Aeronautics: Science and Applications

Aeronautics: Science and Applications

Edited by **Margaret Ziegler**

Willford Press
New York

Published by Willford Press,
118-35 Queens Blvd., Suite 400,
Forest Hills, NY 11375, USA
www.willfordpress.com

Aeronautics: Science and Applications
Edited by Margaret Ziegler

International Standard Book Number: 978-1-68285-226-2 (Hardback)

Printed in China.

Contents

Preface

There have been many scientific and technological innovations in the field of aeronautics in the recent past. This book consists of contributions made by international experts on aeronautics and lucidly discusses significant topics like aerodynamics, navigation, air safety, etc. This text is a vital tool for all researching or studying aeronautics as it gives incredible insights into emerging trends and concepts in this field. It is a ripe text for professionals, students and researchers.

The researches compiled throughout the book are authentic and of high quality, combining several disciplines and from very diverse regions from around the world. Drawing on the contributions of many researchers from diverse countries, the book's objective is to provide the readers with the latest achievements in the area of research. This book will surely be a source of knowledge to all interested and researching the field.

In the end, I would like to express my deep sense of gratitude to all the authors for meeting the set deadlines in completing and submitting their research chapters. I would also like to thank the publisher for the support offered to us throughout the course of the book. Finally, I extend my sincere thanks to my family for being a constant source of inspiration and encouragement.

Editor

1

Verification of Response of Neutron Monitor for In-Flight Neutron Dosimetry

Claudio Antonio Federico[1], Odair Lélis Gonçalez [1], Evaldo Simões da Fonseca[2], Karla Cristina de Souza Patrão[2], Marlon Antonio Pereira[1], Linda Viola Ehlin Caldas[3]

ABSTRACT: In this work, we present the results of verification of special neutron monitoring equipment for use in aircrew dosimetry inside aircraft. The equipment was precalibrated with conventional neutron sources in a National Standard Laboratory of the Institute for Radioprotection and Dosimetry (IRD, Brazil) and verified in a Cosmic Energy Reference Field (CERF), a special field from CERN (Centre Européan de Recherche Nucléaire), a facility that reproduces the shape of neutron field encountered in the atmosphere at aircraft altitudes. The equipment consists of a special neutron probe from Thermo Scientific with response up to 5 GeV neutrons and was verified with respect to ambient dose equivalent rate linearity and angular incidence. The results show the adequacy of the equipment for this type of measurement and the feasibility to use conventional neutron sources to calibrate this specific equipment, in the absence of access to the CERF field.

KEYWORDS: Radiation dose, Aircrew, Neutron dose.

INTRODUCTION

In recent decades, the problem of controlling the level of ionizing radiation dose received by aircrew and sensitive equipment has received great attention and motivated several studies in specialized international literature (FAA, 1990; Wilson *et al.*, 1998; Bartlett, 2004; Edwards *et al.*, 2004; Hajek *et al.*, 2004). In Brazil, these studies were initiated by the Brazilian Air Force (FAB) in conjunction with research institutions from the National Commission on Nuclear Energy (CNEN) (Federico *et al.*, 2010a, 2010b, 2012).

The reason for this concern is the fact that the dose rate arising from cosmic radiation (CR) undergoes a considerable increase with altitude, and consequently, aircrew frequently exceed the annual effective dose limit proposed by international organizations for planned exposures to members of the public (ICRP, 1991), which is 1 mSv.

The International Commission on Radiological Protection (ICRP) recognizes the need to control exposure for the group of professional flight crew and pilots (ICRP, 1998) because this group is exposed to radiation levels that are comparable with or greater than the average levels of radiation received by professionals working with radiation in medicine and technology.

The measurement quantity recommended for aircraft is the ambient dose equivalent (H*(10)), and its employment for purposes of dosimetric control is made by conversion factors for effective dose recommended in the literature, so that the results obtained through the calibrated equipment

1.Instituto de Esudos Avançados – São José dos Campos/SP – Brazil 2.Instituto de Radioproteção e Dosimetria – Rio de Janeiro/RJ – Brazil 3.Instituto de Pesquisas Energéticas e Nucleares – São Paulo/SP – Brazil

Author for correspondence: Claudio Antonio Federico | Instituto de Estudos Avançados | Trevo Coronel Aviador José Alberto Albano do Amarante, 1 | CEP 12.228-001 São José dos Campos/SP – Brazil | Email: claudiofederico@ieav.cta.br can be compared to the limits applicable to the crew of the aircraft.

Due to the large contribution of the neutron component in ambient dose equivalent incident in aircrews and in single-event effects in aircraft avionics, it is necessary to calibrate the response of the instrumentation used in relation to the field of neutrons present at flight altitudes. This calibration is not a conventional procedure due to higher energy of the field and the lack of facilities with this capability.

The neutron radiation field produced by CR covers energy up to the order of hundreds of MeV, presenting more pronounced peaks in the region of thermal neutrons (about 0.023 eV), neutron evaporation (about 1 MeV), and neutrons from spallation processes in atmospheric constituents (peak around 100 MeV). In special, the thermal peak is not very significant in the atmosphere, due to neutron absorption on nitrogen, but can be greatly enhanced by the presence of thermalizing materials within aircraft, such as, the fuel. The characteristics of this type of field are very different from those obtained in neutronic fields emitted by conventional sources of radionuclides, accelerators, or reactors, forcing the use of specific and characterized fields, with broad spectra, created specifically to simulate those obtained at the sites of measurement (Schuhmacher, 2004), for verification or calibration of measuring instruments.

There is a special arrangement in the laboratories of CERN (Centre Européan de Recherche Nucléaire) called CERF (CERN-EU high energy Reference Field), Prevessin, France, designed to reproduce, with reasonable proximity, the existing field on aircraft flight altitudes. While this field cannot be treated as a metrological standard, it is recognized as amongst "the best efforts" in the world to reproduce this kind of neutron field and is widely used to verify the instrumentation used for aeronautical dosimetry purposes.

EXPERIMENTAL PROCEDURE

THE DETECTION SYSTEM

The FH THERMO detection system consists of a Thermo Electron Eberline acquisition, model FH40G-10, with remote programming capability through connection to a computer, coupled to an internal proportional counter, and allows the connection of an external probe with simultaneous acquisition. The internal proportional counter gives the results from low LET components of the field, and the system was designed and calibrated to provide results directly of the operational quantity ambient dose equivalent (H*(10)) and can be operated directly or programmed to acquire measurements of dose rate at time intervals set by the user, which can vary from 1 second to several hours. The system has a volatile internal memory that lets you store the results of up to 256 measuring points, which can be collected by the user later. Each measurement point consists of a record, with information covering the start and finish times of each measurement interval, the result of the proportional internal probe measurement unit, the measurement result of the optional external probe, the measurement unit, the measurement interval (in seconds), and the type of external coupled probe.

For the measurements reported here, the THERMO FHT-762 (WENDI-II) was coupled to the external probe system, which consists of a ^{3}He proportional type neutron probe, covered with a layer of tungsten and an external polyethylene layer, which allows obtaining the response to neutrons with energies up to about 5 GeV, and which differs greatly from conventional neutron detectors that typically have an adequate response to energies up to at most about 20 MeV.

The efficiency curve for ambient dose equivalent as a function of neutron energy is shown in Fig. 1, compared

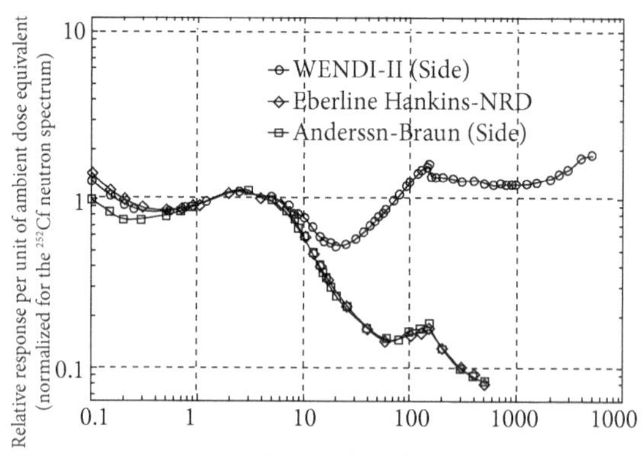

Figure 1. Response curve depending on the incident neutron energy to the probe FHT-762 (WENDI-II), compared with that obtained for other conventional monitor type Eberline Hankins-NRD and Andersson-Braun (extracted from Thermo-Scientific, 2009).

Table 1. Main characteristics of FHT-762 probe (Thermo-Scientific, 2009).

Parameters	Value	Observations
Measurement interval	(0.001 to 1×10^6) μSv/h	
Sensitiveness	0.84 cps/(μSv/h)	Referenced to neutrons of ^{252}Cf
Linearity	-9% to +11%	
Energy interval	25 meV to 5 GeV	Response curve according to ICRP 74 (1997)
Angular dependence	±20%	
Pressure interval	500 to 1500 hPa	

with the curve obtained in other conventional equipment (Thermo-Scientific, 2009). The main features of neutron measurement system are presented in Table 1 and the same is calibrated in the field relative to a neutron source from ^{241}Am/Be.

CONVENTIONAL SOURCES

Initially the neutron probe was recalibrated at the Neutron Laboratory of the Ionizing Radiation Metrology National Laboratory, Brazil, using standard sources of ^{241}Am/Be and ^{252}Cf. The traditional calibration protocol of the laboratory was modified to suit the peculiarity in the use of equipment that operates with low dose rates, present in aircraft flight altitudes. The equipment was exposed to different distances from the standard sources, ranging from about 0.86 m up to 4.7 m, where the ambient dose equivalent of the neutron field is known with a combined average uncertainty varying from around 9 to 2% for sources of ^{252}Cf and ^{241}Am/Be respectively. Such uncertainty is a combination average of the sources of uncertainty given in Table 2.

In the calibration process, we used a shadow cone constructed in accordance with the Safety Report Series 16 (IAEA, 2000), and for each measurement point (distance from source) calibration was performed in two stages, with and without using the shadow cone, so as to eliminate the contribution of the components originating from neutrons scattered in the environment, according to the methodology used in Federico *et al.*, (2010a).

The quantity used for calibration was the ambient dose equivalent, whose conventional true value was obtained

Table 2. Source of uncertainties at the Neutron Laboratory setup of the Ionizing Radiation Metrology National Laboratory.

Factor	Uncertainty (%)	
	^{241}Am/Be	^{252}Cf
Emission rate	1.1	2
Converting coefficient	4	1
Time	1	1
Scattering	1	1
Positioning	0.6	0.6

from the value of the emission rate of the neutron source, determined by means of a bath of manganese sulfate. The bath of manganese sulfate, which is the primary standard for determining the rate of emission of neutron sources, is traceable to International Metrological System through the "K9" key comparison (Roberts *et al.*, 2011), coordinated by the Bureau International des Poids et Mesures (BIPM).

THE CENTRE EUROPÉAN DE RECHERCHE NUCLÉAIRE HIGH ENERGY REFERENCE FIELD

The CERF field is produced by a hadron beam, comprising 34.8% protons, 60.7% pions, and 4.5% kaons, accelerated to an energy of 120 GeV and incident on a copper target (Mitaroff and Silari, 2002). The acceleration of the beam until the energy extraction occurs on an acceleration ring of 6.8 km in circumference, having a 48s cycle between each pulse of particles to be accelerated. Due to the dispersion occurred in the acceleration process, the extracted beam has an average duration of 9s, with the pulse shape shown in Fig. 2 for average dose rates during the pulse varying between 11 and 1300 µSv/h, where the intensity of the pulse can be controlled by the user from the main control room, through a set of collimators that are adjustable remotely.

Two neutron fields are available with different characteristics. The first one uses the neutrons generated on the target and passing through an iron platform according to the schematic drawing shown in Fig. 3.

The distribution of neutron fluence as a function of the energy was evaluated by Mitaroff and Silari (2002) and, for the field in iron platform, the shape of spectra extends from thermal up to 600 MeV, with the predominance of the neutron energy in the range of 0.1–1 MeV.

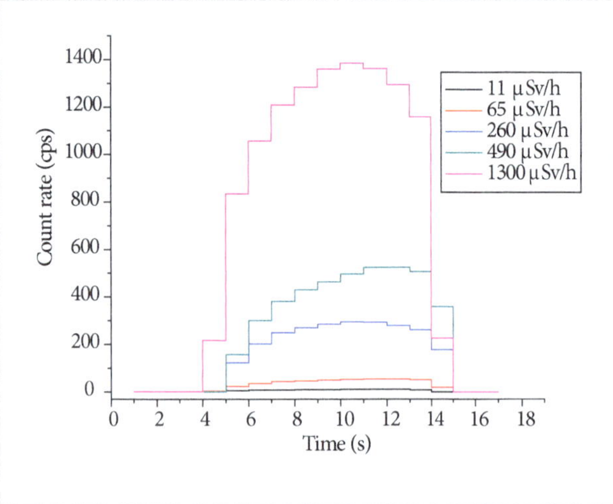

Figure 2. Hadrons pulse shape for dose rates used.

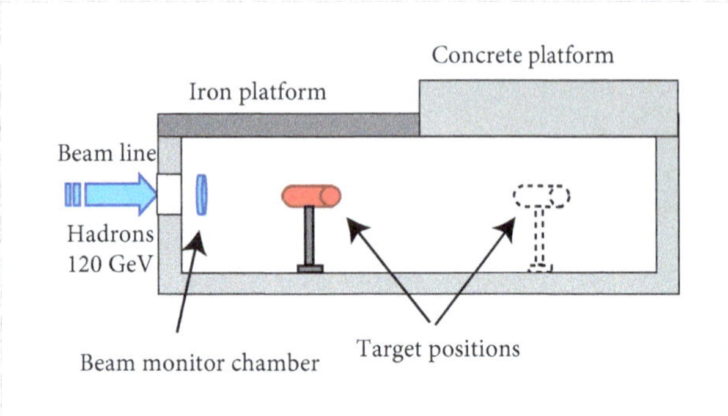

Figure 3. Schematic drawing of the layout of irradiation, with the positions of the beam, target, and measurement sites.

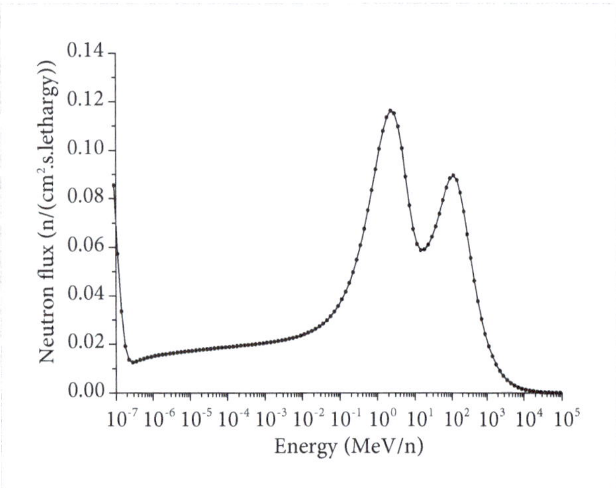

Figure 4. Typical shape of the neutron flux spectrum at 35,000 ft altitude, made using EXPACS code (Sato and Niita, 2006).

The second field, of the most interest to this work, utilizes neutrons generated on the target and passing through a concrete platform, whose spectrum in the position above the concrete platform extends from thermal neutrons up to 500 MeV, with the presence of pronounced peaks characteristic of processes of evaporation and spallation in atomic nuclei (from 1 up to hundreds of MeV) and with great similarity, in shape of the spectrum, with the typical spectrum of neutrons in aircraft flight altitudes (Fig. 4).

The detectors were irradiated at predetermined positions on the platforms of concrete and iron, where the particle flux and spectra can be calculated from the number of counts (PIC counts) obtained in one beam monitor chamber, placed at the exit of the hadrons extraction channel (Fig. 3).

RESULTS

The results of measurements with the source ^{241}Am/Be and ^{252}Cf are shown in Figs. 5 and 6 respectively, along with the lines resulting from a linear fit to experimental data, where the uncertainties in the measurements and in the conventional true value were considered as weight factor in the fitting process. The conventional true value is the best estimative of the true value of the quantity, given by the laboratory maintainer, based on metrological references, as explained previously.

During the verification process of the response of the instrument in the CERF, it was observed that the

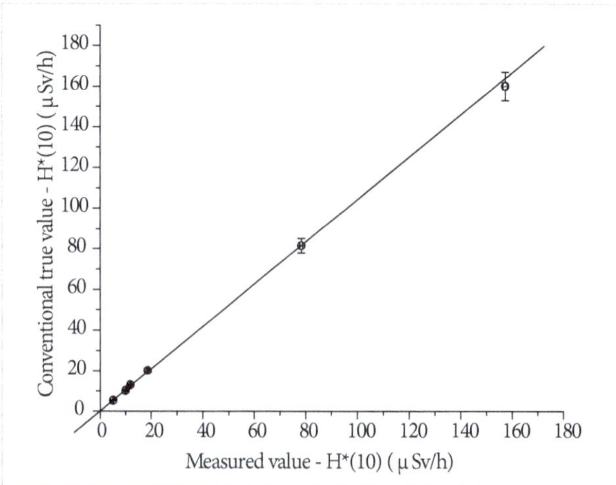

Figure 5. Measurements obtained with the ^{241}Am/Be source, with straight line fitted to experimental data.

background(BG) radiation in the place is variable and depends, among other factors, on the opening of collimators which are used to control the rate of the desired dose. Through the FH40G-10 detector and its neutron probe, we obtained the BG neutron field in some configurations and an empiric function was adjusted to express the BG to be subtracted from experimental measurements, as a function of the opening of collimators for each measurement, as can be seen in Fig. 7. The function type and the coefficients obtained are presented in Table 3.

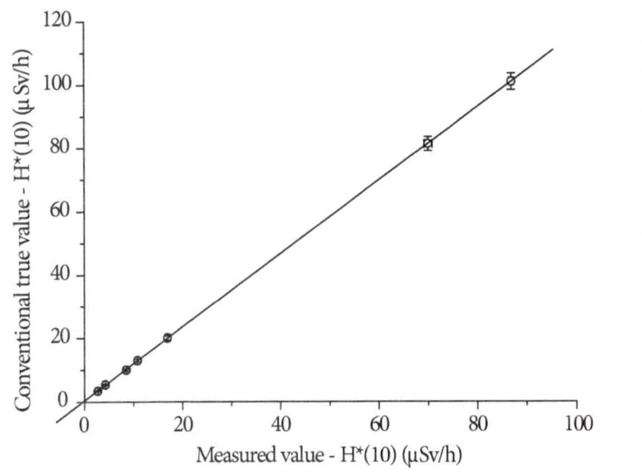

Figure 6. Measurements obtained with the ^{252}Cf source, with straight line fitted to experimental data.

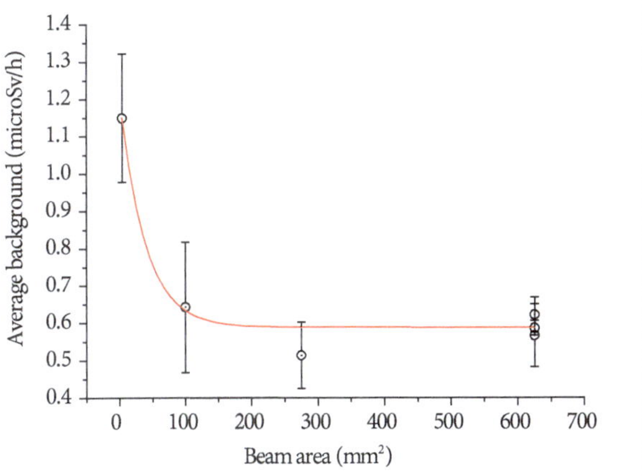

Figure 7. Evaluation of the background component in the neutron field Centre European de Recherche Nucleaire high energy Reference Field.

Table 3. Empirical function and coefficients from background adjust.

Function type	Y=A1*exp(-x/A2) + A3	
Parameters	**Value**	**Uncertainty**
A1	0.62	0.22
A2	38	58
A3	0.587	0.017

The verification of the response of the detector FH40G-10 with the neutron probe (FHT-762) was made by means of comparison with the conventional values of the ambient dose equivalent of the neutron field, for the CERF fields from concrete and iron platforms. The results are shown in Figs. 8 and 9, in which it can

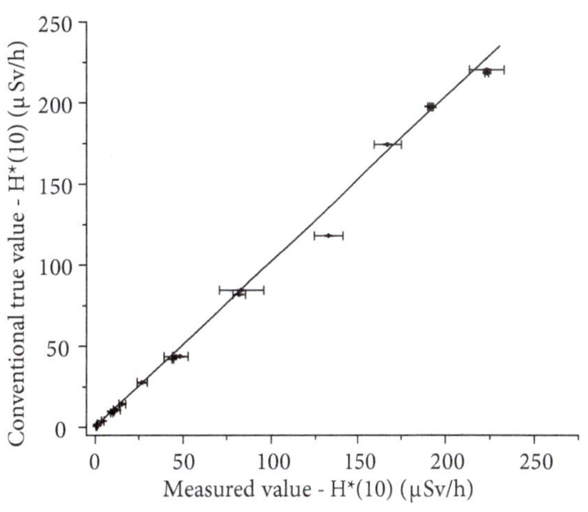

Figure 8. Response curve as a function of dose rate for the probe FHT 762, in the field of concrete, with straight line fitted to experimental data.

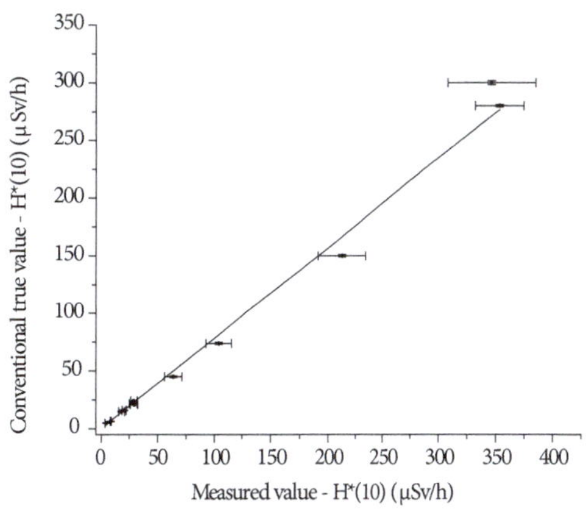

Figure 9. Response curve as a function of the dose rate for the FHT 762 probe, in the field of iron, with straight line fitted to experimental data.

also be observed the lines resulting from linear fittings to the experimental data, where the uncertainties in the measurements were considered as weighting factors in the fitting process. The conventional value of the ambient dose equivalent was previously obtained by Mitaroff and Silari (2002).

Table 4 presents the coefficients obtained for each linear fitting performed by the least squares method, where one can observe that the value of the constant term for the four-neutron radiation fields used are compatible to zero, with a confidence interval of up to two standard deviations. It is observed in the same table an underestimation of 4% on the value indicated by the instrument for ^{241}Am/Be neutron field and 16% for the ^{252}Cf neutron field. This underestimation of the real value can be explained by the dependence of the response to the energy of the neutrons reported by the equipment manufacturer, considering that the average energy of the ^{241}Am/Be neutrons is approximately 4.5 MeV and the ^{252}Cf ones approximately 1 MeV.

We can suppose that, for a good calibration, the relation between the measured values and the reference values is linear, without any constant off-set. Indeed, for the CERF field in concrete platform, which is of particular interest in this work, the constant coefficient obtained in the fitting is compatible to zero, which indicate that there is no systematic deviation of the measured values in relation to the reference values of neutron field. Therefore, a new fitting assuming the constant coefficient as null was accomplished. The uncertainty σ_A obtained previously is now encompassed in the uncertainty of the calibration factor (angular coefficient). This new fitting obtains the value of 1.022±0.005 for the slope, with an adjusted R-square equal to 0.998, which indicates that the instrument readings should be corrected by +2.2% with a statistical uncertainty in coefficient of 0.5%.

Table 4. Results from the linear fitting to the experimental data.

Source	Constant coefficient		Angular coefficient	
	A	σ_A	B	σ_B
^{241}Am/Be	0.14	0.25	1.040	0.020
^{252}Cf	0.33	0.21	1.160	0.022
CERF (iron)	0.05	0.91	0.782	0.018
CERF (concrete)	0.12	0.39	1.021	0.006

This uncertainty, obtained from the linear fitting of the instrument readings on CERN standard field values, is of statistical nature because it is only associated with the instrument reading errors. But, Mitaroff and Silari (2002) reported an uncertainty of 2% in the value of the standard field used in this calibration that affects in the same way all calibration data. Therefore, we must consider this as a systematic error of statistical nature in the calibration procedure, which could be taken into account as a covariance. Considering that the calibration function is linear and that we considered only the variations of statistical nature in obtaining the calibration coefficient by the linear fitting, the uncertainty of the field must be quadratically added to the uncertainty of the calibration coefficient value in order to calculate the standard deviation of the calibration coefficient. Thus, the calibration coefficient for the THERMO FH detector system for the neutron component is 1.022±0.021.

We also evaluated the dependence of the response with the orientation of the detector positioning on the concrete platform of the CERF field, using the positions of standing, lying, and upside down, where differences below 1% were observed for the equipment in question, as can be seen in Fig. 10. We may therefore disregard the influences of the positioning direction.

Comparing the results obtained in this study with measurements made by Mayer *et al.*, (2007) for this same type of equipment, it is observed that the author noticed a difference of -11.9% in the response of the detector to the

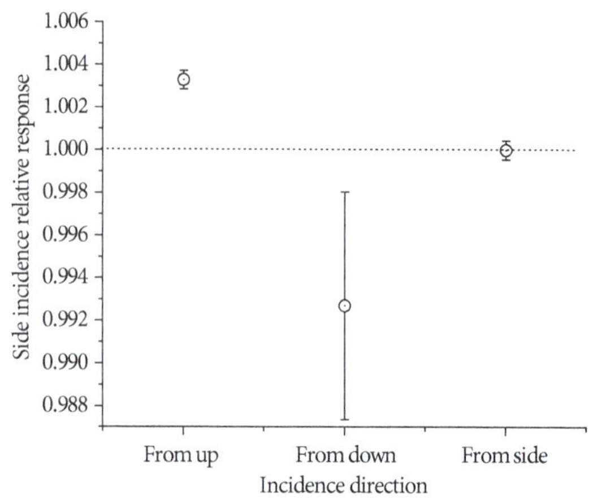

Figure 10. Response curve as a function of preferential direction of incidence for FHT 762 probe.

field in the CERF concrete platform regarding calibration performed for the field of ^{252}Cf. In this study, the difference between the response of the CERF field in concrete platform in relation to the response obtained for the field of ^{252}Cf is about 13.9%, with an uncertainty of about 2.2%, confirming the result obtained by Mayer *et al.,* (2007) within one standard deviation.

Similarly, Yasuda and Yajima (2010) made tests with the same type of monitor, calibrated for neutron from $^{241}Am/Be$ sources and found differences in the H*(10) rate measured in relation to the H*(10) calculated rate of about 3%, consistent with the underestimation of 2.2%, with an uncertainty of about 2.1%, found in this study.

CONCLUSION

The results in terms of rate of environment equivalent dose indicate that the detector responds appropriately to high-energy neutrons, which can be used for radiation measurements on aircraft, with a small correction in the value read from the instrument. By the results obtained in this work and by Yasuda and Yajima (2010), one can observe that the difference between the response of H*(10) of such equipment to the CR field and the response calibrated to conventional sources of $^{241}Am/Be$ is about -2 to -3%, and this type of source can be used for calibration of this instrument, in the case of inability to use the CERF field. This is a very important result due to the difficulties related on support for this type of calibration in CERF field, especially for people from outside of Europe.

Unfortunately, Mayer *et al.,* (2007) show that the same is not the case for other types of conventional neutron detectors whose responses are hardly suitable for energy greater than 16 MeV, being therefore, for the most part, unsuitable for H*(10) measurements of neutrons from CR.

The characterization of this equipment to the field of CR at aircraft flight altitude is an important step in the consolidation of a Brazilian group with training, experience, and proper equipment to perform such measurements in flight, which does not currently exist in Brazil, nor possibly elsewhere in Latin America.

ACKNOWLEDGMENTS

We thank FINEP and CNPQ for partial financial support; to CERN, for allowing the verification of instruments in the CERF field; and to the Brazilian Air Force, which supports this research in IEAv.

REFERENCES

Bartlett, D.T., 2004, "Radiation protection aspects of the cosmic radiation exposure of aircraft crew", Radiation Protection Dosimetry, Vol. 109, No. 4, pp. 349-355.

Edwards, R.E., Normand, E., Dyer, C.S., 2004, "Technical Standards for atmospheric radiation single event effects (SEE) on avionics electronics", IEEE Radiation effects Data Workshop Record, IEEE 04TH8774, 2004.

FAA, 1990, Federal Aviation Administration, U.S. Department of Transportation, Radiation Exposure of Air Carrier Crewmembers. FAA Advisory Circular n. 120-152.

Federico, C.A., Gonçalez, O.L., Fonseca, E.S., Martin, I.M. and Caldas, L.V.E, 2010a, "Neutron Spectra Measurements in the South Atlantic Anomaly Region", Radiation Measurements, Vol. 45, No. 10, pp.1526-1528.

Federico, C.A., Pereira, H.H.C., Pereira, M.A., Gonçalez, O.L. and Caldas, L.V.E., 2010 b, "Estimates of Cosmic Radiation Dose Received By Aircrew of DCTA's Flight Test Special Group", Journal of Aerospace Technology and Management, Vol. 2, pp. 137-144.

Federico, C.A., Gonçalez, O.L., Sordi, G.M.A.A. and Caldas, L.V.E., 2012, "Effects of Cosmic Radiation in Aircrafts: A Discussion about Aircrew over South America", Journal of Aerospace Technology and Management, São José dos Campos, Vol. 4, No. 2, pp. 219-225, Apr-Jun.

Hajek, M., Berger, T. and Vana, N., 2004, "A TLD-based personal dosemeter system for aircrew monitoring", Radiation Protection Dosimetry, Vol. 110, No. 1-4, pp. 337-341.

IAEA, 2000, Calibration of Radiation Protection Monitoring Instruments, Safety Report Series 16, International Atomic Energy Agency, Vienna.

ICRP, 1991, International Commission on Radiological Protection. Recommendations of the international commission on radiological protection, Publication 60, Annals of the ICRP, Vol. 21, No. 1-3.

ICRP, 1997, International Commission on Radiological Protection. Conversion coefficients for use in radiological protection against external radiation, Publication 74, Annals of the ICRP, Vol. 26, No. 3.

ICRP, 1998, International Commission on Radiological Protection. General principles for the radiation protection of workers, Publication 75, Annals of the ICRP, Vol. 27, No. 1.

Mayer, S., Forkel-Wirth, D., Fuerstner, M., Menzel, H.G., Mueller, M.J., Perrin, D., Theis, C. and Vincke, H., 2007, "Response of neutron detectors to high-energy mixed radiation fields", Radiation Protection Dosimetry, Vol. 125, No. 1-4, pp. 289–292.

Mitaroff, A. and Silari, M., 2002, "The CERN–EU High-Energy Reference Field (CERF) Facility for Dosimetry at Commercial Flight Altitudes and in Space", Radiation Protection Dosimetry, Vol. 102, No. 1, pp. 7-22.

Roberts, N.J., Jones, L.N., Králik, M., Park, H., Choi, K.O., Pereira, W.W., da Fonseca, E.S., Cassette, P., Dewey, M.S., Moiseev, N.N. and Kharitonov, I.A., 2011, "International Key Comparison of Measurements of Neutron Source Emission Rate (1999-2005): CCRI(III)-K9.AmBe", Metrologia, 48, Tech. Suppl., 06018.

Sato, T. and Niita, K., 2006, "Analytical Functions to Predict Cosmic-Ray Neutron Spectra in the Atmosphere", Radiation Research, Vol. 166, pp. 544-555.

Schuhmacher, H., 2004, "Neutron Calibration Facilities", Radiation Protection Dosimetry, Vol. 110, No. 1-4, pp. 33-42.

Thermo-Scientific,2009, FHT 762-WENDI-2 Datasheet. (n.42540/85 TD-E).

Wilson, J.W., Goldhagen, P., Maiden, D.L. and Tai, H., 1998, "High altitude radiations relevant to the high speed civil transport (HSCT)". NASA Technical Documentation, NASA.

Yasuda, H. and Yajima, K., 2010, "Characterization of Radiation Instruments at the Summit of Mt. Fuji", Radiation Measurements, Vol. 45, pp. 1600-1604.

2

Experimental Study of Polyurethane-Based Fuels with Addition of Paraffin and Aluminum for Hybrid Rocket Motors

Susane Ribeiro Gomes[1], Leopoldo Rocco Junior[1], José Atílio Fritz Fidel Rocco[1], Koshun Iha[1]

ABSTRACT: Experimental investigation was conducted to determine the relative propulsive and combustion behavior of several polyurethane-based solid-fuel formulations containing 30% w/w of paraffin or 10% of aluminum powder. In total, seven solid-fuel formulations were investigated, four containing 30% of paraffin and three with 10% of aluminum. The polyurethane was synthesized with pre-polymer technology. The oxidizer was gaseous oxygen, which was forced into the combustion chamber with axial and swirl methods. Firing tests with 7 configurations were performed. Thrust measurements indicated that the addition of paraffin increased thrust at about 57% and regression rates at about 70%. No relevant improvement in performance was obtained with aluminum addition. Specific impulse decreased when aluminum particles were added to the fuel. The mixture that produced the best ballistic parameters was polyurethane plasticized with castor oil and 30% w/w of paraffin with gaseous oxygen injected through a swirler.

KEYWORDS: Hybrid rocket motor, Aerospace propulsion, Rocket motors.

INTRODUCTION

Hybrid rocket engine (HRE) is a type of rocket motor that combines the advantages of both solid and liquid fueled rockets and avoids many of the underperformances. In a hybrid, the oxidizer is stored as liquid or gas in tanks and the fuel is stored as solid-fuel grain in the combustion chamber. Oxidizer is injected over the burning fuel surface, and the resulting gases are expelled out of a nozzle to produce thrust. Because the fuel and oxidizer are separate, and cannot easily mix (because they are at different phases), hybrid rockets have very little danger of exploding (Boardman, 2001). Hybrids may provide higher I_{sp} than solid motors, and due to the high-density solid, they may present higher density-specific impulse than liquid engines. Advantages such as fuel versatility — additives can be easily embedded in the fuel grain — potential environmental friendliness, and low cost due to high levels of safety and minimal failure modes are widely acknowledged. Classical hybrid rocket motors also present disadvantages yet to overcome, such as combustion inefficiencies due to poor mixing and mixture ratio shifting and mainly low regression of the fuel grain (Altman, 2001). Several methods for improving regression rates have been proposed and investigated. Approaches to increase regression rates usually employ advanced fuels into three categories:

- Addition of energetic particles: A methodology to increase the regression rate in HREs has been to introduce aluminum into the solid fuel (Chiaverini, 2007). At Penn

1.Instituto Tecnológico de Aeronáutica – São José dos Campos/SP – Brazil

Author for correspondence: Susane Ribeiro Gomes | Instituto Tecnológico de Aeronáutica | Praça Eduardo Gomes, 50 | CEP 12.228-901 São José dos Campos/SP – Brazil | Email: susaneribeiro@gmail.com

State University, nanosized particles of aluminum powder with diameters of 70–150 μm were mixed into hydroxyl-terminated polybutadiene (HTPB) fuel grains and tested (Chiaverini *et al.*, 2000; Evans *et al.*, 2009). The results showed that the regression rates increased by 50% over the HTPB without aluminum. Researchers at University of Pennsylvania (UPENN) reported improvements of 60% on regression rate with the use of embedded nanoparticles (Risha *et al.*, 2007).

- Use of energetic polymers instead of conventional fuels such as HTPB; this polymer is widely used due to its strong mechanical and durability properties and the technology is well known due to the use in solid rocket motors (Altman and Humble, 1995; Altman and Holzman, 2007; Geisler *et al.* 2010).
- Use of paraffin-based fuels due to the inherent mass-transfer mechanism. Paraffin Fuels Researchers at Stanford University (Karabeyoglu *et al.* 2001) discovered that paraffin-based fuels exist which have regression rates that are three to four times than those of conventional hybrid fuels. This is largely due to the development of a thin liquid layer on the fuel grain surface which becomes instable: instability appears due to the incoming oxidizer flow pattern and liquid fuel droplets are injected into the boundary layer (Karabeyoglu *et al.* 2001; Karabeyoglu and Cantwell, 2002). This enhanced mass-transfer mechanism increases fuel mass flow without the blocking effect typical of gaseous fuel blowing.

Paraffin presents poor mechanical properties; a binder is hence desirable to sustain loads on flight condition. Therefore, the purpose of this work was the development of a polyurethane (PU) binder, or polymeric matrix, with total substitution of HTPB pre-polymer for another polyol, or mix of polyols. This product is supposed to be applicable as a fuel binder or solid propellant binder formulations. The synthesized binder is embedded with paraffin and micron-sized aluminum particles. Synthesized fuels are tested to assess the improvement in combustion efficiency and regression rates.

In addition to developing a potential high regression rate fuel, a second methodology was employed to improve combustion efficiency. Swirling type injectors are known to increase heat transfer from the flame to the grain surface. This method has first-order effects on regression rates (Knuth *et al.*, 1998; Carmicino and Sorge, 2005, 2007).

The swirl injector was used in the second series of tests to evaluate the relative improvement in combustion efficiency and regression rates in comparison to the axial injector. The use of paraffin and swirl injectors yielded improvements in thrust up to 57%, and in regression rates up to 70% in relation to the standard PU fuel and axial injector.

METHODOLOGY

This section outlines the steps taken in this work: synthesis of PU pre-polymer, addition of paraffin addition of aluminum particles, and firing tests.

In the binder synthesis, HTPB-hydroxyl-terminated polybutadiene was replaced with a modified polytetramethylene ether glycol (PTMEG) pre-polymer with 4.3% of free NCO. The mixture was cured with a liquid amine (ETHACURE) and plasticized with castor or mineral oil. Finally, PU elastomers were obtained.

A microcrystalline paraffin from petroleum (Petrobras 140/145-1) composed of saturated aliphatic hydrocarbons with melting point of 61.4°C and boiling point of 290°C, at standard conditions, was added to the synthesized PU. In this work, aluminum particles with diameters of 100 μm were added to three chosen fuel combinations in the proportion of 10% w/w. Firing tests with axial and swirl injectors were done and the ballistic results were obtained and evaluated. A side view of swirl injector is shown in Fig. 1, and a scheme of the flow entering the combustion chamber is presented in Fig. 2.

EXPERIMENTAL SETUP

The baseline engine design was developed from a need for simplicity and flexibility; therefore, a modular design was incorporated. The set could be assembled and disassembled in minutes, allowing the practice of several tests per session.

The case and the flanges were made from stainless steel. The case was machined to fit between the steel flanges. The nozzle was adapted in the aft flange, to prevent nozzle escape. A hydraulic system was settled to perform the

Figure 1. Side view of the swirl injector. The holes are the oxidizer hose entrances.

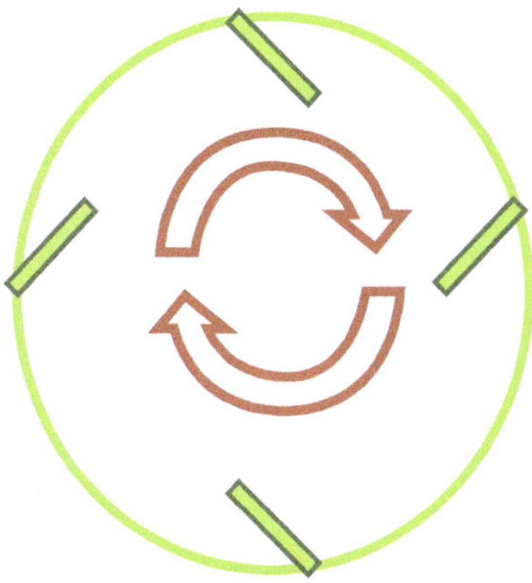

Figure 2. Schematic diagram of the swirl injector cross-section.

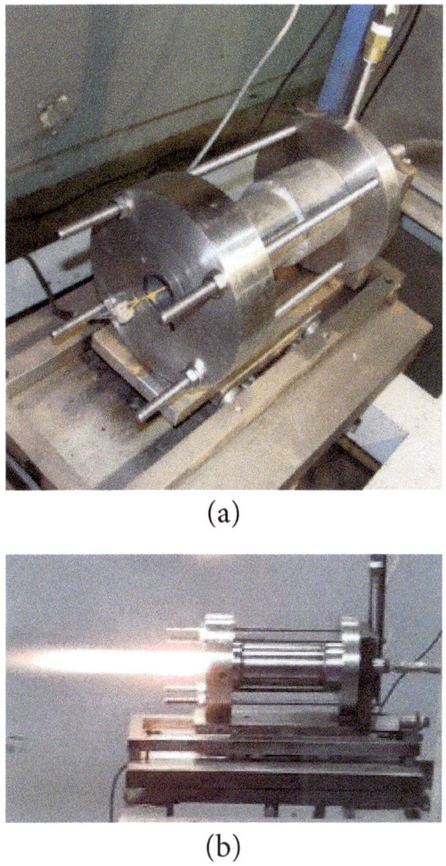

Figure 3. Side view of the labscale hybrid rocket motor on the test bench (a) and during test (b).

thrust measurements. A view of the motor attached to the test bench is shown in Fig. 3. Compression fitting valves were used in the oxygen feed line. A squib was used to achieve ignition.

A schematic diagram of the test facility is exposed in Fig. 4. Compression fitting valves were used in the oxygen feed line. A pyrotechnic method was used to achieve ignition.

The combustion chamber is constructed of 316 L stainless steel, total length of 215 mm, and inner diameter of 68.3 mm. The fuel grain is 185 mm long, inner diameter of 20 mm. Pre-chamber is 15 mm long and post-chamber is 10 mm long. No thermal isolation was used in this project; it is intended to use a high temperature coating in future work. The experiments were performed in ambient conditions, with temperature of approximately 27°C and 1.024 MPA and humidity values varying between 40 and 80%.

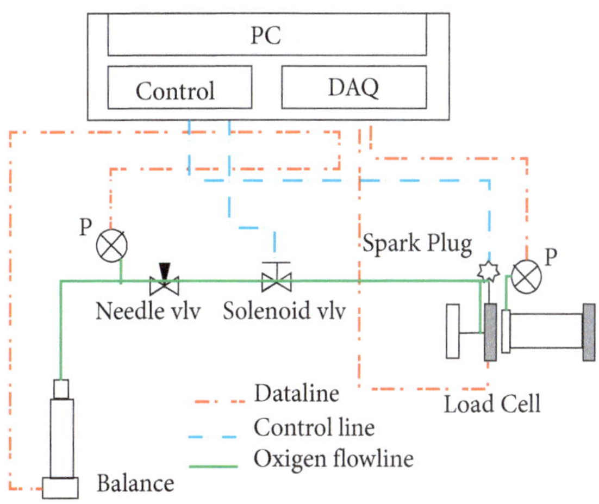

Figure 4. Diagram of the experimental apparatus.

Nozzle is made of graphite, throat diameter of 8 mm, convergence angle of 30° and divergence angle of 15°. The length was calculated accounting for the desired exit pressure as close as possible to the ambient pressure.

RESULTS AND DISCUSSION

Average regression rate is defined in Eq. 1 and oxidizer mass flux is experimentally defined in Eq. 2:

$$\dot{r}_{avg} = a\, G_{ox}^{n} \tag{1}$$

$$G_{ox} = \frac{\dot{m}_{ox}}{\pi\left(\frac{D_i + D_f}{2}\right)^2} \tag{2}$$

Using experimental data, average regression rate is calculated as follows, Eqs. 3 and 4:

$$\Delta m = \frac{\pi}{4}\left(D_2^2 - D_1^2\right)\rho L = \frac{\pi}{4}\left(D_2 + D_1\right)\left(D_2 - D_1\right)\rho L \tag{3}$$

$$\dot{r} = \frac{D_2 - D_1}{\Delta t} = \frac{4\Delta m}{\pi\left(D_2 + D_1\right)\rho L \Delta t} \tag{4}$$

The specific impulse is determined by dividing the total impulse by the weight of the propellant (fuel and oxidizer) consumed. The total impulse is calculated by numerically integrating the thrust over time for the duration of the test using a simple Riemann squares approximation implemented on matlab and then divided by the weight of the propellant used as shown in Eq. 5:

$$I_{sp} = \frac{\sum_{i=0}^{i=t_b} \frac{1}{2}\left(T_{i+\Delta t} + T_i\right)\cdot\left(t_{i+\Delta t} - t_i\right)}{g \cdot m_{prop}} \tag{5}$$

where I_{sp} is the specific impulse, T_i is the current thrust value, $T_{i+\Delta t}$ is the thrust at the next time step, t_i is the current time, $T_{i+\Delta t}$ is the time at the next time step, g is the acceleration due to gravity, and m_{prop} is the mass of the propellant consumed during the burn.

The first series of synthesis provided three fuel formulations, the first was plasticized with mineral oil, the second with castor oil, and the third was plasticized with castor oil and embedded with 30% w/w of paraffin. To establish a baseline, commercial PU was acquired and compared.

A series of tests with each one of the three synthesized fuel and the commercial PU were performed. The method of injection was axial to the grain, and the gaseous oxygen mass flow rate was kept constant at 40 g/s. The thrust results are presented in Fig. 5.

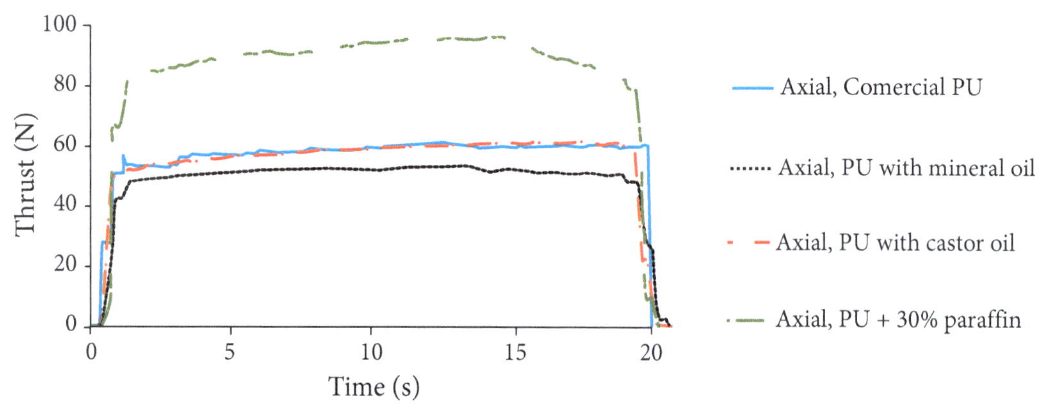

Figure 5. Thrust data of polymeric fuels and paraffin wax mixture with axial injector and 20-second tests.

The graph shows that the PU synthesized with mineral oil yielded the lowest thrust, followed by commercial and PU synthesized with castor oil, both with the approximate thrust of 62 N. Results obtained with the use of paraffin were the most promising reaching almost 100 N. Additional data are summarized in Table 1. Thrust presented the highest uncertainty of 4.9%; other variables showed smaller uncertainties. The chamber pressure was calculated as the mean value of the plateau pattern obtained in the pressure curves.

It was observed that the binder that contained castor oil had better performance and higher stability. It increased I_{sp} in about 15.6% and thrust in about 19.4%. The addition of 30% paraffin increased I_{sp} in about 29.5% and thrust in about 56.9% and the chamber pressure achieved with the same oxidizer mass flow rate is 58.8% higher. Regression rates in relation to the PU synthesized with castor oil increased almost 70%. For the same oxidizer mass flow rate, pressure achieved was at least 70% higher when 30% w/w of paraffin was added.

The next step was increasing the percentage of paraffin. Unfortunately, there were temperature issues on the surface of the axial injector and on the case. The swirl injector was used, and the test duration was reduced to 6 seconds. The second series of tests was performed with PU also synthesized with castor oil in the following proportions:

- PU with 30% of paraffin
- PU with 30% of paraffin and 10% of aluminum (w/w)
- PU with 10% of aluminum

The results were the most promising so far; for more information, see Table 2.

The highest regression rate was achieved in the configuration of 30% paraffin and 10% aluminum; however the highest thrust was obtained when no aluminum was added (PU + 30% paraffin). Specific impulse was the lowest when aluminum powder was present and the highest when no aluminum was added to the fuel.

Previous analysis has shown that addition of aluminum to traditional hybrid solid fuels actually decreases the specific impulse because the molecular weight of the exhaust products is increased, which counteracts the raise in the temperature.

Figure 6 shows the steadiness of thrust achieved with PU with 30% of paraffin tested with the swirl injector. Average thrust values were about 23% higher than both grains with aluminum particles.

PU with 30% of paraffin ignites faster and presents the best performance. The other two cases show roughly the same constant thrust at about 88.9 N. Figure 7 shows a picture of the grain with 30% of paraffin after the firing test.

Paraffin melted and later suffered condensation; one could deduce that a motor with paraffin or any other liquefying fuel will work differently if restarted. Another effect observed in this series of tests was the contraction of the fuel grain after firings, which is a dangerous effect, once combustion can occur at the outer region of the grain and jeopardize the mission.

Table 1. Experimental results for each fuel using of the axial injector.

Fuel	P (atm)	O/F	Thrust (N)	I_{sp} (s)	r (mm/s)	Injector
PU + paraffin 30%	13.98	1.63	98.5	177.9	0.97	Axial
PU w/o castor oil	8.11	3.72	52.7	121.1	0.50	Axial
PU with castor oil	8.94	3.23	62.9	140.0	0.53	Axial
Commercial PU	8.97	2.89	62.6	137.5	0.53	Axial

Table 2. Experimental results for each test configuration.

Fuel	P (atm)	O/F	Thrust (N)	I_{sp} (s)	r (mm/s)	Injector
PU + paraffin 30%	17.65	2.63	108.8	140.0	2.63	Swirl
PU + 10% Al	13.86	3.23	88.9	88.9	3.29	Swirl
PU + 30% paraffin + 10% Al	13.82	3.46	88.4	127.7	2.63	Swirl

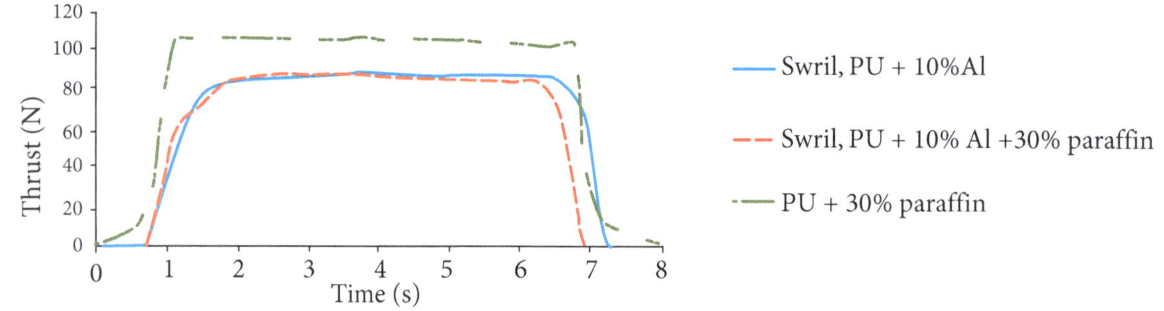

Figure 6. Thrust data of polymeric fuels and paraffin wax mixture with swirl injector and 6-second tests.

ADDITION OF ALUMINUM AND GRAIN ANALYSIS

Figure 8 shows the grain inside the stainless steel case prior to testing. This grain contains 10% of micron-sized aluminum powder and PU synthesized with castor oil.

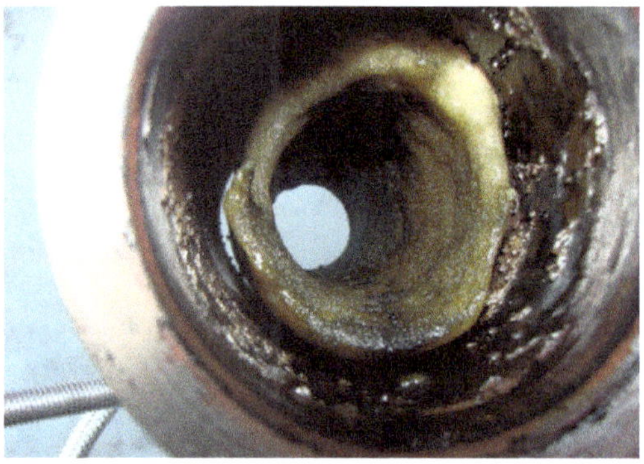

(a)

(b)

Figure 7. PU + 30% of paraffin grain seen from the injector (a) and from the nozzle (b).

This particular grain was hard to ignite; two squibs were necessary to achieve ignition. It is supposed that the oxide layer on the surface is responsible for this problem, once the particles were not stored in an oxygen-free controlled environment.

Figure 9 shows the deposition of metallic aluminum on the rear section of the motor, around the nozzle.

Metallic powder was also deposited on the injector head, as can be seen in Fig. 10.

It is evident that the aluminum deposition on several parts of the engine might cause problems; the most obvious is the accumulation of material close to the nozzle, once a possible throat obstruction is likely to occur. Even though the aluminum amount used was the same with or without the paraffin addition, the deposition when paraffin was present was less significant.

Figure 8. PU with 10% of micron-sized aluminum powder before testing.

(a) (b)

Figure 9. Deposition of aluminum powder around the nozzle.

Figure 10. Deposition of aluminum powder on the injector.

CONCLUSION

The embedding of paraffin increased the overall ballistic parameters of the motor. The use of the binder is essential to structure and strengthen the paraffin to endure flight condition loads. Mixing ratios of 30% w/w of paraffin wax in the binder structure yielded good results, particularly when swirl injectors are used. Several additional tests must be performed to address the best mixing ratio. The aluminum particles probably presented an oxide layer too thick to participate in the combustion process. In future work, aluminum particles with smaller diameters and stored in controlled environments must be used.

ACKNOWLEDGMENTS

The authors gratefully acknowledge financial support from CNPq (Conselho Nacional de Desenvolvimento Científico e Tecnológico) and AEB (Agência Espacial Brasileira).

REFERENCES

Altman, D., 2001, Rocket motors, hybrid. *In* Meyers, R.A. (Ed.), "Encyclopedia of Physical Science and Technology" (3rd ed., Vol. 14, pp. 303-321), Elsevier Science Ltd.

Altman, D. and Holzman, A., 2007, Overview and history of hybrid rocket propulsion, Progress in Astronautics and Aeronautics Series (1st ed., Vol. 218, p. 650). AIAA.

Altman, D. and Humble, R., 1995, "Hybrid Rocket Propulsion Systems", Space Propulsion Analysis and Design", in R. Humble, G. Henry, W. Larson (Eds.). McGraw-Hill, New York, pp. 365-441.

Boardman, T.A., 2001, "Rocket Propulsion Elements", In G.P. Sutton, John Wiley & Sons.

Carmicino, C. and Russo Sorge, A., 2005, "Role of injection in hybrid rockets regression rate behavior", Journal of Propulsion and Power, Vol. 21, No. 4, pp. 606-612.

Carmicino, C. and Russo Sorge, A., 2007, "Performance comparison between two different injector configurations in a hybrid rocket", Aerospace Science and Technology, pp. 61-67.

Chiaverini, M., 2007, Review of solid-fuel regression rate behavior in classical and nonclassical hybrid rocket motors, in M.J. Chiaverini, K. K. Kuo (Eds.), "Progress in Astronautics and Aeronautics" (1st ed., Vol. 218, pp. 37-126), American Institute of Aeronautics and Astronautics.

Chiaverini, M., Serin, N., Johnson, D., Lu, Y.C., Kuo, K. and Risha, G., 2000, "Regression rate behavior of hybrid rocket solid fuels", Journal of Propulsion and Power, Vol. 16, No. 1, pp. 125-132.

Evans, B., Boyer, E., Kuo, K. K. and Risha, G., 2009, "Hybrid Rocket Investigations at Penn State University's High Pressure Combustion Laboratory: Overview and Recent Results", 45th AIAA/ASME/SAE/ASEE Joint Propulsion Conference and Exhibit, AIAA Paper 2009-5349, Denver, Colorado, 2009.

Geisler, R.L., Frederick Jr, R.A. and Giarra, M., 2010, Historical overview and solid rocket motor fundamentals, in R. Blockley, W. Shyy (Eds.), "Encyclopedia of Aerospace Engineering" (Vol. 9), John Wiley & Sons, Ltd. All rights reserved.

Karabeyoglu, M.A. and Cantwell, B.J., 2002, "Combustion of liquefying hybrid propellants: part 2, Stability of liquid films", Journal of Propulsion and Power, Vol. 18, No. 3, pp. 621-630.

Karabeyoglu, M.A., Cantwell, B.J. and Altman, D., 2001, "Development and Testing of Paraffn-Based Hybrid Rocket Fuels", in Proceedings of the 37th AIAA/ASME/SAE/ASEE Joint Propulsion Conference and Exhibit, July 2001, AIAA Paper 2001-4503.

Knuth, W.H., Chiaverini, M.J., Gramer, D.J. and Sauer, J.A., 1998, "Experimental Investigation of a Vortex-Driven High-Regression Rate Hybrid Rocket Engine", 34th AIAA/ASME/SAE/ASEE Joint Propulsion Conference & Exhibit. AIAA.

Risha, G.A., Evans, B.J., Boyer, E. and Kuo, K.K., 2007, Metals, energetic additives, and special binders used in solid fuels for hybrid rockets, in M. J. Chiaverini and K. K. Kuo (Eds.), "Progress in Astronautics and Aeronautics" (1st ed., Vol. 218, pp. 413-456), American Institute of Aeronautics and Astronautics, Inc.

3

Influence of Ethylene Glycol on the Mullite Crystallization Processes Analyzed by Rietveld Refinement

Flaviano Willians Fernandes[1], Tiago Moreira Bastos Campos[1], Luciana de Simone Cividanes[1], João Paulo Barros Machado[1], Evelyn Alves Nunes Simonetti[1], Gilmar Patrocínio Thim[1]

ABSTRACT: Mullite is an excellent structural material due to its high temperature stability, high electrical insulation capabilities and creep resistance. This material has a number of technological applications, such as rocket nozzles used in the aerospace industry. In this work, mullite was obtained by sol-gel process, using silicic sol, aluminum nitrate and ethylene glycol, besides the following volume ratios of silica sol dispersion to ethylene glycol: 1/0; 1/1; 1/2; and 1/3. After drying, the samples were thermal treated at temperatures of 1,000; 1,100; 1,200 and 1,250°C. The samples were characterized by scanning electron microscopy (SEM), X-ray diffraction (XRD) and specific surface area (Bruner-Emmett-Teller – BET). SEM showed that mullite particles are fine and nearly equiaxed. The sample without ethylene glycol showed 3/2 mullite after heat treatment at 1,250°C. The sample with intermediate ethylene glycol concentration presented two crystallization processes: the first at 1,000°C forming mullite and spinel phases, and the second at 1,250°C forming only 3/2 mullite. However, the sample with the highest ethylene glycol concentration crystallized directly to mullite at 1,000°C with the highest yield. There is a strong dependence on the specific surface area with temperature. The Rietveld refinement showed that the a cell lattice of mullite and the Al/Si molar ratio in the mullite formula depend on the ethylene glycol presence and on the calcination temperature. The lattice parameters b and c are not dependent on the alumina content, but the parameter a increases with the increase in the alumina content. Samples prepared with higher ethylene glycol concentrations reached higher mullite yields at lower temperatures.

KEYWORDS: Mullite, Sol-gel, Ethylene glycol, Rietveld refinement.

INTRODUCTION

Mullite has the nominal composition of $3Al_2O_3 . 2SiO_2$ (3:2 mullite) and is the only stable crystalline phase in the SiO_2-Al_2O_3 binary system at atmospheric pressure (Schneider *et al.*, 2008). Mullite is known as an important material for electronic, optical and high-temperature structural applications because of its excellent properties, such as high-temperature strength, creep resistance, low thermal expansion coefficient and good dielectric properties, even at elevated temperatures and high oxidative atmosphere (Chakraborty, 2008; Cividanes *et al.*, 2010a, 2011). Mullite has a number of technological applications, such as rocket nozzles used in the aerospace industry (Schneider *et al.*, 2008).

The mullite structures consist of chains of distorted edge-sharing Al–O octahedra at the corners and center of each unit cell running parallel to the c-axis. The chains are crosslinked by Si–O and Al–O corner-sharing tetrahedrons. The mullite crystal system is orthorhombic - dipyramidal class (H-M Symbol 2/m 2/m 2/m); Space Group: Pbam (Schneider and Komarneni, 2005; Shackelford and Doremus, 2008). Mullite is a solid solution of silica and alumina and its stoichiometry is based on the alumina/silica molecular ratio (Campos *et al.*, 2012; Chakraborty, 2008; Cividanes *et al.*, 2010b; Fischer *et al.*, 1996; Gerardin *et al.*, 1994; Schneider *et al.*, 2008). The chemical formula of mullite is often given by $Al_{4+2x}Si_{2-2x}O_{10-x}$, where x=0, x=0.25 and x=0.4 correspond to sillimanite, 3:2 mullite and 2:1 mullite, respectively (Campos *et al.*, 2012; Chakraborty, 2008; Cividanes *et al.*, 2010a; Fischer *et al.*, 1996). The amount of oxygen vacancies per unit cell is represented by x, which is related to the

1. Instituto Tecnológico de Aeronáutica – São José dos Campos/SP – Brazil

Author for correspondence: Gilmar Patrocínio Thim | Praça Marechal Eduardo Gomes, 50 – Vila das Acácias | CEP 12.228-900 São José dos Campos/SP – Brazil | Email: gpthim@gmail.com replacement of silicon ions (Si^{4+}) by aluminum ions (Al^{3+}) in the tetrahedral sites of the mullite structure (Mazza *et al.*, 2008). With increasing alumina content, the cation Si^{4+} is replaced by the cation Al^{3+} and the anion (oxygen) and the oxygen vacancies are created to maintain charge neutrality (Campos *et al.*, 2012; Cividanes *et al.*, 2011; Okada, 2008). In order to accommodate the structural defects, significant distortions in the aluminum and silicon polyhedra are caused. In mullite, there are three tetrahedral "chains" in the unit cell. Therefore, the distorted alumina tetrahedrons have to be arranged in an oxygen-deficient tri-cluster (three tetrahedrons sharing single corner-bridging oxygen). These clusters constitute a distinctive element of mullite crystal structure (Schneider and Komarneni, 2005; Shackelford and Doremus, 2008). Oxygen vacancies tend to cluster with short-range order along specific crystallographic directions which depend on the alumina content in mullite formula (Cividanes *et al.*, 2010a). Moreover, oxygen vacancies tend to correlate in parallel with the lattice parameter *a* and, in a lesser extent, with *b*, considering that mullite with lower alumina concentration tends to show more random vacancy distribution (Fischer *et al.*, 1996; Gerardin *et al.*, 1994; Yabuki *et al.*, 2002). One can assume that the lattice parameter *a* shows a linear dependence on the Al_2O_3 content (Fischer *et al.*, 1996; Gerardin *et al.*, 1994; Yabuki *et al.*, 2002).

The mullite crystallization sequence depends on the homogeneity of silicon and aluminum in the precursor (Aksaf and Pask, 1975; Campos *et al.*, 2012; Cividanes *et al.*, 2010b; Inoue, 2004; Richardson *et al.*, 1988). When the precursor is homogeneous, the mullite crystallization temperature is lower, and the alumina content in the mullite structure is the same of the starting material. Therefore, the control of the hydrolysis and condensation rate of the starting materials is very important to increase the precursor homogeneity (Chakraborty, 2008). Otherwise, phase segregation can occur, which not only promotes the crystallization of undesirable phases, such as α-alumina and spinel, but also determines a higher mullitization temperature.

In the literature, many ceramic synthesis methods are described, such as: mixtures of solid reagents, coprecipitation of mixed salts, sol-gel, spray pyrolysis etc. (Cividanes *et al.*, 2010a; Schneider *et al.*, 1994). Depending on the application and value of the final product, each method can be justified, since each one has its peculiarity. The sol-gel process has been used for the mullite synthesis to generate products with high purity and homogeneity. In addition, it is a versatile and good reproducible method (Cividanes *et al.*, 2010b). Moreover, the temperature required for the mullite crystallization is low (1,000–1,350°C) compared to traditional methods, such as mixing of reactive powder (1,500–1,700°C) as a consequence of the high homogeneity degree of the precursor (Cividanes *et al.*, 2010b; Hong and Messing, 1998).

The tetraethylorthosilicate (TEOS) in alcoholic medium is one of the most frequently used silica source for synthesizing mullite by sol-gel (Oliveira *et al.*, 2010; Cividanes *et al.*, 2010b). One of the purposes of this study was the replacement of TEOS by silicic acid (aqueous medium), with the aim of reducing costs. The use of silicic acid as silica source leads to higher mullite crystallization temperature. Thus, chemical additives may be used to reduce this temperature, which can overcome the most sol-gel process shortcoming that is to control the hydrolysis and condensation reaction rates of the precursors. This difference in the reactivity of the starting materials can result in phase segregation prior to gelation process. However, the sol homogeneity level can be controlled by the action of chemical additives, such as carboxylic acids, β-diketones or functional alcohols, which act as chelating agents and modify the precursor reactivity (Brandhuber *et al.*, 2005; Campos *et al.*, 2012; Chakraborty, 2008; Inoue, 2004; McMahon *et al.*, 1999; Richardson *et al.*, 1988).

Rietveld proposed a method for refinement of crystal structures based on mathematical algorithms, which consists in comparing the experimental data obtained from X-ray diffraction with calculated data (Rietveld, 1969). This method can be used for the unit cell refinement, crystal structure refinement, microstructure analysis, quantitative analysis of phases and determining the preferred orientation (Le Bail, 2004; Langford *et al.*, 2000). In this work, the Rietveld refinement is used for the determination of *a*, *b* and *c* cell lattices of mullite synthesized at various temperatures. As the parameter *a* is related to the alumina content of mullite formed, the Al/Si molar ratio and the *x* factor of mullite formula can be calculated.

Therefore, in this paper, mullite was obtained by sol-gel process using silicic acid and aluminum nitrite as silica and alumina precursors. Ethylene glycol was used as an agent to control the homogeneity level. The effect of ethylene glycol was analyzed by means of Rietveld refinement method.

EXPERIMENTAL

Silica sol and aluminum nitrate nonahydrate ($Al(NO_3)_3.9H_2O$; Vetec) were used as sources of silica and alumina, in molar ratio Al/Si = 3/1. Ethylene glycol

(Vetec) was used as a chemical additive in four volumetric proportions in relation to silica sol dispersion: 1/0, 1/1, 1/2 and 1/3, and the samples were named as S-0, S-1, S-2 and S-3, respectively. The procedure of obtaining silica sol and the mullite precursor gel/xerogel were described elsewhere (Campos *et al.*, 2012). The xerogels were pre-calcined at 430°C (for removal of organic matter), and then were calcined at 1,000; 1,100; 1,200 and 1,250°C for 5 hours. The amounts of mullite crystallized in these samples were determined by a calibration curve constructed using the X-ray diffraction (XRD) techniques described below.

Calcined samples were analyzed by XRD, in a Philips X-ray diffractometer, PW 1830/1840 model, using CuKα radiation and operating at 40 kV and 25 mA. The XRD analysis was performed between 10° and 90°, with a scan step time of 10.1600 s and with a step size of 0.0170°. The Rietveld refinement was performed with the data obtained from the standard reference material LaB_6. This refinement led to a set of instrumental function parameters and the results are obtained by a validation method using a profile matching method. A modified pseudo-Voigt function (TCHZ) was used to fit the profiles.

A calibration curve for mullite was used to quantify the amount of crystallized mullite in the samples calcined at 1,000; 1,100; 1,200 and 1,250°C. This curve is based on XRD analysis of solid mixtures of pure CaF_2 and pure mullite, where CaF_2 is used as internal standard. The XRD intensities at 26° (related to mullite) and 28° (related to CaF_2) were determined, and the relative yield of mullite crystallization was determined according to Eq. 1:

$$Yield = \frac{I_{26}}{I_{28}} \text{ x } 100 \qquad (1)$$

where I_{26} stands for the intensity of the XRD profile at 26° (2θ) divided by the mass of the calcined sample and I_{28} stands for the intensity of the same XRD profile at 28° (2θ) divided by the mass of CaF_2 introduced into the calcined sample.

Rietveld refinements were used to analyze the XRD data with the GSAS/EXPGUI software, and the *a*, *b* and *c* cell lattices were determined. Then, the molar concentration of alumina in the mullite (*m*) was obtained with the linear relationship between the lattice parameter *a* and the concentration of alumina (*m*), according to Eq. 2 (Fischer *et al.*, 1996):

$$m = 144.5 \text{ x } a\text{-}1029.5 \qquad (2)$$

After calculating the molar concentration of alumina (*m*), the parameter *x* of the mullite equation ($Al_{4+2x}Si_{2-2x}O_{10-x}$) was determined with the Eq. 3 (Fischer *et al.*, 1996; Schneider *et al.*, 2008):

$$x = 10 - 6\left[\frac{m+200}{m+100}\right] \qquad (3)$$

Then, the "concentration" of Al and Si in this material (4+2*x* and 2-2*x*, respectively) was determined. Therefore, the molar ratio Al/Si (mullite stoichiometry) can be obtained.

Scanning electron microscopy (SEM) analyses were made using the Jeol JSM-5310 microscope, in order to observe the influence of heat treatment on the microstructure of ceramic powders. The surface area analysis of the materials calcined at 1,000; 1,100; 1,200 and 1,250°C were made by Bruner-Emmett-Teller (BET) specific surface area measurements, using a Quantachrome NOVA-1200 equipment and nitrogen as the gas.

RESULTS

Figure 1 shows the X-ray diffraction of samples S-0, S-1, S-2 and S-3 calcined at 1,000; 1,100; 1,200 and 1,250°C for 5 hours. Figure 1(a) shows the crystallization of only α-alumina after firing sample S-0 at 1,000°C, while samples S-1, S-2 and S-3 formed mullite and spinel phase after firing at the same temperature. The intensities of mullite peaks at 26° of the samples S-1, S-2 and S-3 are practically the same. Using these peaks as references, one can observe the decrease of the intensity in the spinel peak, in 45°, as the ethylene glycol (EG) content is increased. Therefore, EG should increase the homogeneity of mullite precursors and should make it possible to mullite crystallizes at 1,000°C, which is the mullite crystallization temperature of single phase precursor (the most homogeneous one). This increase in the homogeneity was observed in a previous work of our research group (Campos *et al.*, 2012). This work showed, using images of a scanning electronic microscope with energy dispersive X-ray (EDX-elemental mapping), that the samples prepared with EG are constituted by silicon and aluminum particles whose sizes are much smaller than those present in the sample prepared without EG. This previous work (Campos *et al.*, 2012) also showed that the Al(IV) and Al(V) contents in the mullite precursors can be a decisive factor for crystallizing mullite at 1,000°C.

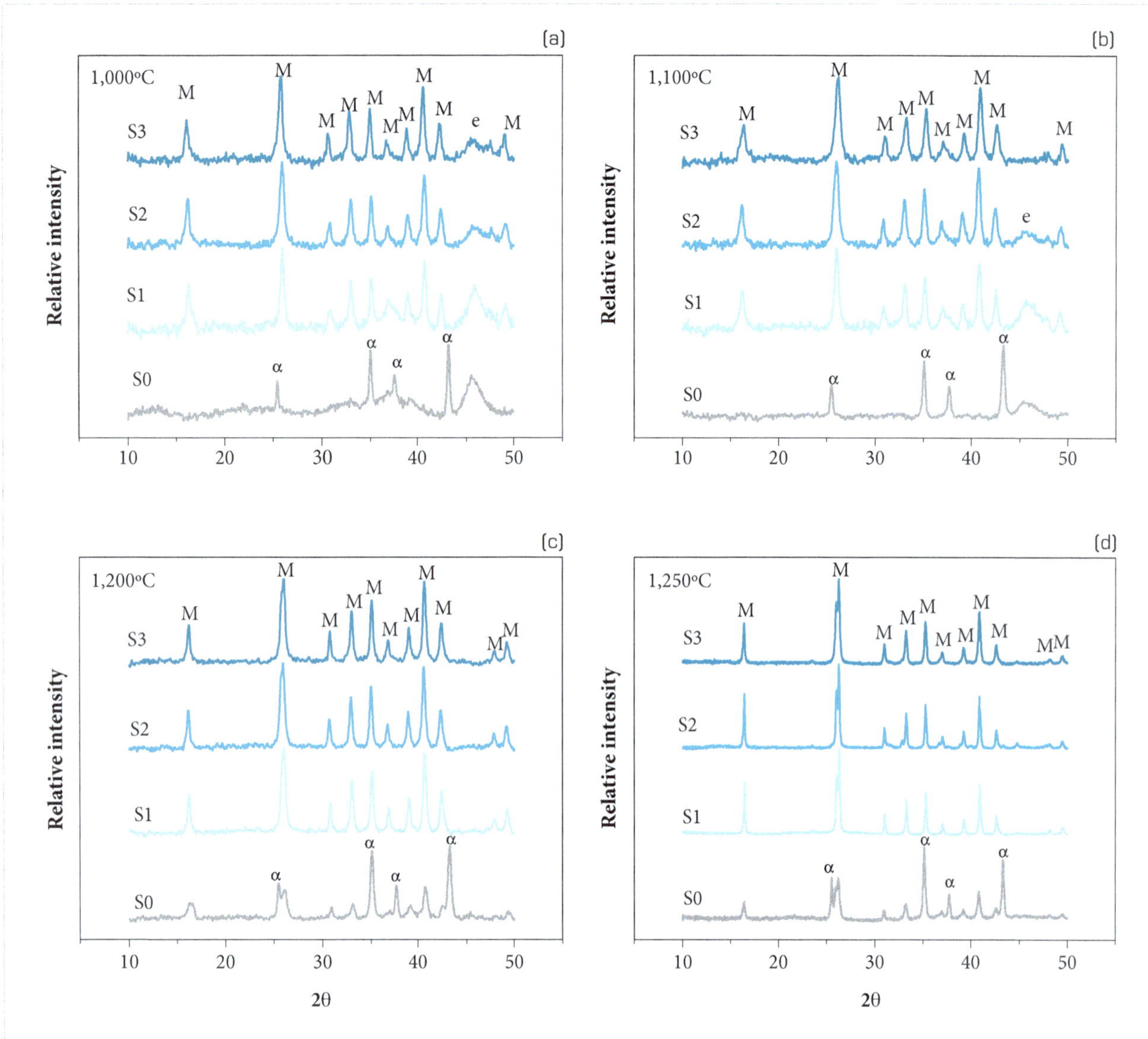

Figure 1. X-ray diffraction patterns of samples S-0, S-1, S-2 and S-3 calcined for 5 hours at: (a) 1,000°C; (b) 1,100°C; (c) 1,200°C and (d) 1,250°C; M: mullite; e: spinel; α: α-alumina.

Sample without EG had low contents of aluminum with these coordinations (Campos *et al.*, 2012).

Figure 1(b) shows that α-alumina is still the only crystalline phase in the sample S-0 and samples prepared with EG crystallized mullite. Figure 1 also shows that the sample with the highest EG concentration, S-3, did not show the presence of spinel phase, but the other two samples prepared with EG did.

Figure 1(c) shows the disappearance of the spinel phase in all samples fired at 1,200°C and the formation of mullite phase. Thus, mullite was formed by consumption of spinel. One can still observe that while the samples S-1, S-2 and S-3 formed only mullite at 1,200°C, the sample S-0 formed mullite and α-alumina. Mullite could be obtained from sample S-3 at 1,200°C, which is close to the crystallization temperature for mullite obtained from TEOS (Oliveira *et al.*, 2010; Ban *et al.*, 1996). Figure 1(d) shows the mullite formation in all samples calcined at 1,250°C, but sample S-0 continues to form the α-alumina phase together with mullite phase.

Table 1 shows the parameters obtained from Rietveld refinements for mullite crystallization at 1,000°C. There are no data for sample S-0 since mullite did not crystallized at this temperature from this sample. One can see that sample S-1 crystallized mullite with the major percentage of alumina (70%) and samples S-2 and S-3 show practically the same percentage (~ 67%). However, the mullite yield is quite different between the samples, where the sample prepared with the highest EG concentration showed the highest yield (52%), while the sample prepared with the lowest EG concentration showed the lowest mullite yield (35%). The cell parameters *b* and *c* do not depend on the alumina content in the mullite formula, as reported in the literature. Therefore, they are virtually constant in all experiments. The parameter *a* determined for mullite from all samples is very similar too.

It is important to note that the alumina content showed in Table 1 is the Al_2O_3 molar content that is present in the mullite structure, according to the chemical formula of mullite: $Al_{4+2x}Si_{2-2x}O_{10-x}$. In this table, 3/2 mullite has 60% of Al_2O_3 molar content, and 2/1 mullite has 67%.

Table 2 shows that alumina contents in all samples treated at 1,100°C are lower than that treated at 1,000°C. Therefore, samples were initially rich in alumina and the thermal treatment at 1,100°C provoked the incorporation of silica into mullite formula. Samples prepared with higher concentration of EG lead to mullite with higher contents of alumina and higher crystallization yields. The lattice parameters *b* and *c* are not depended on the alumina content, but the parameter *a* increases with the increase of alumina content.

Table 3 shows that S-0 crystallized mullite with 53% of alumina, while all samples prepared with EG reach alumina contents closer to mullite 3/2 (64%). However, mullite at this temperature prepared from all samples with EG always formed mullite with virtually the same value of alumina content. One can conclude that the alumina content of mullite depends on the presence of EG, but does not depend on its concentration. However, the mullite yield is strongly dependent on the EG concentration. Samples prepared with higher EG concentration showed higher yield than samples prepared with lower concentration.

Table 4 shows that at 1,250°C all samples crystallize mullite with composition near to Al/Si = 3/2 and all samples that

Table 1. Mullite crystallization for samples calcined at 1,000°C.

Sample	Al_2O_3 (% mol)	*x*	Yield (%)	*a* (Å)	*b* (Å)	*c* (Å)
S-1	70±2	0.48±0.08	35±4	7.61±0.01	7.72±0.01	2.891±0.003
S-2	67±2	0.41±0.08	40±4	7.58±0.01	7.72±0.01	2.888±0.002
S-3	68±2	0.43±0.08	52±4	7.59±0.01	7.72±0.01	2.890±0.002

Table 2. Mullite crystallization for samples calcined at 1,100°C.

Sample	Al_2O_3 (% mol)	*x*	Yield (%)	*a* (Å)	*b* (Å)	*c* (Å)
S-1	62±2	0.29±0.08	37±4	7.552±0.007	7.675±0.005	2.879±0.001
S-2	64±2	0.34±0.08	51±4	7.568±0.005	7.693±0.003	2.887±0.001
S-3	67±2	0.41±0.08	62±4	7.589±0.006	7.673±0.003	2.884±0.001

Table 3. Mullite crystallization for samples calcined at 1,200°C.

Sample	Al_2O_3 (% mol)	*x*	Yield (%)	*a* (Å)	*b* (Å)	*c* (Å)
S-0	53±2	0.1±0.1	41±4	7.493±0.008	7.687±0.007	2.872±0.002
S-1	64±2	0.33±0.08	69±4	7.563±0.002	7.690±0.002	2.8885±0.0007
S-2	64±2	0.35±0.08	81±4	7.568±0.003	7.700±0.002	2.8877±0.0007
S-3	65±2	0.36±0.08	90±4	7.575±0.003	7.699±0.003	2.8891±0.0008

Table 4. Mullite crystallization for samples calcined at 1,250°C.

Sample	Al_2O_3 (% mol)	*x*	Yield (%)	*a* (Å)	*b* (Å)	*c* (Å)
S-0	63±2	0.32±0.08	66±4	7.560±0.004	7.696±0.004	2.886±0.001
S-1	65±2	0.36±0.07	100±4	7.571±0.001	7.696±0.001	2.8886±0.0002
S-2	65±2	0.36±0.07	100±4	7.572±0.001	7.691±0.001	2.8878±0.0003
S-3	64±2	0.33±0.07	100±4	7.563±0.001	7.695±0.01	2.8869±0.0003

were prepared with EG reached a yield of 100%, while the sample prepared without EG reached a yield of 66%. All cell lattices are practically the same in the mullite crystallized at this temperature.

Schneider and Komarneni (2005) showed the influence of the Al_2O_3 content of mullite ceramics on the mechanical properties of the material. They showed that 3/2 mullite (60 mol % Al_2O_3) has about 380 MPa of bending strength at 1,300°C in air, while 2/1 mullite (67 mol % Al_2O_3) has about 240 MPa at the same conditions. They also showed that 3/2 mullite has about 3.4 $MPa.m^{1/2}$ of fracture toughness and 2/1 mullite has about 2.4 $MPa.m^{1/2}$, both at 1,300°C in air. Therefore, one can conclude that 3/2 mullite shows better mechanical properties than 2/1 mullite, which can increase the life time and the reliability of the 3/2 mullite-based pieces for using in aerospace applications. The 2/1 mullite-based pieces can fail during application in a shorter time than 3/2 mullite, because of the Al_2O_3 content. Therefore, Rietveld method can be used as an analytical tool to ensure that the alumina content would not interfere with the mullite properties, since it can determine *a* cell lattice of mullite and this parameter is related to alumina content.

The results obtained by BET for samples S-0, S-1, S-2 and S-3 calcined at 1,000 and 1,200°C are shown in Table 5. Samples S-1, S-2 and S-3 show higher surface area than S-0 when they were fired at 1,000°C. However, these surface areas are practically the same after firing at 1,200°C, since the surface area of samples S-1, S-2 and S-3 decreased substantially when they were fired at 1,200°C. These results can be correlated to the results previously expressed in Tables 1 to 4, which showed that EG increased the homogeneity of the mullite precursors, accelerating the incorporation of silica into mullite formula, decreasing the alumina contents or increasing the mullite yield. EG molecules should act during the sol-gel stage decreasing the polymerization kinetics of silica, resulting in small silica clusters. These small silica clusters may be more homogeneously mixed with those of alumina. Single phase mullite gels are very homogeneous because they have aluminum and silicon mixed in an atomic level.

Therefore, EG increases the homogeneity of the precursor, favoring the overall mullite kinetics. The higher surface area of the samples prepared with EG calcined at 1,000°C should be correlated with the homogeneity of these samples.

Campos *et al.* (2012) also showed that samples prepared with EG were constituted by silicon and aluminum particles whose sizes are much smaller than those present in the sample prepared without EG. Samples prepared with EG also showed good homogeneity, since they crystallizes mullite in lower temperatures compared to the sample prepared without EG.

Figure 2 shows the SEM micrographs of samples S-0, S-1 and S-3 calcined at 1,000°C for 5 hours (Fig. 2a, b and c, respectively) and 1,250°C for 5 hours (Fig. 2d, e and f, respectively). SEM micrograph of sample S-2 is not shown because it was very similar to the samples S-1 and S-3.

It can be seen from the images, and in agreement with the BET results, that samples with the same EG/silanol volumetric proportion and calcined at 1,000°C present smaller particles compared to samples calcined at 1,250°C. Furthermore, for both temperatures (1,000 and 1,250°C), it is clear that S-0 sample presents particles with irregular shapes and samples S-1 and S-3 contain particles with near-angular shapes. This changing in the morphology of the particles is explained by the changing in the phases (from α-alumina to mullite, or from mullite with different alumina contents) present in the samples.

Table 5. Effect of heat treatment on mullite amount and surface area.

1,000°C			
Sample	Al_2O_3 content	Yield % (mol/mol)	Surface area (m^2/g)
S-0	–	–	17
S-1	70	35	30
S-2	67	40	30
S-3	68	52	31
1,200°C			
Sample	Al_2O_3 content	Yield % (mol/mol)	Surface area (m^2/g)
S-0	53	41	6
S-1	64	69	7
S-2	64	81	10
S-3	65	90	10

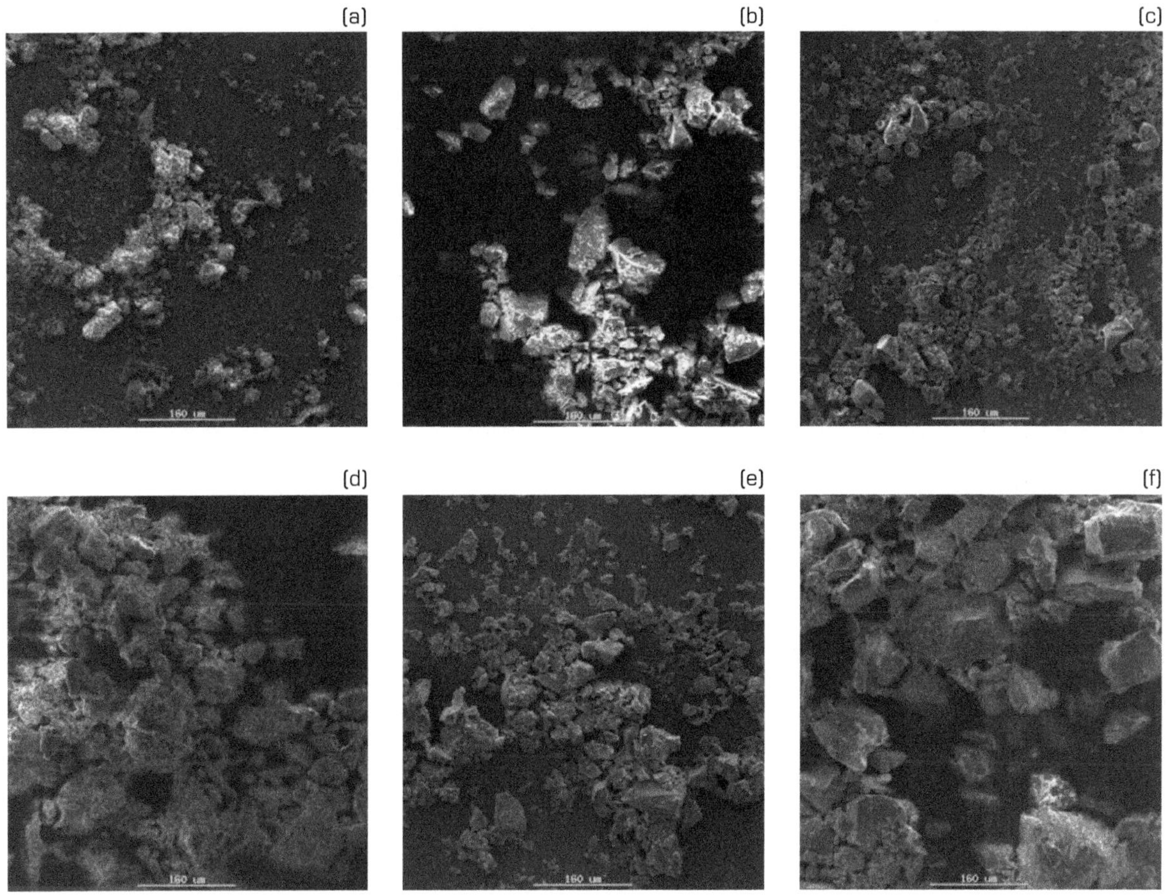

Figure 2. Scanning electron microscopy micrographs of samples S-0, S-1 and S-3 calcined at: 1,000°C for 5 hours (a, b and c, respectively) and 1,250°C for 5 hours (d, e and f, respectively).

CONCLUSIONS

EG has a positive effect in the crystallization of mullite, with the suppression of undesirable phases like α-alumina. The experimental methodology used here showed to be effective for obtaining mullite powder with fine particles, and in the temperature near 1,200°C (near the crystallization temperature when using TEOS as precursor), with a lower cost than the methods that use TEOS as silica source. These fine powders have adequate characteristics to be applied in the sintering process of heat insulation or heat exchanges parts.

The X-ray analysis and the Rietveld refinement showed that *a* cell lattice of mullite and the Al/Si molar ratio in the mullite formula depend on the presence of EG and on the calcination temperature. However, different EG concentrations influence the *a* cell lattice virtually in the same way. The samples prepared with higher concentration of EG lead to mullite with higher crystallization yields at lower temperatures.

The Rietveld method showed to be efficient to analyze the Al_2O_3 content in the mullite, which is associated to the mechanical properties of mullite. Then, this method can be used to ensure that the mullite ceramics will not fail during aerospace application.

ACKNOWLEDGMENTS

The authors gratefully acknowledge CNPq, FAPESP and CAPES for financial support.

REFERENCES

Aksaf, İ.A.and Pask, J.A., 1975, "Stable and Metastable Equilibria in the System SiO_2 -Al_2O_3", Journal of the American Ceramic Society, Vol. 58, No. 11-12, pp. 507-512. doi: 10.1111/j.1151-2916.1975.tb18770.x

Ban, T., Hayashi, S., Yasumori, A., and Okada, K., 1996, "Characterization of Low Temperature Mullitization", Journal of the European Ceramic Society, Vol. 16, No. 2, pp. 127-132. doi: 10.1016/0955-2219(95)00131-X

Brandhuber, D., Torma, V., Raab, C., Peterlik, H., Kulak, A., and Hu, N., 2005, "Glycol-Modified Silanes in the Synthesis of Mesoscopically Organized Silica Monoliths with Hierarchical Porosity", Chemistry of Materials, Vol. 17, No. 16, pp. 4262-4271. doi: 10.1021/cm048483j

Campos, T.M.B., Cividanes, L.S., Brunelli, D.D., Sakane, K.K., and Thim, G.P., 2012, "Effect of ethylene glycol on the mullite crystallization", Journal of the European Ceramic Society, Vol. 32, No. 4, pp. 835-842. doi: 10.1016/j.jeurceramsoc.2011.09.028

Chakraborty, A.K., 2008, "Si-incorporated alumina phases formed out of diphasic mullite gels", Journal of Materials Science, Vol. 43, No. 15, pp. 5313-5324. doi: 10.1007/s10853-008-2775-y

Cividanes, L.S, Brunelli, D.D., Bertran, C.A., Campos, T.M.B., and Thim, G.P., 2011, "Urea effect on the mechanism of mullite crystallization", Journal of Materials Science, Vol. 46, No. 23, pp. 7384-7392. doi: 10.1007/s10853-011-5699-x

Cividanes, L.S, Campos, T.M.B., Rodrigues, L.A., Brunelli, D.D., and Thim, G.P., 2010a, "Review of mullite synthesis routes by sol–gel method", Journal of Sol-Gel Science and Technology, Vol. 55, No. 1, pp. 111-125. doi: 10.1007/s10971-010-2222-9

Cividanes, L.S, Campos, T.M.B., Bertran, C.A., Brunelli, D.D., and Thim, G.P., 2010b, "Effect of urea on the mullite crystallization", Journal of Non-Crystalline Solids, Vol. 356, No. 52-54, pp. 3013-3018. doi: 10.1016/j.jnoncrysol.2010.05.076

Fischer, R.X., Schneider, H. and Voll, D., 1996, "Formation of Aluminum Rich 9:1 Mullite and its Transformation to Low Alumina Mullite upon Heating", Journal of the European Ceramic Society, Vol. 16, No.2, pp. 109-113.

Gerardin, C., Sundaresan, S., Benziger, J., and Navrotsky, A., 1994, "Structural Investigation and Energetics of Mullite Formation from Sol-Gel Precursors", Chemistry of Materials, Vol. 6, No. 2, pp 160-170. doi: 10.1021/cm00038a011

Hong, S.H. and Messing, G.L., 1998, "Anisotropic Grain Growth in Diphasic-Gel-Derived Titania-Doped Mullite", Journal of the American Ceramic Society, Vol. 81, No. 5, pp. 1269-1277. doi: 10.1111/j.1151-2916.1998.tb02478.x

Inoue, M., 2004, "Glycothermal synthesis of metal oxides", Journal of Physics: Condensed Matter, Vol. 16, No. 14, pp. S1291-S1303. doi:10.1088/0953-8984/16/14/042

Langford, J., Louër, D. and Scardi, P., 2000, "Effect of a Crystallite Size Distribution on X-ray Diffraction Line Profiles and Whole-powder-pattern Fitting", Journal of Applied Crystallography, Vol. 33, No. 2, pp. 964-974. doi: 10.1107/S002188980000460X

Le Bail, A., 2004, "Monte Carlo Indexing with McMaille", Powder Diffraction, Vol. 19, No. 3, pp. 249-254. doi: 10.1154/1.1763152

Mazza, D., Ronchetti, S., and Costanzo, A., 2008, "Atomistic simulations on mullite $Al_2(Al_{2+2x}Si_{2-2x})O_{10-x}$ in a variable range of composition", Journal of the European Ceramic Society, Vol. 28, No. 2, pp. 367-370. doi: 10.1016/j.jeurceramsoc.2007.03.003

McMahon, C.N., Alemany, L., Callender, R.L., Bott, S.G. and Barron, A.R., 1999, "Reaction of $Al(^tBu)_3$ with Ethylene Glycol: Intermediates to Aluminum Alkoxide (Alucone) Preceramic Polymers", Chemistry of Materials, Vol. 11, No. 11, pp. 3181-3188. doi: 10.1021/cm990284q

Okada, K., 2008, "Activation Energy of Mullitization from Various Starting Materials", Journal of the European Ceramic Society, Vol. 28, No. 2, pp. 377-382. doi: 10.1016/j.jeurceramsoc.2007.03.015

Oliveira, T.C., Ribeiro, C.A., Brunelli, D.D., Rodrigues, L.A. and Thim, G.P., 2010, "The Kinetic of Mullite Crystallization: Effect of Water Content", Journal of Non-Crystalline Solids, Vol. 356, No. 52-54, pp. 2980-2985. doi: 10.1016/j.jnoncrysol.2010.05.078

Richardson, J.W., Pluth, J.J., Smith, J.V. and Dytrych, W.J., 1988, "Conformation of Ethylene Glycol and Phase Change in Silica Sodalite", The Journal of Physical Chemistry, Vol. 92, No. 1, pp. 243-247. doi: 10.1021/j100312a052

Rietveld, H.M., 1969, "A Profile Refinement Method for Nuclear and Magnetic Structures", Journal of Applied Crystallography, Vol. 2, pp. 65-71. doi:10.1107/S0021889869006558

Shackelford, J.F. and Doremus, R.H., 2008, "Ceramic and Glass Materials", Springer, New York, USA.

Schneider, H., Voll, D., Saruhan, B. and Schmücker, M., 1994, "Constitution of the γ-alumina Phase in Chemically Produced Mullite Precursors", Journal of the European Ceramic Society, Vol. 13, No. 5, pp. 441-448. doi: 10.1016/0955-2219(94)90022-1

Schneider, H. and Komarneni, S., 2005, "Mullite", Wiley-VCH, Weinheim, Germany.

Schneider, H., Schreuer, J. and Hildmann, B., 2008, "Structure and Properties of Mullite – A Review", Journal of the European Ceramic Society, Vol. 28, No. 2, pp. 329-344. doi: 10.1016/j.jeurceramsoc.2007.03.017

Yabuki, M., Takahashi, R., Sato, S., Sodesawa, T., and Ogura, K., 2002, "Silica-Alumina Catalysts Prepared in Sol-Gel Process of TEOS with Organic Additives", Physical Chemistry Chemical Physics, Vol. 4, No. 19, pp. 4830-4837. doi: 10.1039/B205645C

Analysis of Total Ionizing Dose Effects on 0.13 µm Technology-Temperature-Compensated Voltage References

Thiago Hanna Both[1], Dalton Colombo[1], Ricardo Vanni Dallasen[1], Gilson Inácio Wirth[1]

ABSTRACT: The purpose of this work is to briefly discuss the effects of the total ionizing dose (TID) on MOS devices in order to estimate the results of future irradiation tests on temperature-compensated voltage references that are implemented on a mixed-signal chip fabricated using IBM 0.13 µm technology. The analysis will mainly focus on the effects of the parametric variations on different voltage references. Monte-Carlo analyses were performed in order to determine the effects of threshold voltage shifts in each transistor on the output voltage.

KEYWORDS: Ionizing dose, Radiation, TID, Voltage reference.

INTRODUCTION

Space and high-altitude aeronautical applications of electronics are exposed to a continuous action of ionizing radiation. The interaction between radiation and the structure of semiconductor devices results in undesired effects, which may be either transient or accumulated. Predicting these effects and accurately addressing solutions that guarantee operation under radiation is a challenge for the aerospace industry.

The transient effects occur when an energetic particle interacts with a sensitive region of the structure, resulting in a generation of charge that produces an electric pulse at a circuit node (Single Event Effect–SEE). Transient effects may result in hard errors, which are permanent, destructive, faults (Sexton, 2003); soft errors, which are non-destructive faults that affect circuit operation (Karnik and Hazucha, 2004); or disturbances that do not cause a fault.

Accumulated effects can be categorized into displacement damage (DD) and total ionizing dose effects (TID). Displacement damage occurs when the collision of a particle results in non-ionizing energy loss to the medium. This collision displaces an atom from the lattice, damages the crystalline structure of the silicon, and creates a defect that degrades electric parameters of the device (Srour, 2003). The TID usually result from accumulated charge in the dielectric and at the interface between the dielectric and the semiconductor, known, respectively, as oxide-trapped charge and interface-trapped charge.

1. Universidade Federal do Rio Grande do Sul – Porto Alegre/RS – Brazil

Author for correspondence: Thiago Hanna Both | Universidade Federal do Rio Grande do Sul | Avenida Bento Gonçalves, 9500, Campus do Vale, Bloco IV CEP: 91.509-900 Porto Alegre/RS – Brazil | Email: thboth@gmail.com

Ionizing radiation generates electron-hole pairs in the oxides of semiconductor devices. These electron-hole pairs suffer from an initial recombination that is a function of the electric field applied and the energy and type of the incident particle. After this initial recombination, the electrons, which have a higher mobility inside the insulator, escape from the oxide in a few picoseconds. The holes, on the other hand, have lower mobility and are eventually trapped in oxide traps, border traps or interface traps. The detailed mechanism of hole trapping in oxides can be found elsewhere (Oldham and McLean, 2003). In both NMOS and PMOS transistors, oxide-trapped charge is typically net positive, whereas interface-trapped charge is typically net positive for PMOS transistors and negative for NMOS transistors.

The consequence of this trapped charge is the degradation of electric parameters of the semiconductor devices, such as the threshold voltage, the leakage current, and the carrier mobility. It is also reported in the works of Fleetwood and Scofield (1990) and Fleetwood *et al.*, (1994) that TID effects increase the 1/f noise in MOS devices, as shown in Fig. 1. It should be noted that the spikes in Fig.1 are related to the power-line fundamental frequencies and their harmonics.

The 1/f noise in MOS devices is associated with charge carrier trapping near or at the interface between the semiconductor and the insulator. Charge trapping causes carrier number fluctuations and mobility fluctuations in the channel region of transistors. The activity of these traps results in discrete variations of the signal, known as Random Telegraph Signal (RTS), which, in MOS devices, is the major source of 1/f noise. It has also been reported that the annealing of the device after irradiation, for relatively thick oxides, results in a reduction of 1/f noise (Meisenheimer and Fleetwood, 1990), as observed in Fig. 2. The degradation of electric parameters of the device and the increase in the 1/f noise may either result in malfunction or fully disable an electronic system.

The purpose of this work is to review the TID effects on MOS devices in order to estimate the effects of an irradiation test on different temperature-compensated voltage references that are integrated in one chip using IBM 0.13 μm technology. Monte-Carlo analyses were performed in order to identify how parametric variations affect different topologies.

TEMPERATURE-COMPENSATED VOLTAGE REFERENCES

Voltage references are building blocks that are present in a large variety of circuits; for instance, voltage regulators, comparators, and data converters. A voltage reference circuit should provide a stable output voltage despite variations in temperature, power supply, and process. Since voltage reference circuits can limit the accuracy of these applications, it is important to verify the impact of TID on the generated output voltage (V_{REF}).

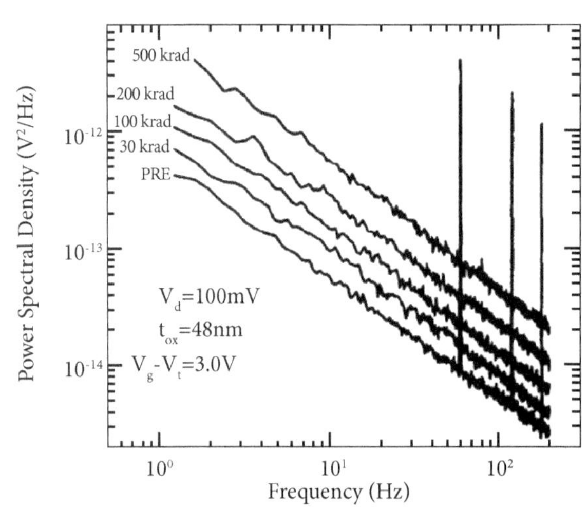

Figure 1. 1/f noise spectra for an irradiated NMOS transistor (oxide thickness of 48 nm). Adapted from Meisenheimer and Fleetwood, 1990.

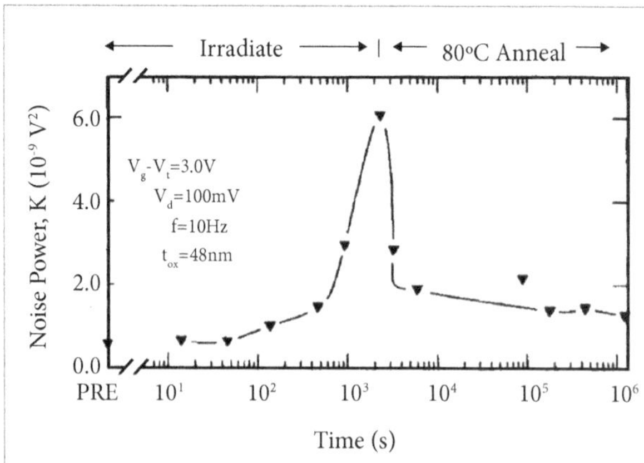

Figure 2. Noise power response during irradiation and annealing of an NMOS transistor (oxide thickness of 48 nm). Adapted from Meisenheimer and Fleetwood, 1990.

The traditional implementation for voltage reference is the bandgap circuit, where the generated V_{REF} is the bandgap voltage of silicon extrapolated to absolute zero Kelvin (i.e., ~1.2 V). Due to its stability regarding temperature and process, bandgap references have been used for the past forty years.

Bandgap references generate a temperature-compensated V_{REF} through a balanced sum of the diode voltage and the thermal voltage. The diode voltage (or base-emitter voltage) has a negative temperature coefficient (TC), while the thermal voltage has a positive TC. The thermal voltage is given by k*T/q, where k is the Boltzmann constant, T is the temperature and q is the electron charge.

Another topology of voltage reference that is widely used in the industry is the threshold-voltage-(V_{TH})-based reference. This topology of reference generates V_{REF} that is equal to the threshold of a transistor extrapolated to absolute zero Kelvin (Colombo *et al.*, 2011). The threshold-based voltage reference has gained increasing importance recently because of its ability to operate under low supply voltages (e.g., less than 1 V). Different from bandgap circuits that use diodes or BJTs, V_{TH}-based references use the gate-source voltage (V_{GS}) of MOS transistors to generate a voltage with negative a TC.

Bandgap and threshold-voltage-based references are two of the most frequently used techniques to generate temperature-compensated V_{REF} and, consequently, they are chosen as the case study for our project. An integrated circuit with five different voltage references was designed. In the fabricated chip, there are four references based on V_{TH}, shown in Figs. 3 to 5. and one bandgap-based reference, shown in Fig. 6.

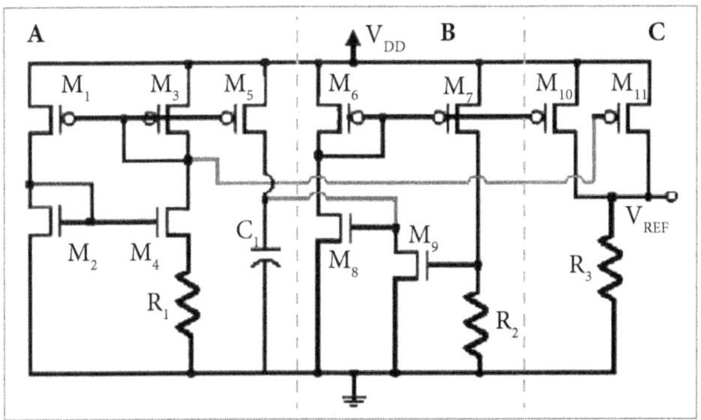

Figure 3. Simulated voltage reference 1 (Colombo *et al.*, 2011).

Reference 2 is the simplest V_{TH}-based voltage reference that can be implemented. Its output voltage is given by the sum of V_{GS} of M6 and the voltage across R2, which is proportional to the temperature. Reference 1 is similar to reference 2, but the temperature compensation is done by adding currents with opposite TC instead of voltages. Reference 3 and 4 are alternative V_{TH}-references that can be implemented with either one or none resistors, respectively. These architectures are appropriate for applications with requirement of a low silicon area. Reference 5 is a bandgap reference that uses a PMOSFET acting as a diode to generate V_{REF}. Further information regarding the designed voltage references can be found elsewhere (Colombo *et al.*, 2012; Banba *et al.*, 1999; Ueno *et al.*, 2009). These circuits were simulated under TID effects.

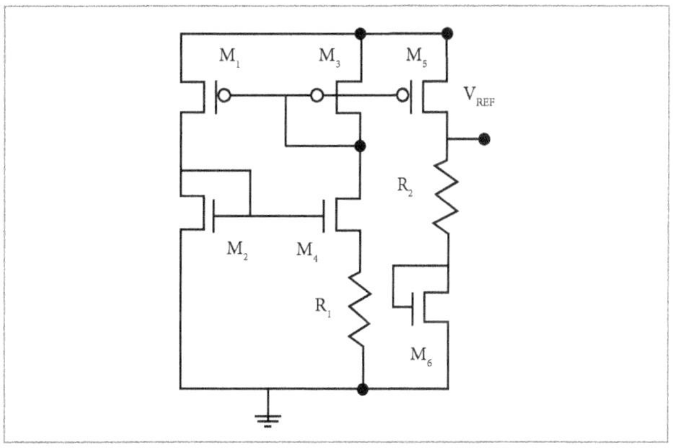

Figure 4. Simulated voltage reference 2 (Colombo *et al.*, 2011).

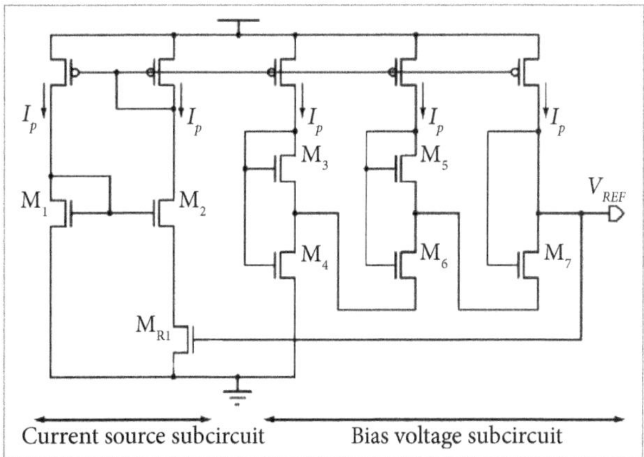

Figure 5. Simulated voltage reference 3 (using a resistor rather than the M_{R1} NMOS transistor) and voltage reference 4 (using the M_{R1} NMOS transistor) (Ueno *et al.*, 2009).

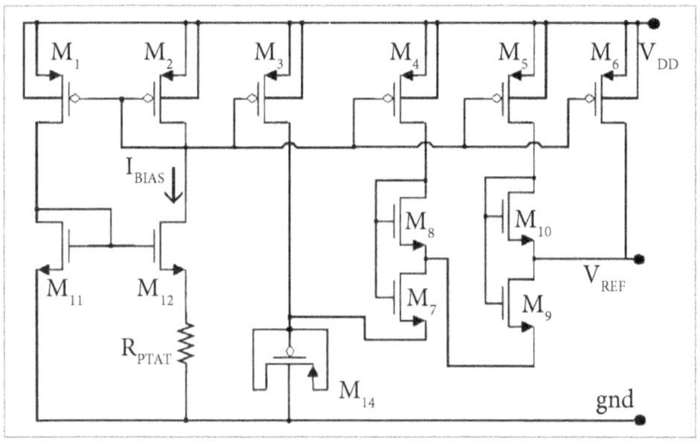

Figure 6. Simulated voltage reference 5 (Colombo *et al.*, 2012).

SIMULATIONS

In order to evaluate total ionizing dose effects on each voltage reference, SPICE simulations were performed, applying threshold voltage shifts to each transistor of the circuit. Data regarding threshold voltage shifts due to total ionizing dose effects on 0.13 μm MOS transistors were obtained from (Haugerud *et al.*, 2005).

The voltage references were simulated for different bias conditions and under different temperature conditions. A total of 10,000 Monte-Carlo simulations were performed for each voltage reference. Process variability, provided in the IBM 0.13 μm commercial design kit, was also included in thesimulations, in an attempt to evaluate TID effects for different circuit conditions. For each of the 10,000 Monte-Carlo simulations, threshold voltage shifts due to process variability and also due to total ionizing dose effects were pseudo-randomly selected from normal distributions and assigned to each transistor of the circuit, meaning that for each simulation every transistor has its own threshold voltage shifts due to TID and also due to variability. It is important to state that data regarding total ionizing dose effects on 0.13 μm technology obtained from (Haugerud, *et al.*, 2005) were not obtained for the 0.13 μm IBM technology used for integration and simulation.

For references 1, 2, 3, and 4, DC simulations were performed for temperature conditions ranging from -40°C to 80°C, at a fixed supply voltage of 1.2 V. For reference 5, DC simulations were performed for temperature conditions ranging from -40 to 80°C, at a fixed supply voltage of 2.5 V. All simulations were performed while considering a pre-irradiation situation (0 krad) and under total doses of 50 krad (Si) and 70 krad (Si).

RESULTS AND DISCUSSION

The output voltage obtained for the five voltage references simulated under 0 krad (Si), 50 krad (Si), and 70 krad (Si) is presented in Figs. 7 to 11. The variation of the standard deviation as a function of temperature and total dose is presented in Tables 1 to 3.

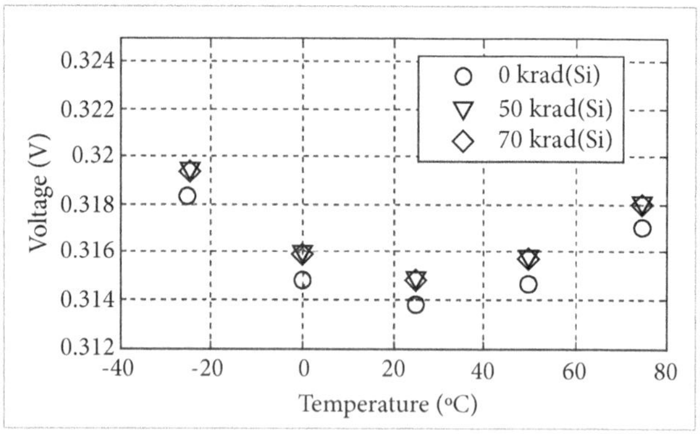

Figure 7. Output voltage for reference 1.

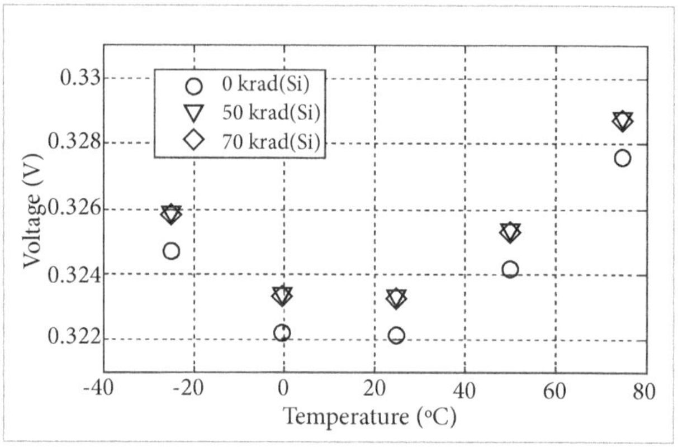

Figure 8. Output voltage for reference 2.

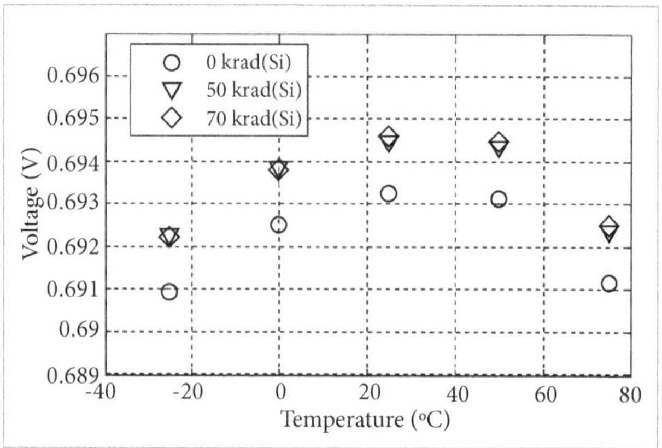

Figure 9. Output voltage for reference 3.

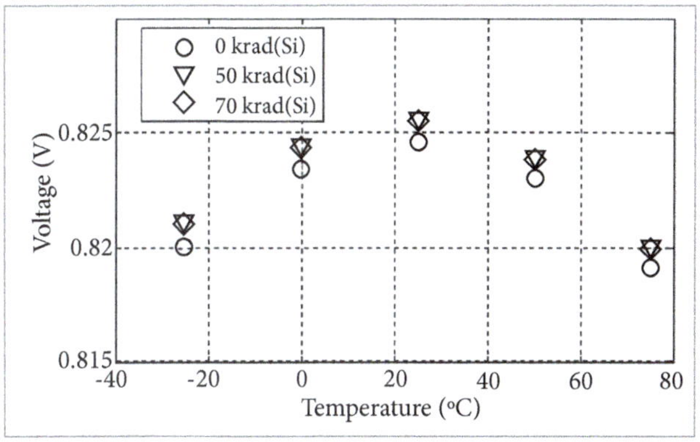

Figure 10. Output voltage for reference 4.

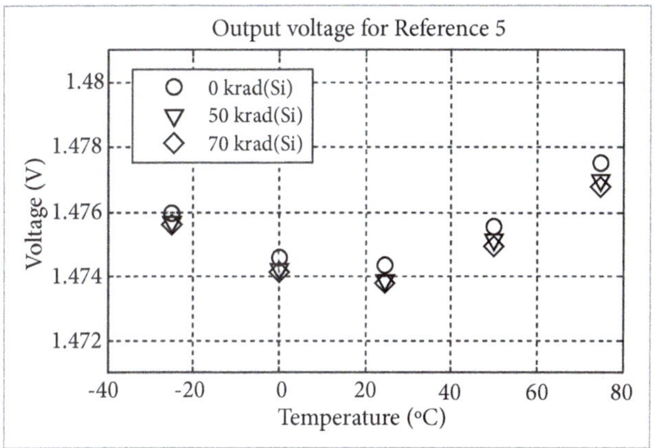

Figure 11. Output voltage for reference 5.

The simulation results indicate that for the 0.13 µm technology voltage references, total ionizing dose is not a major concern. Compared with pre-irradiation simulations, both 50 and 70 krad simulations for the five voltage references showed shifts of a few millivolts in the average output voltage. The standard deviation compared with the pre-irradiation, which considers only process variability, increased by a few millivolts under total ionizing dose for voltage references 1, 2, 3 and 5. On the other hand, voltage reference 4 experienced a decrease of the standard deviation under TID. This reduction was also observed when threshold voltage shifts due to total ionizing dose were applied only to the NMOS transistor M_{R1}, as shown in Fig. 5. The mechanism responsible for this reduction of the standard deviation, however, was not identified.

The observed overall reduction in the standard deviation between 50 krad and 70 krad was expected due to the fact that the standard deviation of the threshold voltage shift data from Haugerud *et al.*, (2005) is reduced between 50 krad and 70 krad, regardless of the fact that the average shift increases.

Table 1. Standard deviation of the output voltage for 0 krad (Si).

Voltage references	Temperature			
	-25°C	0°C	25°C	50°C
1 (mV)	14.7	14.9	15.2	15.4
2 (mV)	14.6	14.9	15.2	15.6
3 (mV)	27.8	28.2	28.8	29.2
4 (mV)	32.0	32.3	32.7	33.1
5 (mV)	23.8	25.5	27.2	29.2

Table 2. Standard deviation of the output voltage for 50 krad (Si).

Voltage references	Temperature			
	-25°C	0°C	25°C	50°C
1 (mV)	18.0	18.2	18.5	18.8
2 (mV)	16.2	16.5	16.9	16.9
3 (mV)	28.1	28.6	29.1	29.6
4 (mV)	27.9	28.2	28.6	29.1
5 (mV)	25.5	21.2	29.0	31.0

Table 3. Standard deviation of the output voltage for 70 krad (Si).

Voltage references	Temperature			
	-25°C	0°C	25°C	50°C
1 (mV)	16.6	16.7	17.0	17.2
2 (mV)	15.6	15.9	16.2	16.5
3 (mV)	28.0	28.5	29.0	29.5
4 (mV)	27.8	28.1	28.5	28.9
5 (mV)	24.7	26.3	28.1	30.0

The average output voltage for all voltage references also presented a small variation between 50 and 70 krad compared with the variation observed between 0 rad and 50 krad. This is expected due to similar threshold voltage shifts for such doses.

The results obtained in this work indicate that these voltage references are functional for doses of approximately 70 krad (Si), and are hence suitable for most space applications. The standard deviation increase of the output voltage due to TID effects, however, should also be considered in order to guarantee proper circuit operation. It is necessary to observe, however, that the data regarding threshold voltage shifts due to TID, employed in simulation, were not obtained for the 0.13 μm IBM technology. Further irradiation tests are necessary to confirm the simulation results.

In addition, the simulations performed in this work did not take circuit layout into consideration. This is an important issue considering that circuit layout techniques, such as the layout of an enclosed gate for NMOS transistors to reduce leakage current beneath the bird's beak region (Mavis and Alexander, 1997), affect radiation tolerance.

CONCLUSIONS

The five voltage references simulated in this work presented tolerance to total ionizing dose effects for doses of approximately 70 krad (Si). For the simulated 0.13 μm technology, parametric shifts due to TID effects resulted in shifts of a few millivolts in the output voltage of voltage references, and in the standard deviation of the output voltage. Despite the positive results in simulations, irradiations tests are required, as circuit layout, for instance, was not accounted for during simulation.

ACKNOWLEDGMENTS

The authors would like to acknowledge the support rendered by CNPq, FAPERGS, and AEB for partially funding this research.

REFERENCES

Banba, H., Shiga, H., Umezawa, A., Miyaba, T., Tanzawa, T., Atsumi, S. and Sakui, K., 1999, "A CMOS bandgap reference circuit with sub-1-V operation", IEEE Journal of Solid-State Circuits, Vol. 34, pp. 670-674.

Colombo, D., Fayomi, C., Nabki, F., Ferreira, L.F., Wirth, G. and Bampi, S., 2011, "A Design Methodology Using the Inversion Coefficient for Low-Voltage Low-Power CMOS Voltage References", Journal of Integrated Circuits and Systems, Vol. 6.

Colombo, D., Werle, F., Wirth, G. and Bampi, S., 2012, "A CMOS 25.3 ppm/°C bandgap voltage reference using self-cascode composite transistor", IEEE Third Latin American Symposium on Circuits and Systems (LASCAS), pp. 1-4.

Fleetwood, D.M. and Scofield, J.H., 1990, "Evidence that similar point defects cause 1/f noise and radiation-induced-hold trapping in metal-oxide-semiconductor transistors", Physical Review Letters, Vol. 64, pp. 579-582.

Fleetwood, D.M. Meisenheimer, T.L. and Scofield, J.H., 1994, "1/f noise and radiation effects in MOS devices", IEEE Transactions on Electron Devices, Vol. 41, pp. 1953-1964.

Haugerud, B.M. Venkataraman, S., Sutton, A.K., Prakash, A.P.G., Cressler, J.D., Guofu Niu, Marshall, P.W. and Joseph, A.J., 2005, "The impact of substrate bias on proton damage in 130 nm CMOS technology", IEEE Radiation Effects Data Workshop Proceedings, pp. 117-121.

Karnik, T. and Hazucha, P., 2004, "Characterization of soft errors caused by single event upsets in CMOS processes", IEEE Transactions on Dependable and Secure Computing, Vol. 1, pp. 128-143.

Mavis, D.G. and Alexander, R., 1997, "Employing radiation hardness by design techniques with commercial integrated circuit processes", Digital Avionics Systems Conference, 16th DASC., AIAA/IEEE, Vol. 1, pp. 15-22.

Meisenheimer, T.L. and Fleetwood, D.M., 1990, "Effect of Radiation-Induced Charge on 1/f Noise in MOS Devices", IEEE Transactions on Nuclear Science, Vol. 37.

Oldham, T.R. and McLean, F.B., 2003, "Total Ionizing dose effects in MOS oxides and devices", IEEE Transactions on Nuclear Science, Vol. 50, pp. 483-499.

Sexton, F.W., 2003, "Destructive single-event effects in semiconductor devices and ICs", IEEE Transactions on Nuclear Science, Vol. 50, pp. 603-621.

Srour, J.R., 2003, "Review of displacement damage effects in silicon devices", IEEE Transactions on Nuclear Science, Vol. 50, pp. 653-670.

Ueno, K., Hirose, T., Asai, T. and Amemiya, Y., 2009, "A 300 nW, 15 ppm/°C, 20 ppm/V CMOS Voltage Reference Circuit Consisting of Subthreshold MOSFETs", IEEE Journal of Solid-State Circuits, Vol. 44, pp. 2047-2054.

5

The Compact Irradiator Modulus Designed for DNA Repair and Mutagenesis Studies in ISS Microgravity Environment Using UVA Emitted by Light-Emitting Diodes

Marcelo Sampaio[1], Heitor Evangelista[2], Roberto d'Amore[3], Nasser Ribeiro Asad[2], Lídia Maria Buarque de Oliveira Asad[2], Adriano Caldeira de Araújo[2], Ana Paula Hagge Brasil[2], Nelson Veissid[1], Valeri Vlassov[1], Alessandra Pacini[4], Monique Thérèze Schulz Fontoura[2]

ABSTRACT: This work presents the design and characteristics of a new compact ultraviolet (UV) irradiator used in a biological onboard space flight experiment. The experiment, called DRM, took place in the International Space Station research facility (ISS-13 expedition), during the Centenary Mission (Russian-Brazil) in March-April 2006. The DRM main objective was to correlate the DNA repair mechanism and mutagenesis with microgravity. A compact irradiator apparatus was designed for DRM to allow *in situ* induced radiation in space. This apparatus, called CIM, uses UV-A Light-emitting diodes (LEDs) with 375 nm wavelength as molecular lesions inducers on four bacterial *E. coli* strains. The manned space mission restrictions were focused on during the CIM main parts design. The ultraviolet dosimetry is also described in this document as DRM experiment results and the CIM operational data are reported to certify the CIM design and DRM protocol compatibility in space application.

KEYWORDS: Ultraviolet radiation, Microgravity, ISS.

INTRODUCTION

During the last decades, experiments using different bacteria (e.g.: *Bacillus subtilis* HA 101, *Echerichia coli* B/r, *Echerichia coli* PQ37, *Echerichia coli* 3 cp, *Deinococcus radiodurans* type R, *Deinococcus radiodurans* type R+rec30) have been conducted and their preliminary conclusions point to no significant statistical difference existing in DNA repair mechanism responses to equivalent radiation doses under microgravity or terrestrial conditions. These experiments have been developed under an approximately common basic protocol: essentially, they have used facilities of irradiation on Earth (e.g.: employing X-ray, gamma ray or UV-C radiation) to achieve a large spectrum of dose responses, followed by cooling and transportation to the microgravity environment, where the samples were incubated under controlled temperatures, ranging normally from + 20°C to + 37°C, depending on the experiment type and organism employed. In these conditions, the biological systems are available to repair damages in the DNA from naturally occurring and artificially induced radiation. After a period of time in space, they were cooled again and sent back to the Earth's surface, where their results were compared with ground controls. One exception was the *in situ* space irradiation experiment, conducted by Pross *et al.* (2000) during the Shuttle Atlantis flight STS-84. The experiment employed a ^{63}Ni beta source to irradiate diploid mutant rad54-3 of the yeast *Saccharomyces*

1.Instituto Nacional de Pesquisas Espaciais – São José dos Campos/SP – Brazil **2.**Universidade do Estado do Rio de Janeiro – Rio de Janeiro/RJ – Brazil **3.**Instituto Tecnológico de Aeronáutica – São José dos Campos/SP – Brazil **4.**Universidade do Vale do Paraíba – São José dos Campos/SP – Brazil.

Author for correspondence: Marcelo Sampaio | Instituto Nacional de Pesquisas Espaciais | Avenida dos Astronautas 1758, São José dos Campos/SP | CEP: 12.227-010 – Brazil | Email: marcelo.sampaio@dge.inpe.br *cerevisiae*. They also concluded that no significant difference was observed for either induction or repair but due to the type of radiation used in the experiment, their conclusion was restricted to double-strand breaks. Table 1 shows a brief compilation of space flight experimental conditions employing studies of DNA repair and mutagenesis.

An alternative method to achieve higher dose levels over the biological material in space flights, without using artificial sources aboard, is the external exposition of the biological experiment, as performed during the EXOBIOLOGIE experiment that belonged to the French PERSEUS mission (1999) onboard the MIR station (Rettberg *et al*., 2002). The authors investigated the ability of microorganisms (spores of the *B. subtilis* DNA repair, wild-type strain (HA 101X) and a DNA-repair deficient mutant strain (TKJ 6312, uvr.4Zf.Js pl-1)) to survive high doses of extraterrestrial solar UV alone or in combination with other space parameters (e.g. vacuum, microgravity, and cosmic radiation). They reported the protective effects over microorganisms of inorganic substances like artificial or real meteorites (in a powdered state).

Other microbiological experiment compilations which discuss ionized radiation conducted in Low Earth Orbit (LEO) can also be found at Olsson-Francis and Cockell (2010). If, on the one hand, there is acceptable evidence that weightlessness alters several cellular functions and effects signaling pathways associated to cell proliferation, differentiation and death (Manti, 2006), its effective influence on DNA repair still raises controversies, despite an apparent trend to invariance between ground and space results. Kiefer and Pross (1999) concluded that the radiation repair is not impaired by the space environment and, based on the premise that the repair mechanism appears to be evolutionarily conserved, they further argued that it also holds true for mammalian cells. Nevertheless, direct comparison of onboard and terrestrial experimental results still presents important constraints: the interaction of the total spectra of cosmic ionizing radiation with an organism is not fully simulated on the ground, especially in the case of HZE particles, the ones with $Z > 2$. They are ~ 1% of the total galactic cosmic radiation with energies high enough to penetrate at least 1 mm of the spacecraft walls. HZE particles may represent up to 50% of the effective dose in space (Ottolenghi *et al*., 2001). Their contribution of the radiation derived from the by-products originating from collisions of the cosmic rays primary beams with the nuclei of the materials which compose the spacecraft shielding.

In the history of radiobiological experiments in space, the use of onboard radiation sources occurred between the Gemini 3 Mission and the Spacelab D1 Mission, the results of which were reported between 1967 and 1986 (Kiefer and Pross, 1999). In this period, the onboard experiments used gamma ray radiation point sources that produced doses varying from 60 to 1,600 Gy, and beta radiation of ^{32}P and ^{85}S". These experiments focused on alterations in the chromosomal or hematopoietic tissue, formed by fibers and cell types that support the blood tissue-forming cells (stem cells). Nevertheless, restrictions were imposed on the use of these manufactured ionizing radiation sources, which may generate high doses and high dose rates onboard during space missions, since their use could result in increased risks to the crew members. The use of ultraviolet lamps also created safety problems due to their toxic vapor contents. More recently, when there has been a surge of techniques that would significantly improve our knowledge of space biology responses to space radiation, through molecular level research, most radiobiological protocols have still been limited to pre-flight irradiation conditions. A promising alternative to that is the use of unmanned, free-flying nanosatellites, like the SESLO experiment during the Space Environment Organism/Organic Exposure to Orbital Stresses (O/OREOS), mission (Nicholson *et al*. 2011). The mission was launched to a 650 km Earth orbit, providing a total dose rate in the experiment 15 times higher than observed at the ISS (exposed to microgravity, ionizing radiation, and heavy-ion bombardment due to its high-inclination orbit), using living organisms (*B. subtillis* wild-type 168 and radiation-sensitive mutant WN1087 strains).

In view of the above limitations, we have developed the DRM (DNA Repair Under Microgravity) experiment to observe the biological response at the molecular level due to the combination of naturally occurring cosmic radiation and induced UV-A. We have developed a portable, hermetic UV-A irradiator, based on UV-A LED technology to conduct the DRM experiment, the Compact Irradiator Modulus (CIM). Here the UV-A irradiation was used in order to just enhance the biological response (not because it is significant for space missions) and observe if any synergistic response would occur in the presence of cosmic radiation. The equipment allowed for the study of the DNA repair and mutagenesis of *E. coli* K12 strains in conditions of irradiation in space (International Space Station-ISS/13th mission), during the Centenary Space Mission (Brazil-USA-Russian) in March-April 2006. Here we describe the CIM apparatus and the DRM experiment procedures and results. We have also characterized the space weather conditions during the experiment.

Table 1. A brief summary of recent radiobiological studies involving DNA repair in bacteria, yeast and human lymphoblastoid during flight missions.

Biological system	Type of radiation	Site of irradiation	Dose	Flight payload; Mission	Reference
Saccharomyces cerevisiae	X-ray, 80 kV	Ground	> 140 Gy	STS-42; IML-1	Pross *et al.*, 1994
Bacillus subtilis	UV-C, 254 nm	Ground	> 335 J m^{-2}	STS-65; IML-2	Horneck *et al.*, 1996
Escherichia coli B/r	X-ray, 150 kV	Ground	120 Gy	STS-65; IML-2	Horneck *et al.*, 1996
Escherichia coli PQ37	Gamma ray, ^{60}Co	Ground	300 Gy	STS-65; IML-2	Horneck *et al.*, 1996
Deinococcus radiodurans	Gamma ray, ^{60}Co	Ground	2-12 kGy	STS -65; IML-2	Kobayashi *et al.*, 1996
Saccharomyces cerevisiae	X-ray, 80 kV	Ground	140 Gy	STS-76; SMM-03	Pross and Kiefer, 1999
Deinococcus radiodurans	Gamma ray, ^{60}Co	Ground	1-12 kGy	STS 79	Kobayashi *et al.*, 2000
Saccharomyces cerevisiae	Beta ray, ^{63}Ni	Space	Not mentioned	STS 84	Pross *et al.*, 2000
Escherichia coli KY396	X-ray, 150 kV and UV-C, 254 nm	Ground	150 Gy and 2 J m^{-2}	STS-91	Takahashi *et al.*, 2001
Saccharomyces cerevisiae	UV-C, 254 nm	Ground	30-300 J m^{-2}	STS-91	Takahashi *et al.*, 2001
Saccharomyces cerevisiae	X-ray, 150 kV	Ground	100-750 Gy	STS-91	Takahashi *et al.*, 2001
Saccharomyces cerevisiae	X-ray, 150 kV	Ground	100-750 Gy	STS-91	Takahashi *et al.*, 2001
Spores of the *Bacillus subtilis* wildtype strain (HA 101X) and DNA-repair deficient mutant strain (TKJ 6312, uvrA10 spl-1)	Full extraterrestrial UV spectrum	Space	36.8-48.7 mGy	MIR; EXOBIOLOGIE/ Perseus	Rettberg *et al.*, 2002
Deinococcus radiodurans and *Bacillus sp.* (PS3D)	Extreme UV, 30.4 nm	Space	$6x10^{16}$ photons m^{-2}	Terrier Black Brant rocket; SERTS	Saffary *et al.*, 2002
Wild type *Saccharomyces cerevisiae* strain (ATCC 18824)	Satellite environment radiation / Not provided	Space	Not mentioned	Practice 8 recoverable satellite	Yi *et al.*, 2011
Human lymphoblastoid cell lines: TSCE5 (wtp53 gene status) and WTK1 (mp53 gene status)	Full extraterrestrial radiation spectrum	Space	71.2 mSv	ISS	Takahashi *et al.*, 2011
Bacillus subtilis (wild-type 168 and and mutant WN1087)	Satellite environment radiation (650 km Earth orbit)	Space	1.0-1.6 Gy	O/OREOS nanosatellite	Nicholson *et al.* 2011
Human lymphoblastoid TK6 cells	Full extra-terrestrial radiation spectrum	Space	54 mSv	ISS	Yatagai *et al.*, 2012

STS: Space Shuttle Missions.

A specific hardware was designed owing to attend the experiment, the specifications and flight restrictions: weight, dimensions, biological barriers protocols and temperature safety limits. The bacterial samples were isolated in slots wile being irradiated. Each slot is submitted to the environmental cosmic radiation and induced UV-A radiation generated by an array of 18 UV-A 375 nm LEDs. The use of UV-A LEDs in compact biological experiments presented several advantages: it promotes a closer contact of the biological material with the irradiation sources improving the dosimetric model; the geometry of each radiation slot as a small rectangular hermetically closed box resulted in a relatively well distributed dose to the biological material inside them; UV-A LED generates 375 nm irradiation with 10 nm bandwidth, at full spectrum half width. They are more resistant to vibration than conventional lamps.

Unlike the previous experiments, the CIM made it possible to irradiate the bacterial strains with a predetermined dose of radiation while in the microgravity of space. Thus, it was possible to carry out both irradiation and cell repair in the same environment without having to manipulate the biological material in the confined environment of the spacecraft. Therefore, it is a new experimental process.

REQUIREMENTS AND TECHNICAL CONSTRAINTS

The CIM apparatus was transported to the ISS by the Soyuz TMA-8 spacecraft. Once onboard the ISS, it was operated in the Russian Segment (RS) of the space station before returning to Earth aboard Soyuz TMA-7 on April 9, 2006.

The requirements and constraints imposed on biological experiments onboard the space shuttles (Cogoli,1996) are basically the same as those of the ISS/Soyuz. The CIM design takes into account all of the safety restrictions of this manned satellite, the Soyus-TMA transport limitations, and the established conditions for the DRM project (SSP41163, 1999; SSP50094, 2000; SSP50146, 1998). The CIM was powered by a DC power supply of the ISS module power supply system (PSS), with a 28.5 ± 0.5 VDC (3A) nominal voltage. The mass was limited to 3 kg for transport aboard the Soyuz to the ISS and to 1.5 kg for the return. Using the appropriate LINS (Line Impedance Stabilization Network), the conducted electromagnetic emissions were found to be between 10 and 100 Hz, and the radiated electromagnetic emissions emitted during CIM operation were between 10 kHz and 1000 MHz. The integrity of the characteristics and the functioning of the CIM system were checked to comply with the expected pressure, temperature, and humidity constraints of the Soyuz and ISS environments. Several Environmental Compatibility requirements had to be met: touch temperature exposed parts; off-gassing (a quantitative analysis of gases that may be generated by non-metallic materials); biological barriers (the guarantee of containment of biological material); external characteristics (no sharp edges, corners and shatterable materials are allowed, in order to avoid any injury during microgravity handlings. The fan blades are protected by 3 mm grid holes); and ergonomy (protection against erroneous operations).

THE COMPACT IRRADIATOR MODULUS (CIM) SYSTEM AND OPERATION

The total CIM hardware is divided into 4 different modules, as shown in Fig. 1 and Fig. 2: the Irradiator Module, with four slots, loaded with biological material that is irradiated during the DRM experiment; the Fan Module, responsible for maintaining the biological material's proper temperature during irradiation period; the Memory Module, which stores all relevant data; and the Electronics Module that controls the proper functioning of all the equipment. Only the Irradiator Module and the Memory Module return to Earth. All CIM electronic components are Commercial Off-The-Shelf (COTS), except for the DC/DC converters, the Input EMI filter, connectors and the DC power cable.

Irradiator Module (IM): The main body of the IM contains the biological material distributed in four 1 mL square wells called "Irradiated slots," where the irradiation takes place; four 25 μL "background slots", where no irradiation occurs; and one fence, designed to contain two thermoluminescent solid state dosimeters (TLD-100 - LiF : Mg,Ti), used to quantify the environmental ionizing radiation. Each one of the 4 irradiated slots has 18 UV-A LEDs (NICHA Co. model NSHU550A UV-A 375 nm) facing its surfaces. The LEDs are distributed over its upper and bottom halves (Fig. 3), containing 9 units on each side. These slots are isolated from the environment by biological barriers that are implemented with silicon and nitrilic rubber o-rings and Scotch Weld® 2216 epoxy adhesive.

The Fan Module (FM): When switched on, each UV-A LED dissipated about 35 mW. Therefore, extra thermal power was generated while a total of 72 LEDs were switched on during the experiment. This process was responsible for a temperature increase of the equipment. The FM was designed to be connected to the irradiation module in order to maintain the biological material temperature never above 39°C; in fact, during the irradiation aboard the ISS, it was always maintained

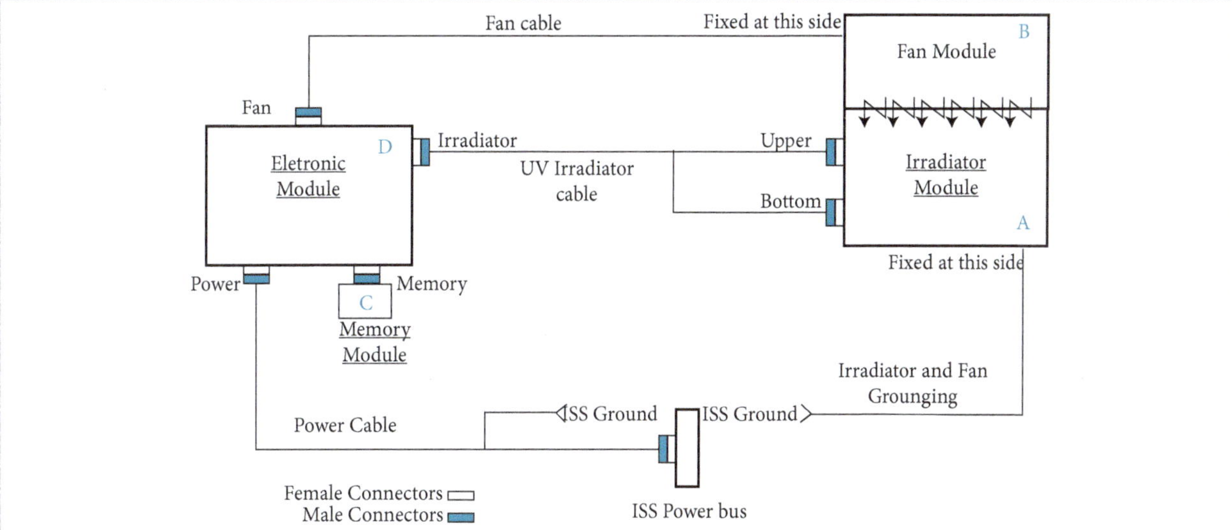

Figure 1. CIM block diagram.

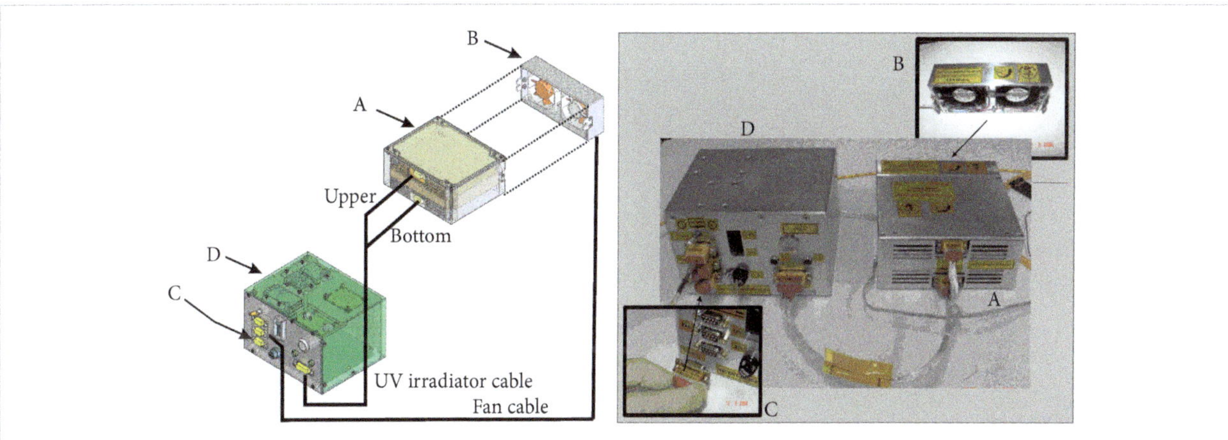

Figure 2. Compact Irradiator Modulus interconnected. (A) Irradiator (IM); (B) Fan (FM); (C) Connector to attach Memory (MM); and (D) Electronic (EM).

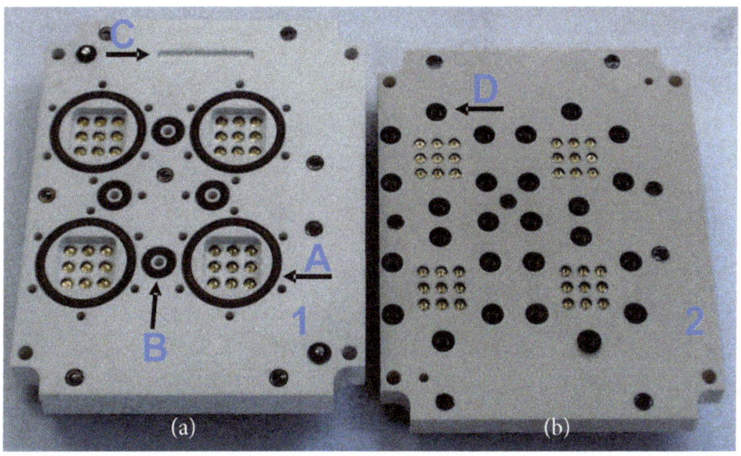

Figure 3. (a) Bottom half; (b) Upper half of inner Irradiator Module. (A) One Irradiated slot with the silicon rubber o'ring; (B) One background slot with the silicon rubber o'ring; (C) Thermoluminescent dosimeter fence; (D) nitrilic rubber o'rings. The screws which fasten the two halves are passed through nitrilic o'rings. The picture presents (4x) 9 UVA LEDs facing the upper half and (4x) 9 LEDs facing the bottom half of the irradiated slot.

below 30°C. Convective cooling was generated by two 60 mm ball bearing type fans (one redundant). Pre-flight temperature controlled tests demonstrated that, while fan cooling was applied, the irradiation slot temperature did not exceed 8°C above room temperature, merely reaching an equilibrium with room temperature.

Memory Module (MM): The MM is composed of two serial EEPROM IC memories (one redundant), mounted in a military/high reliability specification connector and encapsulated with a silicone rubber compound. This module is attached to the Electronic Module by a faced external connector (Fig. 2 c). All relevant experiment data is recorded in it. After integration and initialization of the experiment, this module recorded the 8051 firmware status and the following relevant parameters: (1) the temperature within the slots; (2) the temperature of the irradiation module itself; (3) the irradiation status (on/off) and; (4) the voltage values in the memory module. Time resolution of acquisition data was 1 minute. This data was recovered on Earth at the mission's end.

Electronic Module (EM): This module was composed of the 8051 family microcontroller, which commanded both the IM and FM. The power supply used by the fan module was provided by the ISS DC bus. It had two internal DC/DC converters which provided the DC power to all the other modules of the experiment. The ISS safety requirements specify that maximum temperature of any external surface of an experiment device should not exceed 40°C. To meet this requirement, the aluminum case (alloy AL 6063 T6), used in the EM was deeply anodized in order to increase its heat emissivity (to a depth of not less than 0.82). The less dissipating electronics were placed on the bottom-front panel, while the upper part of the irradiator module was loaded with the 5 most dissipating components: two DC/DC converters, two Voltage Regulators, and one EMI Filter (for a total of 7.6 Watts during the irradiation period).

UV: UV-A dosimetry was conducted by measuring the spatial distribution of light intensity inside all irradiation slots, since the bacteria will occupy this volume randomly during the experiment. The irradiation conditioning was simulated by filling up the 1 mL slot with the same saline solution used in the flight experiment. Parallel planes within the slot composed by the array of 9 LEDs at the upper surface and the same at the bottom of the Irradiator Module were analyzed. UV-A spatial distribution was obtained experimentally by screening the light intensity horizontally and vertically inside the slot, using a photocell covered by a mask with a 1 mm orifice diameter at its center. This photocell was previously calibrated by a UV-A sensor (ORIEL 70260 Radiant Power Meter and ORIEL 70282 Photodiode Absorber Head, that provide sensitivity between 3 mW to 300 mW for a 375 nm wavelength lighting source). Vertical resolution of each step of measurement was 0.1 mm and 0.2 mm for the horizontal axis. Figure 4 depicts the configuration of the isopleths of irradiance at 4 different distances from the LED array (upper or bottom part of the irradiation slot). Taking into account both surfaces with 9 LEDs, we obtain each plane's irradiance distribution. This procedure made it possible to compile a complete description of the UV-A intensity distribution inside the irradiation slot. The statistical analysis of the results is presented in Fig. 5. The median value of this collection is 4.38 $mW.cm^{-2}$, this value was used as total volume slot irradiance (Sampaio, 2007).

Operation and procedures: Aboard the ISS, the astronaut had to follow the unpacking procedures, assemble the Memory and Fan cables; then interconnect the module cables to the EM, and the power cable to the ISS DC bus. Once it was switched on, the EM firmware started a CIM checkup. The Memory and Fan cables were verified and two LEDs on the EM front panel indicated the test condition. After this automatic test, the UV Irradiation was begun by an astronaut command through an EM front panel pushbutton. The IM with FM attached was 154 mm in length, 149 mm wide, and 65 mm high, weighing 1.45 kg. The EM with MM was 149 mm x 130 mm and 100 mm (l,w,h), weighing 1.1 kg. All flight models, including cables, extra fuses and the spare MM, were disposed of in three different Nomex® bags. The overall CIM launch dimensions were 260 mm x 200 mm x 250 mm (l x w x h) and the total mass was 3.08 kg. The return mass was 1.24 kg. During the irradiation period, the maximum CIM consumption was 16.6 W (28 VDC, 0.59 A).

After the irradiation period, the equipment was shut down and disabled. The IM and MM were packaged for return to Earth and all other parts were prepared for disposal.

BACTERIAL STRAINS, TRANSPORT, FLIGHT AND EXPERIMENTAL STRATEGY

The biological material employed in the experiment was composed of bacterial strains derived from *E. coli* K 12 with the following DNA repair phenotypes: AB1157 (wild type), AB2463 (as AB1157, but *recA*13 — deficient in the repair by recombination

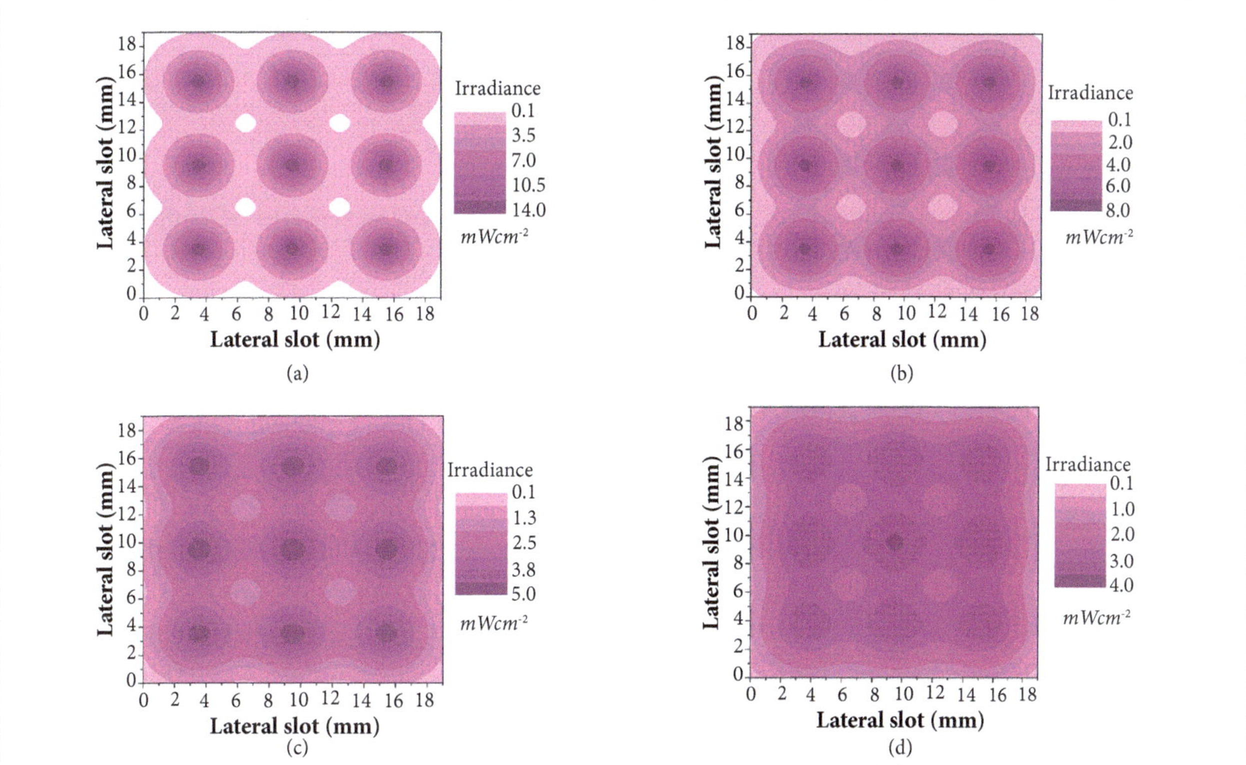

Figure 4. Isopleths plots for layer irradiance distribution inside one slot: horizontal plane at (a) 0.5 mm from LED surface; (b) 1.3 mm; (c) 2.1 mm and (d) 2.9 mm. These layers present partial contribution for one slot surface with 9 LEDs.

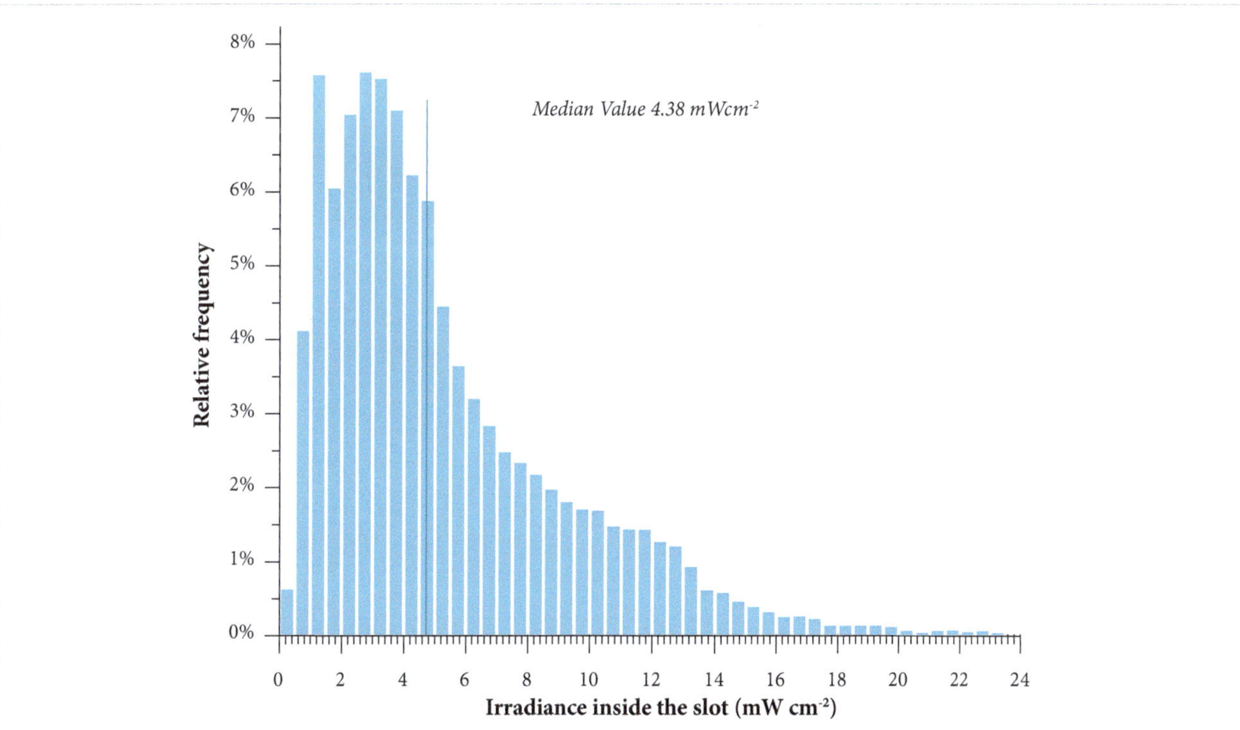

Figure 5. Histogram of Irradiance values inside the irradiated slot. The median value of this collection represents a total slot Irradiance of 4.38 mW.cm^{-2}.

and SOS), AB2480 (as AB1157, but *uvrA-recA* — deficient in nucleotide excision repair, recombination and SOS) and DNA polymerase I deficient strain P3478, a $polA_1$ mutant derivative from *E. coli* wild-type W3110 strain. They were transported from Brazil (Rio de Janeiro State University) directly to the desert of Baykonur (Kazakhstan). The transport was conducted under room temperature and bacterial strains were fixed in Millipore filters of 47 mm diameter, enclosed in a sterilized plastic holder. In Baykonur, a compact biological laboratory was adapted in an installation of the Russian Flight Center, in order to maintain basic conditions concerning the handling and inoculation of the the biological material in the experimental apparatus.

Figure 6 depicts typical experimental setups for pre-irradiation (ground irradiation) and onboard irradiation (mostly aboard the ISS, MIR, nanosatellites and Shuttle missions) as indicated in Table 1. The basic difference in protocol steps is where the irradiation takes place. In both cases, it is assumed that the initial phases of the biological repair mechanism occur in microgravity. The inoculation of the biological material took place 6 hours before the launch of the Soyus TMA-8 spacecraft. On March 29th, the biological material was boarded onto the Reentry Module and on March 31st, 2006 (02:30 h GMT), Soyus was launched from Baykonur. The experimental apparatus was transferred to the ISS on March 31st and the total period of stay aboard the ISS was 9 days and 12 hours. The equipment was installed and operated close to the air conditioning outflow on the "PIRS docking compartment", between the Zarya and Zvezda modules. Indoor temperature at that site was maintained constant, during the 9 days, at ~ +21°C. The apparatus, containing the bacterial strains, returned to Earth on April 9th (GTM: 20:04, April 8th), using the identification "Module of Biological Urgency" and immediately transported to Moscow/Russia where it was analyzed, in the same compact biological laboratory once installed in Baykonur. A detailed chronology of the experiment is presented in Table 2.

EXTERNAL RADIATION DOSE AND SPACE WEATHER CONDITIONS DURING THE EXPERIMENT

Even in low Earth orbits, such as the orbital altitude of the ISS, any biological material is under the influence of Solar Cosmic Rays (SCR) and Galactic Cosmic Rays (GCR), which are formed by particles whose energies vary from 10^{-1} to 10^{8} MeV. Besides these contributions, trapped particles, especially protons (Reitz *et al.*, 2005), by the Earth's magnetosphere in the Van

Table 2. Chronology of the experiment during the Centenary Mission (March-April 2006)

Experimental procedure	Date	Time GMT	Experiment chronology	Temperature amplitude
Inoculation of bacterial strains in the irradiator in Baykonur	March, 29th	06:00 h	00:00 h	~ +25°C
Pre-flight tests	March, 29th	06:30 h	00:30 h	~ +21°C
Refrigeration of the Irradiator with the bacterial strains	March, 29th	07:00 h	01:00 h	~ +10°C
Boarding the apparatus in the Soyus TMA-8	March, 29th	12:20 h	06:20 h	~ +22 to +23°C
Lauching of Soyus TMA-8	March, 30th	02:30 h	20:30 h	~ +22 to +23°C
Soyus TMA 8 docked to the ISS	April, 1st	04:18 h	2 days + 22:18 h	~ +21°C
Begin of UV-A irradiation	April, 7th	13:40 h	9 days + 07:40 h	~ +21°C *
End of UV-A irradiation	April, 7th	17:40 h	9 days + 11:40 h	~ +21°C *
Soyus TMA 7 undocked from the ISS	April, 8th	20:23 h	10 days + 14:23 h	~ +21°C
Reentry at Earth's atmosphere	April, 8th	23:47 h	10 days + 17:47 h	~ +25°C
Delivery of the apparatus to the research team at Moscow	April, 9th	08:00 h	11 days + 02:00 h	~ +10 to +25°C
Begin of survival and mutagenesis tests	April, 9th	12:40 h	11 days + 06:40 h	~ +25°C

(*) a maximum, of short duration, +30°C was reached inside the irradiation module, during the 4 hours of irradiation.

Small Solid Propellant Launch Vehicle Mixed Design Optimization Approach

Fredy Marcell Villanueva[1], He Linshu[1], Xu Dajun[1]

ABSTRACT: For a small country with limited research budget and lack of advanced space technology, it is imperative to find new approaches for the development of low-cost launch vehicles (LV), which is, among all possibilities, an interesting option for rapid access to space, focused on integration of acquired components complemented with indigenously developed subsystems. This approach requires the cooperation of developed countries with huge experience and knowledge in LV development and operations. The main objective is to develop a small three stage solid propellant LV capable of delivering a payload of 100 kg to a circular low earth orbit of 600 km altitude, with the first and second stage solid rocket motors (SRM) hypothetically acquired from different countries and the third one designed and produced domestically in accordance with the production and technological capability. This approach provides main advantages such as: reduction in total time to access the space and to master the basic knowledge of launch operations. For this purpose, an integer continuous genetic algorithm global optimization method was selected and implemented, the SRM characteristics of the first and second stage were considered as integer variables, whereas the design variables of the third stage SRM and the trajectory variable were considered as continuous. A multi discipline feasible (MDF) framework was implemented along with the propulsion, aerodynamic, mass and trajectory models. Despite their particular characteristics and constraints, the results show highly acceptable values, and the approach proved to be reliable for conceptual design level.

KEYWORDS: Launch vehicle, Mixed design optimization, Solid propellant.

INTRODUCTION

The last decade may be characterized by an increased number of small satellites delivered into the low earth orbit (LEO), and this tendency will be dominant in the coming years.

Small satellites have a reduced manufacturing cost, and are relatively easy to operate and maintain. Furthermore, the miniaturization of technology makes possible its delivery into space by using small cost effective launch vehicles (LV).

Small countries generally have a limited research budget oriented to space technology development, however, nowadays it is possible to deliver a small satellite into orbit with a reasonable budget, considering the cooperation with technologically more advanced countries.

This research was focused on finding a way to have rapid access into space and to master the basic knowledge of space development and operations. In such a way, several options had been analyzed, among them the most suitable alternative in terms of economic investment and development time resulted in a small solid propellant LV with mixed design configuration, involving a strong cooperation with different countries.

The strategy considered here prioritizes the technology integration over expensive and time consuming new development, this means that complex and advanced devices were acquired and complemented with indigenous manufactured devices using available resources and technology.

As a result, a three stage solid propellant LV was configured, where the first and second stage solid rocket motors (SRM) were acquired from different providers, complemented with a locally developed third stage SRM, which was designed and optimized to accomplish the specific mission.

1.School of Astronautics – Beihang University – Beijing – China

Author for correspondence: Fredy Marcell Villanueva | Beihang University | School of Astronautics | Haidian District, 37 Xueyuan Road – New Main Building, B923 | Beijing 100191 – China | Email: marcell385@ yahoo.com

In our research, a mixed integer continuous variables genetic algorithm (GA) method has been used in order to optimize the overall configuration of the LV.

LAUNCH VEHICLE MODEL

LAUNCH VEHICLE DEFINITION

A small three stage solid propellant LV in tandem configuration is considered for this research. The mission is to deliver a 100 kg payload to a circular LEO of 600 km of altitude. The payload volume requirements and the instrument module weight were specified beforehand in mission definition analysis and are shown in Table 1.

Table 1. Launch Vehicle data.

Variables	Units	Value
Payload	kg	100
Fairing mass	kg	50
Instrument module	kg	50
Payload deployment module	kg	50

CONSIDERED SOLID ROCKET MOTORS

The considered SRM are listed in the Table 2 and were selected based on the variety of design characteristics. However, it is possible to add additional parameters such as cost, availability, technology complexity, country of origin among others.

PROPULSION ANALYSIS

The propulsion analysis has been conducted for all three stages of the LV, using the classical approach presented in Sutton and Biblarz (2001) and He (2004a; 2004b). For the third stage SRM, a detailed analysis was conducted, considering the properties of the propellant. In this analysis, the burning surface is considered constant by introducing a grain geometry shape coefficient, k_s, the burning surface of the grain S_b can be calculated as:

$$S_b = k_s \, D_m \, L_m \tag{1}$$

where, L_m is the rocket motor cylindrical length and D_m the diameter.

The burning time t_b, grain mass m_{gn}, and mass flow rate $\dot{m}_{gn}$ of the grain are calculated as:

$$m_{gn} = \frac{\pi}{4} \rho_{gn} \eta_v \lambda_{gn} D_m^3 \tag{2}$$

Table 2. Selected solid rocket motors.

	SRM	Grain mass (kg)	SRM mass (kg)	Diameter (m)	Length (m)	Specific impulse (N.s/kg)	Burn time (s)	Mass flow (kg/s)	Thrust (N)
Stage1	11	18400	20791	1990	4.80	2364	65	283.08	669194
	12	15000	16779	1390	7.25	2314	74	202.70	469054
	13	9950	11281	1390	5.20	2280	62	160.48	365903
	14	4530	5207	0.98	4.60	2265	70	64.71	146578
Stage2	21	9800	10950	1990	3.62	2805	65	150.77	422908
	22	5080	5607	1390	3.10	2776	64	79.38	220345
	23	4138	4412	1390	2.86	2746	65	63.66	174815
	24	3700	4190	1390	2.40	2754	68	54.41	149850
	25	3300	3650	1390	2.63	2824	55	60.00	169440
	26	1760	1949	0.98	1.85	2776	46	38.26	106212
	27	650	719	0.85	1.50	2849	43	15.12	43066

$$t_b = \frac{\pi \eta_v D_m}{4 u k_s} \quad (3)$$

$$\dot{m}_{gn} = \rho_{gn} u\ S_b = \rho_{gn} u\ k_s \lambda_{gn} D_m^{\ 2} \quad (4)$$

$$\lambda_{gn} = \frac{L_{gn}}{D_{gn}} \quad (5)$$

where, u is the burning rate of propellant, ρ_{gn} density of the grain, $L_{gn} = L_m + 0.314D_m$ length of the grain, $D_{gn} = D_m$ diameter of the grain, λ_{gn} fineness ratio of the grain (grain length/diameter), and η_v the grain volumetric loading fraction.

The expansion ratio ε, nozzle throat area A_t, and nozzle exit area A_e are calculated as:

$$A_t = \frac{\rho_{gn} u\ S_b}{\Gamma_0 p_{c\max}} \sqrt{R_c T_c} \quad (6)$$

$$\varepsilon = \frac{\Gamma_0}{\left(\frac{p_e}{p_c}\right)^{\frac{1}{k}} \sqrt{\frac{2k}{k-1}\left[1-\left(\frac{p_e}{p_c}\right)^{\frac{k-1}{k}}\right]}} \quad (7)$$

$$A_e = A_t \varepsilon \quad (8)$$

$$\Gamma_0 = \sqrt{k}\left(\frac{2}{k+1}\right)^{\frac{k+1}{2(k-1)}} \quad (9)$$

where, P_c is the chamber pressure, p_e exit pressure, $R_c = 296$ *J/(kg.K)* gas constant, $T_c = 3300$ *K* temperature in the combustion chamber, $P_{c\max} = 1.1P_c$ maximum value of chamber pressure, and $k = 1.2$ the specific heat ratio of gas.

The specific impulse I_{sp}, and the thrust T can be calculated as:

$$I_{sp} = I_{sp}^a + \left(\frac{p_e}{p_c}\right)^{\frac{k-1}{k}} \frac{R_c T_c}{g^2 I_{sp}^a} \quad (10)$$

$$T = I_{sp} \dot{m}_{gn} - p_a A_e \quad (11)$$

where, p_a is the atmospheric pressure, I_{sp}^a average specific impulse, g acceleration due to gravity, and A_e the nozzle exit area.

MASS ANALYSIS

The mass analysis was conducted for the entire LV, and is represented by the following equations:

$$m_{LV} = m_{01} + m_{02} + m_{03} + m_{Inert} \quad (12)$$

$$m_{03} = m_{st} + m_{gn} \quad (13)$$

$$m_{Inert} = m_{IM} + m_{PDM} + m_{PAY} \quad (14)$$

where, m_{LV} is the LV gross mass, m_{01} first stage mass, m_{02} second stage mass, m_{03} third stage solid rocket mass, m_{IM} instrument module mass, m_{PDM} payload deployment module mass, m_{PAY} payload mass, and m_{st} the structural mass of the third stage SRM.

He (2004a; 2004b) provided a methodology and a detailed calculation of the third stage SRM structural mass. This design consisted in a classical metallic case made of high strength steel, ethylene propylene diene monomer (EPDM) for chamber insulation, and carbon phenolic for the nozzle.

AERODYNAMIC ANALYSIS

The aerodynamic coefficients were estimated using the Missile DATCOM 1997 digital (Blake, 1998). This software is easy to use and implemented, and accurate enough for the conceptual design phase. Qazi and He (2005) and Villanueva *et al.* (2013) applied DATCOM in LV aerodynamics analysis. The lift and drag forces were calculated using the following relations:

$$L = C_L q S_{ref} \quad (15)$$

$$D = C_D q S_{ref} \quad (16)$$

where, q is the dynamic pressure, D drag force, L lift force, S_{ref} vehicle reference area, C_L lift coefficient, and C_D the drag coefficient.

The aerodynamic coefficients were calculated repeatedly for each LV configuration, the selected Mach ranged from 0 to 8 and the angle of attack from −8 to +1 degrees.

TRAJECTORY ANALYSIS

The trajectory analysis considers a 3 degree of freedom (3DOF) model, which has been modeled in SIMULINK (Zipfel, 2007; Fleeman, 2001). The previously calculated aerodynamic coefficients, the mass and the propulsion are the input parameters. In order to obtain a quick result, a 2D coordinate system was adopted, the LV flies as a point mass in a non rotating earth model. Figure 1 illustrates the forces acting on a LV and below a set of governing equations of motion (Xiao, 2001). The LV is flying in an inertial reference coordinate system XOY, with

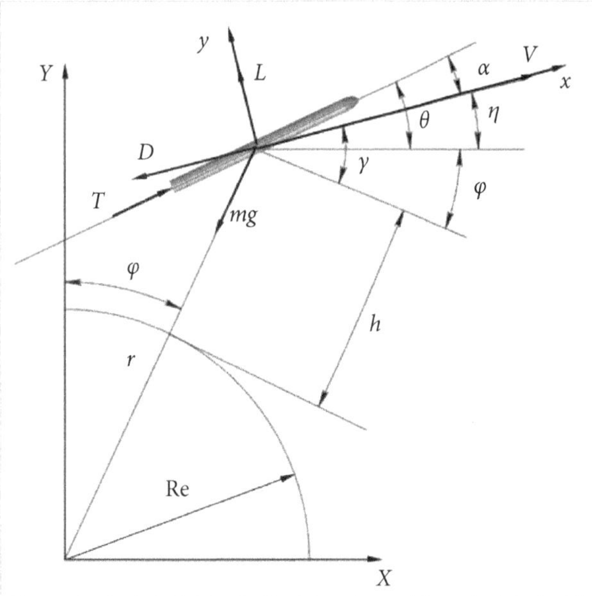

Figure 1. Forces acting on a launch vehicle.

its origin located in the center of the earth. Furthermore, all forces applied to the LV were considered in relation to the body centered velocity coordinate systems xoy as shown in Fig. 1.

$$\begin{aligned} \frac{dV}{dt} &= \frac{T\cos\alpha - D}{m} - g\sin\gamma \\ \frac{d\gamma}{dt} &= \frac{T\sin\alpha + L}{mV} - \frac{g\cos\gamma}{V} + \frac{V\cos\gamma}{r} \\ \frac{dh}{dt} &= V\sin\gamma \\ \frac{d\varphi}{dt} &= \frac{V\cos\gamma}{r} \\ \alpha &= \theta + \varphi - \gamma \\ \frac{dm}{dt} &= -\frac{T}{I_{sp}} \\ \alpha &= \alpha_{prog}(t) \end{aligned} \tag{17}$$

where, V is the velocity, m vehicle mass, θ pitch angle, η trajectory angle, γ flight path angle, φ range angle, h height above ground, α angle of attack, and α_{prog} *(t)* is the programmed angle of attack.

The axial and normal overload coefficients ensure the integrity of the LV in all phases of flight, and were calculated in a body centered velocity coordinate systems (xoy), as follows:

$$n_x = \frac{T + L\sin\alpha - D\cos\alpha}{mg} \leq n_{x\max} \tag{18}$$

$$n_y = \frac{T\sin\alpha + L}{mg} \leq n_{y\max} \tag{19}$$

The thrust to weight ratio gives an important value to evaluate the liftoff characteristics of the LV:

$$\nu = \frac{T}{m\,g} \tag{20}$$

The density variation with altitude can be calculated as:

$$\rho = \rho_0 e^{(-h/\beta)} \tag{21}$$

The gravity varies with altitude and can be represented as:

$$g = \frac{\mu}{(R_e + h)^2} \tag{22}$$

where, ρ_0 is the sea level density, R_e radius of earth, β density scale height, and μ the earth gravitational parameter.

The mission requires to deliver the payload to an altitude h_f with a circular orbital insertion velocity V_f:

$$V_f = \sqrt{\frac{\mu}{h_f + R_e}} \tag{23}$$

TRAJECTORY PHASES

The trajectory of the LV can be described as a composition of several phases, as presented by He (2004a), Qazi and He (2005) and Villanueva *et al.* (2013). For the present research, the trajectory was sectioned in seven phases, as shown in Fig. 2. Each phase has a specific flight characteristic as described next:

- Vertical launch phase: This phase starts from the time of ignition of the first stage SRM until the end of vertical flight time t_v ($t_v = t_1$ in Fig. 3), during this time the LV flies vertically with a flight path angle equal to 90 degrees.
- Pitch over phase: During this phase, the LV maneuver with a negative angle of attack until the transonic velocity is reached. In this point, the angle of attack should approaches zero degrees.
- Powered first stage phase: This phase lasts until the end of the burning time of the first stage SRM. The angle of attack should be kept at zero during the stage separation process.
- Coasting phase 1: The LV flies with no thrust until the second stage ignites.
- Powered second stage phase: The duration of this phase starts with the ignition of the second stage SRM and is equal to its burning time.

- Coasting phase 2: This phase is characterized by a prolonged ballistic free flight approaching the target altitude.
- Kick phase: This phase starts with the ignition of the third stage SRM until the insertion altitude at the required orbital velocity and flight path angle.

FLIGHT PROGRAM FORMULATION

The flight profile defines the performance and loads acting on the LV Consequently, its selection should be integrated in the optimization process. Figure 3 explains the variation of the angle of attack during the pitch over phase (He, 2004a; Xiao, 2001):

$$\alpha_{prog}(t) = -\alpha_{\max} \sin^2 f(t) \tag{24}$$

$$f(t) = \frac{\pi(t - t_1)}{a_m(t_2 - t) + (t - t_1)} \tag{25}$$

$$a_m = \frac{t_a - t_1}{t_2 - t_a} \tag{26}$$

where, $\alpha_{\max}$ is the maximum angle of attack, a_m launch maneuver variable, t_a time corresponding to maximum angle of attack, t time of flight, and t_1 the time of start of pitch over phase, coincident in value with time t_v.

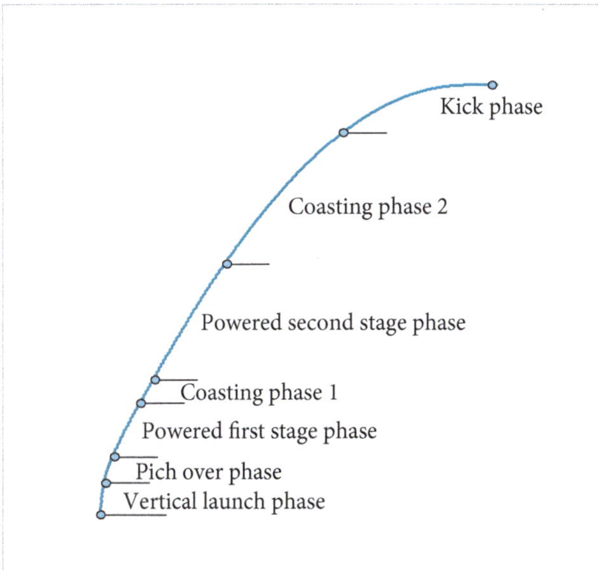

Figure 2. Trajectory phases of launch vehicle.

DESIGN OPTIMIZATION PROBLEM

OBJECTIVE FUNCTION

There can be different objective functions for LV optimization problem, such as minimization of the LV cost, which can be obtained knowing the cost of the first and second stage SRMs and the development cost of the third stage, and also the minimization of the development time, knowing the availability of the first and second stage SRM and the development time of the third stage SRM. However, this analysis considers the minimization of the gross launch mass (m_{LV}). The mathematical description of design objective is as follows:

$$\min m_{LV} = f(X) \tag{27}$$

$$g_j(X) \leq 0 \tag{28}$$

$$h_k(X) = 0 \tag{29}$$

$$X_{lb} \leq X_i \leq X_{ub} \tag{30}$$

where, g_j is the inequality constraints, h_k the equality constraints, X the set of variables, X_{lb} the lower bound of variables and X_{ub} the upper bound of variables.

DESIGN VARIABLES

The design variables are composed from integer (first and second stage SRMs), and continuous third stage SRM and trajectory variables. They are listed in Table 3 and can be represented as:

$$X = [X_{SRM}, X_{SRM3}, X_{Trajectory}] \tag{31}$$

$$X_{SRM} = [srm_1, srm_2] \tag{32}$$

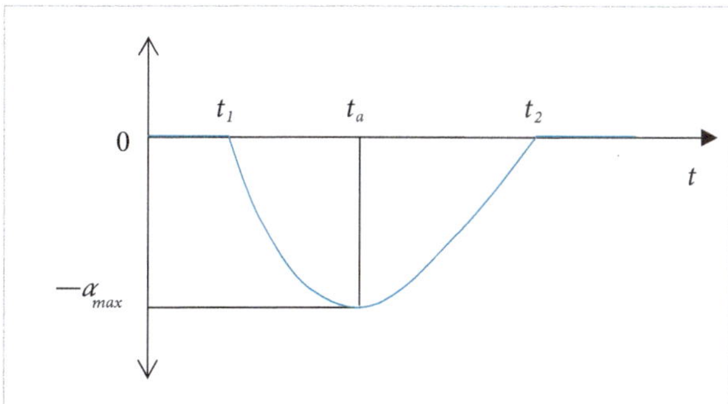

Figure 3. Pitch over ascent phase of launch vehicle.

$$X_{SRM3} = [L_{m3}, D_{m3}, p_{c3}, p_{e3}, k_{s3}, u_3, \rho_{gn3}] \quad (33)$$

$$X_{Trajectory} = [t_v, t_m, t_{c1}, t_{c2}, \alpha_{max}, a_m] \quad (34)$$

DESIGN CONSTRAINTS

The selections of constraints were oriented in order to satisfy the mission, to prevent any failure during flight, and to consider the limitation of the third stage manufacturing technology. They are listed in Table 4:

OPTIMIZATION STRATEGY

INTEGER CONTINUOUS OPTIMIZATION APPROACH

The particularity of our problem deals with integer and continuous variables simultaneously. The selection of SRM type for first and second stages were considered as integer variables. Meanwhile, the trajectory and design parameters of the third stage SRM were considered as continuous.

Several engineering applications of mixed integer continuous optimization approach were presented by Haupt *et al.* (2009), Faustino *et al.* (2006), as well as detailed explanation in Yeniay (2005) and Gantovnik *et al.* (2005).

Garfield and Allen (1995) used integer optimization applied to the configuration of LVs, Johnson (2002) conducted a screening process of booster for hypersonic vehicles, Calabro *et al.* (2002) presented the optimization of the propulsion for multistage LVs, and Bhatnagar *et al.* (2012) solved the mass distribution problem under restrictive condition.

Hartfield *et al.* (2004) have shown the application of GA in finding the global optimum in ramjet propulsion. Bayley and Hartfield (2007) used GA for LV multidisciplinary design optimization with emphasis on minimum cost.

GA has been effectively applied to solve the problem of liquid propellant based LV (Riddle *et al.* 2007), as well as

Table 3. Design variables.

	Variables	Symbol	Units
X1	Stage 1	srm_1	
X2	Stage 2	srm_2	
X3	Rocket motor cylindrical length	L_{m3}	m
X4	Rocket motor diameter	D_{m3}	m
X5	Chamber pressure	P_{c3}	Pa
X6	Nozzle exit pressure	P_{e3}	Pa
X7	Coefficient of grain shape	k_{s3}	
X8	Grain burning rate	u_3	m/s
X9	Grain density	ρ_{gn3}	kg/m^3
X10	Vertical flight time	t_v	s
X11	Time to pitch over	t_m	s
X12	Coasting time 1 (between 1st and 2nd stage)	t_{c1}	s
X13	Coasting time 2 (between 2nd and 3rd stage)	t_{c2}	s
X14	Maximum angle of attack (absolute)	α_{max}	deg
X15	Launch maneuver variable	a_m	

Table 4. Design constraints.

	Constraints	Value	Units
C1	Orbit insertion velocity	$V_f \geq 7560$	m/s
C2	Final altitude	$V_f \geq 600$	km
C3	Axial overload	$n_x \leq 14$	
C4	Normal overload	$n_y \leq 2$	
C5	Maximum dynamic pressure	$q_{max} \leq 85$	kPa
C6	Angle of attack (0.8 ≤ M≤ 1.3)	$\alpha = 0$	deg
C7	Orbit insertion angle	$\gamma = 0 \pm 0.2$	deg
C8	Rocket motor diameter	$D_{m1} \geq D_{m2}$	m
C9	Rocket motor mass	$m_{SRM1} \geq m_{SRM2}$	kg
C10	Total LV length	$L_{LV} \leq 18$	m
C11	Grain fineness	$\lambda_{gn3} < 2$	
C12	Thrust to weight ratio	$v_3 \geq 1.8$	
C13	Nozzle exit diameter	$d_{e3} \leq 0.9D_{m3}$	m
C14	Burning time	$t_{b3} \leq 65$	s

solid propellant LVs (Bayley *et al.* 2008). Rafique *et al.* (2009) and Goldberg (1989) provides detailed and comprehensive implementation of GA in solving complex problems.

OPTIMIZATION METHOD

The adopted and implemented GA optimization method is shown in Fig. 4, where a set of input design variables (SRM type, trajectory and third stage), as well as the lower and upper bounds, are passed to the main loop, where an initial population is randomly created. Furthermore, the selection, the crossover and the mutation operations are performed until the stopping criteria is achieved. The constraints were calculated and handled by external penalty function, as presented in Deb (2000) and detailed and explained in Coello (1999) and Kramer (2010). At each routine, the propulsion, mass, aerodynamics and trajectory analysis were performed.

The main characteristics of GA are presented in Table 5.

OPTIMIZATION FRAMEWORK

The optimization framework considered for this research is based on the multi-discipline feasible (MDF) design, which allows an easy and accurate result (Qazi and He L, 2006), as shown in Fig. 5.

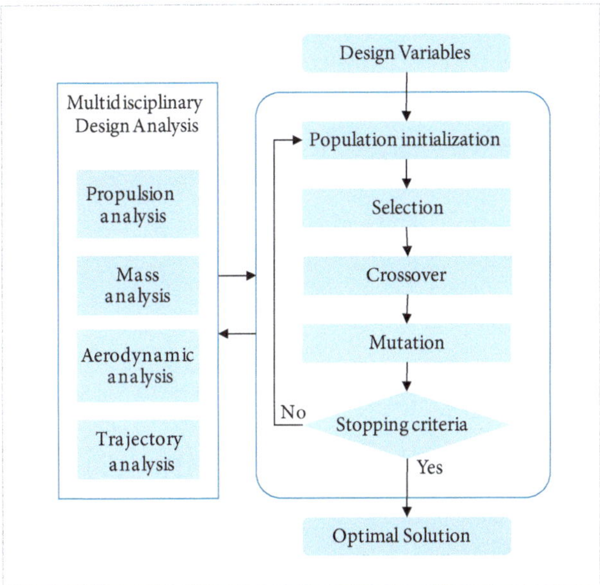

Figure 4. Genetic algorithm optimization approach.

Table 5. Genetic algorithm characteristics.

Variables	Characteristics
Generations	200
Population size	100
Stopping criteria	Function tolerance 10e-6
Population type	Double vector
Selection	Stochastic uniform
Crossover	Single point pc = 0.8
Mutation	Uniform pm = 0.2564
Reproduction	Elite count = 2
Function evaluation	2000

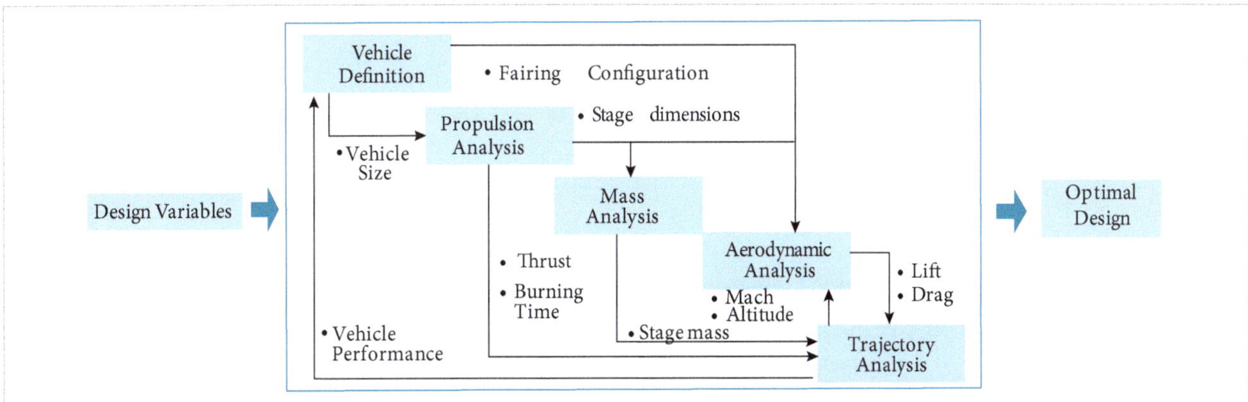

Figure 5. Multidisciplinary design optimization.

OPTIMIZATION RESULT

The results show that the considered mixed integer-continuous GA optimization approach successfully reached the objective function. The optimized LV has a total mass of 23,530 kg and a 16.12 m of length. Table 6 shows the optimized value of variables and in Table 7 the main parameters of the LV third stage are listed.

The first and second stages SRM design type (SRM 12 and SRM 22), had been optimized and selected from Table 2. Both SRMs have the same diameter but different length. As it is represented in Fig. 6, the shroud design is configured with the same diameter as the third stage, in order to reduce the aerodynamics forces and interferences.

The performance characteristics of the LV, shown in Fig. 7, demonstrates the capability of the three stage solid propellant LV to place a small payload into the LEO orbit maintaining its main parameters inside its limit values, furthermore, the overall design configuration facilitates its launch operations.

CONCLUSION

A small three-stage solid propellant LV was configured and optimized using a mixed integer-continuous GA optimization method. The first and second stages SRM types were considered as integer variables, whereas the third stage SRM and trajectory as continuous. The main advantage of using GA relies on its independency of initial values to start the optimization, and the ability to handle integer variables. The propulsion, mass, aerodynamic and dynamic models were developed and integrated in a MDF framework.

Table 6. Optimum values of variables.

Variables		Lower Bound	Upper Bound	Optimized Value
X1	srm_1	11	14	12
X2	srm_2	21	27	22
X3	L_{m3} (m)	0.80	1.20	0.90
X4	D_{m3} (m)	0.80	1.20	0.83
X5	P_{c3} (Pa)	70e5	80e5	77.42e5
X6	P_{e3} (Pa)	0.05e5	0.15e5	0.133e5
X7	k_{s3}	1.10	1.60	1.14
X8	u_3 (m/s)	6.0e-3	8.0e-3	6.71e-3
X9	ρ_{gn3} (kg/m^3)	1650	1740	1683.1
X10	t_v (s)	3.0	6.0	3.01
X11	t_m (s)	18.0	25.0	21.57
X12	t_{c1} (s)	2.0	8.0	4.46
X13	t_{c2} (s)	360	400	372.61
X14	α_{max} (deg)	3.0	6.0	5.731
X15	a_m	0.28	0.42	0.319

Table 7. Parameters of launch vehicle third stage.

Parameters	Value (Units)
Stage mass	893.6 (kg)
Propellant mass	762.1 (kg)
Stage dry mass	131.2 (kg)
Propellant mass fraction	0.854
Average Thrust	33.82 (kN)
Specific Impulse vac	2702.8 (N.s/kg)
Nozzle expansion ratio	48.01
Thrust to weight ratio	3.69
Burning time	60.9 (s)

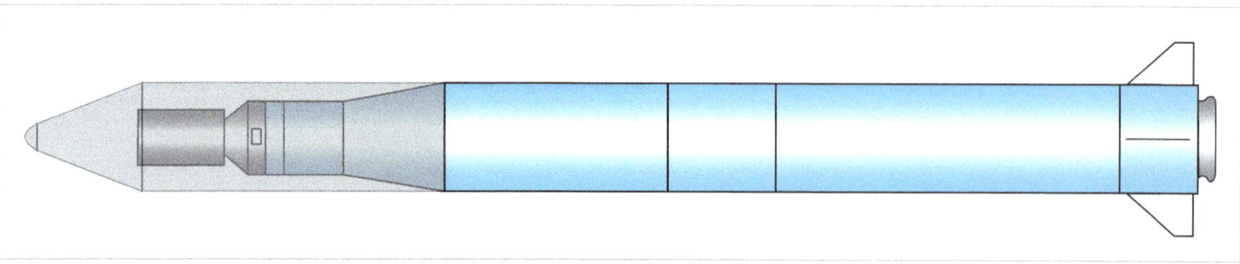

Figure 6. Optimized three stage launch vehicle.

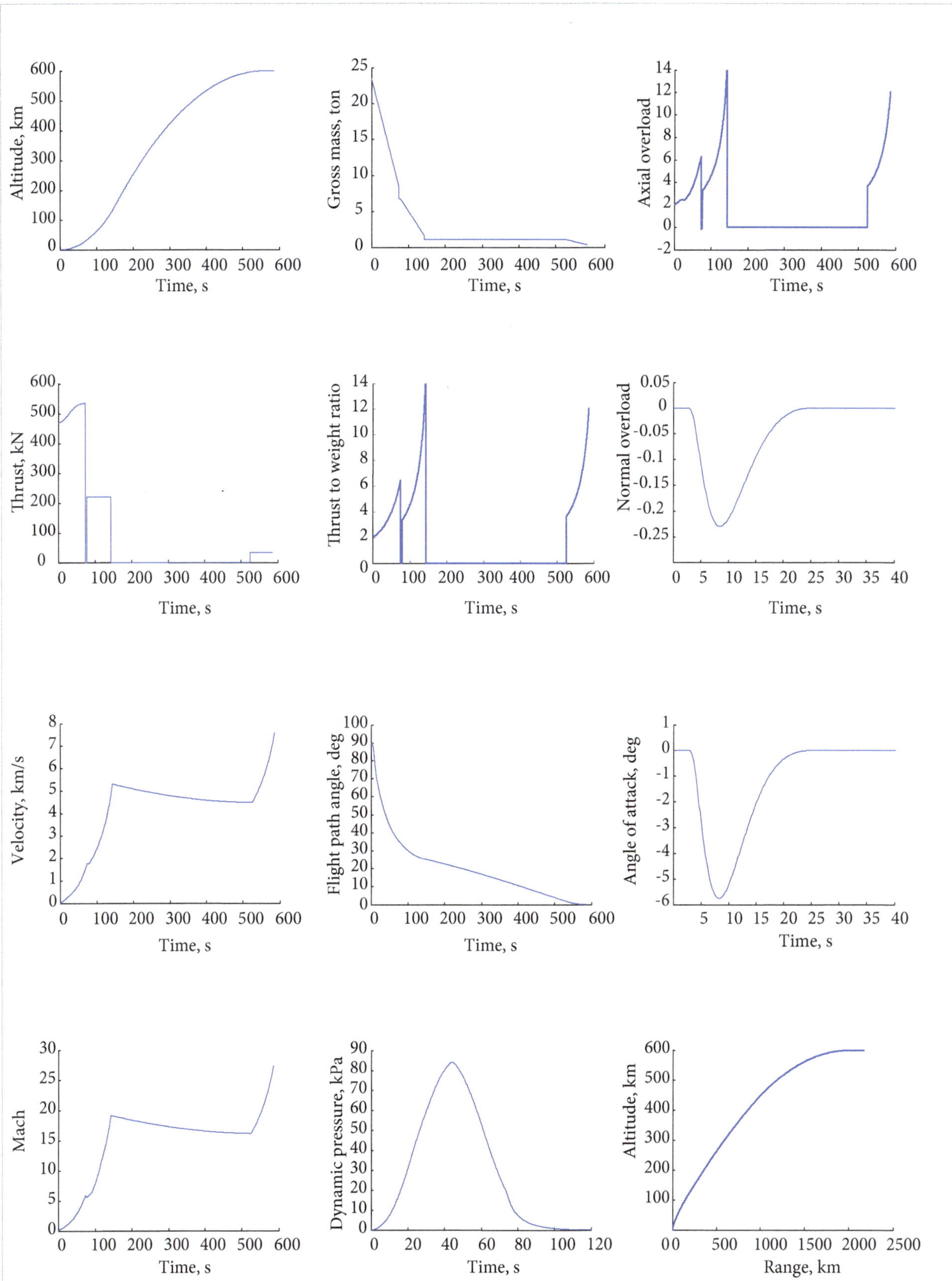

Figure 7. Performance characteristics of launch vehicle.

An important contribution of this research is the approach in finding the best LV design configuration to rapid access to space with limited research budget, relied mainly on international cooperation and complemented with the indigenous aerospace manufacturing technology capability.

ACKNOWLEDGMENT

Fredy Villanueva wishes to thank Beihang University and China Scholarship Council (CSC), for their financial support.

REFERENCES

Bayley, D.J. and Hartfield, R.J., 2007, "Design Optimization of a Space Launch Vehicles for Minimum Cost Using a Genetic Algorithm", AIAA 43rd AIAA/ASME/SAE/ASEE Joint Propulsion Conference and Exhibit, AIAA 2007-5852.

Bayley, D.J., Hartfield, R.J., Burkhalter, J.E. and Jenkins R.M., 2008, "Design Optimization of a Space Launch Vehicle Using a Genetic Algorithm", Journal of Spacecrafts and Rockets, Vol. 45, No. 4, pp. 733–740.

Bhatnagar, P., Rajan, S. and Saxena, D., 2012, "Study on Optimization Problem of Propellant Mass Distribution Under Restrictive Condition in Multistage Rocket", International Conference on Advances in Computer Applications (ICACA), No. 1, pp. 27-29.

Blake, W.B., 1998, "Missile DATCOM: User's Manual-1997 FORTRAN 90 Revision", Wright-Patterson AFB, Oklahoma.

Calabro, M., Dufour, A. and Macaire, A., 2002, "Optimization of the Propulsion of Multistage Solid Rocket Motor Launcher", Acta Astronautica, Vol. 50, No. 4, pp. 201–208. doi: 10.1016/S0094-5765(01)00164-3.

Coello, C.A., 1999, "A Survey of Constraint Handling Techniques Used with Evolutionary Algorithm", Technical Report Lania-RI-99-04, Laboratorio Nacional de Informatica Avanzada, Rebamen 80, Xalapa, Veracruz 91090, Mexico.

Deb, K., 2000, "An Efficient Constraint Handling Method for Genetic Algorithm", Computer Methods in Applied Mechanics and Engineering, Vol. 186, No. 2-4, pp.311-338. doi: 10.1016/S0045-7825(99)00389-8.

Faustino, A.M., Judice, J.J., Ribeiro, I.M. and Neves, A.S., 2006, "An Integer Programming Model for Truss Topology Optimization", Investigação Operacional, Vol. 26, No. 1, pp. 111-127.

Fleeman, E.L., 2001, "Tactical Missile Design", AIAA, Reston.

Gantovnik, V.B., Gurdal, Z., Watson, L.T. and Anderson-Cook, C.M., 2005, "Genetic Algorithm for Mixed Nonlinear Programming Problems Using Separate Constraint Approximations", 31st AIAA Journal, Vol. 43, No. 8, pp. 1844-1849.

Garfield, J.R. and Allen, B.D., 1995, "Screening Studies and Techniques for All-Solids Space Launch Vehicles", 31st AIAA/ASME/SAE/ASEE Joint Propulsion Conference and Exhibit.

Goldberg, D.E., 1989, "Genetic Algorithms in Search, Optimization, and Machine Learning", Addison-Wesley.

Hartfield, R.J., Jenkins, R.M. and Burkhalter, J.E., 2004, "Ramjet Powered missile design using a genetic algorithm", 42nd AIAA Aerospace Sciences Meeting and Exhibit, AIAA 2004-0451.

Haupt, S.E., Haupt, R.I. and Young, G.S., 2011, "A Mixed Genetic Algorithm Used in Biological and Chemical Defense Applications", Software and Computing, Vol. 15, pp. 51-59. doi: 10.1007/s00500-009-0516-z.

He, L., 2004a, "Launch Vehicle Design", Beijing University of Aeronautics and Astronautics Press, Beijing.

He, L., 2004b, "Solid Ballistic Missile Design", Beijing University of Aeronautics and Astronautics Press, Beijing.

Johnson, D.B., 2002, "Screening Process for Boosters for Hypersonic Vehicles", AIAA/AAAF 11th International Space Plane and Hypersonic Systems and technologies, AIAA 2002-5218.

Kramer, O., 2010, "A Review of Constraint-handling Techniques for Evolution Strategies", Applied Computational Intelligence and Soft Computing, Vol. 2010, Article ID 185063, pp. 11. doi: 10.1155/2010/185063.

Qazi, M. and He, L., 2005, ""Rapid Trajectory Optimization Using Computational Intelligence for Guidance and Conceptual Design of Multistage Space Launch Vehicle", AIAA Guidance, Navigation, and Control Conference, AIAA 2005-6062.

Qazi, M. and He, L., 2006, "Nearly Orthogonal Sampling and Neural Network Metamodel Driven Conceptual Design of Multistage Space Launch Vehicle", Journal of Computer Aided Design, Vol. 38, No. 6, pp. 595-607. doi: 10.1016/j.cad.2006.02.001.

Rafique, A.F., He, L., Zeeshan, Q., Kamran, A., Nisar, K. and Wang Xiaowei, 2009 "Integrated System Design of Air Launched Small Space Launch Vehicle Using Genetic Algorithm", 45th AIAA/ASME/SAE/ASEE Joint Propulsion Conference and Exhibit, AIAA-2009-5506.

Riddle, D.B., Hartfield, R.J., Burkhalter, J.E. and Jenkins, R.M., 2007, "Genetic Algorithm Optimization of Liquid Propellant Missile Systems", 45th AIAA Aerospace Sciences Meeting and Exhibit, AIAA 2007-0362.

Sutton, G.P. and Biblarz, O., 2001, "Rocket Propulsion Elements", 7th edition, Wiley-Interscience, New York.

Villanueva, F.M., He, L., Rafique, A.F. and Rahman T., 2013, "Small Launch Vehicle Trajectory Profile Optimization Using Hybrid Algorithm", International Bhurban Conference on Applied Science and Technology, Pakistan. doi: 10.1109/IBCAST.2013.6512150.

Xiao, Y., 2001, "Rocket Ballistics and Dynamics", Postgraduate Course, Beihang University.

Yeniay, O., 2005, "Penalty Function Methods for Constrained Optimization with Genetic Algorithms", Mathematical and Computational Applications, Vol. 10, No. 1, pp. 45-56.

Zipfel, P.H., 2007, "Modelling and Simulation of Aerospace Vehicle Dynamics", AIAA, Reston.

7

Theoretical Optimal Trajectories for Reducing the Environmental Impact of Commercial Aircraft Operations

Cesar Celis[1], Vishal Sethi[2], David Zammit-Mangion[2], Riti Singh[2], Pericles Pilidis[2]

ABSTRACT: This work describes initial results obtained from an ongoing research involving the development of optimization algorithms which are capable of performing multi-disciplinary aircraft trajectory optimization processes. A short description of both the rationale behind the initial selection of a suitable optimization technique and the status of the optimization algorithms is firstly presented. The optimization algorithms developed are subsequently utilized to analyze different case studies involving one or more flight phases present in actual aircraft flight profiles. Several optimization processes focusing on the minimization of total flight time, fuel burned and oxides of nitrogen (NO_x) emissions are carried out and their results are presented and discussed. When compared with others obtained using commercially available optimizers, results of these optimization processes show satisfactory level of accuracy (average discrepancies ~2%). It is expected that these optimization algorithms can be utilized in future to efficiently compute realistic, optimal and 'greener' aircraft trajectories, thereby minimizing the environmental impact of commercial aircraft operations.

KEYWORDS: Trajectory optimization, Aircraft emissions, Environmental impact.

INTRODUCTION

In this globalized world, where the efficient transportation of people and goods greatly contributes to the development of a given region or country, the aviation industry has found the ideal conditions for its development. These conditions have made the aviation industry one of the fastest growing economic sectors during the last decades. The growth in the aviation industry is reflected in the increase in air transport, expressed in terms of Revenue Passenger-Kilometers (RPKs), which has risen in an average annual rate of around 5% over the past 20 years (Boeing Commercial Airplanes, 2009). Market projections associated with this industry indicate that this growth will continue over the following years.

Environmental issues associated with aircraft operations are currently one of the most critical aspects of commercial aviation (Green, 2003; Clarke, 2003; Brooker, 2006; Riddlebaugh, 2007). These are due to both the continuing growth in air traffic and the increasing public awareness about anthropogenic contribution to global warming. The critical nature of this problem means that currently several organizations worldwide are focusing their efforts towards large collaborative projects whose main objective is to identify the best alternatives or routes to reduce the environmental impact of aircraft operations. Particular examples of these projects are Partnership for AiR Transportation Noise and Emissions Reduction (PARTNER) project (PARTNER, 2003) and European Clean Sky JTI (Joint Technology Initiative) project (Clean Sky JTI, 2008). The Clean Sky JTI project has been demonstrating and validating different technologies, thereby making a major move towards achieving

1. Pontifícia Universidade Católica do Rio de Janeiro – Rio de Janeiro/RJ – Brazil. 2. Cranfield University, Cranfield – United Kingdom.

Author for correspondence: Cesar Celis | Departamento de Engenharia Mecânica | Pontifícia Universidade Católica do Rio de Janeiro – Rua Marquês de São Vicente 225, Rio de Janeiro/RJ | CEP: 22453-900 – Brazil | Email: cesar.celis@puc-rio.br the environmental goals set by Advisory Council for Aeronautics Research in Europe (ACARE). These targets for 2020 include reductions in carbon dioxide (CO_2) and NO_X emissions by 50% and 80%, respectively.

Cranfield University (CU) and other partners of the European Aviation Industry are collaboratively participating in several areas of Clean Sky JTI, including the Systems for Green Operations (SGO) Integrated Technology Demonstrator (ITD). The SGO ITD concentrates on two key areas: (1) Management of Aircraft Energy (MAE) and (2) Management of Trajectory and Mission (MTM). One of the main contributions of CU to the SGO ITD is the development of suitable computational algorithms for the management of aircraft trajectory and mission, in particular, for carrying out aircraft trajectory optimization processes. This paper focuses on the initial stages of the development of these optimization algorithms and their applications for determining theoretical optimum aircraft trajectories. The optimization algorithms developed are a part of an optimization suite known as 'Polyphemus' (oPtimisatiOn aLgorithms librarY for PHysical complEx MUlti-objective problemS).

As a part of the ongoing research about trajectory optimization, a methodology for optimizing aircraft trajectories has been initially devised. Then, Polyphemus has been developed and/or adapted for carrying out these aircraft trajectory optimization processes. Computational models simulating different disciplines, such as aircraft performance, engine performance, and formation of pollutants, have also been selected or developed as required. Simplified aircraft trajectory optimization processes have finally been carried out to evaluate the mathematical performance of Polyphemus primarily. Main results of these optimization processes are summarized in this paper.

AIRCRAFT TRAJECTORY OPTIMIZATION

Optimization can be defined as the science of determining the best solutions for certain mathematically defined problems which are often representations of physical reality (Fletcher, 1987). There are several criteria and methodologies for classifying and solving optimization problems, respectively (Walsh, 1975; Schwefel, 1981; Bunday, 1984; Everitt, 1987; Krotov, 1996; Rao, 1996). Thus, aircraft trajectory optimization problems can be mainly classified as constrained, dynamic, optimal control, nonlinear – *the functions relating inputs (design variables) and outputs (objective functions) are unknown in this work and they are presumed to be non-linear, non-smooth, and non-differentiable* – real-valued (mostly), deterministic (mostly), multimodal, multidimensional, and multiobjective. A number of optimization methods have been developed in the past, many of which are customized for a specific problem. Most important optimization methods can be grouped under three broad categories (Schwefel, 1981): (1) hill climbing methods (direct search methods, gradient methods and Newton methods); (2) random search methods; and (3) evolutionary methods. A detailed review of these methods can be found in Celis et al. (2009) and Celis (2010).

Evolutionary methods are inspired by nature, biological structures and processes that can be observed in natural environments for solving technical problems. They are based on Darwin's principles of species evolution reproduction cycle, natural selection and diversity by variation (Quagliarella, 1998). Most important evolutionary methods are evolutionary programming, evolution strategies, genetic programming and genetic algorithms (GAs). Among all evolutionary techniques, GAs are most widely used, and they have had a significant impact on optimization (Russell and Norvig, 2003). Like other evolutionary techniques, GAs are based on the principles of natural genetics and natural selection. Thus, basic elements of natural genetics (reproduction, crossover and mutation) are used in the genetic search procedure. Generally, evolutionary methods, in particular GAs, are robust, which help to solve problems in which the functions relating inputs to outputs are unknown and may have an unexpected behavior. In these situations, standard nonlinear programming techniques would be inefficient, computationally expensive, and in most cases, find a relative optimum that is the closest to the starting point (Rao, 1996). It has been argued (Betts, 1998) that evolutionary methods (including GAs and other techniques involving some sort of stochasticity during the optimization process) are not adequate to solve trajectory optimization problems and are computationally inferior when compared to methods that use gradient information. This inadequacy argument is originated from considering that trajectory optimization problems are not characterized by discrete variables. However, the results shown in this work highlight the fact that GAs are indeed suitable for this class of problems. Even more, for aircraft trajectory optimization involving multimodel integration, where the characteristics of the functions relating inputs to outputs are unknown, algorithms of this type appear to be the only practical alternative. A number of reasons that help support this point of view are as follows:

- GAs do not use specific knowledge of the optimization problem domain. Instead of using previously known domain-specific information to guide each step, they make random changes in their candidate solutions and then use the fitness function to determine whether those changes result in an improvement. As GAs optimization routines are both model- and problem-independent, and they allow the users to (simultaneously) run different models for simulating different disciplines, they appear to be the ideal methods.
- GAs are well-suited to solve problems where the fitness landscape is complex (discontinuous and multimodal), number of constraints and objectives are involved and the space of all potential solutions is large (particular characteristic of nonlinear problems).
- GAs make use of a parallel process of search for the optimum, which means that they can explore the solution space in multiple directions at once. If one path turns out to be a dead end, they can easily eliminate it and progress in more promising directions, thereby increasing the chance of finding the optimal solution.

From four main evolutionary algorithms, GAs have been initially chosen because of their large number of previous successful applications, worldwide. However, it is important to highlight that the hybridization of GAs with other optimization techniques has also been considered. This is due to fact that although GAs are an extremely efficient optimization technique, they are not the most efficient for the entire search phases (Rogero, 2002). Thus, hybrid optimization methods will be developed in future, as they have the potential to improve the performance in a given search phase; for example, GAs techniques involving the use of both a random search phase during the beginning of the optimization process (to increase the quality of the initial population) and a hill climbing phase at the end of the optimization (to refine the quality of the optimum point once the global optimum region has been found).

STATUS OF OPTIMIZATION ALGORITHMS (POLYPHEMUS)

Different numerical methods that could be used for solving the aircraft trajectory optimization problem were firstly reviewed, and a suitable optimization technique was initially selected (see "Aircraft Trajectory Optimization" section). The next step in the development of Polyphemus was reviewing the track record of optimizers developed by CU for a range of applications, and identifying a candidate which could be used as a suitable 'starting point'. This resulted in the decision to use GA-based optimization routines developed by Rogero (2002) as the basis for the development of Polyphemus. Rogero's optimizer already includes several algorithms for each of the main phases involved in a GA-based optimization process; however, there are additional enhancements that can be introduced to further improve the quality of the optimizer. These improvements include the use of adaptive GAs (e.g., 'master-slave' configurations), which would allow using optimum GA parameters (e.g., population size, crossover ratio, mutation ratio, etc.) during the optimization processes; and also inclusion of the concept of Pareto optimality (Pareto fronts), which would improve its capabilities when performing multiobjective optimization processes. These improvements can be made based on successful past experiences of these concepts as part of previous optimizers (Gulati, 2001; Sampath, 2003) developed by CU.

As the development of Polyphemus is continuous, only a brief description of the main aspects characterizing its current status is presented here. Polyphemus has been implemented using Java as the main programming language. Its core has been developed based on the basic structure of 'SGA Java V1.03' from Hartley (1998), which involves a Java implementation of the 'simple GA' (SGA) from Goldberg (1989). However, the original model has been recoded and extensively modified to both adapt to engineering design optimization problems and to maximize its performance. The main modifications made help to improve both the optimization performance, through an adaptation to the application domain, and the technique and genetic operators utilized during the optimization process. The optimizer application domain considered was engineering design. Thus, the chromosome modules have been developed in a way to support real-number parameter encoding in conjunction with a defined allowable range for the parameters (genes). In addition, algorithms for keeping a historical record of all created chromosomes and for preventing the creation of duplicate ones have been implemented. In order to improve the GA technique, concepts such as elitism (preservation of the genetic material of the best members through generations), steady-state replacement (partial replacement of the newly-generated chromosomes to avoid loss of potentially good genetic material), and fitness scaling (trade-off between premature convergence and genetic drift by keeping the selection pressure relatively constant along the whole optimization process) have been introduced.

Another phase of the optimization performance improvement involved the implementation of more advanced and efficient GA operators (crossover, mutation and selection) (Rogero, 2002). Thus, several crossover techniques suitable for real-number encoding have been implemented, including weighted averaging crossover method, blend crossover BLX-a method and simulated binary crossover SBX method. The simulated binary crossover SBX method involves the creation of solutions within the whole search space. For instance, for a problem with 'k' design variables (or genes), a real-number vector (chromosome) can be given by:

$$X = \{x_1, x_2, x_3, \ldots, x_k\} \tag{1}$$

Then, one of the most simple ways of performing a crossover operation using any two parent chromosomes, X_1 and X_2, involves the combination of the two vectors representing them, which is as follows:

$$\begin{aligned} X_1' &= \lambda_1 X_1 + \lambda_2 X_2 \\ X_2' &= \lambda_1 X_2 + \lambda_2 X_1 \end{aligned} \tag{2}$$

Here, the multipliers λ_1 and λ_2 (subject to the condition $\lambda_1 + \lambda_2 = 1$) represent the weights randomly selected during the crossover process, and X_1' and X_2' are the child chromosomes. Depending on the permissible values of the multipliers λ_1 and λ_2, different subtypes of crossover methods can be derived. The (weighted) averaging crossover corresponds to the special case, in which $\lambda_1 = \lambda_2 = 0.5$. The averaging crossover suffers from contraction effects because it allows the creation of offspring only along the line generated between the two parental chromosomes. This problem is solved to some extent in the blend crossover BLX-a method, which uses an exploration factor (α) to increase the exploration capability of the crossover operator.

In addition to the standard random mutation operator, others such as creep mutation, with and without decay, and Dynamic Vectored Mutation (DVM) - have also been implemented. In general, for a given parent chromosome X, Eq. (1), if its element (gene) x_i is selected for mutation, a (random) change in the value of this selected gene within its domain, given by a lower LB_i and upper UB_i bound, will result in the following transformation:

$$\{x_1, x_2, x_3, \ldots, x_i, \ldots, x_k\} \rightarrow \{x_1, x_2, x_3, \ldots, x_i', \ldots, x_k\},$$
$$x_i, x_i' \in [LB_i, UB_i] \tag{3}$$

As creep mutation is basically operated by adding or subtracting a random number to a gene of the chromosome selected for mutation, the mutation of a given gene, x_i, using this method is limited to a creep range centered on its original value (Davis, 1991). A creep mutated gene, x_i', is then computed as follows:

$$x_i' = x_i + (2r - 1)\Delta_{max}, \text{ where} \tag{4}$$

$$\Delta_{max} = \delta.(UB_i - LB_i) \tag{5}$$

In Eqs. (4) and (5), Δ_{max} is the maximum size used for the creep mutation, δ is the range ratio, and r is a random number from [0,1]. The level of disruption produced by the mutation process is controlled by the creep size δ. In the creep mutation with decay method, the creep size is altered as a function of the stage of the search process as follows:

$$\delta_{t+1} = \delta_t (1 - \gamma) \tag{6}$$

In Eq. (6), γ represents the creep decay rate and t is the generation number. This type of implementation allows the use of large values of δ in the beginning of the search process and small values at the end; the exploration and exploitation capabilities required during the process are balanced in this way. Details about the DVM method can be found in Rogero (2002).

Selection operators implemented in the optimizer include a modified roulette wheel selection operator (limiting the number of chromosome instances) and the Stochastic Universal Sampling (SUS) technique. Roulette wheel selection, in which an area proportional to its fitness is allocated to each chromosome on a virtual roulette wheel, is the best known selection method. The selection process is carried out by spinning the wheel a number of times equal to the number of chromosomes to be selected (each time a single chromosome is selected). One drawback associated with this selection method is that it has a tendency to select a large number of copies of the best chromosome, which can lead to loss of diversity. This problem can be solved to some extent by using the SUS selection method. In this method, like in the roulette wheel selection, a chromosome occupies on the wheel an area proportional to its fitness. However, instead of spinning the wheel several times for selecting chromosomes, a single spin of the wheel identifies all parent selections simultaneously. This is possible because there is another wheel on the outside of the roulette wheel containing a number of equally spaced pointers equal to the number of chromosomes to be selected (Callan, 2003). With regard to replacement operators, tournament replacement and ranked replacement have been improved and implemented as replacement operators.

Finally, Polyphemus uses a unique optimization method based on Wienke's idea of target vector optimization (Wienke *et al.*, 1992). In this method, designers can define, for each parameter, a target to be attained, a range within which this parameter should remain, and the requirement to maximize or minimize the given parameter. Accordingly, the quality of the design is determined by the level of both the achievement of the targets and the violation of the parameters ranges. This approach enables designers to have total control over the optimization process with neither having to know much about the optimization algorithms, nor having to devise a fitness function (Rogero and Rubini, 2003). The optimization results presented in the following sections were obtained using the current version of Polyphemus, whose main characteristics have been summarized above.

TRAJECTORY OPTIMIZATION CASE STUDIES

In this section, computational models utilized, flight profiles optimized and the methodology followed for optimizing these flight profiles have been particularly emphasized.

Computational Models

In the aircraft trajectory optimization processes, three computational models, i.e., aircraft performance simulation model (APM), engine performance simulation model (TurboMatch) and emissions prediction model (Hephaestus), have been utilized. Figure 1 illustrates the general arrangement of these models, as well as different parameters exchanged among them. The APM (Long, 2009) is a generic tool that determines flight path performance for a given aircraft design. It uses steady-state performance equations to resolve aerodynamic lift and drag and to determine the thrust required for a given kinematic flight state. In order to easily identify the behavior of Polyphemus, airspeed limitations – such as critical Mach number (M), never-exceed speed and wave drag at transonic M – have not been implemented in the model. As APM uses endpoints to compute performance, the user must declare a trajectory segment in terms of ground range and altitude intervals, whereby a constant flight path angle is then defined. Flight conditions are then assumed to be constant over that segment. The aircraft modeled in this work corresponds to a typical mid-sized, single-aisle, twin turbofan airliner with a maximum takeoff weight (MTOW) of about 72,000 kg and a seating capacity of about 150 passengers.

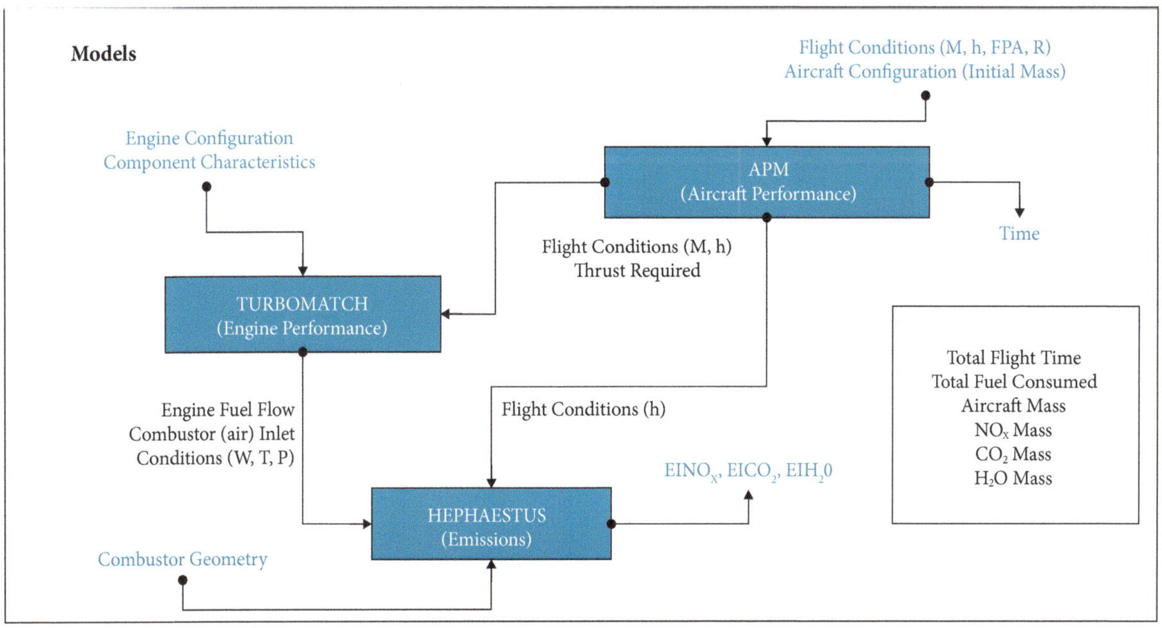

Figure 1. Computational models configuration and exchange of parameters.

The performance of the engines was simulated using TurboMatch (Palmer, 1999), which is the in-house CU gas turbine performance code that has been developed and refined over a number of decades. TurboMatch performance simulations range from simple steady-state (design and off-design point) to complex transient performance computations. Finally, the gaseous emission predictions have been performed using the CU emissions prediction software, Hephaestus. An integral part of Hephaestus constitutes the emissions prediction model described in Celis *et al.* (2009), which follows an approach based on the use of a number of stirred reactors for modeling combustion chambers and estimating the level of pollutants emitted from them. Additional details of these computational models can be found in Celis *et al.* (2009) and Celis (2010).

Flight Profiles

It is clear that in order to demonstrate the suitability of an optimizer for optimizing aircraft trajectories, an extensive validation process of the algorithms that are implemented needs to be carried out using different analytical problems with known optimal values. In the case of Polyphemus, this part of the validation process has already been performed (Rogero, 2002) and is therefore not repeated here. In order to provide insight into the results that can be expected using Polyphemus, simplified aircraft trajectory optimization processes using this optimizer have been performed. It is relevant to note that the main objective of these processes was evaluation of the mathematical performance of Polyphemus rather than the generation of realistic aircraft trajectories. Consequently, simplifications (in terms of number of flight segments, design variables, constraints and objective functions, etc.) have been introduced when optimizing the aircraft flight profiles. Indeed, the results discussed in this work correspond to single-objective optimization processes only, which means that the determination of non-dominated or Pareto optimal solutions that characterize multi-objective optimization processes is out of the scope of this work.

In this work, the aircraft flight profiles have been divided into only a small number of segments, as illustrated in Fig. 2. This helps getting a greater visibility on the characteristics of the Polyphemus performance when assessing results. This would have been more difficult if the trajectory had been divided into a greater number of segments. These hypotheses are a simplification of real cases but provide numerical solutions that are used to commission the methodology. In order to obtain meaningful results in terms of actual optimum trajectories, the flight path needs to be divided into a much larger number of segments, each small enough so that the errors associated with the assumptions made within each segment will be cumulatively insignificant. All the optimization processes carried out involved only vertical profiles. Therefore, only three parameters have been used to define a given aircraft trajectory: (1) flight altitude (*h*); (2) aircraft speed: true airspeed (TAS), equivalent airspeed (EAS) or Mach number (*M*); and (3) range (*R*): the horizontal distance flown by the aircraft. One of the main uses of Polyphemus involves the optimization of aircraft trajectories between city pairs. Thus, *R* has been usually kept constant during the optimizations, and only altitude and aircraft speed vary (i.e., used as design variables) to compute optimum aircraft trajectories that minimize, separately, total flight time, fuel burned and NO_x emissions.

Several aircraft flight profiles have been optimized in order to assess the mathematical performance of Polyphemus. The optimization results associated with three of these flight profiles are summarized in this paper. A brief description of these profiles, which were analyzed as part of three separate case studies, is presented as follows:

- *Case 1: Simple Climb Profile Optimization.* (i) Flight profile has been divided into four segments (Fig. 2). (ii) Climb segments have been defined by arbitrary segment lengths (range, *R*). (iii) Overall climb has been defined by the cumulative range, start and end altitudes, and Mach numbers. (iv) Variation in intermediate Mach numbers (initial *M* in segments 2 and 3) and altitudes (initial altitude in segments 2, 3 and 4) has been allowed during the optimization processes. (v) Only explicit constraints have been utilized, i.e., range of permissible values of the design variables (*h* and *M*) are limited. (vi) Lower and upper bounds for these permissible ranges have been set at 457 m (1,500 ft) and 10,668 m (35,000 ft), respectively, for *h* (profile start and end altitudes); and 0.38 and 0.80, respectively, for *M*. (vii) International Standard Atmosphere

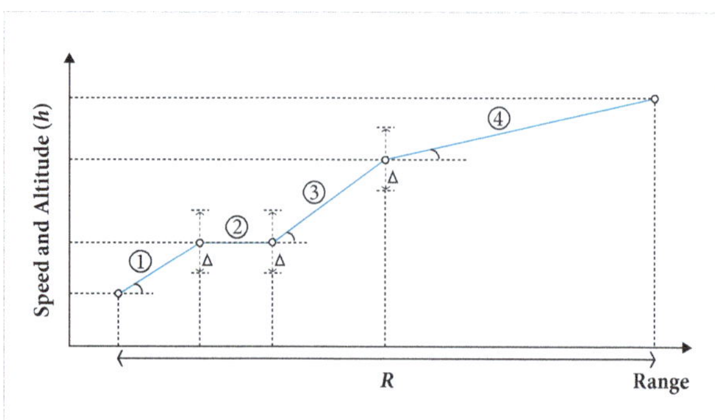

Figure 2. Generic aircraft flight profile.

(ISA) conditions are assumed. (viii) Range of flight path angle (FPA) allowed: [0, 7.5] deg. Data corresponding to this particular profile optimization are summarized in Table 1.

- *Case 2: Implicitly Constrained Climb Profile Optimization.* (i) Flight profile is similar to the profile used in Case 1; however, aircraft speeds have been specified at the start and end of each climb segment allowing continuity in aircraft speed. (ii) Climb schedule has been described as follows; 1st seg.: climb at constant EAS from 1,500 ft (457 m) up to 10,000 ft (3,048 m); 2nd seg.: EAS acceleration at 10,000 ft (level flight); 3rd seg.: climb at constant EAS up to a segment final altitude where (cruise) M is about 0.8; 4th seg.: climb at constant M from this altitude up to 35,000 ft (10,668 m). (iii) Design variables and their range of permissible values: initial EAS in segment 1 (EAS_{1i} [89.0, 128.6] m/s), final EAS in segment 2 (EAS_{2f} [133.8, 221.2] m/s), initial altitude in segment 3 (h_{3i} [3048, 4400] m), and initial altitude in segment 4 (h_{4i} [3048,10668] m). (iv) Implicit constraint: initial M in segment 4 (M_{4i}) – allowable range ±0.5% of its nominal value, 0.8. (v) ISA conditions have been assumed. (vi) FPA range allowed: [0, 7.5] deg. Additional details about this case study are shown in Table 2.
- *Case 3: Quasi-Full Flight Profile Optimization.* (i) Flight profile (involving climb, cruise and descent) has been divided into eight segments. (ii) Profile has been defined following a similar approach to that used in Case 2. (iii) Flight schedule has been described as follows; 1st seg.: climb at constant *EAS* from 1,500 ft (457 m) up to 10,000 ft (3,048 m); 2nd seg.: EAS acceleration at 10,000 ft (level flight); 3rd seg.: climb at constant *EAS* up to a segment final altitude where (cruise) M is about 0.8; 4th and 5th seg.: level flight cruise at constant M; 6th seg.: descent at constant *EAS* to 10,000 ft (3,048 m); 7th seg.: EAS deceleration at 10,000 ft (level flight); and 8th seg.: descent at constant *EAS* from 10,000 ft (3,048 m) to 1,500 ft (457 m). (iv) Design variables and their range of permissible values: initial *EAS* in segment 1 (EAS_{1i} [89.0, 128.6] m/s), final EAS in segment 2 (EAS_{2f} [117.1, 184.6] m/s), initial altitude in segment 3 (h_{3i} [3048, 4400] m), initial altitude in segment 4 (h_{4i} [6096, 12192] m), initial altitude in segment 7 (h_{7i} [3048, 4400] m), and initial EAS in segment 8 (EAS_{8i} [89.0, 128.6] m/s). (v) Implicit constraint: initial M in segment 4 (M_{4i}) – allowable range ±0.5% of its nominal value, 0.8. (vi) ISA conditions have been assumed. (vii) FPA range allowed: [0, 7.5] deg during climb and cruise, and [-7.5, 0] deg during descent. Table 3 summarizes the data associated with this flight profile optimization.

The lower and upper bounds of the range of permissible values of the design variables were in general defined in a way to reduce the computational time of the optimization processes, to take into account typical air traffic control (ATC) restrictions, and/or to avoid the aircraft losing (gaining) altitude during climb (descent) processes. For instance, below 10,000 ft, the *EAS* lower and upper bounds usually correspond to, respectively, the aircraft stall speed (89.0 m/s

Table 1. Case 1 (simple climb profile) – Baseline trajectory and design variables.

Seg. No.	h_i (m)	h_f (m)	M	R (km)	Design variables
1	457	3048	0.38	20	–
2	3048	3048	0.46	10	$0.38 \le M_i \le 0.80$; $457 \le h_i \le 10668$
3	3048	7000	0.58	60	$0.38 \le M_i \le 0.80$; $457 \le h_i \le 10668$
4	7000	10668	0.80	100	$457 \le h_i \le 10668$

Table 2. Case 2 (implicitly constrained climb profile) – Baseline trajectory and design variables.

Seg. No.	h_i (m)	h_f (m)	M_i	M_f	EAS_i (m/s)	EAS_f (m/s)	R (km)	Design variables
1	457	3048	–	–	128.6	128.6	20	$89.0 \le EAS_i \le 128.6$
2	3048	3048	–	–	128.6	164.6	10	$133.8 \le EAS_f \le 221.2$
3	3048	7724	–	–	164.6	164.6	60	$3048 \le h_i \le 4400$
4	7724	10668	0.80	0.80	–	–	100	$3048 \le h_i \le 10668$

Table 3. Case 3 (quasi-full flight profile) – Baseline trajectory and design variables.

Seg. No.	h_i (m)	h_f (m)	M_i	M_f	EAS_i (m/s)	EAS_f (m/s)	R (km)	Design variables
1	457	3048	–	–	128.6	128.6	20	$89.0 \leq EAS_i \leq 128.6$
2	3048	3048	–	–	128.6	164.6	10	$117.1 \leq EAS_f \leq 184.6$
3	3048	7724	–	–	164.6	164.6	160	$3048 \leq h_i \leq 4400$
4	7724	7724	0.80	0.80	–	–	230	$6096 \leq h_i \leq 12192$
5	7724	7724	0.80	0.80	–	–	230	–
6	7724	3048	–	–	164.6	164.6	140	–
7	3048	3048	–	–	164.6	128.6	20	$3048 \leq h_i \leq 4400$
8	3048	457	–	–	128.6	128.6	70	$89.0 \leq EAS_i \leq 128.6$

EAS for the particular aircraft modeled) and the maximum *EAS* permissible below this altitude (according to ATC restrictions), i.e., 250 kts *EAS* or 128.6 m/s. In Case 3, in particular, the range of values in which the initial altitude in segment 4 can vary was established in a way to allow the aircraft cruising at altitudes between 20,000 ft (6,096 m) and 40,000 ft (12,192 m). Thus, EAS_{2f} permissible values were limited to those speeds that yield Mach numbers of about 0.8 at these cruise altitudes. Similar considerations were made in the other case studies analyzed in this work.

Optimization Process

According to the methodology followed in this work for optimizing a given aircraft trajectory, Polyphemus first randomly changes the values of the design variables (altitude and/or aircraft speed in one or more trajectory segments) in order to create a group of potential solutions. For a given potential solution, by making use of the initial aircraft weight (aircraft empty weight plus fuel on-board, constant), the APM carries out the computations related to the first segment of the aircraft trajectory, and determines the thrust required, flight time, etc. (Fig. 1). TurboMatch subsequently uses the flight conditions and the thrust required to determine the engine operating point, thereby establishing the engine fuel flow and the combustor inlet conditions among others. Hephaestus then makes use of the combustor inlet conditions and combustor geometric parameters to calculate the emission indices for the main pollutants. Based on the fuel flow and flight time, the fuel burned during the first trajectory segment and the new aircraft weight (i.e. the initial weight less fuel burned) are calculated. Computations continue in a similar fashion for all the remaining trajectory segments. When all the segments have been computed, among other calculations, the total flight time, fuel burned and gaseous emissions produced during the whole aircraft trajectory are also computed. This process is repeated for all the potential solutions, and for all generations of potential solutions that Polyphemus utilizes in order to determine an optimum trajectory according to given criteria initially specified by the designer. The results were obtained following a procedure similar to that described before.

RESULTS AND DISCUSSIONS

Main results of the optimization processes corresponding to three case studies indicated above are summarized in this section. In these processes, the minimization of total flight time, fuel burned, and NO_X emissions have been considered as the objective functions.

Case 1: Simple Climb Profile Optimization

The baseline climb profile for this case study as well as the optimum trajectories computed using Polyphemus and two commercial optimizers (MATLAB®, 2008) are illustrated in Fig. 3. Two different approaches used within the commercial package were (i) a pattern search algorithm called mesh adaptive search (MADS), and (ii) GAs. Both Polyphemus and the commercial optimizers yielded very similar results (Fig. 3a). Even though this first optimization case study (climb profile) corresponded to a hypothetical one, the reasonable agreement among the optimizers (average discrepancies ~2%) confirmed the validity of the approach. Figure 3c shows that in order to minimize the time spent during climb, Polyphemus suggests a solution where the aircraft flies at the highest *M* permissible, which was fixed at 0.38 and 0.80 in the first and fourth segment, respectively, and free to rise to 0.8 in the remaining middle two. Polyphemus also

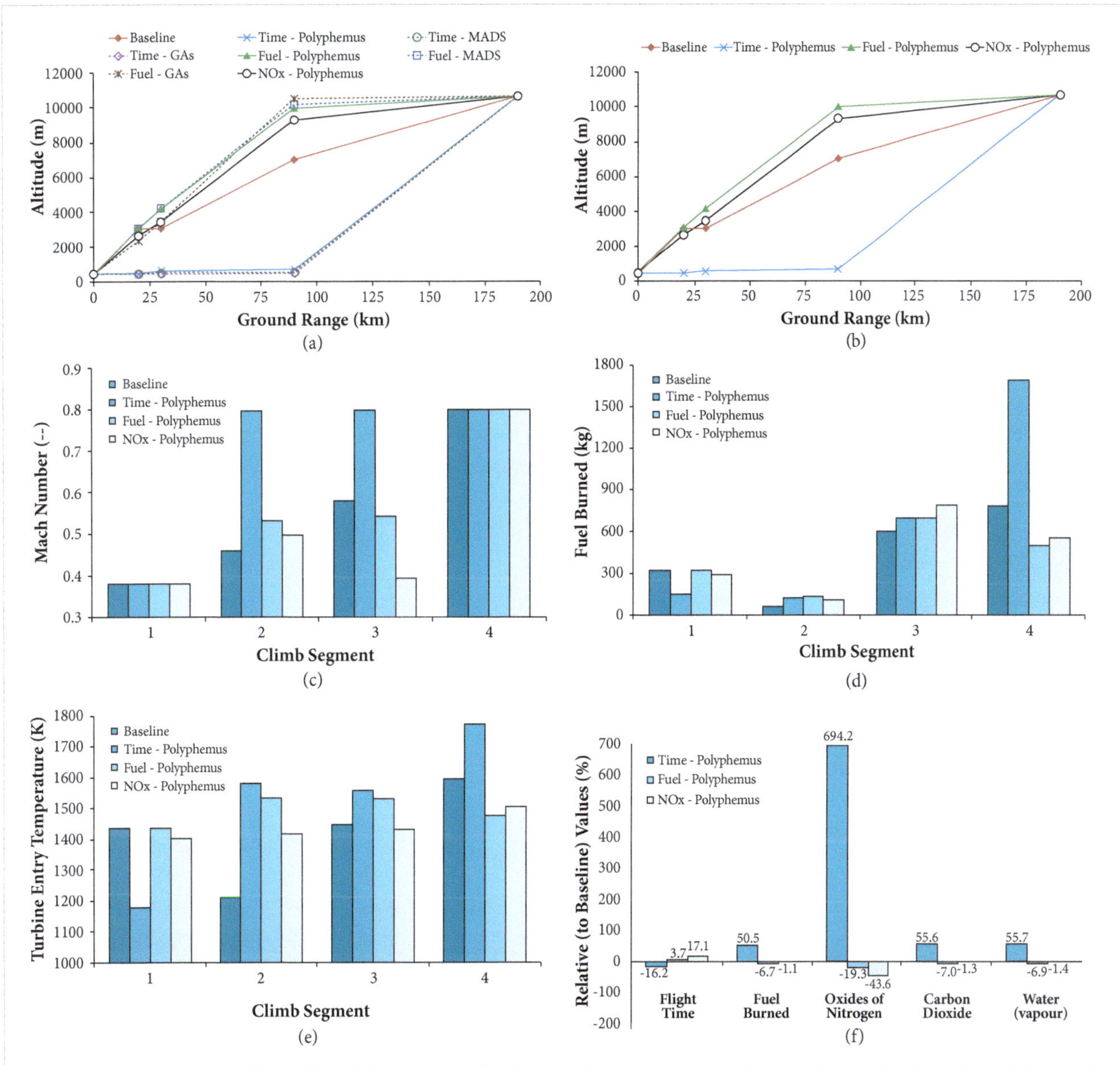

Figure 3. Case 1 – Simple climb profile optimization results.

suggests that the aircraft should fly at low altitudes for as long as possible before climbing rapidly to the target end altitude (Fig. 3b). This is mathematically correct because the speed of sound is the highest at sea level, thus enabling the aircraft to fly faster (maximization of TAS) if it could actually achieve *M* 0.8 at this level. This solution, however, does not represent practical flight profiles because never exceed speed (V_{NE}) is much lower than *M* 0.8 at sea level, thus restricting large transport category aircraft from approaching such high Mach numbers. Nevertheless, it is an interesting solution, confirming that the optimizer is working correctly in the absence of *M* (or TAS) constraints.

Figure 3 also illustrates that in order to reduce fuel burn, the optimizer suggests flying slower (Fig. 3c) and higher (Fig. 3b) than the reference trajectory (segment 3). This is again conceptually correct given the current reference trajectory. It is interesting to note that the fuel optimized trajectory proposes second and third segments affording a greater fuel burn (relative to the baseline) (Fig. 3d) in order to gain height (Fig. 3b), which then subsequently yields a lower fuel burn in the last segment and an overall lower fuel burn for the climb profile as a whole. In terms of flight profile, one could conclude from Fig. 3b that the trajectories optimized for minimum fuel burned and NO_X emissions are similar. However, there are significant differences between these two trajectories. The main difference is related to the fact that the NO_X emissions optimized trajectory is flown at relatively lower Mach numbers than the fuel burned optimized trajectory (Fig. 3c). These lower Mach numbers result in lower engine thrust settings, i.e., the thrust required to fly a given segment is lower, which in turn results in lower engine turbine entry temperature (TET) values (Fig. 3e). Consequently, as one of the main factors determining the level of NO_X emissions produced (besides the fuel burned) is TET, the trajectory optimized for minimum NO_X emissions produces a significant reduction in the amount of NO_X emitted (~-43%). Interestingly, Fig. 3e shows that in order to minimize NO_X emissions, the optimizer proposes a trajectory in which the engine TET remains almost constant (~1,400–1,500 K) for the entire climb profile. It is relevant to note in this discussion that the level of NO_X formed at temperatures near to and above 1,700–1,800 K increases exponentially with temperature.

An aspect to be highlighted in Fig. 3f is the level of gaseous emissions (NO_X, CO_2 and H_2O) associated with the optimum trajectories relative to the reference climb trajectory. As expected, variations in CO_2 and H_2O are directly proportional to the variations in the amount of fuel burned (species in chemical equilibrium). However, the aircraft trajectory optimized for total flight time significantly increases the amount of NO_X emissions. One of the main factors responsible for this significant increase in NO_X emissions (besides the increase in fuel burn) is the increase in TET resulting from the higher thrust settings. Fig. 3f also illustrates the increase in total flight time associated with the trajectory optimized for minimum NO_X emissions. Although this parameter increases, the total fuel burned slightly decreases as a consequence of the lower thrust settings (i.e., lower engine fuel flow relative to the baseline trajectory). Additional details about the results analyzed in this first case study can be found in Celis *et al.* (2009). In the following two case studies, complexities (in terms of operational constraints, number of segments, number of trajectory flight phases, etc.) were included gradually. This gradual approach afforded greater visibility of the mathematical performance of Polyphemus when assessing results, which would have been more difficult if the analysis had been initiated with very complex trajectories.

Case 2: Implicitly Constrained Climb Profile Optimization

Results obtained in the second case study (Fig. 4) are in general similar to those obtained in the first case study. Thus, when minimizing the time spent during climb, i.e., maximizing TAS, Polyphemus suggests a solution where the aircraft flies the first segment at the highest *EAS* permissible (fixed at 128.6 m/s) (Fig. 4c). This is conceptually correct because in the first segment, since the flight altitude is fixed, TAS increases with the increase in *EAS*. The optimizer also suggests that the aircraft should accelerate in the second segment to the highest *EAS* permissible (fixed at 221.2 m/s), and fly the following segments at low levels (Fig. 4b) as long as possible before climbing rapidly to the target end altitude. This is again mathematically correct because, firstly, as previously indicated, once the flight altitude has been established, the TAS increases with increase in *EAS*; and, secondly, for a given *M*, TAS increases with the decrease in altitude (speed of sound is the highest at sea level). Clearly, the influence of the third and fourth segments on the total climb time is more important than the corresponding second segment. Otherwise, the initial altitude in segment 3 would be the highest permissible.

Figure 4 also shows that in order to reduce the climb fuel burned, Polyphemus suggests flying mostly slower (Fig. 4c) and higher (Fig. 4b) than the reference trajectory. In particular, it suggests flying the first segment at the highest EAS permissible. It is clear that in order to minimize the total amount of fuel burned, the total energy required by an aircraft to describe a given flight profile (aircraft energy change plus path-dependent energy required to impart that change) must be minimized. Thus, in this particular case, the total energy required to climb must also be minimized. It means that the total aircraft kinematic energy change needs to be minimized. The aircraft kinetic energy change is minimized when the initial kinetic energy is maximized. It implies, in turn, maximization of the initial aircraft speed, i.e., TAS. This TAS maximization results in the maximization of the EAS at the first segment, as highlighted. One of the main factors driving the minimization of the fuel burned during a given flight profile is the aircraft mass change. Different factors affect the fuel burned and, consequently, the changes in the aircraft mass. The aircraft speed and flight altitude constitute two of these main factors. Reducing the speed and increasing the altitude reduce drag and, consequently, the thrust required to fly a given segment. This lower thrust requirement translates into a lower engine thrust setting, and, consequently, a lower fuel burn. However, neither altitude nor speed can be increased or decreased arbitrarily. Speed reductions imply in general, an increase in flight time, which can negatively affect the total fuel burned. In addition, in order to quickly achieve higher altitudes, higher engine thrusts (i.e. higher thrust settings) are required. These higher thrust settings require higher fuel flow, which

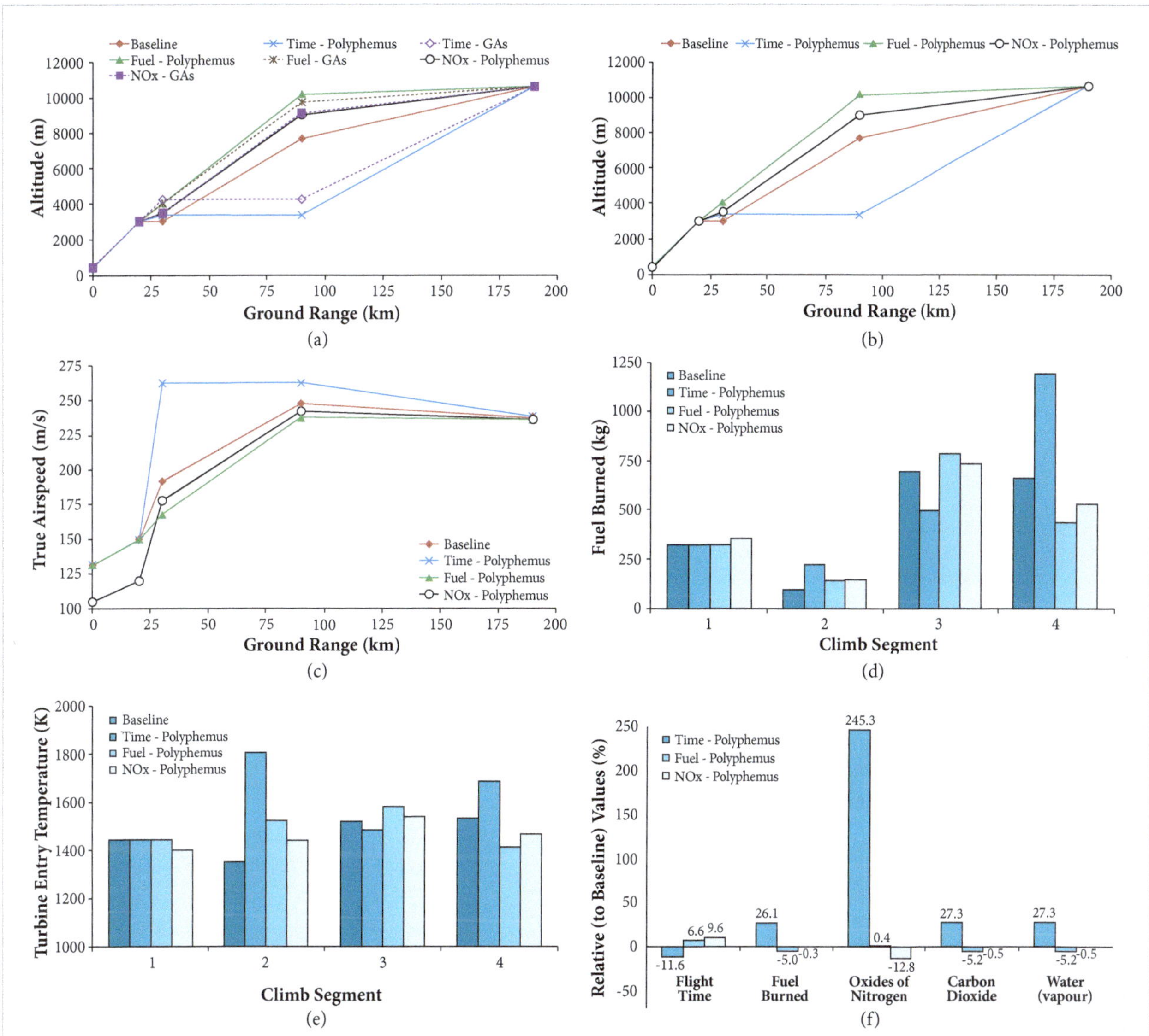

Figure 4. Case 2 – Implicitly constrained climb profile optimisation results.

negatively affects the fuel burned during the process. Therefore, a compromise between aircraft flight altitude and speed, which directly affect the changes in the aircraft mass, needs to be achieved at some stage. The fuel-optimized trajectory computed is a typical example of the referred compromise. It is interesting to note in Fig. 4d that this fuel-optimized trajectory proposes second and third segments affording a greater fuel burn (relative to the baseline) in order to gain height (Fig. 4b), which, then, is translated into a lower fuel burn in the last segment and an overall lower fuel burn for the whole climb profile.

Regarding the trajectory optimized for minimum NO_X emissions, the results show that, similar to the fuel optimized one, this trajectory is flown mostly slower (Fig. 4c) and higher (Fig. 4b) than the baseline trajectory. In general, lower speed and higher altitude lead to a reduction in the thrust required to fly the climb segments. These lower thrust requirements are in turn translated into lower engine TET values (Fig. 4e), which in turn result in a reduction in the level of NO_X emissions (Fig. 4f). As in the first case study, Fig. 4e illustrates that in order to minimize NO_X emissions, the aircraft describes a trajectory in such a way that the engine TET remains almost constant along the whole climb profile. As before, CO_2 and H_2O vary proportionally to the variations of fuel burned (Fig. 4f). Even though the NO_X emission-optimized trajectory increases total flight time, as a consequence of the lower engine thrust settings utilized, the total amount of fuel burned is slightly reduced. In Fig. 4f, it should also be noticed that the aircraft trajectory optimized for minimum flight time significantly increases the amount of NO_X emissions. This is partially because of the large amount of thrust required to increase both aircraft kinetic energy in segment 2 and potential energy in segment 4. This higher engine thrust requirement is translated into higher TET values (Fig. 4e), and consequently into a significant increase in the level of NO_X emissions.

Case 3: Quasi-Full Flight Profile Optimization

As highlighted before and illustrated in Fig. 5, which shows the main results obtained for this particular case study, minimization of total flight time implies maximization of TAS (Fig. 5c). Thus, when determining the minimum flight time-optimized trajectory, Polyphemus suggests a solution in this case, where the aircraft flies the first and the last segments at the highest EAS permissible (fixed at 128.6 m/s). This is conceptually correct because the first and last segments are flown at fixed altitudes, where TAS increases with the increase in EAS. Polyphemus also suggests that the aircraft should accelerate in the second segment to the highest EAS permissible (fixed at 184.6 m/s), and start the third segment as high as possible, and the fourth one as low as possible. This is again mathematically correct because, firstly, TAS increases with the increase in both flight altitude and EAS (segments 2 and 3); and, secondly, for a given Mach number, the TAS increases with the decrease in altitude (segments 4 and 5).

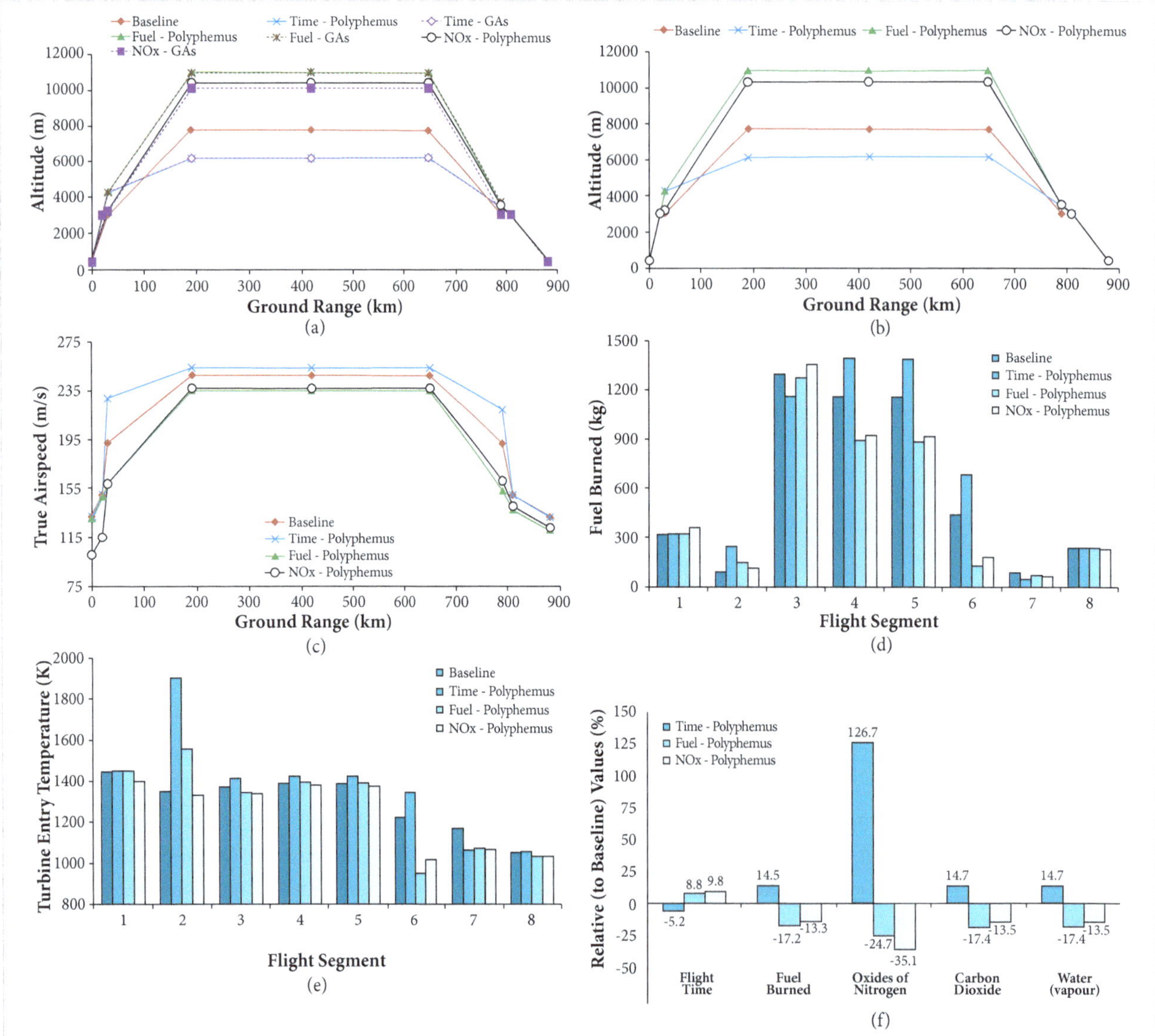

Figure 5. Case 3 – Quasi-full flight profile optimization results.

As a consequence of the larger distance covered by the cruise segments 4 and 5, their influence on the total flight time is more important than that associated with the third and sixth segments. This is emphasized by the fact that the aircraft has a tendency to cruise at low altitude levels as observed in Fig. 5b.

Regarding the fuel-optimised trajectory, it is observed that in order to reduce the total fuel burned, the optimiser suggests flying mostly slower (Fig. 5c) and higher (Fig. 5b) than the reference trajectory. In particular, it suggests flying the first segment at the highest EAS permissible. This situation is similar to that encountered in the second case study. In order to minimize the total fuel burned during the flight profile, the total energy required during the process must be minimized. In this case, it implies, in turn, minimization of the aircraft kinetic energy change; or, more specifically, maximization of the initial aircraft speed and minimization of the final one (in terms of TAS). As in the second case study, this TAS maximization makes the aircraft fly the first segment at the highest EAS permissible. Results also show that even though the aircraft arrives to the endpoint at a low speed, this does not correspond to the lowest EAS permissible (fixed at 89.0 m/s), as it could be expected. It is believed that one aspect that might be influencing this particular result is the path-dependent energy, which has a direct relationship with the aircraft speed and also needs to be minimum. In the foregoing analysis, the aircraft mass changes were not considered. These mass changes cannot be ignored however because in reality they are one of the main factors driving the minimization of the fuel burned during the optimization process. As highlighted before, there are two main parameters that affect the fuel burned and, consequently, the changes in the aircraft mass: the aircraft speed and the aircraft flight altitude. These two parameters directly or indirectly affect, in turn, other parameters, such as drag, thrust required, flight time and engine thrust setting (consequently, fuel flow, TET, etc.), among others. It implies that a fuel-optimized trajectory represents in fact a tradeoff among all these parameters, some of which conflict with each other.

The flight profile optimized for minimum NO_x emissions is flown similar to the fuel-optimized one, i.e., mostly slower (Fig. 5c) and higher (Fig. 5b) than the baseline trajectory utilized. In general, the relative lower speed and higher altitude utilized to fly this trajectory lead to a reduction in the thrust required to fly the trajectory segments. These lower thrust requirements are in turn translated into lower engine TET values (Fig. 5), which ultimately result in a reduction in the level of NO_X emissions produced (Fig. 5f). Fig. 5e shows, in particular, that from all TET values, those corresponding to the NO_X emissions, optimised trajectory values are the lowest ones. This is expected, of course, because this parameter has a direct influence on the level of NO_X emissions produced. In Fig. 5f, it can also be observed that the changes in CO_2 and H_2O present the expected behavior, and, even though the NO_X emissions optimized trajectory increases the total flight time, the total amount of fuel burned is largely reduced. This is a consequence of the lower engine thrust settings utilized to fly this trajectory. As in the first two case studies, the aircraft trajectory optimized for minimum flight time significantly increases the amount of NO_x emissions (Fig. 5f). This is partially due to the large amount of thrust required to fly some of the segments of this particular optimized trajectory.

CONCLUSIONS

Initial results obtained using optimization algorithms (i.e., Polyphemus) capable of performing multidisciplinary aircraft trajectory optimization processes were described in this work. A short description of both the rationale behind the initial selection of a suitable optimization technique and the status of Polyphemus was firstly presented. The Polyphemus optimizer was subsequently utilized to analyze three different case studies involving the optimization of one or more phases of actual aircraft flight profiles. Results related to optimum trajectories obtained using Polyphemus, minimization of total flight time, fuel burned and NO_X emissions were presented and discussed. The results obtained using Polyphemus and other commercially available optimization algorithms presented a satisfactory level of agreement (average discrepancies ~2%). Further developments on Polyphemus are currently underway in order to identify and efficiently compute optimum and 'greener' aircraft trajectories, which help minimize the impact of commercial aircraft operations on the environment. It is worth emphasizing that the main objective of different case studies analyzed was the evaluation of the mathematical performance of Polyphemus rather than the generation of realistic aircraft trajectories. As, in general, these different case studies provided mathematically and conceptually correct solutions, it is concluded that the approach utilized in this work for carrying out the aircraft trajectory optimization processes is a valid one. This, of course, provides the necessary motivation for continuing with the development of the Polyphemus optimizer.

ACKNOWLEDGMENTS

During this work, Cesar Celis was partially supported by the Programme Alβan, the European Union Programme of High Level Scholarships for Latin America, Scholarship No. E07D400097BR. Source of financing: European Union.

The authors would like to extend their gratitude to both Mr. Hasan Zolata, who using commercially available algorithms produced some of the results mention here, and Dr. Jean-Michel Rogero, whose PhD research at CU has provided a foundation for the Polyphemus optimizer used in this work. Special thanks also to Mr. Richard Long, who developed the aircraft performance model utilized.

REFERENCES

Betts, J., 1998, "Survey of Numerical Methods for Trajectory Optimization", Journal of Guidance, Control, and Dynamics, Vol. 21, No. 2, pp. 193–207.

Boeing Commercial Airplanes, "Market Analysis, Boeing Current Market Outlook 2009 to 2028", Available from: www.boeing.com/cmo. Accessed on: 16 August 2009.

Brooker, P., 2006, "Civil Aircraft Design Priorities: Air Quality? Climate Change? Noise?", The Aeronautical Journal, Vol. 110, No. 1110, pp. 517–532.

Bunday, B.D., 1984, "Basic Optimisation Methods", Edward Arnold, London, UK.

Callan, R., 2003, "Artificial Intelligence", Palgrave Macmillan, New York, US.

Celis, C., 2010, "Evaluation and Optimisation of Environmentally Friendly Aircraft Propulsion Systems", Ph.D. Thesis, School of Engineering, Cranfield University.

Celis, C., Long, R., Sethi, V. and Zammit-Mangion, D., 2009, "On Trajectory Optimisation for Reducing the Impact of Commercial Aircraft Operations on the Environment", 19th Conference of the International Society for Air Breathing Engines, ISABE-2009, Montréal, Canada.

Celis, C., Moss, B. and Pilidis, P., 2009, "Emissions Modelling for the Optimisation of Greener Aircraft Operations", Proceedings of GT2009, ASME Turbo Expo 2009, Power for Land, Sea and Air, Orlando, Florida, USA.

Clarke, J.P., 2003, "The Role of Advanced Air Traffic Management in Reducing the Impact of Aircraft Noise and Enabling Aviation Growth", Journal of Air Transport Management, Vol. 9, No. 3, pp. 161–165.

Clean Sky JTI (Joint Technology Initiative), 2009, Available from: www.cleansky.eu. Accessed on: 18 August 2009.

Davis, L. (editor), 1991, "Handbook of Genetic Algorithms", Van Nostrand Reinhold, New York, US.

Everitt, B., 1987, "Introduction to Optimization Methods and their Application in Statistics", Chapman and Hall, London, UK.

Fletcher, R., 1987, "Practical Methods of Optimization", 2nd Edition, John Wiley, Chichester, UK.

Goldberg, D.E., 1989, "Genetic Algorithms in Search, Optimization and Machine Learning", Addison-Wesley, Reading, MA, US.

Green, J. E., 2003, "Civil Aviation and the Environmental Challenge", The Aeronautical Journal, Vol. 107, No. 1072, pp. 281–299.

Gulati, A., 2001, "An Optimization Tool for Gas Turbine Engine Diagnostics", Ph.D. Thesis, School of Engineering, Cranfield University.

Hartley, S.J., 1998, "Concurrent Programming: The Java Programming Language", Oxford University Press, New York, US.

Krotov, V.F., 1996, "Global Methods in Optimal Control Theory", Marcel Dekker, New York, US.

Long, R.F., 2009, "An Aircraft Performance Model for Trajectory Optimisation", School of Engineering, Cranfield University, UK (unpublished).

MATLAB®, 2008, "The Language of Technical Computing, Version 7.7 (R2008b)", The MathWorks, Inc. Available from: www.mathworks.com.

Palmer, J.R., 1999, "The TurboMatch Scheme for Gas-Turbine Performance Calculations, User's Guide", Cranfield University, Cranfield, UK.

PARTNER, 2009, "Partnership for AiR Transportation Noise and Emissions Reduction", Available from: web.mit.edu/aeroastro/partner/. Accessed on: 18 August 2009.

Quagliarella, D., 1998, "Genetic Algorithms and Evolution Strategy in Engineering and Computer Science, Recent Advances and Industrial Applications", John Wiley & Sons, Ltd., Chichester, UK.

Rao, S.S., 1996, "Engineering Optimization: Theory and Practice", 3rd Edition, John Wiley, New York, US.

Riddlebaugh, S.M. (editor), 2007, "Research & Technology 2006, NASA/TM - 2007-214479", NASA Glenn Research Center, Cleveland, Ohio, USA.

Rogero, J.M., 2002, "A Genetic Algorithms-based Optimisation Tool for the Preliminary Design of Gas Turbine Combustors", Ph.D. Thesis, School of Mechanical Engineering, Cranfield University.

Rogero, J.M. and Rubini, P.A., 2003, "Optimisation of Combustor Wall Heat Transfer and Pollutant Emissions for Preliminary Design Using Evolutionary Techniques", Proceedings of the Institution of Mechanical Engineers, Part A: Journal of Power and Energy, Vol. 217, No. 6,pp. 605–614.

Russell, S. and Norvig, P., 2003, "Artificial Intelligence: A Modern Approach", 2nd Edition, Prentice Hall, New Jersey, US.

Sampath, S., 2003, "Fault Diagnostics for Advanced Cycle Marine Gas Turbine Using Genetic Algorithms", Ph.D. Thesis, School of Engineering, Cranfield University.

Schwefel, H.P., 1981, "Numerical Optimization of Computer Models", John Wiley, Chichester, UK.

Walsh, G.R., 1975, "Methods of Optimization", John Wiley, London, UK.

Wienke, D., Lucasius, C.B. and Kateman, G., 1992, "Multicriteria Target Vector Optimization of Analytical Procedures Using a Genetic Algorithm. Part I. Theory, Numerical Simulation and Application to Atomic Emission Spectroscopy", Analytica Chimica Acta, Vol. 265, No. 2, pp. 211–225.

Disaster Monitoring Constellation Using Nanosatellites

Mohamed Kameche[1], Haider Benzeniar[1], Ayhane Bey Benbouzid[1], Redha Amri[1], Nadir Bouanani[1]

ABSTRACT: In this paper, a constellation of four low earth orbit nanosatellites for disaster monitoring is presented; their small size, low cost, and short time development give them the opportunity to be widely selected. The space segment is based on existing subsystems that are assembled on the bottom part of a developed structure and present acceptable performance for the mission. The payload is a multispectral camera which fulfills the mission's requirements (remote sensing). The ground segment is based on an existing modular ground station, used for the ALSAT-1 DMC, which will be adapted to fit with the specifications of the mission.

KEYWORDS: Nanosatellite, Constellation, Earth observation, LEO orbit.

INTRODUCTION

African and Asian developing countries need support to face natural disasters such as floods, forest fires, desertification, and locust monitoring. The problems of desertification and locust are especially related to African countries, which are not able to handle them.

This paper describes a mission idea, to develop a low cost disaster monitoring constellation using four Nanosatellites, which allows different Asian and African developing countries to access satellite images, in order to face different natural disasters. The designed system consists on a space segment which is composed by two main parts. The bottom part of the structure – also called platform – has the same dimensions used for the U6 Cubesat platform structure to hold the existing different functions – Attitude and Orbit Control Systems (AOCS), Radio Frequency (RF), power, thermal, propulsion, On-Board Data Handling (OBDH), etc. The upper part of the structure – also called payload structure – is used in order to hold the camera (optic and electronic subsystems). The ground segment is based on an existing modular ground station, used for the ALSAT-1 DMC, which will be adapted to fit with the mission requirements.

The paper is subdivided into six sections. In the first section, the Mission Objectives are presented, followed by the Concept of Operations, and then the mission Key Performance are highlighted.

The Space Segment, Orbit and Constellation are described further on. The last section of this work is dedicated to the Implementation Plan.

1. Centre de Développement des Satellites – Bir Eldjir/Oran – Algeria

Author for correspondence: Mohamed Kameche | Centre de Développement des Satellites Agence Spatiale Algérienne | BP 4065, Ibn Rochd USTO | 31130 | Oran/Algeria | Email: mkameche@asal.dz

MISSION OBJECTIVES

The main objective is to launch the first low cost disaster monitoring constellation, which allows different Asian and African developing countries to access the satellite images, in order to face different natural disasters.

- Floods: analysis and assessment of damage caused by floods, ability to restore decisions of the authorities with the necessary key information for emergency measures to be undertaken;
- Locust biotopes monitoring: analysis of ecological conditions in the regions of desert locust breeding;
- Forest fires: characterization of forest, prevention against forest fires, and aid for decision taking by the authorities;
- Desertification: tracing of different maps of desertification sensitivity for many African countries.

The other main objective of this mission is to encourage the African developing countries to acquire the space technology, to develop low cost constellation for their economic development and to obtain a multidisciplinary team which, by the end of the project, shall be able to design, to integrate and to operate a full nanosatellite mission.

In absence of disaster, the constellation is dedicated for educative and scientific purposes.

CONCEPT OF OPERATIONS

The mission consists of a constellation of four Nanosatellites, and a ground segment for commands, telemetry and monitoring and for image processing. The launcher, to be selected for the constellation, is Polar Satellite Launch Vehicle (PSLV), which presents the advantage of its low cost, regarding other launchers.

The space segment is a nanosatellite, consisting of two main parts: platform and payload, with a total mass below 12 Kg (around 6 kg for each part). The developed structure is designed to have the same internal dimensions to support modules already used on ISIS 6U platform (CubeSat Kit, 2011). The payload is a multispectral pushbroom imager, suitable for wide range of missions in Low Earth Orbit (LEO). It is encapsulated in a compact unit and provides three multispectral channels, with an effective sensor length of 2098 pixels per channel.

To support the four nanosatellites of the constellation, an existing modular ground station used for the ALSAT-1 DMC will be adapted to fit the requirement of the mission (Amri, 2002). The existing ground station supports the communication in VHF/UHF/S-band (Fig. 1). It is implemented in the "Centre of space techniques building", located in Arzew, west of Algeria. It includes Yagi-Uda antennas and radio for the bidirectional UHF communications and for receiving the VHF beacon (Fig. 2) and a 3.7 m parabolic dish and radio, for receiving and transmitting the S-band signal (Fig. 3). The ground station software will be modified and adapted to meet the mission's requirements.

Regarding the operational concept, three operation phases are summarized below:

BEFORE NOMINAL OPERATION

The Launch and Early Operations Phase (LEOP) starts after the satellite separation from the launcher. This phase includes

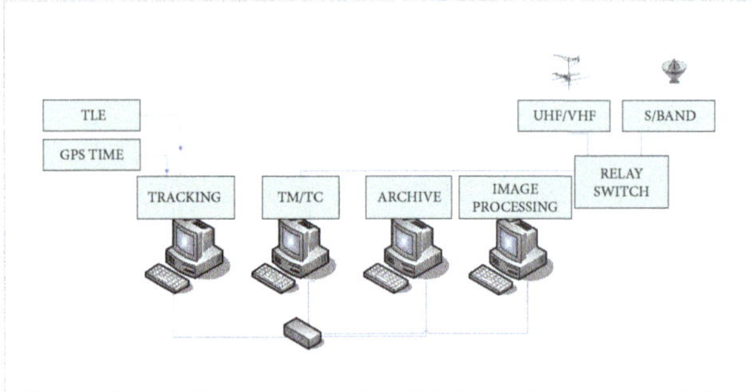

Figure 1. Ground station architecture diagram.

Figure 2. VHF/UHF antenna structure.

Figure 3. S-band antenna structure.

the antenna's deployment, the initial satellite acquisition (beacon signal), the validation of the satellite-ground station link, the switching of the equipments on, the acquisition, and the transmission of the first image to check that the satellite is not damaged during launch. After the first image, a phase of orbit correction will be performed to position each satellite in its mission orbit. The duration of LEOP phase depends strongly on the injection conditions (dispersions, initial injection velocity of each nanosatellite, etc). The period of corrections depends on the accuracy of injection by the launcher. Once the satellite is placed on its mission orbit, a commissioning phase starts. During this phase, the performances of the satellite will be tested and validated. The results will be used to correct and calibrate the on-ground satellite models.

NOMINAL OPERATIONS

During nominal operation phase, the satellite will be fully operational and shall perform the Earth observation mission in accordance with the defined timeline. This phase starts from the end of the commissioning phase up to 24 months after launch.

CONTINGENCY OPERATIONS

In order to avoid surprises during different phases (in particular the LEOP phase) and to prepare the solutions in advance, to predict possible failure scenarios and recovery plans will be elaborated.

KEY PERFORMANCE PARAMETERS

For each nanosatellite, the camera used gives a 146-kilometer field of view at a 62-meter spatial resolution in three spectral bands, the Red, the Green and the Near Infra-Red. This field of view allows the constellation to cover the whole Earth within 72 hours and to communicate with the main ground station, located in Algeria or other (so-called) auxiliary ground stations situated in other developing countries. The band spectrums selected for this mission are the Red: [0.22-0.69] mm, Green: [0.52-0.60] mm, and Near Infra-Red (NIR): [0.77-0.90] mm.

The nanosatellite is three axis stabilized when imaging and evolves in a Barbeque Mode (BBQ) mode out of imaging time. The Attitude Determination and Control Subsystem give attitude pitch/roll/yaw stability during imaging of less than 0.2 deg. The imaging system allows windowing, and it is supported by a total storage capacity of 2 Gbytes. The mission is designed to record the maximum of 16 elementary images of 2098 x 2098 pixels per day, which could be downloaded to a ground station at a 1 Mbps data rate, within four visibilities per day, with a total duration of about 32 min.

In order to assure the circularization of the constellation satellites, the station acquisition (which allows the satellites to equally separate from each other, so the daily global coverage is respected), and the station keeping of the four satellites, each nanosatellite of the constellation is equipped with a propulsion system of 50mN thrust and a 0.9-liter-capacity tank. The estimated preliminary delta V total is about 25 m/s and it can reach up to 30 m/s.

The mission lifetime for each nanosatellite is 18 months and it can reach 24 months or more, by introducing added protections against radiations (shielding on the sensitive equipments).

SPACE SEGMENT DESCRIPTION

The satellite is equipped with camera in order to take images with a 146-kilometer field of view and a 62-meter spatial resolution in three spectral bands, the Red, the Green and the Near Infra-Red ones. It will be powered by a battery which is charged by solar arrays on the outside of the structure. A Data handling payload will process, store and send the images to the S-band transmitter, for downloading them into the ground station. The actions of the satellite will be controlled by an active attitude control system, and will be oriented based on the Earth's magnetic field. The structure will be designed to meet its mission objective and to meet the specifications of the interface system that will be used for launching vehicle integration (Fig. 4).

The Nano-satellite specifications state the maximum mass allocation at 15 Kg, since it is the preliminary design, and the mass budget is based on the primary design, it will be updated in the following phases. The overall mass of the satellite is about 11.5 Kg, which means a margin of about 23.34%.

In the choice of the components, the tolerance to radiation had to be considered towards the harsh LEO environment. The anti-fuse rad-hard Field-Programmable Gate Array (FPGA) for Payload data handling, the anti-latch-up function in the command data handling system, and the experience learning from similar Cubesat based mission flying with analogue chip, reveal over estimated effective mission lifespan as Libertad-1 and Delfi-C3. Also, the interior arrangement of sub-systems boards had to be done according to former Cubesat based missions, by setting the On-Board Computer (OBC) between the other boards. The Commercial Off-The-Shelf (COTS) selected have tolerance up to 500krad of radiation and temperature tolerance in the range of - 40 to 85°C.

PAYLOAD

The objective of the mission is to take, process, store, and transfer, during one visibility, up to three or four elementary images of 2098 x 2098 pixels (15.75 Mbytes). To achieve this, we have selected a line scan camera system, designed to provide medium resolution of 62m, high dynamic range and low noise imagery. The image sensor selected is KODAK KLI-2113 with an imager size of 29.37 mm x 0.24 mm, which offers high sensitivity, high data rate and low noise. For each sensor, only one line among three is derived to obtain one active Charged-Coupled Device (CCD) line for each multispectral band.

To meet the mission requirements (low cost and time development), the sensor element will be requested without the RGB (red-green-blue) organic dye filters. A specific filter is procured from BARR to achieve the required spectral bands. The lens is a Schneider Apo-Componon. It's a flight proven lens (flown onboard Tsinghua-1 and Alsat-1). This commercial lens has been subject to a set of tests, prior to use. The camera board consists in a three video chain, composed by each of a pre-amplifier for CCD signal and a correlated double sampling, and a 10 bit A/D converter function, represented

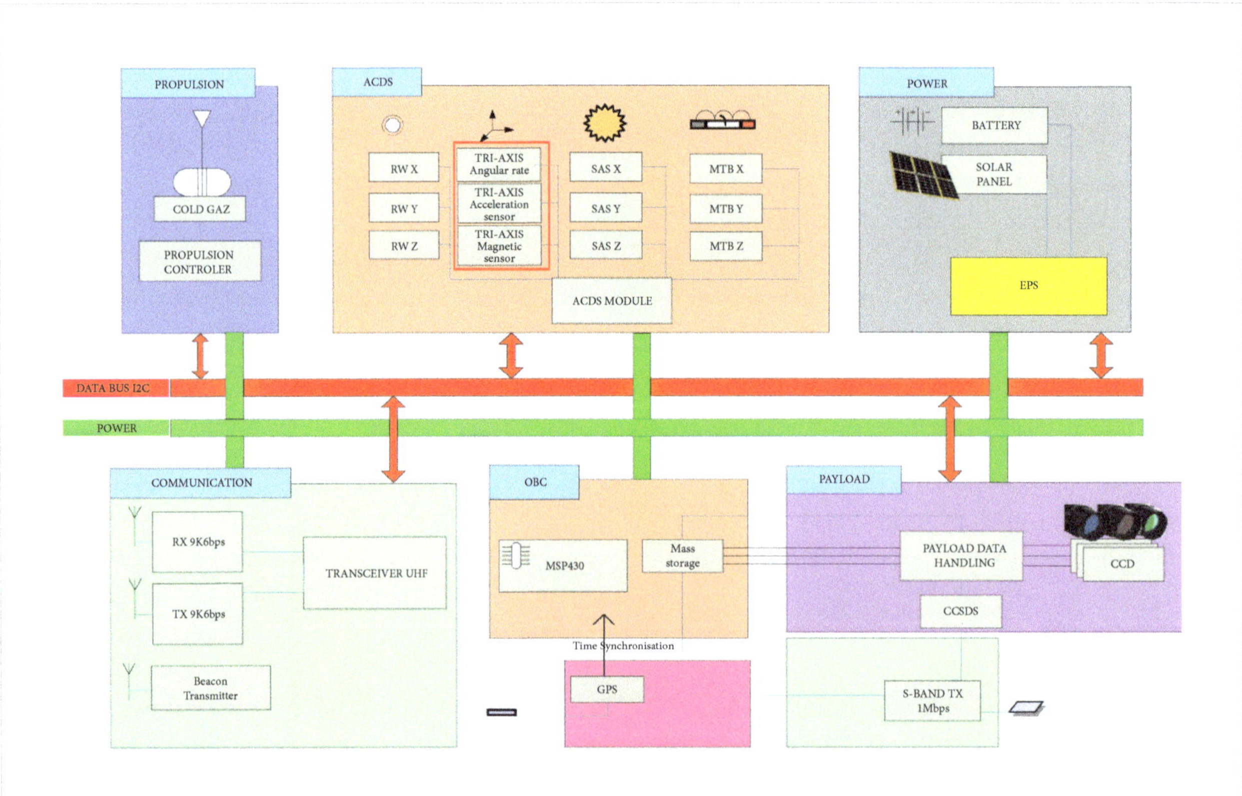

Figure 4. Nanosatellite block diagram.

in Analog device 9840A chip. Meanwhile, a high density FPGA is used to interface, via the used bus, between the video chain, the mass memory hosted by the on-board data handling FM430 and the TX sub system. A dedicated Actel FPGA RT54SX is used to packetize the mission data in The Consultative Committee for Space Data Systems (CCSDS) format before transmission.

ATTITUDE CONTROL SYSTEM

The main goal of the attitude determination and control system (ADCS) is to guarantee an accurate Earth pointing when imaging. This can be achieved using sensors for acquiring spacecraft attitude and position, and after performing on-board calculation, or receiving ground commands the spacecraft makes, using actuators to achieve the desired attitude or trajectories. Once the spacecraft achieves a certain desired attitude or orbit, the role of the ADCS is not over, because of the continuous disturbing forces acting on the spacecraft: aerodynamic drag, solar radiation pressure, gravity gradient torques and genetic field torques.

The proposed ADCS is based on the ADACS IMI200 integrated system from Maryland Aerospace, Inc, that uses three-axis active controls with three reaction wheels and three magnetorquers, providing a 0.02Nms momentum and 5mNm as maximum torque. In addition to that, three sun sensors, triaxial gyroscope, triaxial accelerometer, and triaxial magnetometer are also used to enhance the attitude determination, for accuracy of about 0.2°.

The miniaturized sun sensors are provided from SFL/Sinclair interplanetary Company, and the triaxial inertial sensor with analog magnetometer is from ANALOG DEVICES, Inc. A fixed nadir pointing is set as nominal pointing, for both imaging mission and in eclipse, since the payload is a multispectral camera. The ADCS works with three modes, the detumbling mode – applied after launch separation using magnetorquers, the imaging mode – with nadir pointing, and the BBQ mode – for thermal equilibrium. In this mode, the satellite is spinning around the yaw axis, which is pointing towards Earth. The imaging and the BBQ modes are achieved by the reaction wheels and the magnetorquers for momentum dumping. However, when proceeding to orbital maneuver, the satellite can slew to some degrees to perform the required maneuver.

The GPS receiver selected is the SGR-05U from SSTL, used for an accurate timing and synchronization.

ELECTRICAL AND POWER SYSTEMS

The preliminary satellite power budget estimation in Table 1 shows that the estimated required average power for the imaging and reading mode simultaneously, or only reading mode during night visibilities, are respectively 22 and 15 W, provided by 58 GaAs junctions, placed on the four lateral sides of the structure, and a Lithium Polymer battery (CS-SBAT2-30) with a capacity of 30 W.h and a depth of discharge of 20%. In order to improve the total power provided by solar arrays and the thermal protection of the satellite, the solar arrays will be

Table 1. Nanosatellite power budget estimation.

Equipment	Peak power (W)	Power (imaging mode) (W)	Power (orbital manoeuvre) (W)	Power (BBQ mode) (W)	Power (imaging + read mode) (W)	Power (read mode) during night visibilities (W)
ADCS	4.925	4.925	4.925	4.925	4.925	4.925
Payload (imager + electronic parts)	7	7	Off	Off	7	Off
Command & Data Handling	0.22	0.22	0.22	0.22	0.22	0.22
S-band Transmitter	5	Off	Off	Off	5	5
VHF Beacon Transmitter	1.5	1.5	1.5	1.5	1.5	1.5
Transceiver UHF TM/TC Data	2.71	2.71	0.000066	2.71	2.71	2.71
GPS Receiver	1	1	1	1	1	1
Propulsion System	10	Off	10	Off	Off	Off
Electrical Power System	0.1	0.1	0.1	0.1	0.1	0.1
Total Power (W)	32.455	17.455	17.745	10.455	22.455	15.455

specified in accordance with the exact dimensions of each lateral panel. The battery is required for the mission because the solar cells alone cannot produce enough power when peak power is needed, and it provides power to subsystems that cannot be turned off while the satellite is in eclipse. The power and the electrical subsystems have to accommodate a variety of power needs, as required by the mission timeline. If the satellite is in sunlight and it takes and transmit images to the ground station, the payload and both transmitters (UHF and S-band transmitter) will all need full power to operate. In these conditions, the satellite acquires supplementary power from the battery.

Regarding the calculated peak power (all equipments On) shown in Table 1, it is recommended that no mission (neither imaging nor data reading) will be performed the day of orbital maneuvers, because 10 W is required for the heater and the pipe heater, to evaporate the liquid butane before injection through the thruster.

COMMUNICATIONS

The communication to and from the satellite (control, command, and monitoring) is achieved using a commercial UHF transceiver (NanoCom U480), operating in the amateur radio band. The images data will be transmitted by using an S-band transmitter, operating in the commercial band (2.2-2.29 GHz), with a data rate of 1Mbps and an output power of 1 W. Also, a dedicated transmitter operating in VHF band will be used to send a beacon signal containing satellite health to aid in satellite identification, tracking and monitoring, particularly in the days following launch. Monopole and patch antennas are used for uplink and downlink communications. The uplink and the downlink

Table 2. S-band Telemetry link budget.

S-BAND TELEMETRY LINK BUDGET						**V1.0 of the 07/12/10**
Frequency	2260.00 MHz	***Link data***		***Station***		
Polarisation	C	Orbit Altitude	680 Km	Name	ARZEW	
		Availability %	99.00	Latitude	35.8585000 °	
Data rate - to download images data	1.00 Mbps			Longitude	-0.3139000 °	
Modulation	QPSK			Height	0.260 Km	
Coding	Conv. (R=1/2)			Antenna Diameter	3.7 m	
				Antenna Gain	25.8 dbi	

		Nominal Mode				**Comments**
SL to G/S site angle		5 °	10 °	45 °	89 °	
SL - Station distance		2517 Km	2112 Km	919 Km	680 Km	
On-Board field of view		64.19 °	62.86 °	39.72 °	0.90 °	
Transmitter output power	dBW	1.00	1.00	1.00	1.00	
Pertes harnais	dB	-1.00	-1.00	-1.00	-1.00	
On-board antenna gain	dBi	4.00	4.00	4.00	4.00	
S/C EIRP	**dBW**	**4,00**	**4,00**	**4,00**	**4,00**	
Space losses	dB	-139.01	-137.48	-130.26	-127.64	
Gaz losses	dB	0.41	0.20	0.05	0.04	
Scintillation losses	dB	0.82	0.36	0.07	0.04	
Cloud losses	dB	0.02	0.01	0.00	0.01	
Rain losses	dB	0.00	0.00	0.00	0.00	
Total atmospheric losses	dB	-1.23	-0.56	-0.12	-0.08	
Flux density at G/S	**dBW/m2**	**-136,24**	**-134,05**	**-126,38**	**-123,72**	
Isotropic area	dBm2	-28.54	-28.54	-28.54	-28.54	
Ground station G/T	dB/K	11.10	11.40	11.90	11.90	
Pointing losses	dB	-0.50	-0.50	-0.50	-0.50	
Polarisation losses	dB	-1.00	-1.00	-1.00	-1.00	
G/T degradation due to rain	dB	-0.18	-0.09	-0.03	-0.02	
Received C/N0	dBHz	73.25	75.82	84.06	86.73	
Rx + Tx quality losses	dB	-2.00	-2.00	-2.00	-2.00	
Useful data rate	Mbps	1.00	1.00	1.00	1.00	
Eb/N0	**dB**	**11,25**	**13,82**	**22,06**	**24,73**	
Required Eb/N0	dB	6.35	6.35	6.35	6.35	
Margin	**dB**	**4,90**	**7,47**	**15,71**	**18,38**	

budgets show very good margins. Table 2 illustrates the S-band Telemetry link budget.

COMPUTER

The OBC used in the design consists on a single board computer flight module FM430, designed by Pumpkin, for low cost and low power consumption embedded applications, built around a Texas Instrument 16-bit TI MSP 430 microcontroller, made for harsh environment. This choice offers an operational reliability (Libertad-1, Delfi-C3). The flight Module computer is charged to monitor satellite and payload housekeeping parameters, and manage bidirectional communications with the mission's ground station, handling data storage as well as retrieving satellite control. The OBC is the master of the I2C-bus on the satellite, and the other sub-systems are slaves.

SOFTWARE

Salvo Pro RTOS is used to handle all software tasks. Highly configurable Pumpkin product of low cost , it can run multiple prioritized tasks and work with a cooperative scheduler. Ideally suitable for MSP430's 2K RAM, it allows application to sleep at all times, waking only for activity when specified events occur. Well documented, it can be developed under Non-intrusive environment and be easily debuged. The mission specifications will be developed under Salvo Pro RTOS environment.

STRUCTURE

The proposed structure of the Nanosatellite must hold all the components and fit the specifications of launch vehicles. The structure is made of Aluminum and based on two main parts, the platform structure which has the same dimensions of an U6 Cubesat platform to hold the components, and the payload structure for the camera. Four lateral panels are attached to the structure, to support the solar cells. The camera is placed on the bottom panel in order to point towards Earth (Fig. 5). The satellite envelope is 340 x 360 x 280 mm^3, and the satellite total mass will be less than 12 kg. The POD-types separation systems cannot be used because of the satellite envelope. For our mission, the AxelShooter separation system is selected, which ensures protrusions such as antennas and camera, and allows the visual inspection and operation check after the attachment (Axelspace, 2011).

THERMAL

A passive thermal protection system consisting of proper insulation will be used to provide proper protection from radiation and heat fluxes. This insulation will consist of a layer of Kapton outside the panels, which will also act as an adhesive for the solar panels, and a Multi-Layer Insulation (MLI) inside the structure. Since solar panels will be covering most of the panels, the layer will be acting as an additional insulation. Additional thermal covering will be used around the most thermally sensitive components.

PROPULSION

A butane cold gas propulsion system, utilizing 460 g of propellants to meet the mission's requirements (ΔV of about 25 m/s up to 30 m/s), will be developed. One tank, with capacity for 0.9 L, will be used for the storage of butane at maximum pressure of 4 bars (Fig. 6). The tank will be connected to the thrusters, using pipe heater to evaporate the liquid butane before injection through the thruster (Gibbon et al, 2001). The propulsion system delivers approximately 50 mN thrust.

Figure 5. Outside satellite configuration.

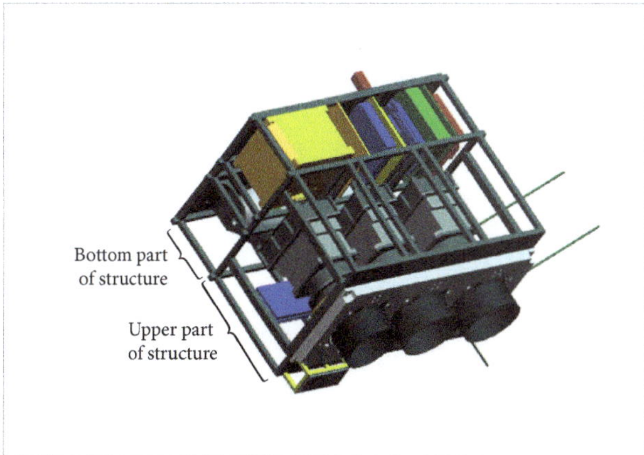

Figure 6. Preliminary satellite configuration.

ORBIT AND CONSTELLATION DESCRIPTION

Since many Nanosatellites are typically launched as a secondary payload, its precise orbital trajectory is dictated by the requirements of the primary mission. Based on previous Cubesat launch trends, a sun-synchronous reference orbit has been chosen with an inclination of 98.08 degrees, an altitude of 670 km over the Equator, and a period of 98.25 mn. The local solar time used in the design is 10h +/- 30 mn, and it will be adjusted when the launch opportunity is defined. This orbit allows the obtaining of an image swath of 146 km, which allows the four satellites of the constellation to cover the whole earth within 96 hours.

For a reference orbit with a repetitivity of (14 + 19/29), the position of the four satellites in the constellation will be: Nanosat 1 is defined as the reference position; Nanosat 2 is located at about 86.9 degrees from Nanosat 1; Nanosat 3 is located at about 86.9 degrees from Nanosat 2; and finally, Nanosat 4 is phased 86.9 degrees from Nanosat 3or 99.3 degrees from Nanosat 1, in opposite direction (Fig. 7).

It also allows the ground station to communicate with the satellites four times per day, with an average duration of 8 min per visibility. The minimum elevation considered for VHF/UHF/S-Band communications is 5 deg.

This reference orbit is a design guide built from historical averages, and additional worst-case figures must be considered whenever possible throughout the mission's design process. In order to have different earth coverage by each satellite, and to avoid simultaneous visibility of two nanosatellites or more from the main ground station, the reference mission orbit of each satellite, and especially the argument of latitude, will be carefully defined. The description of the position of the nanosatellites constellation is done according to Fig. 7.

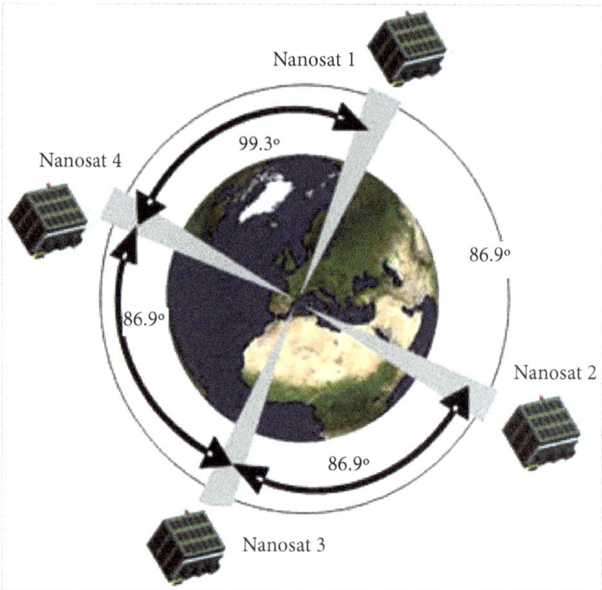

Figure 7. Nanosatellite constellation phasing.

IMPLEMENTATION AND CONCLUSION

The Satellite Development Centre is the technical entity, working with the Algerian Space Agency on the conception, realization and operations of different satellite missions (Earth observation, communication). To date, two Earth observation microsatellites have been successfully launched (ALSAT-1 and ALSAT-2A) and a third satellite ALSAT-2B is planned to be launched within the two following years. The two first satellites have been successfully accomplished with the cooperation of the SSTL for ALSAT-1 and the EADS-Astrium for ALSAT-2A.

For this mission, the conception, the accomplishing, the integration, and the tests of at least one nanosatellite of the constellation, will be performed in the Centre of Space Techniques CDS facilities. The CDS is a new centre located in Oran city (Algeria). It has been inaugurated in February 2012. It contains a clean room, vacuum, acoustics, vibration and RF facilities to perform environmental tests.

REFERENCES

Amri, R., 2002, "ALSAT-1 documentation", ALSAT-1 Project Documentation.

Axelspace, 2011, "Nano-Satellite Separation Mechanism AxelShooter", Retrieved on 2011, from http://www.axelspace.com/product/AxelShooter_e.pdf

CubeSat Kit, 2011, "Begin your CubeSat Mission with the CubeSat Kit", Retrieved on 2011, from http://www.cubesatkit.com

Gibbon, D., Underwood, C., Sweeting, S. and Amri, R., 2001, "Cost effective propulsion systems for small satellites using butane propellant", 52nd International Astronautical Congress, Toulouse, France.

9

Nanoelectronic Devices Based on Carbon Nanotubes

Victor Dmitriev[1], Fernando Gomes[1], Clerisson Nascimento[1]

ABSTRACT: Carbon nanotubes are possible building blocks in the development of new generation of electronic devices. The carbon nanotubes allow one fabrication of devices using nanometric scales. They can be used in projects of a wide range of electronic and optoelectronic components such as diodes, transistors and interconnection elements, among others. This paper presents the state-of-the-art of this area, emphasizing the influence of quantum effects on the characteristics of such components.

KEYWORDS: Nanoelectronics, Carbon nanotubes, Nanotransistors.

INTRODUCTION

Carbon nanotubes (CNT) with different configurations are the most studied nanomaterials since their discovery in 1991 by S. Iijima (1991). They present a single hollow cylinder with one or several layers of carbon atoms. The CNTs behave as one-dimensional (1-D) systems since their diameters are on the nanometric scale and are much smaller as compared to their lengths. 1-D nature of CNT defines their special physical properties, which have been used in design of novel electronic devices. It is expected that the nanocomponents based on CNT will have some of electrical and optical parameters far superior than the existing microelectronic components based on Si, Ge or GaAs semiconductors (Dragoman and Dragoman, 2006).

There is a relentless pursuit of getting smaller and smaller devices with low energy consumption and high processing speed in many areas, including aerospace technology. In this sense, single-wall carbon nanotubes (SWCNT) have been very promising because, depending on their chirality and diameter, they may exhibit metal or semiconducting behavior (Yao *et al.*, 1999). This allows one to project good conducting wires and p-n junction in nanometric scale.

In this paper, we present a review of electrical properties of some components based on SWCNT. In particular, interconnecting elements, Schottky diodes and field-effect transistors are discussed. We make a comparison of the nanoelectronic components based on CNT and the corresponding microelectronic devices and discuss the principal problems in their modelling.

1. Universidade Federal do Pará – Belém/PA – Brazil.

Author for correspondence: Fernando Gomes | Departamento de Engenharia Elétrica, Universidade Federal do Pará | Avenida Augusto Correa, 1 | CEP: 66.075-110 Belém/PA – Brazil | Email: fgomes@ufpa.br

NANOELECTRONIC COMPONENTS BASED ON CNT

CNT AS INTERCONNECTION ELEMENTS

One of the first ideas for applications of SWCNT with metallic characteristics is the use of them as interconnectors in integrated circuits. A great advantage of carbon nanotubes is a possibility of ballistic transport over long length. The regime of ballistic transport, in contrast to diffusive one, is characterized by absence of scattering or collision of moving electrons in the CNT structure (Wong and Akinwande, 2011).

Integrated circuits with a large-scale integration can have billions of components. Before CNT can be used as interconnection wires, there are some challenges that must be overcome. All tracks that connect internal components of the integrated circuits suffer some unwanted phenomena such as delays, heat dissipation, noise, jitter and electromigration. The metallic CNT can serve as an alternative to overcome these challenges. They are characterized by a relatively long mean free path, high current density and resistance to electromigration (Wong and Akinwande, 2011; Javey and Kong, 2010).

Below, we describe the interconnection elements based on metallic SWCNT as a possible replacement of copper tracks. Due to 1-D nature of SWCNT, for a given applied external electric field, only a small number of channels or sub-bands are involved in the process of charge transport. The finite number of states means discretization of these states' energy levels. This leads to some quantum effects that are described by a quantum resistance, quantum capacitance and kinetic inductance, which are very different from the corresponding classical concepts of resistance, capacitance and inductance (Wong and Akinwande, 2011).

Quantum Resistance

Ideally, the good quality interconnection resistance must be negligible when compared to the resistance of the devices to be interconnected. The resistance that appears at the interface between a metal contact and a metallic SWCNT is called the quantum resistance. In a metallic SWCNT at low energies, *i.e.* energy of the order of a few electron volts, only two 1-D sub-bands are involved in the electrical transport. If the transportation is ballistic, the value of the quantum resistance at the interface between the metal contacts and the metallic SWCNT is $R_q = 6.5\ K\Omega$ (Wong and Akinwande, 2011; Javey and Kong, 2010).

Quantum Capacitance

As far as the quantum capacitance's effect is concerned, it appears also only in systems with finite number of states (Wong and Akinwande, 2011; John *et al.*, 2004). This quantum capacitance (C_q) is defined by the additional energy required to release charge carriers in systems due to quantization of the energy levels of quantum states. This capacitance is only noticeable in systems with a very low density of states (DOS). The value of the C_q is much smaller than the value of the classical (geometric) electrostatic capacitance (C_{es}). As the C_q is in series with the C_{es}, the result of this combination is the dominance of the C_q. However, in structures with a large number of free electrons, the C_q is very large when compared with the C_{es}.As a result, there is a predominance of the C_{es}. The value of C_q of metallic CNT is around of 310 $\alpha F \mu m^{-1}$ (Wong and Akinwande, 2011).

Kinetic Inductance

The classical magnetic inductance (L_m) is associated with the magnetic field and the corresponding magnetic energy due to charges motion. The kinetic inductance (L_k) is associated with the kinetic energy of the electrons in motion. In case of ballistic transport in the CNT, the kinetic inductance is relatively large since the kinetic energy is not dissipated due to scattering. The L_k is in series with the classical L_m. Due to the relatively long mean free path, which is characteristic of the CNT, the L_k for signal frequencies above 100 GHz becomes appreciable, which is of great importance in high-frequency electronic applications. The L_k value for metallic CNT is about 3,2 $\eta H \mu m^{-1}$ (Wong and Akinwande, 2011). This value takes into account the degeneracy of spin.

Degeneracy occurs when there are two or more quantum states with the same energy. The Pauli exclusion principle states that two electrons can not be in the same state. So, in order to accommodate two electrons in one state, the spins of the electrons have to be different, *i.e.* one electron with spin up and another electron with spin down. This is degeneracy of spin.

Transmission Line Model of CNT

A model of metallic CNT as a transmission line with distributed elements is presented in Fig. 1. In this equivalent scheme, C_q and L_k effects were taken into account (Wong and Akinwande, 2011). In Fig. 1, $R_c/2$ is the input and output contact resistance of line, *i.e.* we deals with two equal metal contacts, and the R_c must be divided into two. R is the resistance per unit length of the metallic CNT, excluding R_c This term describes ohmic losses in diffusive regime. Δx represents the incremental length of line and ΔL is the length of the transmission line.

The transmission lines are characterized by the characteristic impedance (Z_0) and the propagation constant γ (Sadiku, 2004).

The propagation constant is in general a complex number. The real part (α) and the imaginary part (β) represent the attenuation and phase constant of the transmission line, respectively. With the given parameters of the transmission line in Fig. 1, Z_0 and γ can be determined using the classical transmission line theory.

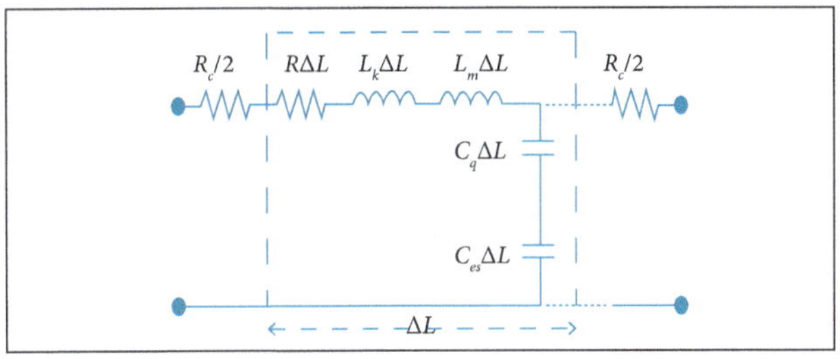

Figure 1. Transmission line model of CNT with distributed elements.

The transmission line can be considered lossy when the signal frequency is much below the critical frequency and lossless when the signal frequency is much above the critical frequency. This critical frequency is usually called zero frequency (ω_z) and it is defined as the ratio between R and L_k of the transmission line, *i.e.* $\omega_z = R/L_k$ (Wong and Akinwande, 2011). For a high-quality SWCNT at room temperature and with the effective mean free path of 1 μm, $\omega_z \approx 318$ *GHz* (Wong and Akinwande, 2011).

In the lossless (ballistic) transmission regime, the propagation constant γ of the line is a pure imaginary number which is proportional to the frequency of the signal, *i.e.* $\gamma = j\beta = jg\omega\sqrt{L_k C_q}$. In this equation, g is a parameter that quantifies the electron-electron interaction, j means that the variable is a complex number and ω is the angular frequency. When $g \sim 1$, the interaction is weak; and when $g \sim 0$, the interaction is strong (Wong and Akinwande, 2011). The line's Z_0, which, in this case, is given by $Z_0 = 1/g\sqrt{L_k C_q}$, is a pure real number. It does not depend on the frequency. This means that a signal that propagates along this line does not suffer attenuation; besides, the voltage and the current of the line are in phase. The group velocity $v_g = \vartheta\omega/\vartheta\beta$ is constant, *i.e.* all the frequency components of the signal propagate along the line with the same velocity v_g given by $v_g = v_F/g$, where is the Fermi velocity: $v_F \approx 10^8$ $cm.s^{-1}$ (Wong and Akinwande, 2011).

In a lossy transmission line (diffusive regime), the attenuation is due to ohmic losses represented in Fig. 1 by R. The characteristic impedance is no longer a pure real number and is given by $Z_0 = 1/g\sqrt{R/2\omega C_{tot}}\, e^{j\Pi/4}$ (Wong and Akinwande, 2011). Analyzing this expression, one can conclude that, for the metallic CNT, the voltage is delayed with respect to the current by $\Pi/4$, presenting, thus, partially capacitive behavior. In addition, the amplitude of the signal becomes dependent upon the frequency (ω). The term C_{tot} is the resulting capacitance of C_q and C_{es} in series, according to the model presented in Fig. 1, and g is computed as $g = \sqrt{C_{tot}/C_q}$.

The group velocity is given by $v_g = 2\sqrt{2\omega/RC_{tot}}$, which also depends on ω. This dependence causes undesirable signal dispersion along the metal CNT. In order to reduce the effect of attenuation and scattering of the signal, one should keep both the length and the C_{es} of the metallic CNT as small as possible. To decrease the influence of ω in the v_g, the C_{es} must be much smaller than the C_q (Wong and Akinwande, 2011).

In general, CNT as metallic interconnecting elements will have an advantage over the copper tracks when the CNT are in ballistic regime, as one can see in Fig. 2 (Wong and Akinwande, 2011; Javey and Kong, 2010).

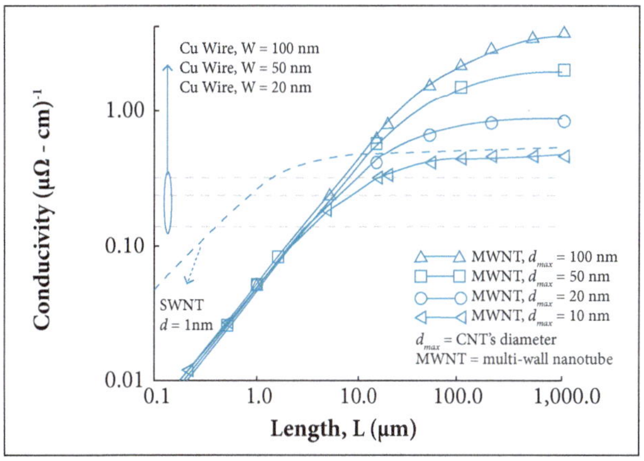

Figure 2. Conductivity for SWCNT and MWNT nanotubes as a function of length (Wong and Akinwande, 2011).

NANODIODES WITH CNT

In order to obtain rectifying p-n junctions, one can use different methods in particular, doping impurities or intramolecular junctions. Intramolecular junctions in principle can be obtained by introducing defects in a CNT, using two different mechanisms. Firstly, by a mechanical deformation, which consists in bending of a CNT with respect to its axis. At the spot of deformation, some defects are introduced, which, in this case, are pentagon-heptagon pairs, so that, in one side of the bending, there is a metallic CNT and, in the other side, a semiconductor CNT (Fig. 3a). Secondly, one can use a junction of two CNT with different diameters. One of the CNT behaves as a metal and the other one, as a semiconductor. At the point of junction, pentagon-heptagon pairs in hexagonal structure of the CNT are also introduced. Figure 3b shows details of the junction (Yao *et al.*, 1999; D'yachkov, 2006).

According to Yao *et al.* (1999), the carbon nanotube shown in Fig. 4a is placed directly on three electrodes. The electrodes are over a film of silicon dioxide (SiO_2), which, in turn, is on a silicon (Si) wafer that has the function of a gate. At room temperature, the resistance of the carbon nanotube's upper segment between the two electrodes is 110 KΩ, shown in detail in Fig. 5a. This thread has metallic behavior and is independent of the applied gate voltage. On the other hand, the current-voltage characteristic of the carbon nanotube bend, also shown in Fig. 5a, has a highly non-linear and asymmetric behavior, resembling the current-voltage characteristics of a diode rectifier. We can observe a very high increase in the current when applying a positive voltage on the top electrode when the voltage exceeds the threshold voltage, which is about 2 V (Fig. 5a). There is a very small increase in current for the reverse bias. This junction has an extremely high impedance for the voltages around 0 V.

Figure 5b represents current-voltage characteristics of metal-semiconductor junction in Fig. 4a. The electrodes are made of gold (Au) and titanium (Ti) and have dimensions of 250 nm wide, 20 nm in thickness and less than 1 nm in height, which minimizes deformation of the CNT. The curve in Fig. 5a is obtained with the grounded gate. In Fig. 5b, the gate voltage from left to right ranges from -4 V to +2 V.

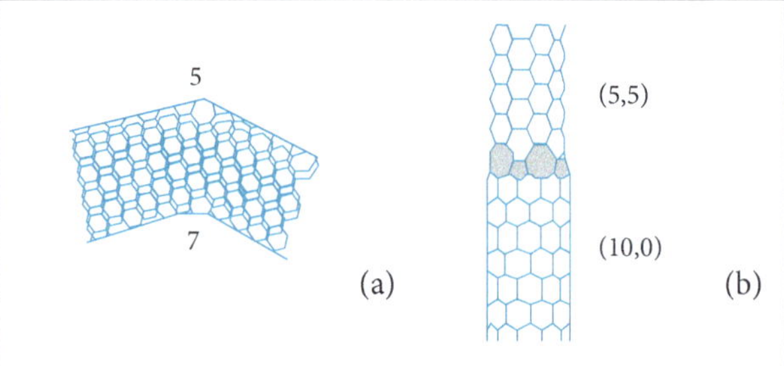

Figure 3. Examples of rectifying p-n junction (D'yachkov, 2006).

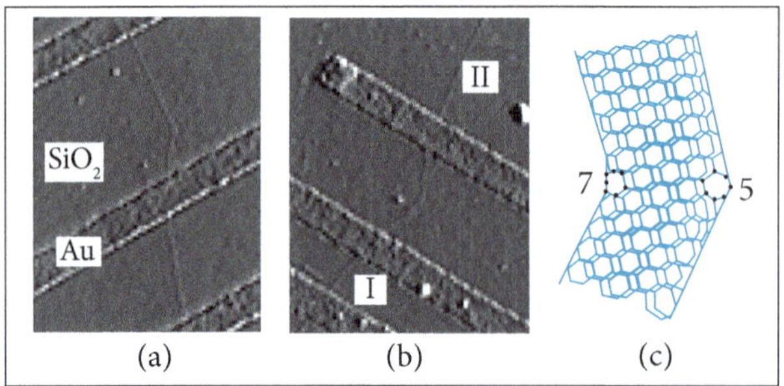

Figure 4. Sample image junction obtained by atomic force microscope (Yao *et al.*, 1999): (a) CNT placed directly on 3 electrodes, (b) CNT folded that presents the effect of rectification, (c) the detail of the fold, in which, on one side of the fold, the CNT is metallic and, on the other one, it is semiconductor.

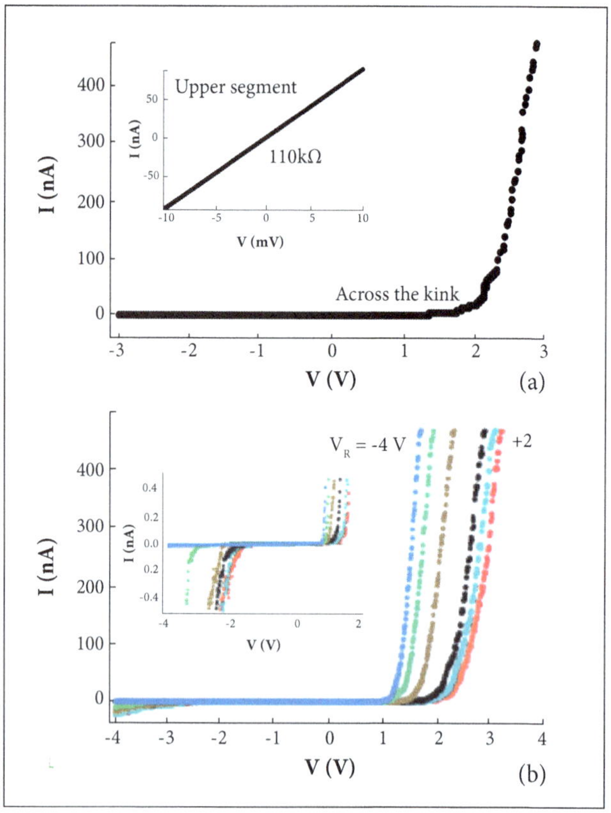

Figure 5. Current-voltage characteristic of metal-semiconductor junction in Fig. 4a.

The Schottky diode is another type of diodes of particular interest for applications in the range of 30 GHz to 3 THz. The diodes that operate in this range are commonly used as detectors, mixers and frequency multipliers in the submillimeter wave region at room temperatures. The CNT diodes are characterized by their high switching speed and inherent suitability for applications at low voltages and high currents (Manohara *et al.*, 2005; Huang *et al.*, 2010).

In devices based on microelectronics in frequencies above a few GHz, there are the inherent parasitic capacitance and resistance. The capacitance is proportional to the area of the junction and the resistance is inversely proportional to the mobility of the electrons. These undesirable capacitances and resistances are mainly due to the limitations of the manufacturing process of the device, as well as to the properties of the materials. These drawbacks severely limit the sensitivity of signal detection and signal generation in THz range. Typical values of signal power are of the order of a few microwatts for frequencies up to 1.5 THz (Manohara *et al.*, 2005).

From the standpoint of new materials, the CNT offer an excellent alternative in the design of high-frequency devices in THz range because of small areas of the junctions, which usually have a maximum of 2 nm in diameter. This leads to low capacitances of the order of tens $aF/\mu m$. Besides, the CNT possess high electron mobility, reaching up to 200.000 $cm^2/V.s$ (Manohara *et al.*, 2005).

Unlike planar microelectronic Schottky diode junctions, the Fermi contact level does not control the properties of devices with CNT. As a result, the best threshold voltage for optimal performance of the device can be chosen (Manohara *et al.*, 2005; Huang *et al.*, 2010).

Figure 6 shows the basic schematic representation of a Schottky diode, whose design employs two different metals at the ends of the p-type CNT. The utilized metals have different work functions (Φ); one of them has a lower work function than the CNT in order to have the Schottky contact; the other one has a higher work function than the CNT in order to obtain the ohmic contact (Manohara *et al.*, 2005).

In Fig. 6a, the Schottky contact is obtained with the use of Ti, and $\Phi_{Ti} = 4.33$ eV $< \Phi_{NTC,}$ with $\Phi_{NTC} \approx 4.9$ eV. The ohmic contact is obtained with the use of platinum (Pt) possessing, and $\Phi_{Pt} = 5.65$ eV $> \Phi_{NTC}$ (Manohara *et al.*, 2005). There is a layer of SiO_2 between the Si substrate and the CNT of approximately 400 nm of thickness.

Figure 7 shows the direct (DC) current-voltage characteristics of four different types of Schottky diode manufactured on the same substrate and in the same batch with the zero gate voltage. All these diodes contain a single CNT with semi-conductive characteristic between the two contacts of different metals formed by Ti and Pt (Manohara *et al.*, 2005). We can see in Fig. 7 that the current-voltage characteristics for the four diodes are quite different. These differences can be attributed to the characteristics of the CNT, such as, for example, its overall length and the overlap that it makes with the metallic contacts (Yang *et al.*, 2005).

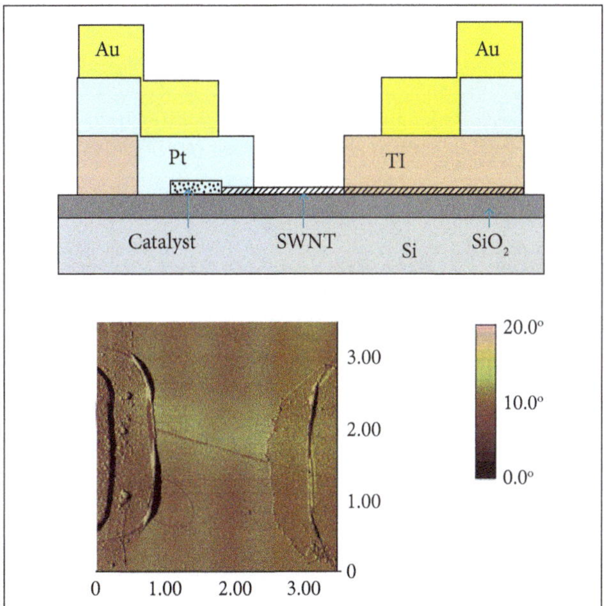

Figure 6. (a) Schottky diode's schematic representation, (b) typical image obtained by atomic force microscope (Manohara *et al.*, 2005).

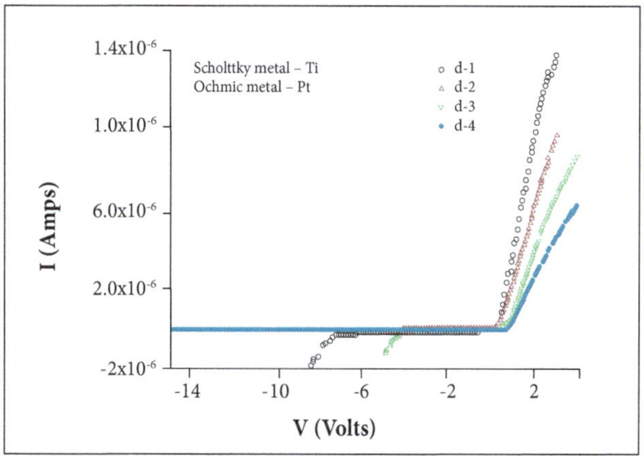

Figure 7. Current-voltage characteristics of four Schottky diodes (Manohara *et al.*, 2005).

There is another type of diode, also for high frequency, which uses multiple CNT in parallel between the metal contacts of the device (Manohara *et al.*, 2005). Each device has from 8 to 10 CNT. These nanotubes may have either metallic or a semi-conductor behavior and this property is difficult to control. The presence of only one carbon nanotube with metal characteristics destroys the desired effect of rectification. Therefore, the CNT with metallic behavior were literally burned by a high current in the technological process that is described in Collins *et al.* (2001) so that all the remained tubes had semi-conductor properties. Figure 8 shows the change in current-voltage characteristic before and after the selective burning of metallic CNT. The inset chart in this figure demonstrates the curve for the rectifier for low bias voltages.

Figure 9 shows the current-voltage curves of symmetric devices, *i.e.* those whose metallic contacts are made of the same material. All electrical measurements were made at room temperature in the air. In Fig. 9a, palladium (Pd) was used as contact metal, resulting in a behavior of a p-type semi-conductor with on-off current ratio of 10^6. In Fig. 9a, it is shown in detail the image that was obtained through a scanning electronic microscope. In Fig. 9b, Ti was used as the metal contact, resulting in an ambivalent behavior, *i.e.* for negative bias voltages, the device features a semi-conductor p-type behavior and, for positive bias voltages, the same device shows a semi-conductor n-type behavior. The on-off ratio is about 10^4. In Fig. 9c, aluminum (Al) was used as the metal contact, resulting in an on-off ratio of 10^3 for negative gate bias voltages and an on-off ratio of 10^6 for positive bias voltages.

As seen in Fig. 9, the simple change of the type of material used as electrical contact modifies the current-voltage characteristic of the device, and it can present both a p-type semi-conductor behavior and an n-type semi-conductor behavior or even ambivalent characteristics.

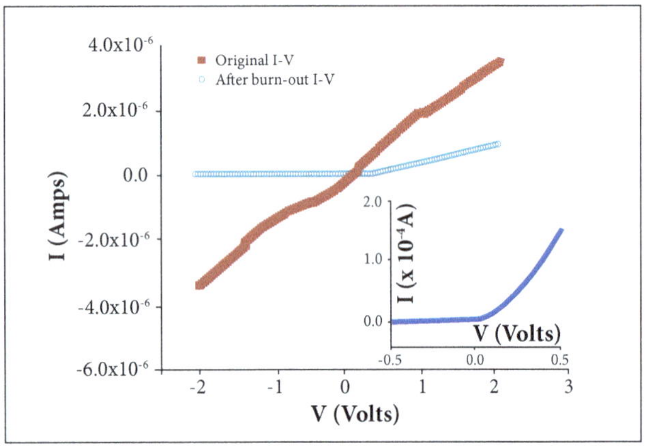

Figure 8. Current-voltage curve of Schottky diode with multiple CNT (Manohara *et al.*, 2005).

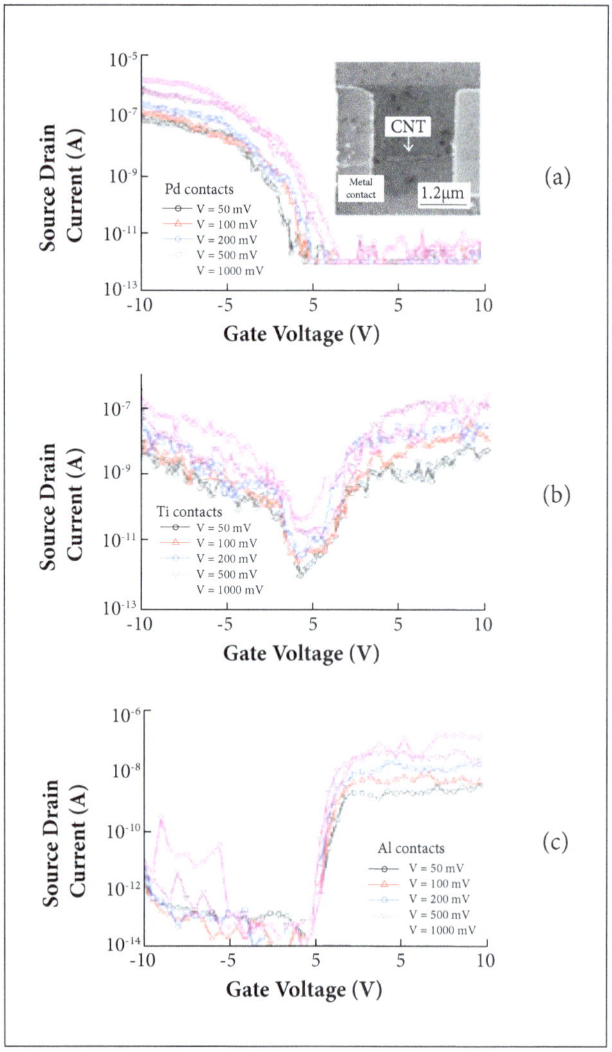

Figure 9. Current-voltage curves of a symmetric Schottky diode with three different metals as contact (Avouris, 2002).

NANOTRANSISTORS BASED ON CNT

In microelectronics, the basic structure of a field-effect transistor (FET) of solid state is composed of two metal electrodes called source and drain, which are connected by a channel in a semi-conductor. There is a third electrode, called the gate, which is isolated from the channel by a thin layer of SiO_2. By varying the channel width, the gate bias voltage controls the current flow between the source and drain terminals.

In FET with CNT (CNFET), the Si semi-conductor channel is replaced by a SWCNT, which is inserted over the metal electrodes with high work function $\Phi > 5$ eV. These electrodes are typically made of Au or Pt (Avouris, 2002).

Back in 1998, there were the first demonstrations that CNT could be used in the design of FET (Adel and Smith, 2007). An example of this transistor is shown in Fig. 10. In this picture, a carbon nanotube is positioned at the top, interconnecting thus the two electrodes of noble metals. These electrodes are the source and drain terminals of the FET, which were fabricated

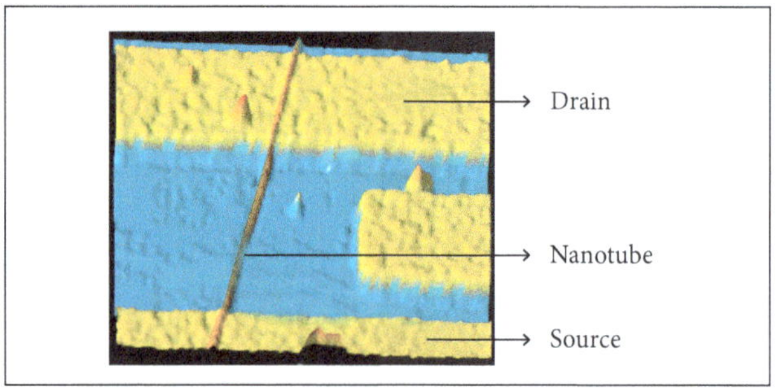

Figure 10. Image of field-effect transistor with CNT obtained by atomic force microscope (Adel and Smith, 2007).

at the top of a SiO_2 film, which, in turn, was deposited on a Si wafer. This Si wafer operates as a third terminal, which has the function of gate. These devices have the properties of a p-type

FET with the ratio of on-off current of the order of 10^5. They are characterized by a high parasitic resistance of contact, which is generally around 1 MΩ, a low current between source and drain and a low transconductance of the order of 10^{-9} A/V.

In order to reduce the contact resistance and improve the features of the CNFET, the scheme shown in Fig. 10 can be modified as proposed in Adel and Smith (2007). In this assembly, there is an oxidized Si wafer placed on the SiO_2, in which the SWCNT is dispersed. The source and drain terminals are fabricated on top of a SWCNT semi-conductor, according to the scheme in Fig. 11. The terminals are made of Ti or cobalt (Co). The curve of Fig. 12 was obtained at room temperature, with the drain, source and gate terminals made of Ti. The thickness of the SiO_2 under the gate is 15 nm. A reduction of the parasitic resistance is obtained by heat treatment between the metal and the SWCNT. As a result, the resistance was reduced from 1 MΩ to 30 KΩ, approximately. Figure 12 shows the output characteristics of this CNFET Adel and Smith (2007). The device in Fig. 11 can also be switched using a Si wafer as a back-gate.

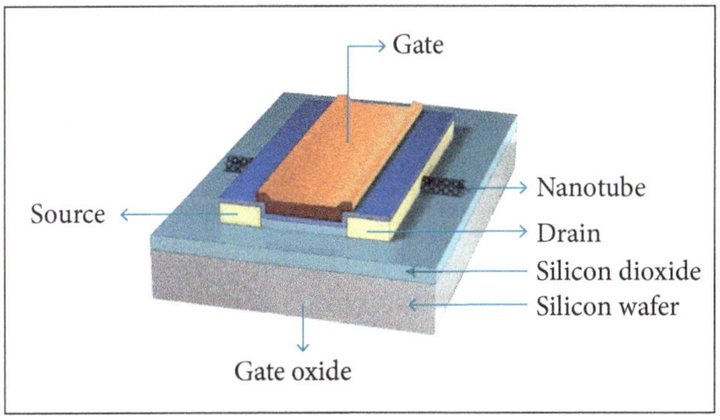

Figure 11. Schematic representation of field-effect transistor with CNT with top-gate (Adel and Smith, 2007).

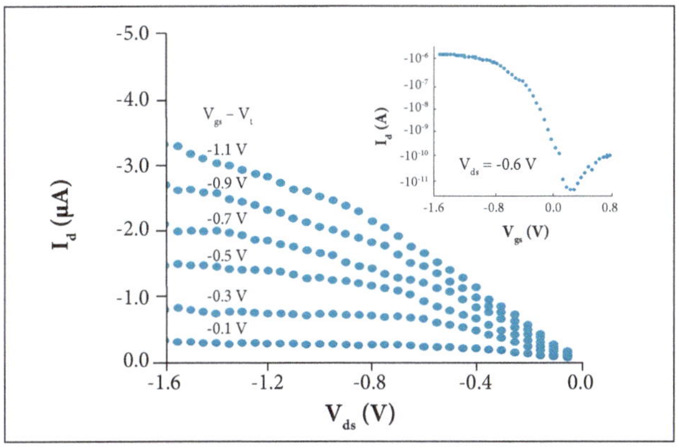

Figure 12. Electrical output characteristics of top-gate field-effect transistor with CNT with source, drain and gate terminals of titanium at room temperature (Adel and Smith, 2007).

Currently, there are several geometric configurations that can be used in implementation of the CNFET. The most commonly used are shown in Fig. 13 (Wong and Akinwande, 2011).

In the settings of Fig. 13, all of them retain some features in common with each other, such as the use of intrinsic CNT semi-conductors with the channel function, the use of metal contacts to the terminals of the source, drain and gate. The gate can be both top-gate and back-gate and, as a common use, all configurations make use of a dielectric or an oxide in order to isolate the gate terminal of the CNT.

The dynamics of CNFET operation is simple to understand. When a bias voltage is applied at the gate terminal of the CNFET, a perpendicular electric field appears in the channel of this device. This electric field controls the amount of charge carriers in the channel. The control can be fulfilled in two different ways: by top-gate, when the gate terminal is placed over the CNT, or by back-gate, when the gate terminal is underneath the Si wafer, which is heavily doped.

The top-gate controls the channel that is right below the gate terminal and the back-gate does the same control, but

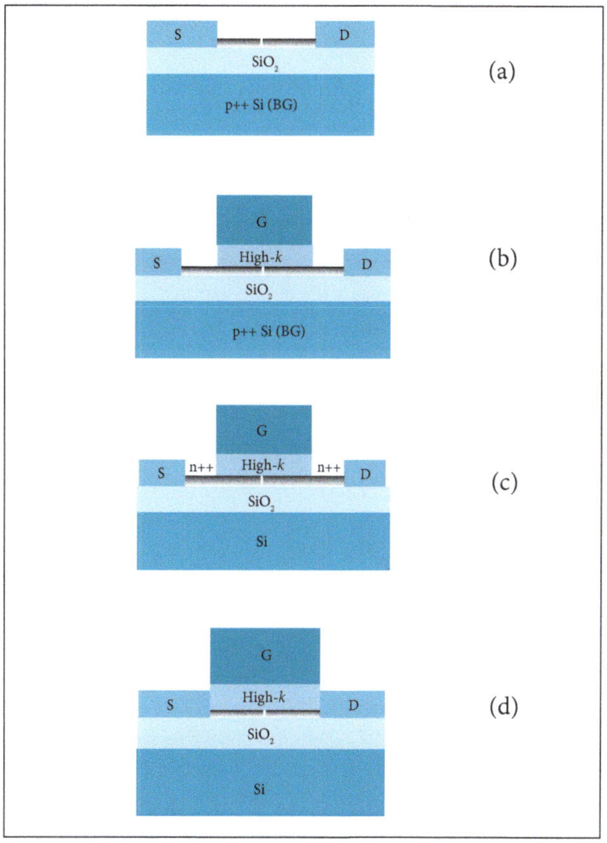

Figure 13. Geometric configurations commonly used in field-effect transistor with CNT (Wong and Akinwande, 2011).

now, through an electrostatic doping in a region of the CNT near of the drain and source terminals in order to decrease the contact resistance between these terminals. This doping is done by electrostatic application of vertical electric field that induces electrons or holes near the region of the drain source terminals. Regardless of the type of control used, whether by a top-gate or a back-gate, there is an electric field due to the applied bias between drain and source. This electric field is responsible for the displacement of charge carriers between the drain and source terminals, whose intensity is controlled by the gate terminal.

Four basic geometry settings of CNFET are shown in Fig. 13. Figure 13a depicts the scheme of the device of Fig. 10, in which the Si substrate is heavily doped and plays the role of back-gate. The difference between Fig. 13b and Fig. 13a is that there is an additional electrode G which plays the role of top-gate. The top-gate is separated from the CNT by a high-k dielectric (the high-k dielectric has the dielectric constant k several times much greater than that of SiO_2). Therefore, in this geometry, one can control the current flow between source and drain terminals, both through the top-gate and by the back-gate. Figure 13c uses the chemical doping in the regions close to the source and drain terminals, and Fig. 13d depicts the geometry of a top-gate with a self-aligned CNFET, which implies a perfect alignment, *i.e.* no space between the gate, source and drain metal contacts. Controlled devices through the top-gate are preferred when a greater control of the channel is desired, as well as the top-gate allows the control of individual gates on a device with multiple transistors. Notice that the charge carriers in CNFET can be both electrons and holes (Wong and Akinwande, 2011).

In Table 1, the four charge transport schemes for CNFET are represented. These schemes are independent on the geometric setting. They are defined by the length, the mean free path of the CNT and the type of contact between the terminals, *i.e.* the ohmic or Schottky barrier. In Table 1, L is the length and L_m is the mean free path of the CNT.

In the projects of CNFET devices, one needs to optimize the current between source and drain terminals. In this sense,

Table 1. The four transport regime of field-effect transistor with CNT (Wong and Akinwande, 2011).

Type of contact between metal-CNT	$L < l_m$	$L > l_m$
Ohmic	Ballistic	Diffusive
Schottky barrier	Ballistic	Diffusive

projects with transparent contacts (*i.e.* ohmic contacts) and ballistic transport offer the best performance.

DIFFERENCES BETWEEN 2-D METAL-OXIDE-SEMICONDUCTOR FIELD-EFFECT TRANSISTOR AND 1-D CNFET

In Table 2, a comparison between the 2-D metal-oxide-semiconductor field-effect transistor (MOSFET) and 1-D CNFET is given. Due to the 1-D nature of CNT, in the linear region, the electrons can move only in one direction, *i.e.* either forward or backward and the charge transport can be ballistic. As for the MOSFET, the electrons can move in the plane and the electrons transport is diffusive. In the active regions of the MOSFET, the current of saturation is due to the pinch-off that occurs in the channel and, as a consequence, the drift rate no longer depends on the drain voltage (Martel *et al.*, 2001). As for CNFET, the current of saturation occurs due to disappearance of the charge carriers from the drain terminal. This implies that the current does not depend on the drain voltage, unlike of what occurs with the MOSFET (Wong and Akinwande, 2011).

Table 2 shows explicitly that the device current in the on state (I_{ON}) for the MOSFET is controlled mainly by the drift velocity (u_{eff}), which depends on the properties of the material. Whereas for CNFET this current mainly depends on the physical properties of states of the electrons in the solid, such as the band-gap structure (E_g), the Boltzmann constant (k_s) depends on the temperature (T) and the quantum resistance (R_q) (Wong and Akinwande, 2011). The parameter u_{eff} is the drift velocity, which is an indicative of the average scattering that the loads suffer in the solid; L and W are the length and width of the channel, respectively; C_{ox} is the capacitance of the insulating placed between the top-gate and the channel; φ_s is

Table 2. Difference between 2-D MOSFET and 1-D CNFET (Wong and Akinwande, 2011).

2-D MOSFET $I_{ON} \approx u_{eff} \frac{C_{ox} W}{2L} (V_G - V_T)^2$		1-D CNFET $I_{ON} \approx \frac{k_B T}{eRq} ln\left(1 + e^{\frac{2e\varphi_s - E_g}{2k_B T}}\right)$	
Parameter	**Origins**	**Parameter**	**Origins**
(i) u_{eff}	Drift velocity	(i) φ_s	Gate coupling
(ii) I/L	Drift velocity	(ii) E_g	Band structure
(iii) C_{ox}/W	Charge sheet	(iii) R_q	Quantum transport
(iv) V^2	Drift velocity and charge sheet	(iv) $k_g T$	Fermi-Dirac distribution

the electrostatic surface potential; V_G is the gate voltage and V_T is the threshold voltage. The other parameters of Table 2 have already been described in the text.

A more detail information about the topics of this paper can be found in references below.

CNT DEVICE SIMULATIONS

In the nanoscale simulations, one needs to consider the effects of a single atom on the characteristics of the device. In the atomic scale, the effects of the individual vacancies and impurities start to have a real influence in the performance of the device.

There are two main steps in the nanoelectronic studies. The first one is a solution of the electronic structure problem that deals with the behavior of electrons' band energy inside of the material as well the DOS in such energy levels. In practice, this study gives us the information about the band gap and the energy levels where the device will work properly. Two categories of methods are used in practice: *ab initio* approaches to solve the Schrödinger equation based on first principles. An example of such methods is density-functional theory (DFT). The second category is semi-empirical methods where the electronic structure is calculated by using adjustable parameters obtained, for example, from experiments.

The second step is the study of electronic transport, *i.e.* the behavior of the electronic current inside the material. The problem here is usually to obtain the I-V (current-voltage characteristic) curves. In this step, it is necessary to consider the electrons as wave functions instead of particles. That is, it is important to take into account the quantum effects that appear in the atomic scales. At these scales, the channel of the device (the active region of interest) is frequently shorter than the mean-free path of the material. Thus, one has to take into account the potential profile of the central region; has to consider quantum mechanical equations and approximation, such as Schrödinger equation; needs to apply, in the right way, the boundary conditions; and has to consider the non-equilibrium situation because the transmission property of the electron is itself bias dependent — that is the problem is non-linear and, for each value of applied voltage, one has a different transmission curve. A combination of the Landauer approach with the non-equilibrium Green function (NEGF) method is now widely used in the analysis and design of nanoscale devices. Quantum transport in nanotubes and other nanoelectronic devices is described in more details in (Wong and Akinwande, 2011; Javey and Kong, 2010; Stokbro *et al.*, 2010; Blom and Stokbro, 2011; Datta, 2005; Nazarov and Blanter, 2009).

CONCLUSION

Some special electrical, optical and magnetic properties, the mechanical strength and chemical stability make the CNT one of the promising nanomaterials to be used in nanoelectronics.

Based on the available literature, this study shows that the devices with CNT are very promising for replacing microelectronic devices that employ Si, germanium (Ge) or GaAs as semi-conductor elements. The use of CNT as a metal to interconnect elements in order to replace the copper tracks is advantageous when the mean free path is smaller than the link track.

The use of CNT in the projects of diodes and FETs allows one to increase the frequency range of these devices, sometimes even to THz. Possible applications of such devices are detectors, mixers, frequency multipliers, among others.

However, there are still some challenges to be overcome in order to raise the manufacturing process of these CNT devices to a commercial scale, such as control of the physical and electronic properties in the CNT growth process. The production of CNFET is still complicated, expensive and requires a lot of time.

REFERENCES

Adel, S. and Smith, K.C., 2007, "Microeletrônica", Fifth Ed., Prentice Hall, São Paulo, Brazil.

Avouris, P., 2002, "Molecular Electronics with Carbon Nanotubes", Accounts of Chemical Research, Vol. 35, No. 12, pp. 1026-1034. doi: 10.1021/ar010152e

Blom, A. and Stokbro, K., 2011, "Towards Realistic Atomic-Scale Modeling of Nanosacale Devices", IEEE Conference on Nanotechnology, pp. 1487-1492. doi: 10.1109/NANO.2011.6144584

Collins, P.G., Arnold, M.S. and Avouris, P., 2001, "Engineering Carbon Nanotubes and Nanotube Circuits Using Electrical Breakdown", Science, Vol. 292, No. 5517, pp. 706-709. doi: 10.1126/science.1058782

D'yachkov, P.N., 2006, "Carbon Nanotubes: Structure, Properties and Applications", BINOM, Moscow, Russia.

Datta, S., 2005, "Quantum transport. Atom to transistor", Cambridge University Press, Cambridge, UK. 404p.

Dragoman, M. and Dragoman, D., 2006, "Nanoelectronics: Principles and Devices", Artech House, Boston, USA/London, UK. 426p.

Huang, L., Chor, E.F., Wu, Y. and Guo, Z., 2010, "Fabrication of Single-Walled Carbon Nanotube Schottky Diode with Gold Contacts Modified by Asymmetric Thiolate Molecules", Carbon, Vol. 48, No. 4, pp. 1298-1304. doi:10.1016/j.carbon.2009.12.003

Iijima, S., 1991, "Helical Microtubules of Graphitic Carbon", Nature, Vol. 354, pp. 56-58,. doi: 10.1038/354056a0

Javey, A. and Kong, J., 2010, "Carbon Nanotube Electronics", Springer, New York, USA.

John, D.L., Castro, L.C. and Pulfrey, D.L., 2004, "Quantum Capacitance in Nanoscale Device Modeling", Journal of Applied Physics, Vol. 96, No. 9, pp. 5180-5184.

Manohara, H.M., Wong, E.W., Schlecht, E., Hunt, B.D. and Siegel, P.H., 2005, "Carbon Nanotube Schottky Diodes Using Ti-Schottky and Pt-Ohmic Contacts for High Frequency Applications", Nano Letters, Vol. 5, No. 7, pp. 1469-1474.

Martel, R., Derycke, V., Lavoie, C., Appenzeller, J.and Chan, K.K., 2001, "Ambipolar Electrical Transport in Semiconducting Single-Wall Carbon Nanotubes", Physical Review Letters, Vol. 87, No. 25, pp. 256805. doi: 10.1103/PhysRevLett.87.256805

Nazarov, Y. and Blanter, Y., 2009, "Quantum Transport. Introduction to Nanoscience", Cambridge University Press, Cambridge, UK.

Sadiku, M.N.O., 2004, "Elementos de Eletromagnetismo", Third Ed., Bookman, Porto Alegre, Brazil.

Stokbro, K., Peterson, D., Smidstrup, S., Blom, A., Ipsen, M. and Kaasbjerg, K., 2010, "Semiempirical Model for Nanoscale Device Simulations", Physical Review B, Vol. 82, No. 7, 075420. doi: 10.1103/PhysRevB.82.075420

Wong, H.S.P. and Akinwande, D., 2011, "Carbon Nanotube and Graphene Device Physics", Cambrigde University Press, Cambrigde, UK. 262p.

Yang, M.H., Teo, K.B.K., Milne, W.I. and Hasko, D.G., 2005, "Carbon Nanotube Schottky Diode and Directionally Dependent Field-Effect Transistor Using Asymmetrical Contacts", Applied Physics Letters, Vol. 87, 253116.

Yao, Z., Postma, H.W.C., Balents, L., and Dekker, C., 1999, "Carbon Nanotube Intramolecular Junctions", Nature, Vol. 402, pp. 273-276. doi:10.1038/46241

10

Gas Turbine Course's Teaching Process at Instituto Tecnológico de Aeronáutica: Theory and Laboratory

Cleverson Bringhenti[1], Jesuíno Takachi Tomita[1], João Roberto Barbosa[1]

ABSTRACT: The Instituto Tecnológico de Aeronáutica (ITA) is an Engineering school maintained by the Air Force Command, Ministry of Defense. The aim of the Turbomachines Department at ITA is the human resources training for design and development of aeronautical and industrial gas turbines, necessary for the Gas Turbine Program of the Departamento de Ciência e Tecnologia Aeroespacial (DCTA). The human resources training is carried out in undergraduate and graduate courses at ITA, where topics in gas turbine and turbomachinery are taught. The gas turbine topic is taught in the undergraduate degree, in Mechanical-Aeronautical Engineering course, and focuses on gas turbines' performance for different configurations (turboshaft, turbojet and turbofan). Lecture notes containing the essential elements of the course are made available for the students, addressing the basic theory of the gas turbines required for the performance study at the design and off-design point. The technological aspects are presented and discussed during detailed studies of the actual cycle. Simple gas turbines and more sophisticated ones are studied, for both aeronautical and industrial application. Performance calculations at design and off-design point of the main engine's components and the cycle are done manually, encouraging students to develop spreadsheets. The theory is complemented with laboratory classes and technical visits, when the practicalities involving gas turbines operation and tests are presented to the students. As an activity laboratory class, the students perform disassembly-assembly of a small industrial gas turbine.

KEYWORDS: Teaching, Gas turbine, Performance, Preliminary design, Numerical simulation.

INTRODUCTION

Instituto Tecnológico de Aeronáutica (ITA) is a technological school focused on Engineering courses: aeronautical, mechanical-aeronautical, aerospace, electronics, computation and civil. Its mission and main aspects of original educational model is to provide research and Engineering education in support to the interests of the Brazilian defense and aerospace industry and related institutions. Academic excellence, self-discipline, citizenship development, full time students and professors, students living in campus and student-centered policy are some specific aspects of the ITA.

The Engineering courses last five years (two fundamental years and three professional years). The undergraduate gas turbines lectures are offered to the students of mechanical-aeronautical Engineering course in the second semester of the second professional year.

The students enter the gas turbine course with good knowledge of thermodynamics, fluid mechanics, heat transfer and flow machines. They had already been introduced to the operational aspects of the rotating machines from the flow machines course (compressors, turbines, pumps, velocity diagrams, losses, surge, choke etc.).

The gas turbines course (identified as MMT02) is taught during the 2nd semester of the 2nd professional year and comprises 48 lecture-hours of theory/exercises and 16 lecture-hours of laboratory. The theory classes are given in two weekdays (two hours of theory and one hour of exercise); the labs are given every other week, for two groups of students, therefore decreasing

1. Instituto Tecnológico de Aeronáutica – São José dos Campos/SP – Brazil.

Author for correspondence: Cleverson Bringhenti | Instituto Tecnológico de Aeronáutica | Departamento de Turbomáquinas | Pr. Marechal Eduardo Gomes, 50 Vila das Acácias | CEP: 12.228-900 – São José dos Campos/SP – Brazil | Email: cleverson@ita.br the number of students in the lab. Lab classes are for hands-on activities and visits to industries whose activities are linked to the gas turbines — Propulsion Division of Instituto de Aeronáutica e Espaço (IAE) at Departamento de Ciência e Tecnologia Aeroespacial (DCTA) and GE-Celma in Rio de Janeiro.

There is indication that the DCTA started activities on gas turbine area at the beginning of its existence. Plans were drawn to build a gas turbine laboratory comprising compressor, combustor, turbine and turboshaft test rigs at the PMO building (engines project), which belongs to ITA. The gas turbine activities at DCTA were discontinued and no records are now available to evaluate the extent of those activities. In 1975, DCTA restarted the gas turbine activities under a gas turbine project. Consultancy from Cranfield Institute of Technology (now Cranfield University) to establish a gas turbine group indicated that ITA should support the base of human resources training, while Cranfield would contribute essentially with specialized courses. In that period, ITA was involved with activities on gas turbines to support PMO's needs, with extra-curricular courses.

In the 1990's, ITA resumed its activities on gas turbines area, creating a new course called Gas Turbines, offered to the Mechanical Engineering undergraduate students, with focus on gas turbine performance calculations. It was decided to concentrate ITA's gas turbine activity in the graduate course due to the diversity and extent of the disciplines involved. Nevertheless, undergraduates are encouraged to take part in projects developed at the Turbomachines Department. Course projects and introduction to scientific research projects have been attracting undergraduate students, as indicated in this article.

The activities in the Turbomachines Department have been directed towards the design and development of a turbojet engine, named TAPP (aero gas turbine of small size), nowadays under development tests at APA (formerly PMO), the aero propulsion department of IAE. TAPP is financed by a consortium comprising Financiadora de Estudos e Projetos (FINEP), DCTA (ITA and IAE) and TGM Turbinas. TAPP was designed based on engine requirements of a local industry, but its gas generator was chosen for both turbojet and turboshaft versions of a 5 kN thrust and a 1.2 MW turboshaft. It is believed that such expertise can be used to improve the quality of the teaching, enrolling instructors who have practical skills acquired during the design and development of TAPP and other projects carried out at the Turbomachines Department (Bringhenti and Barbosa, 2008; Bringhenti *et al.*, 2006).

A study of the engine thermodynamics cycle is already being done (Bringhenti and Barbosa, 2004a; Silva, 2009). The engine will continue the development aiming at performance improvement by increasing pressure ratio and maximum cycle temperature.

At the time of submission of this paper, the Turbomachines Department had four staff with doctoral degrees, as well as seven PhD and ten MSc students. The Turbomachines Department is located in the building E-0028 of ITA and accommodates all the staff and full time students. Lecture rooms and offices are available for all.

Depending on the on-going projects, researchers from other institutions may work collaboratively. A few post-doctoral assignments may join the Department's workforce. It is the Department's policy to develop all computational tools needed for the design of the gas turbine's major components. Several computational codes have been specially developed for such purpose, covering design and performance calculations of both engine and components, flow calculation using Computational Fluid Dynamics (CFD) and lateral and torsional vibration (Barbosa, 1987; Bringhenti, 2003; Alves, 2003; Tomita, 2003, 2009; Tomita *et al.*, 2012). Such in-house developed codes are also used for teaching support, as presented by Tomita and Barbosa (2012) and by Barbosa and Bringhenti (2000). Current software development effort concentrates on the CFD code for turbomachines, with incorporation of chemical reactions, compressor design optimization, multi-fuel combustor design, gas turbine performance deterioration and noise prediction. Major international congresses are attended by the Department's staff and students as a means to exchange ideas and provide update. Several papers have been written as part of the project developed in the Department.

GAS TURBINE COURSE

THEORY

The course's contents and supporting literature are listed in Table 1.

In addition to the three references, the handout lists other books as well as special reports and the students are encouraged to be familiar with them. All the references are made available to the students at the ITA's library. The classes are given in Portuguese, but English may be used by invited specialists.

The course starts with an overview of gas turbine applications (industrial, aeronautical, marine, army) and configurations (single and multi-spool, low and high turbofan bypass ratios, military and civil engines) and it is stressed that physics is associated to the production of power and thrust. Cycles for production of shaft power are studied in details, followed by the cycles for aeronautical propulsion.

Table 1. Course's contents and supporting literature.

MMT-02 Gas Turbines	
Course's contents	Description, classification and applications: turboshafts, turboprops, turbojets, turbofans and ramjets. Main components and their performance characteristics: compressors, combustors, turbines, ducts, intakes and exhausts, propelling nozzles and heat exchangers. Ideal and real cycles. Enthalpy-entropy diagrams. Shaft power cycles. Cycles for aeronautical propulsion. Design and off-design performance. Performance curves. Engine decks.
Required	Flow Machines
Weekly hours	3 hours for theory and exercises, 1 hour for lab and 4 hours for home study.
References	Barbosa, J.R., 2001, "Desempenho de Turbinas a Gás", ITA coursepack; Saravanamuttoo, H.I.H, Cohen, H., Rogers, G.F.C. and Straznick, P., 2009, "Gas Turbine Theory", Sixth Edition, Prentice Hall, USA; Walsh, P.P. and Fletcher, P., 1998, "Gas Turbine Performance", Pennwell Books, Tulsa, USA, 656p.

Design and off-design performance calculations are explained in details, followed by calculations of more complex engines using in-house developed computer programs. The Gas Turbine Performance Program (named GTAnalysis), developed to assist the student in complex cycle calculation, is based on modular engine blocks build-up (Bringhenti, 2003; Barbosa and Bringhenti, 2000; Bringhenti and Barbosa, 2004a, 2004b, 2004c, 2008; Bringhenti *et al.*, 2006; Silva, 2009). The interacting blocks describe the actual component performance to produce high-fidelity gas turbine model and actual engines are quite well modeled. Combining the blocks adequately, it is possible to numerically simulate almost all types of gas turbines (turboshafts, turbojets and turbofans). The program is equipped with libraries of compressors, combustion chambers and turbine maps, which are selected to better represent the studied engine. The program's capabilities have been disclosed by means of publications (Barbosa and Bringhenti, 1999; Bringhenti, 1999; Bringhenti and Barbosa, 2002, 2004a, 2004b, 2004c, 2004d; Barbosa *et al.*, 2011; Tomita *et al.*, 2005; Bringhenti *et al.*, 2010; Silva, 2007; Silva *et al.*, 2005). In addition to the steady state design point performance calculation, the program is able to simulate gas turbines' performance at off-design points, transient operation, using nozzle, compressor and turbine variable geometry, as well as to predict the engine's noise and performance deterioration. If it is necessary to evaluate a specific engine to be designed, the major component maps (compressor, combustor and turbine) can be generated using in-house developed software (Barbosa, 1987; Bringhenti, 2003; Tomita, 2003, 2009). The students have access to the program to evaluate: gas turbine's performance configurations; the importance of variation in main parameters; the best cycle parameters and configurations for a given application; engine degradation. The students are also encouraged to develop their own programs during the course.

Engine decks, like the ones used by the aeronautical industry, are discussed and the in- house developed high-fidelity gas turbine performance calculation is presented. Gas turbine modeling and performance calculations are carried out by the students. Phenomena like choke, stall and surge are presented. The program developed can also be used as a supporting tool in undergraduate and graduate courses. It will allow the discussion of more complex exercise and examples during class. The "student version" is an executable file through which the student has access only to the information he/she needs for his/her particular study. Due to the complexities of the calculations involved, only simplified problems are approached in classroom. More complex calculations are just introduced to the students in a quantitative way. Due to this simplification, important passages in the resolution of exercises are omitted.

The commercially available computer programs were developed for specific tasks and settings and, therefore, cannot be used for general purpose, as well as for any configuration and application. As the programs are developed in this manner, in most cases, not flexibly, they cannot be used for a particular case of study, if it was not previously foreseen. These programs, for example, do not allow users — even advanced ones — to have access to the source code, allowing them to modify the code for a specific case study, which prevents certain types of studies/application.

Several universities, industries and researchers are devoting time to develop gas turbine performance simulation programs around the world, among them, McKinney (1967), that developed SIMulation Of Turbofan Engine (SMOTE); NASA, which developed the GENeralized ENGine (GENENG); Koenig and Fishback (1972), Fishback and Koenig (1972), that developed an advanced version of the GENENG, named GENENG II. Besides, Szuk (1974) developed a computer program named HYDES, a hybrid, analogical and digital program. Sellers and Daniele (1975) developed the DYNGEN, a program for calculating steady-state and transient performance of turbojet and turbofan engines. Wittenberg (1976), OTAN, developed a computer program for off-design performance calculation of turbojet and turbofan. At Cranfield University, Palmer (1967,

1983, 1990) and Palmer and Yang (1982) started to develop gas turbine performance simulations programs, TURBOMATCH and TURBOTRANS. Flack (1990), Stamatis *et al.* (1990), Ismail and Bhinder (1991) also contributed to the developments in this area. TESS (University of Toledo, Reed and Afjeh, 1994); ON-X/OFF-X (Mattingly, 1996). The industry contributed to the development of SOAPP (P&W), CWS/ICS (GE), GECAT/NEPP (SRS Technologies, 2007), TERMAP (2000) (Allison/USAF), RRAP (Rolls Royce), JANUS (Snecma), FAST (Honeywell Allied Signal), ATEST (AEDC, Chappell and McLaughlin, 1993), GASTURB (2015) and GSP (2015).

Invited lecturers may contribute to the course, giving their lectures in other languages, mainly in English. The classes follow the traditional teaching methods, with support of computational resources. The students must attend the lectures and, if they fail to do so, they must justify to their counselor and submit written justification to ITA. Depending on the justification, it is made available to the student a special psychological support.

The course's targets are set in the first lecture, specifying the minimum knowledge expected from the student after each chapter is concluded. For example, at the end of the fundamentals of gas turbine performance, the students must master:

- Description of a gas turbine working at off-design condition.
- Calculation of steady-state operating lines for the different types of gas turbines.
- Utilization of computer programs for performance calculations.
- Correct utilization of engine decks.

Assessment is made based on these targets through two bi-monthly tests, which include theoretical and practical exercises, as well as a final exam. The minimum final score is 65%. If the student gets a score below 50%, his/her enrollment with ITA is finished.

Use is made of a PT6 Pratt & Whitney turboshaft engine to show the students the major engine components and internal flow path. Figure 1 shows details of the engine.

Figure 1. PT6 sectional view.

ENGINEERING EDUCATION METHODS SOUGHT AT THE TURBOMACHINE DEPARTMENT

New Engineering education methodologies have been changing and implemented based on several experiences using different teaching manners to deliver good lectures, facilitating the transfer of information (Mazur, 2012). Generally, the proposals for the culture change in Engineering education modify the course's syllabus and, sometimes, a huge course reconstruction is necessary (Karltun, 2013; Kamp and Klaassen, 2013).

Some of the habits to improve the Engineering education have to be updated and this is not a trivial process because it depends on a common action and culture change. Goldberg (2013) related some important aspects and described some key-points about three habits that must be rethought:

- Noticing, Listening and Questioning: this is applied for both instructors and students.
- Dot Connecting: the need for changing the Engineering education at so many levels as departments, systems, society and schools.
- Collaborative Disruption: the competition between different institutes should be analyzed. Generally, the competition for innovation brings advantages for one institute or another. The open innovation movement sounds a good way to be applied among different schools.

Actually, one of the most common methodologies is presented in Darmofal (2012), in which the Concept, Design, Implement and Operate (CDIO) was applied in Aero Astro at MIT. The change in course's syllabus and research cycle are presented and explained based on lessons learned. There are several CDIO implementations and results in the literature; it is interesting to understand them and to find out what are the experiences and format to adopt for each particular case or course (Karltun, 2013; Al-Atabi, 2013; Koster *et al.*, 2013; Malmqvist *et al.*, 2010; Friesel, 2013; Hoffman, 2013; Lassudrie and Kontio, 2013; The Royal Academy of Engineering and Massachusetts Institute of Technology, 2012). The need for accomplishing a good matching between practice and science can be achieved using CDIO education process as presented in Fig. 2.

The new engineer's generation should be flexible and adapted to the modern world. Personal values, different skills, social and mutual interaction are important items to reach success in a globalization era (Friesel, 2013; Mäkimurto-Koivumaa *et al.*, 2013). Figure 3 shows the coupling requirements to be an entrepreneurial professional (Mäkimurto- Koivumaa *et al.*, 2013).

To achieve these goals, it is highly necessary to have a solid communication between instructor and student. A collaborative class makes big and profound differences to educational process. An interesting structure was studied and presented in Connor (2013), including many students' testimonials, how to manage tasks, team work, preconceptions and so on.

The methodology called ICED (Connor, 2013) brings an idea to infuse creativity and innovative education process based on a team-oriented and problem-based learning program. This involves since science, technology, Engineering and Mathematics (STEM) until science and technology (S&T) focuses. The purpose is to integrate basic Engineering concepts with problem-solving techniques involving real world problems. A schematic diagram of ICED is shown in Fig. 4.

The general ideas of each Engineering education methods can be found in Chubik and Zamyatina (2013). The authors mention the implementation of Elite Engineering Education Program

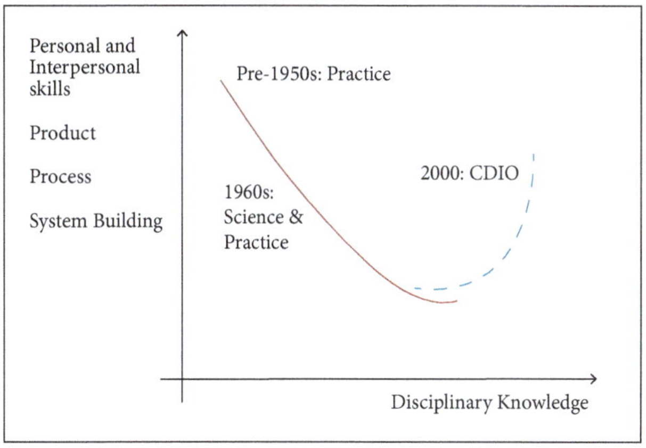

Figure 2. Improvements desired in education applying CDIO.

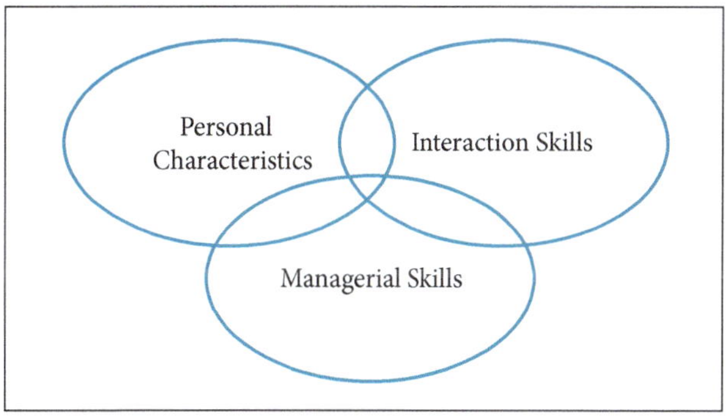

Figure 3. Entrepreneurial behavior and mindset.

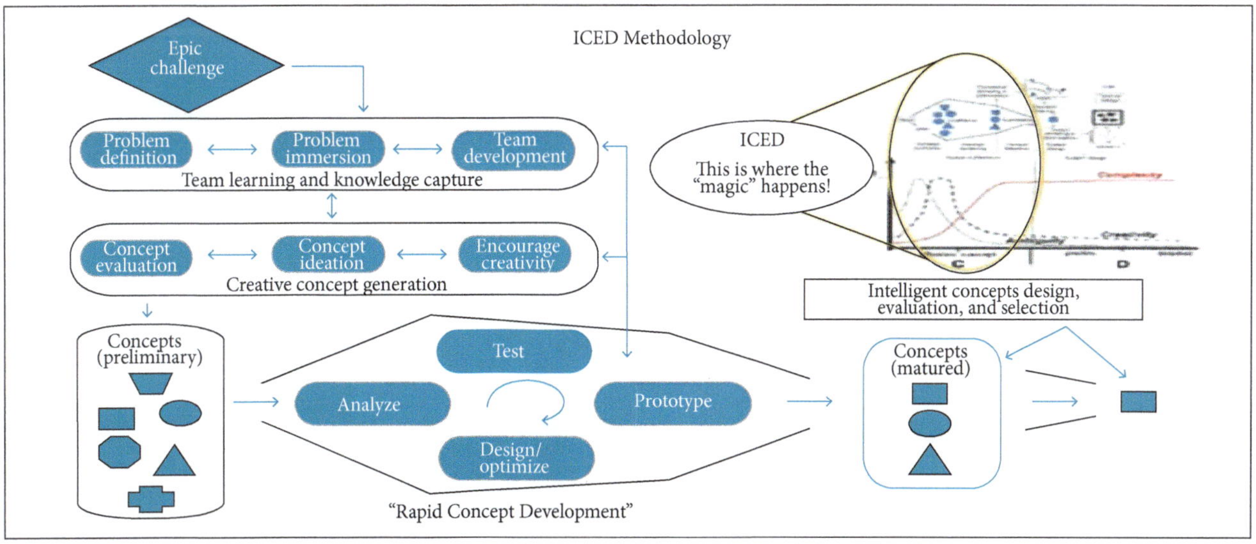

Figure 4. The concept of ICED methodology (Camarda et al., 2013 - courtesy of the authors).

(EEEP) and its capabilities to create competencies with profound fundamental knowledge in Physics, Mathematics and Economics, leadership and team-building qualities, including participation in project-oriented activities. The need for innovation and competition strategies of the industry and society are described and discussed.

What is in common between these methodologies and techniques? All of them mention the importance of Engineering education for interdisciplinary future (Yang, 2009). The characteristics of new professionals should be highly inter and multidisciplinary. The student's skills must be vastly explored. These abilities start in the classroom with adequate and effective education process.

LABORATORY

A gas turbine laboratory requires special and expensive infrastructure. ITA's Gas Turbine Laboratory uses ITA's and IAE's facilities. ITA has two Rover turboshaft engines, model IS60 (45 kW). This engine runs using kerosene and its power is measured through a dynamometer. The former is used only for hands-on training activity (disassembling-assembling). IAE has a 320 N turbojet engine of its own design and development, which runs on liquid (kerosene, diesel, ethyl alcohol and biofuels) and natural gas fuels, and a gas turbine test stand kit, whose details are described below.

The Gas Turbine Laboratory course is based on the following activities:

- Engines at lab: the students are presented to the Rolls-Royce Viper engine, located at the lab. The engine had parts of its body removed, so that the major components can be seen, as shown in Fig. 5. The students, who have already learnt the component associated theory, are induced to answer simple questions about the engine operational characteristics.

The Wright Cyclone engine is used to teach at the lab. Although this engine is very old and is a piston engine, it is used to show the engine's evolution, quoting the major modifications along the time. This engine is coupled to an electrical motor that makes it to work.

In Fig. 6, it is shown one of the first gas turbine engines produced by Rolls-Royce the Derwent. This engine is used as a tool to help to explain the gas turbine's evolution, major modifications and applications of each engine type.

- Visit to GE, Rio de Janeiro, Brazil: it is an engine overhaul company. In this visit, the students have contact with medium and large aero gas turbine engines. All areas are visited by students and detailed information of the activities are explained by GE's staff, associating lecture information with hardware: high-performance fan blades, compressor blade airfoils, Low-Pressure Compressor (LPC), High-Pressure Compressor (HPC), Low-Pressure Turbine (LPT), High-Pressure Turbine (HPT), rotating groups, cooled blades, wall treatments, bypass duct, cold and hot parts, filters, pumps, lubrication systems, secondary flow path, Full Authority Digital Engine Control (FADEC), bearings and its installation positions, blade overhauling, blade material analysis, corrosion, balancing and engine's assembly area.
- Visit to Propulsion Division of the IAE: The DCTA gas turbine project is housed in this institute. Some publications made by the Gas Turbine Group at

Figure 5. Viper engine with its main parts exposed.

Figure 6. Rolls-Royce Derwent.

ITA/IAE (DCTA) (Bringhenti and Barbosa, 2003, 2004c, 2004d; Barbosa *et al.*, 2011, 2012; Tomita *et al.*, 2005; Bringhenti *et al.*, 2010; Silva, 2007; Silva *et al.*, 2005; Creci Filho *et al.*, 2009, 2011) indicate the capability of gas turbines' design and development. In this visit, the students are presented to test facilities, including one of the piston engines. The 15 kN turbojet test facility, which is used to test the 5 kN turbojet under development, is presented in details to the students.

- Small gas turbine engine's disassembly and assembly: usually performed with a small gas turbine, namely the 45 kW Rover IS/60 engine. Firstly, the students are briefed about the engine configuration (compressor type, combustion chamber type, dynamometer, operational envelope, gearbox, among others) with focus on the engine's components. Then, the 60 kW Rover IS/90 engine, which is installed in a dynamometer, is presented to the students, together with significant engine's drawings to support the disassembly. Figure 7 shows some of such drawings that help the students to find the adequate order in which each component should be removed.

This is particularly interesting for understanding the way the engine's components are linked to each other, besides details of couplings and bearings and their locations in the engine. Although this is a simple engine, it is possible to deduce what are the parts and their role in more sophisticated gas turbines.

Figure 8 shows the Rover gas turbine disassembled at lab. The engine is used by the students to assemble and disassemble, which approaches them to the real engine's components and

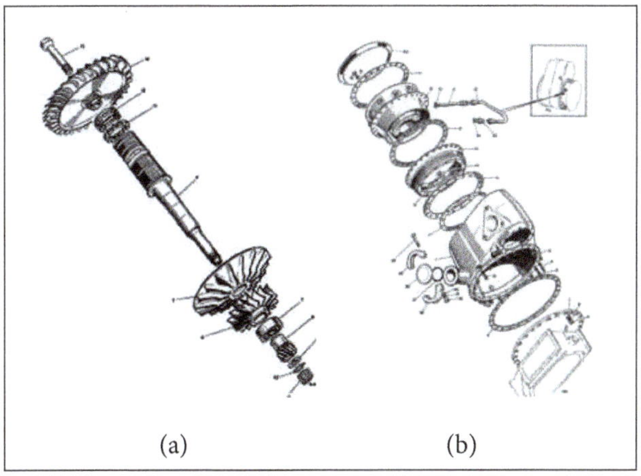

Figura 7. (a) Rotating group (centrifugal compressor and axial turbine), (b) Engine's external parts (casing, seals, ducts, connections etc.).

Figure 8. Rover gas turbine's components disassembled.

stimulates them to make questions. The procedure is accompanied by the professor, while the students make questions about engine and components.

- Engine's run and data acquisition: the students follow an engine's run, from the test preparation to the data acquisition. They are briefed about the test rig and its hydraulic dyno. The engine is the Rover IS/60 turboshaft engine, running with kerosene (Jet-A). A technical report must be submitted by the students, containing the engine's performance calculation obtained from data reduction acquired during the lab class.

Figure 9 shows view of the gas turbine's installation. Recorded data are inlet temperature, fuel pump pressure, fuel consumption, engine's rotational speed, applied load, temperatures and pressures at compressor and turbine's inlet and outlet. Required results are the engine's performance showing fuel consumption, efficiency, pressure ratio and shaft power.

Figure 10 shows the new gas turbine demonstrator purchased for educational purpose and housed at IAE. The Mini Gas Turbine Power System (http://www.turbinetechnologies.com) is a self-contained jet engine laboratory, the SR-30 Turbojet engine. All aspects of the gas turbine's theory, flowpath and thermodynamics are easily demonstrated and readily explored.

Figure 9. Lateral view of the Rover IS/60 gas turbine.

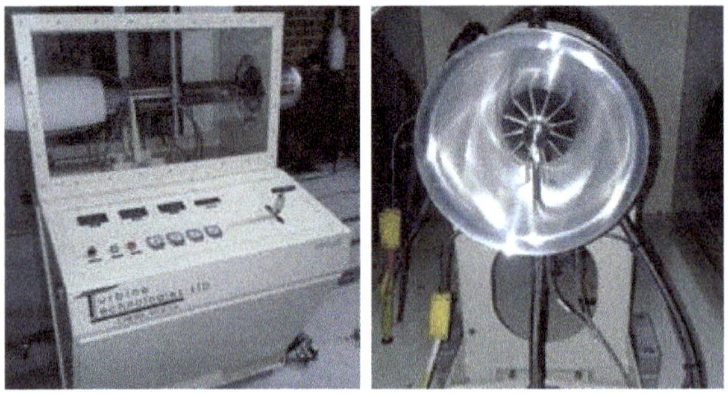

Figure 10. Mini gas turbine's power system.

RESEARCH CARRIED OUT BY UNDERGRADUATE STUDENTS (FINAL PROJECT AND INTRODUCTION TO SCIENTIFIC RESEARCH)

The Turbomachines Department encourages undergraduate students to start their end-of-course projects (monograph of Engineering undergraduate course) at their fourth year, to give them time to go deeper in their researches. The studies made by Sabóia Filho (2011), Ando (2010), Martins (2011), Müller (2011), Ibler (2011) and Carvalho (2012) show some of the research carried out by undergraduate students, in support to the gas turbine activities of the Department.

INTEGRATED ENGINEERING-MASTER PROGRAM

High-end undergraduate students may follow the Integrated Engineering-Master's Program (PIGM), enrolling to post-graduate courses concomitantly with their undergraduate courses, during the fifth year. This would enable them to complete the Master program in one additional year, that is, the student may leave ITA with the Engineering and the Master's degrees in six years. It is strongly recommended that the Master's dissertation be associated to the end-of-course project. As a result, the student that intends to continue with a doctoral program would spend one less year if he/she joins the PIGM program.

LESSONS LEARNED, COMMENTS AND CONCLUSIONS

The Engineering education process has been changing during the last years. Several new techniques and methodologies were created and evaluated to measure the teaching quality and effectiveness. Methodologies as CDIO, used at Chalmers University, MIT and Linköping University (Malmqvist *et al.*, 2010), present the importance of updating the education's development process.

At ITA, the Gas Turbines course is currently focused on the traditional methodology, associating theory and lab classes, but the introduction of hands-on on more complex Engineering problems is being progressively implemented, mainly in the graduate courses. For example, the students are requested to follow the whole design process of a gas turbine's part/component and its manufacturing process (machining) to get acquainted with real life procedures.

The coupling of theory and lab has proved to be very important so far. The students confirm that they understand more easily the subjects taught during the lectures when they see and feel the hardware. Moreover, lab exercises complete the theory, since, generally, the book's exercises have little of practical content.

The lecturers of Turbomachines Department have plans to implement the use of lab classes, despite the high costs. For example, the IAE's Gas Turbine Laboratory is being improved with engine test stand, compressor and overspeed test rigs, so that the students will have new test facilities to exercise their abilities. The test stand is being finalized, where the 5-kN gas turbine is being tested this year.

Taking advantage of the development of computational tools and manufacturing processes for gas turbine's components, a small gas turbine project is being developed at DCTA. A 5-kN/1.2-MW turbojet engine is already in the turbojet's test bed (not shown here) for the performance assessment, after a series of cold runs to check lubrication and vibration. ITA, IAE (DCTA) and TGM Turbinas, a Brazilian steam-turbine industry, are working in this project. The developed gas turbine is a 5:1 pressure ratio, 5-stage transonic axial compressor, single-stage axial turbine and multifuel annular combustion chamber.

Students have been involved in current developments and it is expected that the lessons learned can provide the basis for a new Engineering education methodology and improvement. Effort is being made to work with other universities in order to learn their experiences and incorporate them to our courses (Konstantinos *et al.*, 2012).

The result of the work being done by the Turbomachines Department in undergraduate, graduate (Professional Master, MSc and PhD) as well as in training courses has been helping the industry to hire professionals qualified in turbomachines.

Recently, the DCTA/IAE/TGM absorbed students that finished the Professional Master program at ITA to work in their gas turbine development programs, test bench development and tests. The gas turbine under development (Fig. 11) at DCTA by IAE/TGM is also supported by ITA's staff.

The Turbomachines Department contributes to training professionals who are already working at the industry, as DCTA/IAE, EMBRAER, PETROBRAS, TGM Turbinas and Thermal Power Plants. The training process is done through short courses or through integration of these professionals in Master's and PhD programs at ITA.

Figure 11. Gas turbine designed by IAE, TGM and ITA.

ITA has agreements with universities abroad, thus students who complete the Master's degree in Brazil can do PhD or sandwich PhD abroad. The goal is: when they return to Brazil, they initiate other gas turbine research groups.

ACKNOWLEDGEMENTS

Financiadora de Estudos e Projetos (FINEP), TGM Turbinas, Fundação de Amparo à Pesquisa do Estado de São Paulo (FAPESP), Conselho Nacional de Desenvolvimento Científico e Tecnológico (CNPq) and Coordenação de Aperfeiçoamento de Pessoal de Nível Superior (CAPES). IAE/APA are acknowledged for their support to the research carried out at the Department of Turbomachines (ITA).

REFERENCES

Al-Atabi, M., 2013, "Grand Challenges for Engineering and Development of CDIO Skills", Proceedings of the 9th CDIO Conference, MIT and Harvard University School of Engineering and Applied Sciences, Cambridge, USA.

Alves, M.A.C., 2003, "Non-Adiabatic Transient of Gas Turbines" (in Portuguese), Ph.D. Thesis, Technological Institute of Aeronautics, São José dos Campos, Brazil, 2003.

Ando, V.F., 2010, "Project Optimisation of High-Performance Axial Flow Compressor" (in Portuguese), Final Project, Technological Institute of Aeronautics, São José dos Campos, 2010.

Barbosa, J.R., 1987, "A Streamline Curvature Computer Programme for Performance Prediction of Axial Compressors", Ph.D. Thesis, Cranfield Institute of Technology, Cranfield, UK.

Barbosa, J.R. and Bringhenti, C., 1999, "Gas Turbine Performance Numerical Simulation", Proceedings of Congresso Brasileiro de Engenharia Mecânica (COBEM), Águas de Lindóia, Brazil.

Barbosa, J.R. and Bringhenti, C., 2000, "Numerical Simulation in Classroom", Proceedings of Congresso Nacional de Engenharia Mecânica (CONEM), Natal, Brazil.

Barbosa, J.R., Bringhenti, C. and Tomita, J.T., 2012, "Gas Turbine Transients with Controlled Variable Geometry", ASME Turbo Expo 2012, Copenhagen, Denmark.

Barbosa, J.R., Jefferds, F.S.S., Tomita, J.T. and Bringhenti, C., 2011, "Influence of Variable Geometry Transients on Gas Turbine Performance", Proceedings of ASME TURBO EXPO 2011, Vancouver, Canada.

Bringhenti, C., 1999, "Steady State Gas Turbine Performance Analysis" (in Portuguese), Master Thesis, Technological Institute of Aeronautics, São José dos Campos, Brazil.

Bringhenti, C., 2003, "Variable Geometry Gas Turbine Performance Analysis", Ph.D. Thesis, Technological Institute of Aeronautics, São José dos Campos, Brazil.

Bringhenti, C. and Barbosa, J.R., 2002, "Study of an Industrial Gas Turbine with Turbine Stators Variable Geometry", 9th Brazilian Congress of Thermal Engineering and Sciences, Caxambú, Brazil.

Bringhenti, C. and Barbosa, J.R., 2003, "Analysis of Gas Turbine Off-Design Safe Operation Using Variable Geometry Compressor", Proceedings of International Congress of Mechanical Engineering, São Paulo, Brazil.

Bringhenti, C. and Barbosa, J.R., 2004a, "Gas Generator Study for Use in Turboshaft and Turbojet", Proceedings of Congresso Nacional de Engenharia Mecânica (CONEM), Belém, Brazil.

Bringhenti, C. and Barbosa, J.R., 2004b, "Part-Load Versus Downrated Industrial Gas Turbine Performance", Proceedings of ASME Turbo Expo 2004: Power for Land, Sea, and Air, Vienna, Austria.

Bringhenti, C. and Barbosa, J.R., 2004c, "Performance Study of a 1 MW Gas Turbine", Proceedings of the 10th Brazilian Congress of Thermal Sciences and Engineering, Rio de Janeiro, Brazil.

Bringhenti, C. and Barbosa, J.R., 2004d, "A Methodology for Gas Turbine Performance Improvement Using Variable Geometry Compressors and Turbines", Journal of Power and Energy (Part A), Vol. 218, pp. 541-549.

Bringhenti, C., and Barbosa, J.R., 2008, "Effects of Turbine Tip Clearance on Gas Turbine Performance", Proceedings of ASME Turbo Expo 2008: Power for Land, Sea and Air, Berlin, Germany.

Bringhenti, C., Tomita, J.T. and Barbosa, J.R., 2010, "Performance Study of a 1 MW Gas Turbine Using Variable Geometry Compressor and Turbine Blade Cooling", Proceedings of ASME Turbo Expo 2010: Power for Land, Sea, and Air, Glasgow, UK.

Bringhenti, C., Tomita, J.T., Souza Júnior, F. and Barbosa, J.R., 2006, "Gas Turbine Performance Simulation Using an Optimized Axial Flow Compressor", Proceedings of ASME Turbo Expo 2006: Power for Land, Sea, and Air, Barcelona, Spain.

Camarda, C.J., Weck, O. and Do, S., 2013, "Innovative Conceptual Engineering Design (ICED): Creativity and Innovation in a CDIO-Like Curriculum", Proceedings of the 9th CDIO Conference, MIT and Harvard University School of Engineering and Applied Sciences, Cambridge, USA.

Carvalho, D.S., 2012, "Determinação das Características de Operação de Uma Turbina de Fluxo Radial Utilizando a Técnica da Dinâmica dos Fluidos Computacional" (in Portuguese), Final Project, Technological Institute of Aeronautics, São José dos Campos, Brazil.

Chappell, M.A. and McLaughlin, P.W., 1993, "Approach to Modeling Continuous Turbine Engine Operation from Startup to Shutdown", Journal of Propulsion and Power, Vol. 9, No. 3, pp. 466-471.

Chubik, P.S. and Zamyatina, O.M., 2013, "Training Elite Specialists in Engineering and Technologies", Proceedings of the 9th CDIO Conference, MIT and Harvard University School of Engineering and Applied Sciences, Cambridge, USA.

Connor, J., 2013, "Listening by Design: Teaching Students to Be More Collaborative", Proceedings of the 9th CDIO Conference, MIT and Harvard University School of Engineering and Applied Sciences, Cambridge, USA.

Creci Filho, G., Menezes, J.C. and Barbosa, J.R., 2009, "Numerical Modal Analyses of Blades and Rotors in a Single Spool Gas Turbine", Proceedings of the International Symposium on Dynamic Problems of Mechanics, Angra dos Reis, Brazil.

Creci Filho, G., Menezes, J.C. and Barbosa, J.R., 2011, "Rotordynamic Analysis of a 5- Kilonewton Thrust Gas Turbine by Considering Bearing Dynamics", Journal of Propulsion and Power, Vol. 27, No. 2, pp. 330-336.

Darmofal, D., 2012, "Rethinking Engineering Education: An Overview of the CDIO Approach", LASPAU/Brazil Program, Harvard University, Cambridge, USA.

Fishback, L.H. and Koenig, R.W., 1972, "GENENG II: A Program for Calculating Design and Off-design Performance of Two-and Three-Spool Turbofans with as Many as Three Nozzles", NASA, Washington, DC, USA.

Flack, R.D., 1990, "Analysis and Matching of Gas Turbine Components", International Journal of Turbo and Jet Engines, Vol. 7, pp. 217-226.

Friesel, A., 2013, "Engineering Education Resulting in Skilled, Inventive and Innovative Professionals, How?", Proceedings of the 9th CDIO Conference, MIT and Harvard University School of Engineering and Applied Sciences, Cambridge, USA.

GASTURB, 2015, "Gas turbine performance… has to be understood by all concerned with gas turbine technology", Retrieved in January 23, 2015, from http://www.gasturb.de/

Goldberg, D., 2013, "Three Habits at Three Levels for Improved Engineering Education, Huff Post, Education", Retrieved in January 23, 2015, from http://www.huffingtonpost.com/david-goldberg/3-habits-at-3-levels-for-_b_1751699.html.

GSP, 2015, "Gas Turbine Simulation Program", Retrieved in February 01, 2015, from http://www.gspteam.com/home

Hoffman, J.A., 2013, "Evolution of MIT's 'Introduction to Aerospace Engineering and Design' Course", Proceedings of the 9th CDIO Conference, MIT and Harvard University School of Engineering and Applied Sciences, Cambridge, USA.

Ibler, F., 2011, "Flow Calculations Within Axial Compressor Based on CFD Techniques and Its Comparison with Streamline Curvature Results" (in Portuguese), Final Project, Technological Institute of Aeronautics, São José dos Campos, Brazil.

Ismail, I.H. and Bhinder, F.S., 1991, "Simulation of Aircraft Gas Turbine Engines", Journal of Engineering for Gas Turbines and Power, Vol. 113, No. 1, pp. 95-99. doi: 10.1115/1.2906536

Kamp, A. and Klaassen, R., 2013, "Making Curricular Change: Case Report of a Radical Reconstruction Process", Proceedings of the 9th CDIO Conference, MIT and Harvard University School of Engineering and Applied Sciences, Cambridge, USA.

Karltun, J., 2013, "Renewal of a Local Concept for Engineering Education Including CDIO", Proceedings of the 9th CDIO Conference, MIT and Harvard University School of Engineering and Applied Sciences, Cambridge, USA.

Koenig, R.W. and Fishback, L.H., 1972, "GENENG: a Program for Calculating Design and Off-design Performance for Turbojet and Turbofan Engines", NASA,Washington, DC, USA.

Konstantinos, K., Gronstedt, T. and Barbosa, J.R., 2012, "Gas Turbine Lessons Learned from the Development of Courses on Gas Turbine Multi-Disciplinary Conceptual Design", ASME Turbo Expo 2012, Copenhagen, Denmark.

Koster, J.N., Uhmeyer, K.L. and Soin, G.S., 2013, "Designing a Blended Wing Body Aircraft Globally", Proceedings of the 9th CDIO

Conference, MIT and Harvard University School of Engineering and Applied Sciences, Cambridge, USA.

Lassudrie, C. and Kontio, J., 2013, "Managing the Continuous Improvement Loop of Educational Systems: Students as Key Actors in Program Evaluation", Proceedings of the 9th CDIO Conference, MIT and Harvard University School of Engineering and Applied Sciences, Cambridge, USA.

Mäkimurto-Koivumaa, S., Väänänen, M. and Belt, P., 2013, "Developing Engineering Education to Support Entrepreneurial Behavior", Proceedings of the 9th CDIO Conference, MIT and Harvard University School of Engineering and Applied Sciences, Cambridge, USA.

Malmqvist, J., Bankel, J., Enelund, M., Gustafsson, G. and Wedel, M.K., 2010, "Ten Years of CDIO - Experiences from a Long-Term Education Development Process", Proceedings of the 6th International CDIO Conference, École Polytechnique, Montréal, Canada.

Martins, V.A.C., 2011, "Projeto Preliminar de uma Turbina Axial para Uso em Turbinas a Gás de Pequena Potência" (in Portuguese), Final Project, Technological Institute of Aeronautics, São José dos Campos, Brazil.

Mattingly, J.D., 1996, "Elements of Gas Turbine Propulsion", McGraw-Hill, New York, USA.

Mazur, E., 2012, "Educating the Innovators of the 21st Century", LASPAU/Brazil Program, Harvard University, Cambridge, USA.

McKinney, J.S., 1967, "Simulation of Turbofan Engine - SMOTE: Description of Method and Balancing Technique", pt. 1-2, Air Force Aero Propulsion Lab., s.l..

Müller, L., 2011, "Cálculo de Escoamento Turbulento em Uma Turbina Axial de Simples Estágio Aplicando a Técnica da Dinâmica dos Fluidos Computacional" (in Portuguese), Final Project, Technological Institute of Aeronautics, São José dos Campos, Brazil.

Palmer, J.R., 1967, "The 'TURBOCODE' Scheme for the Programming of Thermodynamic Cycle Calculations on an Electronic Digital-Computer", CoA Report Aero 198, Cranfield Institute of Technology, Cranfield, UK.

Palmer, J.R., 1990, "The TURBOMATCH Scheme for Aero/Industrial Gas Turbine Engine Design Point/Off Design Performance Calculation", Thermal Power Group, Cranfield University, Cranfield, UK.

Palmer, J.R., 1983, "The 'TURBOMATCH' Scheme for Gas Turbine Performance Calculations, User's Guide", Cranfield Institute of Technology, Cranfield, UK.

Palmer, J.R. and Yan, C.Z., 1982, "TURBOTRANS: a Programming Language for the Performance Simulation of Arbitrary Gas Turbine Engines with Arbitrary Control Systems", s.n., s.l.

Reed, J.A. and Afjeh, A.A., 1994, "Development of an Interactive Graphical Propulsion System Simulator", June 1994.

Sabóia Filho, F.A.T., 2011, "Turbojet: an Overview on Design, Off-design and Transient Performance", Monograph, Technological Institute of Aeronautics, São José dos Campos, Brazil.

Sellers, J. and Daniele, C., 1975, "Dyngen: a Program for Calculating Steady-State and Transient Performance of Turbojet and Turbofan Engines", NASA, Washington, DC, USA.

Silva, F.A., 2009, "Preliminary Design and Analysis of a One Stage Axial Flow Turbine for Conversion of a Turbojet into a Turboshaft" (in Portuguese), Master Thesis, Technological Institute of Aeronautics, São José dos Campos, Brazil.

Silva, F.J.S., 2007, "Gas Turbines Transient Performance Study for Axial Compressor Operation Characteristics", Proceedings of the 19th International Congress of Mechanical Engineering, Brasília, Brazil.

Silva, F.J.S., Bringhenti, C. and Barbosa, J.R., 2005, "Transient Performance of Gas Turbine, Proceedings of the 18th International Congress Mechanical Engineering, Ouro Preto, Brazil.

SRS Technologies, 2007, "Graphical Engine Cycle Analysis Tool, GECAT", www.stg.srs.com.

Stamatis, A., Mathioudakis, K. and Papailiou, K.D., 1990, "Adaptive Simulation of Gas Turbine Performance", Journal of Engineering for Gas Turbines and Power, Vol. 112, No. 2, pp. 168-175.

Szuk, J.R., 1974, "Hydes: a Generalized Hybrid Computer Program for Studying Turbojet or Turbofan Engine Dynamics", NASA, Washington, DC, USA.

TERMAP, 2000, "Innovative Visual Modeling Environment for Turbine Engine Reverse Modeling Aid Program (TERMAP)", SRS Technologies, www. stg.srs.com

The Royal Academy of Engineering and Massachusetts Institute of Technology, 2012, "Achieving Excellence in Engineering Education: The Ingredients of Successful Change", Retrieved in January 23, 2015, from http://www.raeng.org.uk/publications/reports/achieving-excellence-in-engineering-education

Tomita, J.T., 2003, "Numerical Simulation of Axial Flow Compressors", Master Thesis, Technological Institute of Aeronautics, São José dos Campos, Brazil.

Tomita, J.T., 2009, "Three-Dimensional Flow Calculations of Axial Compressors and Turbines Using CFD Techniques", Ph.D. Thesis, Technological Institute of Aeronautics, São José dos Campos, Brazil.

Tomita, J.T. and Barbosa, J.R., 2012, "Numerical Tools for High Performance Axial Compressor Design for Teaching Purpose", Proceedings of the ASME Turbo Expo 2012, Copenhagen, Denmark.

Tomita, J.T., Bringhenti, C. and Barbosa, J.R., 2005, "Study of the Air Bleed Influence in the Industrial Gas Turbine Performance", Proceedings of the 18th International Congress of Mechanical Engineering, Ouro Preto, Brazil.

Tomita, J.T., Bringhenti, C., Barbosa, J.R. and Carlesse, V.M., 2012, "Performance Evaluation of a 5kN Gas Turbine Based on Specially Designed Components", ASME Turbo Expo 2012, Copenhagen, Denmark.

Wittenberg, H., 1976, "Prediction of Off-Design Performance of Turbojet and Turbofan Engines", s.n., s.l..

Yang, H.T., 2009, "Engineering Education and Research for an Interdisciplinary Future", University of California, Santa Barbara, USA.

11

Application of the Prado - Project Management Maturity Model at a R&D Institution of the Brazilian Federal Government

Luiz Aldo Leite das Neves[1,2], Luiz Eduardo Nicolini do Patrocínio Nunes[1], Valesca Alves Corrêa[1], Mirabel Cerqueira Rezende[2]

ABSTRACT: Government institutions have sought to improve their processes in project management in an effort to elevate their maturity levels using models that clearly identify the weaknesses in the management of their projects. This article aims to show a case study by applying the Project Management Maturity Model (Prado - PMMM), developed by Darci Prado, in a Research and Development (R&D) Institution of the Brazilian Federal Government. The scores show that the maturity level in project management is weak (institutional level equal to 2.47). The main causes for this score are attributed to the lack of knowledge and the unpreparedness of some sectors and project managers. It was also observed that the dimension named Technical Competence presents the highest value (46%), considered a good score. On the other hand, the dimension involving Behavioral Competence presents the lowest value (9%), which, according to the used methodology, is considered weak, indicating that investments must be made to enhance this dimension.

KEYWORDS: Maturity in management, Project management, Maturity models, Growth plan.

INTRODUCTION

Managing projects is not as simple a task as many may think. Besides having well-defined beginning and end, it involves a series of steps with specific goals, which require financial, human, as well as material resources. It is also complicated by the fact that many institutions are structured as a matrix, where projects permeate several departments within.

The concept of project management has gone through an evolution in the past few years, where the traditional idea of the project manager being a specialist, usually hired by engineering companies, is now considered limited. Nowadays, the concept of project management is viewed under a new perspective inside organizations, which includes more inter-relations that take into consideration varied profiles in medium and high level positions in the organizational charts (Barber, 2004).

In order to remain competitive, organizations have been trying to improve their process management aiming for the success of their projects by adopting project management practices. The demand for qualified professionals in this area has increased since the project manager has become essential by providing an advantage to both public and private institutions.

Public organizations are not different from their private counterparts in terms of their complexity. They are both going through the conflict arising from the changes and innovations required by the current environment against the bureaucratic dynamics of their organizational culture (Pires and Macêdo, 2006). Due to this complexity, which is a characteristic of modern organizations, studies in the area of project management have intensified, with emphasis on project handling, project

1.Universidade de Taubaté – Taubaté/SP – Brazil 2.Instituto de Aeronáutica e Espaço – São José dos Campos/SP – Brazil

Author for correspondence: Luiz Aldo Leite das Neves | Praça Marechal Eduardo Gomes, 50 – Vila das Acácias | CEP 12.228-904 São José dos Campos/SP – Brazil | Email: luiz_aldo@ig.com.br management offices, maturity, strategic alignment, life cycle and risk management (Prado, 2010).

In the 1990's, Brazil saw the beginning of actions meant to create a greater alignment with the public administration management movement, gaining strength after the establishment, in 1995, of Ministry of Federal Administration and State Reform (MARE). At the time, the Brazilian government was becoming more concerned with such issues as administrative efficiency, contract handling, public service, and accountability; in other words, providing a better quality service to the most interested party, its citizens. In this context, government institutions are intensively utilizing tools that measure the maturity and handle the management of its projects with the purpose of directly contributing to the satisfaction of its customers. Using an informed diagnosis, public project managers can create guidelines to improve the administration of projects, which in turn increases their success rate at the end. These tools are being used to enable the evaluation of maturity in project management, taking into consideration the context and the management procedures in above mentioned institutions. The outcome of these evaluations is used to create a plan of action for short and long term growth, as well as to guide the management of programs and projects with the purpose of increasing their success rate.

Prado and Archibald (2009) did some research with public, private, and third sector (public non-profit civil institutions) institutions and confirmed the low rate of project management maturity in public organizations of both direct and indirect type administration. In other words, the study observed that these institutions were less efficient and effective in project management than their private counterparts.

In an effort to contribute to this discussion, this article aims to show the present level of project management maturity of a Research and Development (R&D) institution of the Brazilian Federal Government, using Prado-Project Management Maturity Model (Prado - PMMM) and based on the scores, make some suggestions to advance the project management maturity of the institution in question.

PROJECT MANAGEMENT MATURITY

Beginning in the 1990's, various models were developed to evaluate the maturity of organizations in managing projects, almost all of them inspired by the maturity model in software development, Capability Maturity Model Integration (CMMI), created by Carnegie Mellon University in a partnership with the Systems Engineering Institute (SEI) (Prado, 2010). According to Kerzner (2002), Maturity in Project Management is represented by specifically designed systems and processes, which are characterized by repetitiveness, increasing the probability of success, though not guaranteeing it, in spite of this increase in probability being its main characteristic.

Globalization and the changes in the economic and business scenario forced organizations to change its way of thinking in regards to project management, since they depend on their projects, together with technological advances, to guarantee their competitiveness and survival. Therefore, Project Management Maturity has advanced since it shows that the company that makes use of it has the tools, capability, and the needs to manage its projects. The advantages of Project Management have been largely advertised and the application of their methods is ever more common in companies, especially those that need to provide a swift and effective response to the current organizational and environmental issues (Carvalho and Rabechini, 2006).

In this area of study, the best recommended practices are grouped in the maturity models, which try to identify the present level of maturity in organizations through the use of assessments and subsequently propose improvements by showing which practices are useful to advance their level of maturity in project management. This procedure allows for future assessments using the previous score as reference as well as the growth data of the organization during the period, increasing the chance of success in project execution.

Among the models developed in the 1990's and beyond, we have the Prado - PMMM, based on the experience of its author, Darci Prado, in the implementation of project management in dozens of Brazilian institutions (Prado, 2010).

PRADO - PMMM

The Prado - PMMM used in this case study is characterized by the simplicity of its questionnaire, the practical way of obtaining scores, the applicability to the various sectors of an organization as well as to the organization as a whole. It is also in alignment with our culture, since it has been used in many Brazilian institutions, and is available online.

This model developed by consultant Darci Prado is comprised of 5 levels of maturity and 6 dimensions, as shown in Fig. 1. It takes into account the areas related to processes, people, technology, and strategies spread through the 5 levels, in accordance with the terminology used by the Project Management Body of Knowledge (PMBOK) Guide.

MATURITY LEVELS

According to Prado's model, Level 1 represents the initial stage, where the department hasn't made any coordinated effort to implement project management. In other words, in this stage projects are executed based on intuition, individual effort, and good will. There is usually no planning or control, since there are no standard procedures, and consequently there is the possibility of delay and cost overrun, and most likely the technical specifications will not be met. The first level clearly shows the total disengagement between those involved in the project and the practices of Project Management.

Level 2 "Known" demonstrates that the organization regularly invests in training and has acquired project management software. There may be isolated initiatives to standardize procedures, but their use is limited because there needs to be widespread standardization in order to facilitate project planning and control. In spite of that, failures are still frequent because the lack of standards results in a diluted use of knowledge.

Level 3 "Standardized" has seen the implementation of the Project Management Office, which has standardized the use of procedures that require the utilization of planning and control processes, which in turn demand more dedication on the part of those involved in the project. We can observe manager's improvements in terms of technical, behavioral, and contextual competency. The problems affecting project performance are known but haven't yet been resolved and it is obvious that improvements are needed.

Level 4 "Managed" shows that investments in behavioral competency are efficient because the project managers are better prepared to handle the behavioral aspects of their teams, such as human relationships, conflicts, and negotiations. At this level, the practice of improvement is intensified in order to boost knowledge through an emphasis in advanced course participation (such as MBA's in project management) and visits to other organizations that have consolidated project management processes (benchmarking).

Finally, **Level 5 "Optimized"** indicates that the company has reached a high level of project management understanding; the processes are optimized therefore accomplished in less time, at less cost, but with quality scores, due to the wide experience, knowledge, and attitude of the people involved. That is obtained through the harmonization of the Project Management Model and the Organizational Structure, which are in complete alignment with corporate business.

THE DIMENSIONS

The correlation between the six dimensions and the five maturity levels shows how mature the project management of an institution is. Under this perspective, Prado (2010) takes into consideration this correlation, which brings into evidence the most important characteristics of each of the maturity level of the model, as well as expected success rate of the projects.

The six dimensions of the Prado - PMMM are shown in Fig. 2. In broad terms, the dimensions, or maturity factors, appear in

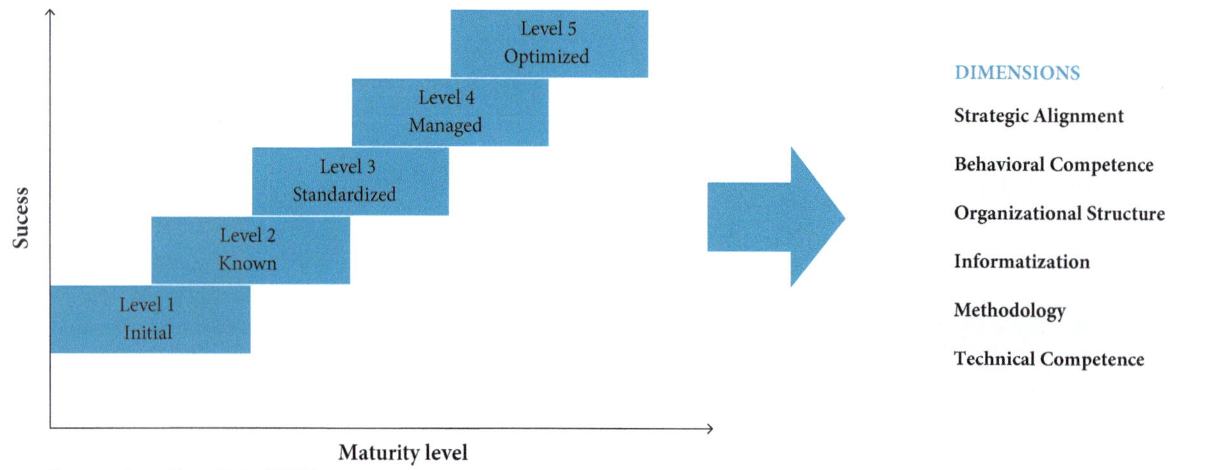

Source: adapted from Prado (2010).

Figure 1. Maturity Level and Dimensions – Prado - PMMM.

each level with more or less intensity, depending on the moment where the peak of maturity occurs in a specific dimension.

The first dimension — **Technical Competency** — addresses the knowledge base of project management, which must be widespread among the project management sectors. In other words, project managers and others involved should strengthen their technical knowledge not only in their specific areas but also other aspects of project management.

The second dimension — **Methodology** — means that the use of a unique methodology in the whole company is highly recommended. It should also allow for small variations among the different sectors, since the correct application of methods, techniques, and tools is guaranteed by carrying out a series of steps.

The third dimension — **Informatization** — addresses the use of project management software since many aspects of the methodology need to be computerized to be used by a number of people as standard procedure.

The fourth dimension — **Organizational Structure** — concentrates on the way a company or its sectors are organized to execute its projects. The most commonly used structures are functional, matrix, and divisional. These also allow for some variations and may coexist with each other or with complementary structures, such as in a Project Management Office.

The fifth dimension — **Behavioral Competence** — recognizes that the execution of the work depends on the people therefore, it is essential that they are motivated to perform their tasks to the best of their ability. Conflicts among those in a team are usually detrimental to the project and should be avoided by managers.

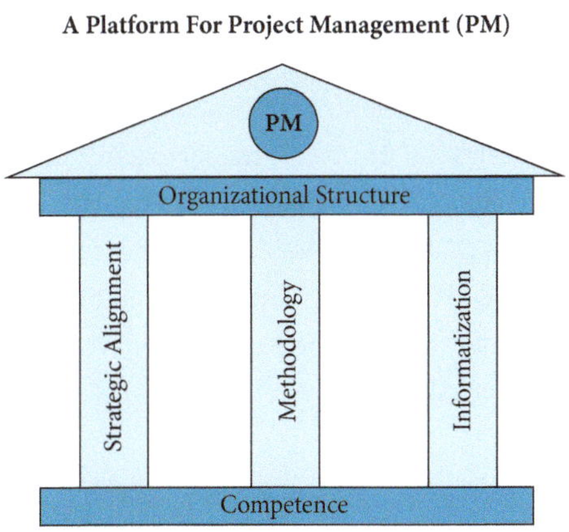

Source: Adapted Prado (2010).

Figure 2. Dimensions of Prado-Project Management Maturity Model – PMMM.

The sixth and last dimension — **Strategic Alignment** — considers it fundamental to have the projects aligned with the business of the organization so that they are adequately planned and executed. Each project must have been assessed according to certain criteria, such as technical and financial evaluation, and risk analysis.

ASSESSMENT

The Prado - PMMM allows for the maturity assessment to be obtained through the application of a 40-item questionnaire. According to the Prado - PMMM (Prado, 2010), the final score of the maturity assessment is obtained from the answers and data of this test. This final score is given on a scale of 1 through 5, which can be interpreted according to illustration in Fig. 3.

The total points obtained from the answers at each level determine the placement in a level. Each answer is measured from 0 to 100 points, or the equivalent percentage, as shown in Fig. 4.

Based on these scores, the weak and strong points rated in each question can be analyzed, and a plan of action for the short, medium, and long term can be established.

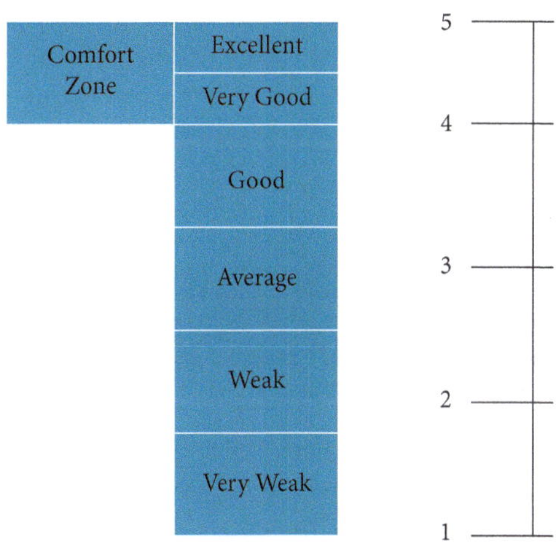

Up to 1.60 .... Very Weak

Between 1.60 and 2.60 .... Weak

Between 2.60 and 3.20 .... Average

Between 3.20 and 4.00 .... Good

Between 4.00 and 4.60 .... Very Good

Above 4.60 .... Excellent

Adapted from Prado (2010).

Figure 3. Management Maturity Scale.

Up to 20%	Weak Adherence
Up to 40%	Average Adherence
Up to 70%	Good Adherence
Up to 90%	Very Good Adherence
Up to 100%	Excellent Adherence

Source: Prado (2010).

Figure 4. Standards for Level Adherence Assessment.

CASE STUDY

This case study was performed at a R&D Institution of the Brazilian Federal Government. This institution has a variety of projects in its portfolio, each with its own characteristics and different funding resources. Its goal is research and development directed at increasing technical and scientific knowledge in order to provide technical solutions to strengthen the country's industrial sector, and therefore contribute to national sovereignty through research, development, and innovation.

The first step in the present research was choosing the model to evaluate project management maturity, in this case the Prado - PMMM. This choice was based on the advantages of this model as explained earlier. The next step was choosing the target audience to apply the Prado - PMMM questionnaire, involving 19 sectors of the chosen organization. The work related to this study started with sector meetings to demonstrate the importance of this research as well as the benefits it would bring to the organization.

The target audience is very important to the organization's projects due to their technical knowledge, their experience in project management, or for their part in their project's team. The selected audience included upper managers, managers, future project managers, internal service suppliers, and others involved in the projects, mainly for their position and job description in each sector of the organizational structure.

The tool used to collect data for this study was a questionnaire based on the Prado - PMMM developed by consultant Darci Prado, available online at "www.maturityresearch.com". This model was chosen for its simplicity, small number of questions, and alignment with our culture. The questionnaire is comprised of 40 questions, addressing four levels of maturity (from 2 to 5) and each level has 10 questions with 5 choices (A, B, C, D, and E), which weigh 10, 7, 4, 2, and 0, respectively. The questionnaires were handed out individually to each respondent in his own sector in order to prove once again the importance of the research. A total of 78 questionnaires were given out.

RESULTS AND DISCUSSION

In order to analyze the scores of this research, some factors were taken into consideration, such as the respondents' profile in regards to technical knowledge, professional experience, participation in project management courses, management position, and length of service in the organization. Respondents' profile: 100% are college graduates, 90% have a doctorate, masters, or a specialty degree, but only 30% have done training in project management.

During the course of this research, most of the respondents (90%) completed the questionnaire. This positive outcome may be attributed to the direct approach used for its distribution, when each could raise concerns regarding the model being used and have these promptly addressed. Of the 78 questionnaires given out, 72 were turned in. The data collected was consolidated and analyzed using the formulas provided by the Prado's model to calculate maturity levels (Prado, 2010).

The respondents who took part in this research work at 23 different sectors of the institution under study. Its organizational structure is considered a "weak" matrix, and since it is a hierarchy, managers' autonomy is limited. Not only do the sectors where the respondents work have different characteristics, the projects within a sector differ as well; some sectors concentrate on R&D, others work directly with the execution of projects, still others provide services, besides project department itself, and upper management.

Based on the analysis of the respondents' as well as the institutional profile, it is possible to improve our comprehension of the final score obtained by the institution in terms of management maturity, adherence to the levels, and adherence to the dimensions.

The institutional Final Maturity Score (FMS) was calculated using Eq. 1 (Prado, 2010). The score obtained was 2.47, which is considered a weak level of project management maturity according to the reference levels proposed by the model (Fig. 3). According to Prado (2010), the FMS obtained by this institution demonstrates a fairly good level of knowledge, and there are isolated individual efforts; it also shows that they have initiated the implementation of a standard project management platform.

$$Institutional\ FMS = \frac{1}{23}\left(\sum_{n=1}^{23} FMS_{sector\ n}\right) \tag{1}$$

$$Institutional\ FMS = \frac{56.88}{23}$$

$$Institutional\ FMS = 2.47$$

Adherence to the four levels proposed by the model used in the research is shown in Fig. 5. An analysis of this figure shows that Level 2, also called "Known", shows the most adherence (62%). According to Prado (2010), this is a good score, since it demonstrates isolated attempts to standardize procedures as well as create a common language. Level 3, on the other hand, also called "Standardized", shows a 48% adherence, considered average. This level is considered very important since it is where the implementation of a project management platform is verified. Nevertheless, according to Prado (2010), even adherence levels close to 100% are no guarantee for consistent and lasting scores. Finally, Levels 4 ("Managed") and 5 ("Optimized") show 30% and 7% adherence levels, respectively, which are considered weak profiles by Prado's model standards.

The assessed score showing institutional adherence by levels indicate that the institution needs to improve its methods in

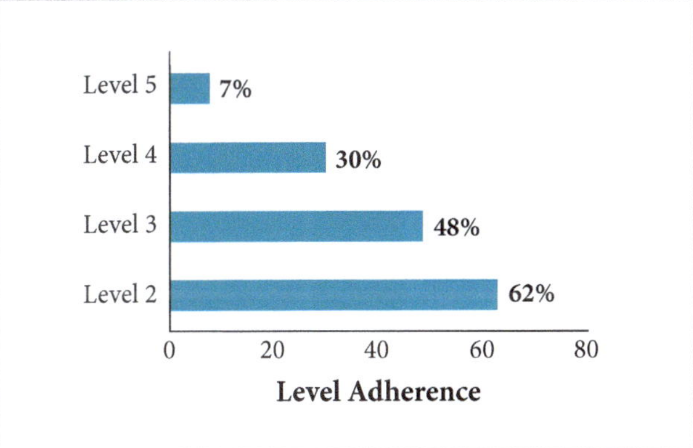

Figure 5. Subject's Percent of Adherence to the Levels of Institutional Maturity.

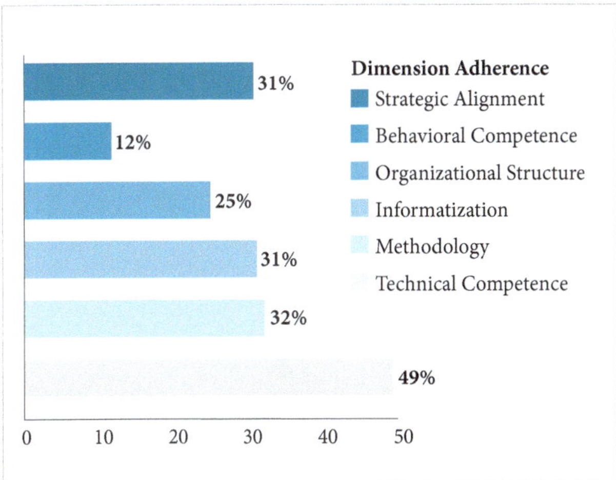

Figure 6. Subject's percent of Adherence to the Dimensions of Institutional Maturity.

the project management area. It is also worth pointing out that the assessed scores may have been influenced by the type of organizational structure. Generally speaking, it has been noticed that the institution has fairly good adherence in certain levels based on Prado's model standards, in spite of not having a centralized model of project management.

Figure 6 represents the institutional adherence to the dimensions based on the data collected from the questionnaires. An analysis of this figure shows that the first dimension — Technical Competence — is the highest at 49%, considered good according to the Prado - PMMM (Prado, 2010). On the other hand, the second (Methodology), the third (Informatization), and the sixth (Strategic Alignment) dimensions show values considered average, i.e. 29%, 29 and 27%, respectively. The fourth dimension — Organizational Structure — has a value of 15%. Finally, the Behavioral Competence (fifth dimension: Behavioral Competence) shows a value of 9%, which is considered weak according to the methodology used.

It can be observed from these scores that the institution presents a profile typical of Level 2, where the organization invests in training and has acquired project management software, but the values in Dimensions 4 and 5 indicate that the institution has to implement actions to enhance these dimensions.

CONCLUSION

The scores obtained through this research, based on the Prado - PMMM, allowed for the identification of the level of Project Management Maturity of the subject institution. Comparing the score obtained through this case study (2.47) with the overall average scores of Brazilian companies in terms of R&D by the survey Brazil Research 2012 (available at www.maturityresearch.com), we see that the subject institution is in line with the national average of institutions in the same category (2.60). It must be emphasized that this is a low score according to the reference levels of the model used in this research (Prado - PMMM).

One of the contributing factors to this score is the lack of tools for project control and follow-up, added to poor communication, which had a significant impact on the final score, since many projects are structured as a matrix therefore permeating various sectors of the institution. In addition to that, it is known that this organization has a "weak" matrix structure, i.e. it is a hierarchical institution where managers' decision making is limited, having a direct effect on the outcome of projects. Nevertheless, this research has identified the following as the main contributors to the score: lack of knowledge, inexperience of some project managers, the way the institution is structured, and the way resources are distributed.

This organization has been going through positive changes in the past few years, such as internal restructuring, improvement in its strategic planning, better quality management. It can be asserted that this first assessment of their project management maturity is consistent with these improvements although not the object of such changes.

One of the positive aspects of this research was pointing out the current level of maturity of the organization and which areas need improvement. The analysis of the dimension adherence scores will be used by the Project Office to prioritize improvements, especially in the areas where the dimension level was low.

In this case study, although the FMS is considered "weak", we can classify the institution as a Level 2 "Known" in maturity, whose final score shows a 60% adherence level, which is considered "good" according to the standards of the Prado model. This means that there are isolated initiatives towards the use of good Project Management practices, but there is a lack of a methodology, which leads to a dilution of the acquired knowledge as evidenced by levels 4 and 5, where adherence was "Average" and "Weak", respectively.

As far as dimension adherence is concerned, the organization is inadequate in all dimensions, in spite of the 49% achievement in "Technical Competence", which is considered "Good" according to the reference scale, because the project managers are not knowledgeable in the methodologies used in project management. This behavior is made worse by the "Behavioral Competence" dimension, since managers are not encouraged to participate in conflict management training, and the "Organizational Structure" dimension, which is hierarchical. The scores point to the need of corrective measures to improve the level of project management maturity.

In order to better understand the score achieved by the institution, the offices were divided into groups based on the type of activity they perform, which made it possible to evaluate the differences among them in regards to project management maturity (Management Group, Research Group, and Service Group). The scores clearly show the respondents' view of project management.

The Project Management Office of this institution is still in its initial stage, which hinders the implementation of a "Methodology". Only through its empowerment and the support of upper management can the Project Management Office improve the administration of processes and prevent the use of isolated software whose sole purpose is managing deadlines. Software to manage risk, cost, and resources is not offered by the institution, though it would help improve the "Organizational Structure" as well as the "Informatization" dimensions.

Based on the scores, a plan of action aiming to develop the institution's project management processes has been proposed, to be coordinated by the Project Management Office. The initiatives shall concentrate on how manager's profile and organizational structure affect project management in order to develop directives meant to solve the problems affecting the performance of projects.

It is important to promote awareness among upper management, project managers, and all others involved in projects, thus creating a culture that values the use of a corporate tool for the management of projects, be it through lectures, training, or even visits to other institutions included in benchmark, as well as intensive training for the present and future project managers.

ACKNOWLEDGMENTS

The authors would like to thank the Project Office of the institution for their support throughout this study as well as all the respondents who promptly contributed to this research.

REFERENCES

Barber, E., 2004, "Benchmarking the Management of Projects: a Review of Current Thinking", International Journal of Project Management, Vol. 22, No. 4, pp. 301-307.

Carvalho, M.M. and Rabechini, R., 2006, "Construindo Competências para Gerenciar Projetos: Teoria e Casos", Atlas, São Paulo, Brazil.

Kerzner, H., 2002, "Gestão de Projetos: As Melhores Práticas", Bookman, Porto Alegre, Brazil.

Pires, J.C.S. and Macêdo, K.B., 2006, "Cultura Organizacional em Organizações Públicas no Brasil", Revista de Administração Pública, Vol. 40, No. 1, pp. 81-105.

Prado, D., 2010, "Maturidade em Gerenciamento de Projetos", INDG Tecnologia e Serviços Ltda., Nova Lima, Brazil.

Prado, D. and Archibald, R., 2009, "Gerenciamento de Projetos para Executivos", INDG Tecnologia e Serviços Ltda., Nova Lima, Brazil.

12

Nonlinear Characteristic of Spherical Joints with Clearance

Zhang Jing[1], Guo Hong-Wei[2], Liu Rong-Qiang[2], Deng Zong-Quan[2]

ABSTRACT: Joints in deployable structures can degrade the stiffness and the stability of spacecraft. In this study, the nonlinear stiffness of spherical joints is investigated. The traditional contact model of spherical joints based on non-conforming contact assumption is presented. A new contact model for spherical joints based on the Winkler model and geometric constraints is established to calculate the stiffness of spherical joints with small clearances. The finite element model (FEM) of spherical joints is built to evaluate the accuracy of the theoretical model. The effects of the clearance and the contact force of spherical joints on the deformation of joints are investigated. When compared with FEM results, the error of the new spherical model is smaller than that of the traditional contact model when the clearance of spherical joint is not excessively large. The new contact model for spherical joint is more accurate than the traditional contact model when the clearance of spherical joint is large.

KEYWORDS: Deployable structure, Nonlinear characteristic, Joint, Stiffness.

INTRODUCTION

Given the need of sending large payloads and high volume of launchers in space missions, deployable structures as supporting ones have been widely used in spacecraft (Puig *et al.*, 2010; Gruber *et al.*, 2007). The deployable mast for solar sails has reached the range of 100 – 1,000 m (Brown, 2011). The large deployable structures usually contain a large number of joints to achieve the smooth rotation of the structure parts. Joint clearance often generates nonlinear characteristics, such as impact or contact, which affects the stability and the accuracy of deployable structures (Qi *et al.*, 2010).

The dynamic performance of deployable structures is important for space structure design. In the first type of dynamic analysis, joint clearance is neglected. Shin and Lee (2000) proposed an improved approach to update the initial finite element model (FEM) for space structures. The second type of analysis considers the impact and free-play generated by the joint clearance. Sarkar modeled the space trusses with nonlinear force caused by joint that is expressed by harmonic function (Sarkar *et al.*, 2004). (Li *et al.*, 2011) investigated the transient dynamic response of plane mechanics and analyzed the effect of clearance on the dynamic performance of structures by utilizing gap element to model clearance joint in ANSYS software. Joint clearance is considered not only in dynamic analysis of space trusses, but also in mechanism analysis. Bauchau and Rodriguez (2002) considered three-dimensional configuration of spherical joint and studied the effect of clearance, lubrication and friction phenomena on the dynamic response of flexible multibody systems.

The contact model of a joint is mainly based on elastic theory and finite element theory. Hertz law is proposed to model the contact of joints (Johnson, 1985). Lankarani and Nikravesh (1990) presented a modified Hertzian-based contact

1.Taiyuan University of Technology – College of Mechanical Engineering – Taiyuan – China 2.Harbin Institute of Technology – School of Mechatronics Engineering Harbin – China.

Author for correspondence: Zhang Jing | Taiyuan University of Technology – College of Mechanical Engineering | 79 – Yingze West Street – Wanbolin District | Zip Code: 030024 – Taiyuan – China | Email: free1985216@163.com law, which includes a damping term to account for the energy dissipation. Tian *et al.* (2009) calculated the contact force of spherical joint in spatial flexible multibody systems by using the continuous contact model proposed by Lankarani and Nikravesh. Boltachev *et al.* (2012) studied the elastic interaction of spherical particles whose stress distribution is determined by a rod model, which was applied in a more wide range of deformation calculation when compared with the classical Hertz law. Liu *et al.* (2006) built an approximate contact model of spherical joints with clearance that is based on distributed elastic forces. Fang *et al.* (2015) proposed an accurate theoretical solution to calculate the conformal contact pressure distribution in spherical plain bearings through the equivalent force principle. Bai and Zhao (2013) presented a new force model of the revolute joint with clearance for planar mechanical systems, which is a hybrid model of the Lankarani-Nikravesh and the improved elastic foundation models. Rajaei and Ahmadian (2014) used the Iwan model to simulate stick-micro/macro slip behavior involved in a friction contact. Pereira *et al.* (2014) estimated the applicability of the cylindrical contact models proposed in previous studies. The research on contact model is mainly based on the Hertz theory. Although energy dissipation, slip, and friction in the contact solids are considered, the accurate of the contact models usually used in mechanism simulation is not completely studied. The models based on Hertz model are not suitable for the conforming contact of joint because Hertz model is based on non-conforming assumption.

Finite element method is the other method to investigate the contact of joints. Cannon and Howell (2005) used pseudo-rigid-body model to predict the behavior of the conforming contact-aided revolute joint and validated the model by finite element analysis. Liu *et al.* (2007) established an approximate model for the contact problem of cylindrical joints by introducing appropriate assumptions and by analyzing the FEM numerical results. Because the elastic contact calculation by implementing finite element software depends on the user's choice in some parameters such as normal contact stiffness and penetration limit, Hattori and Serpa (2015) predicted the values of some parameters by investigating the artificial neural networks. Big-Alabo *et al.* (2015) conducted elastoplastic analysis for a spherical impactor. Finite element method can be used to analyze various contact conditions. However, the major limitation of this type of analysis is time-consuming. This method can hardly be combined with the simulation of mechanism. Thus, an accurate conforming model for a joint is important to mechanism simulation.

In this paper, the classic contact model for spherical joint is presented. The following four assumptions are used as basis: (i) the surfaces are continuous and non-conforming; (ii) the strains are small; (iii) each solid can be considered as an elastic half-space; (iv) the surfaces are frictionless. The non-conforming in the assumptions means that the contact width is significantly smaller than the equivalent radius. The non-conforming assumption for Hertz model is not suitable for the spherical joint with small clearance, because, in general, the contact width of the joint is not significantly smaller than equivalent radius. Therefore, the classic contact model cannot be adequately used to calculate the stiffness of spherical joints in deployable structures, because the clearance is usually small for satisfying the high accuracy of the supporting structures. Based on the geometric constraint model of revolute joint, the geometric model for spherical joint is given. The contact model for spherical joints is built based on Winkler model. The FEM for spherical joint is established to evaluate the accuracy of the proposed contact model.

CONTACT MODEL FOR SPHERICAL JOINTS

CLASSICAL CONTACT MODEL

The contact area of rotation is a circle. The four assumptions cited before exist in the classical contact model. When a normal load P is applied, two spherical solids are shown in the cross section after deformation in Fig. 1. R_1 and R_2 are the radius of the two spheres; $u_z(r)$ is the total deformation of the two spheres on the contact circle of radius r. The total deformation of the two contact solids is $\delta = \delta_1 + \delta_2 = u_z(0)$, where δ_1 and δ_2 are the deformation of the N$^{o.}$ 1 solid and the N$^{o.}$ 2 solid on the centre of the contact area.

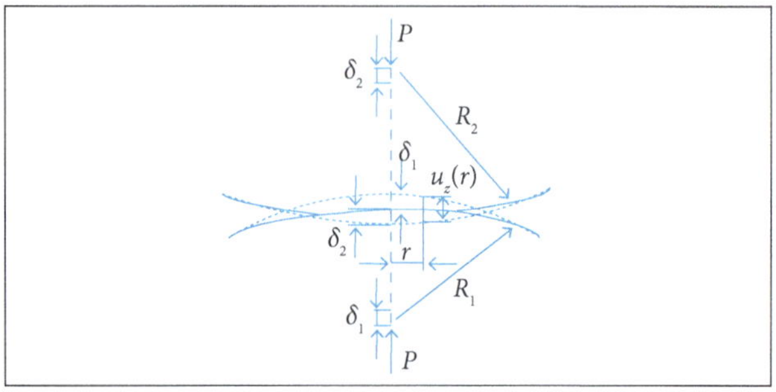

Figure 1. Hertz model for two spherical solids.

The total deformation of the two spheres on the contact circle of radius r can be expressed as:

$$u_z(r) = \delta - \frac{1}{2R} r^2 \tag{1}$$

where:

R: equivalent radius; $1/R = 1/R_1 + 1/R_2$ when the two solids are convex.

The Hertz contact force of spherical joint satisfies:

$$p(r) = p_0 \sqrt{1 - (\frac{r}{a})^2} \tag{2}$$

where:

a : the contact width at the center of contact area; p_0: contact pressure at the center of contact area.

The relationship between the contact deflection of the sphere and the contact force is presented as (Johnson, 1985):

$$\delta = (\frac{9P^2}{16RE^2})^{\frac{1}{3}} \tag{3}$$

where:

E: composite modulus $1/E = (1 - v_1^2)/E_1 + (1 - v_2^2)/E_2$; E_1 and E_2: the elastic modulus of the two contact solids; v_1 and v_1: the Poisson ratios of the two contact solids.

Based on the simplification of Eq. 3, the contact force can be given by:

$$P = \frac{4E\delta}{3} \sqrt{\delta R} \tag{4}$$

CONTACT MODEL OF SPHERICAL JOINT BASED ON THE WINKLER MODEL AND GEOMETRIC CONSTRAINTS

The Hertz contact model is rarely used to obtain high-accuracy contact stiffness of the spherical joint with small clearance because the non-conforming contact is assumed in Hertz contact model of sphere. The spherical joint and the contact model of this joint are shown in Fig. 2; r can vary from 0 to a.

The elastic half-space can be replaced with Winkler elastic base. The simplified model is shown in Fig. 3. The contact forces of the points on the contact circle of radius r can be expressed as (Johnson, 1985):

$$p(r) = \frac{E}{h} u_z(r) \tag{5}$$

where:

h: thickness of the spherical shell.

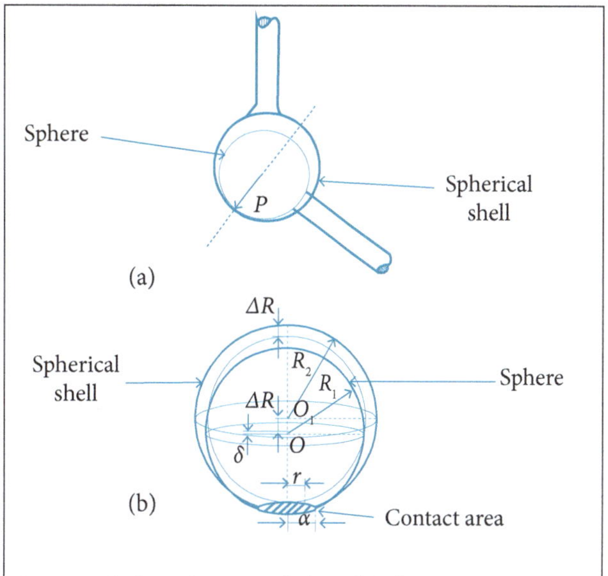

Figure 2. Contact of spherical joint. (a) Spherical joint; (b) Contact model for spherical joint.

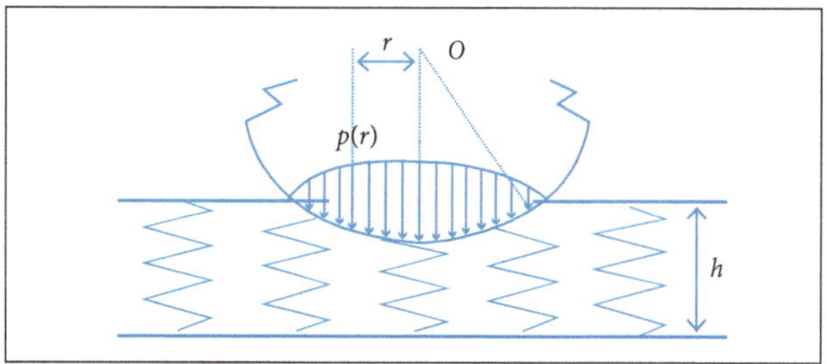

Figure 3. Simplified contact model for spherical joints with clearance.

The contact force of the central point of the contact area can be expressed as:

$$p_0 = \frac{E}{h} \delta \tag{6}$$

The contact force given by traditional Winkler model is expressed as:

$$P = \frac{\pi E \delta^2 R}{h} \tag{7}$$

The pressure distribution of the contact of revolution solids is presented in Hertz model, which is validated. Based on the distribution of the pressure given by Hertz, the total contact force can be expressed as:

$$P = \int_0^a p(r)2\pi r \mathrm{d}r = \frac{2}{3} p_0 \pi a^2 \quad (8)$$

Based on Eqs. 6 and 8, the relationship among the contact force, contact width, and the contact deflection can be given as:

$$P = \frac{2\pi E\delta}{3h} a^2 \quad (9)$$

The contact width and the contact deflection in any rotation plane plotted by dashed line in Fig. 2b satisfy the relationship between contact deflection and contact width of revolute joint described in Liu *et al.* (2013). Taking the contact width expression into Eq. 9, the relationship between contact force and contact deflection can be given by:

$$P = k_j\delta = \delta \frac{2\pi E}{3h}\left[R_2^2 - \frac{1}{4}\left(\frac{R_2^2 - R_1^2}{\delta + \Delta R} + \delta + \Delta R\right)^2\right] \quad (10)$$

VERIFICATION OF THE PROPOSED CONTACT MODEL

The FEM of spherical joint is built to verify the established contact model of spherical joint. The FEM is shown in Fig. 4, which is established by ANSYS. The type of the element is SOLID 45. The contact models of spherical joints with different clearances are established, where $R_2 = 5$ mm, $E = 2.06 \times 10^5$ Mpa and $h = 5$ mm.

The comparison between theoretical model and the FEM for spherical joint is given in Fig. 5, which plots the relationship

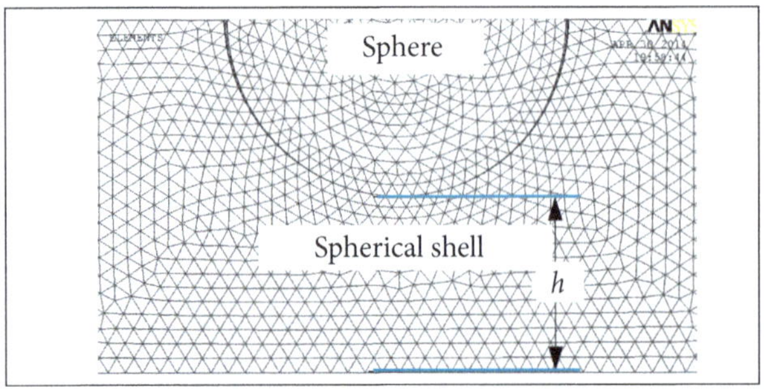

Figure 4. FEM for spherical joint.

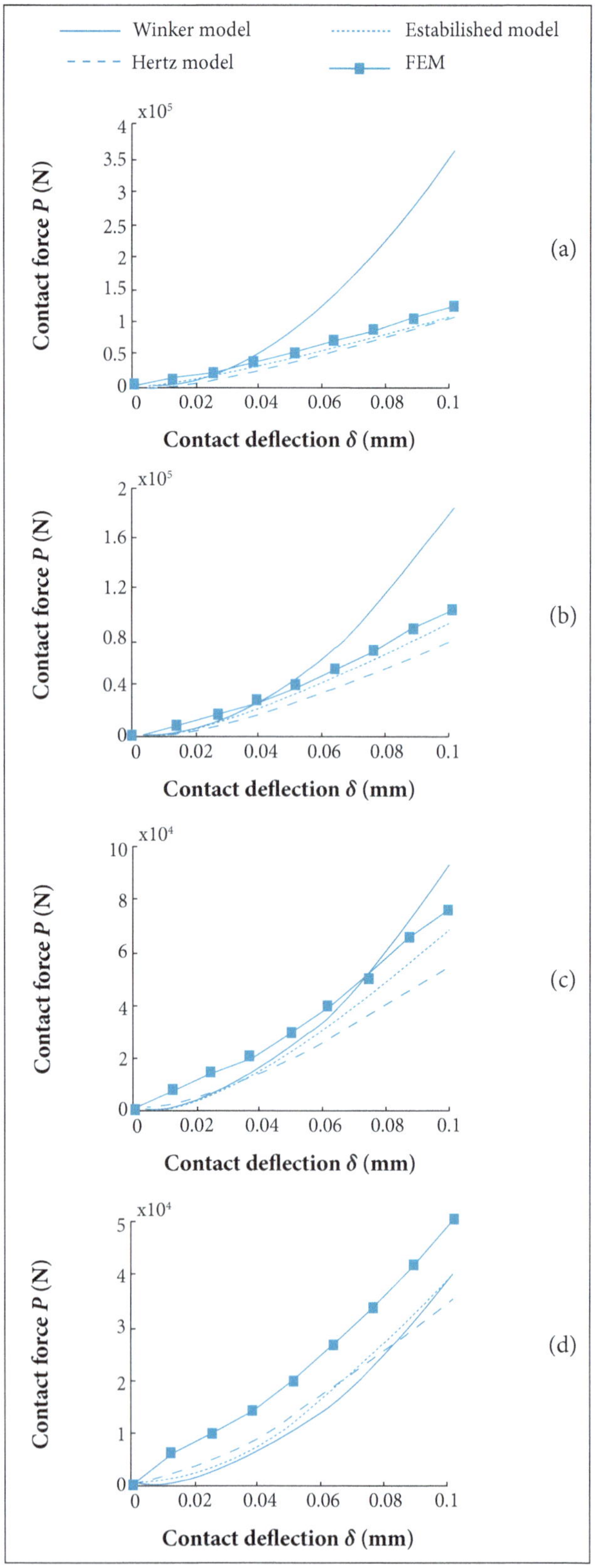

Figure 5. Comparison of the contact models for spherical joint. (a) $\Delta R = 0.05$ mm; (b) $\Delta R = 0.1$ mm; (c) $\Delta R = 0.2$ mm; (d) $\Delta R = 0.5$ mm.

between the contact force and the contact deflection of spherical joint. The stiffness of the joint decreases with the joint clearance. When the contact deflection is small, the error of the theory model is smaller than that of Hertz model. With the increase of the clearance of spherical joint, the error of the theory model in total contact deflection range is closed to that of the Hertz model. When the clearances of the spherical joints are 0.05; 0.1; 0.2 and 0.5 mm, respectively, the errors of the Hertz contact model compared with FEM are 94.1; 43.3; 29.0 and 49.5%. The errors of the Hertz contact model compared with FEM are 30.5, 35.6, 38.6, and 40.4%, respectively. Moreover, the errors of the theory model compared with FEM are 22.4, 23.1, 30.1, and 44.2%, respectively. Thus, the established spherical model is accurate in terms of calculating the stiffness of the spherical joint contact.

CONCLUSIONS

Spherical joints are investigated because of their importance in the application of deployable structures in space missions. The nonlinear characteristic of spherical joints can affect the dynamic response of deployable structures. First, the traditional contact model of spherical joints is presented. The contact model is based on four assumptions. The model is suitable for joints with large clearance. Second, a new contact model for spherical joints is modeled based on the Winkler model and geometric constraints given in literature. Third, finite element models for spherical joint with different clearances are built. The simulation results are compared with the results of traditional model and those of the new model. The error of the new contact model is close to that of the traditional model when the clearance is small. The error of the new contact model is closed to that of the traditional model when the clearance is large. Therefore, the new model can be accurate in calculating the stiffness of spherical joints with clearance.

ACKNOWLEDGEMENTS

This research is financially supported by the Fundamental Research Funds for Central Universities (Grant No. HIT. NSRIF.2015050), Self-Planned Task (No. SKLRS201401A02) of State Key Laboratory of Robotics and System (HIT), Shanxi Provincial Higher School Outstanding Innovative Team (2011), The Qualified Personnel Foundation of Taiyuan University of Technology (No. tyut-rc201448a), and The Special/Youth Foundation of Taiyuan University of Technology (No. 2014TD040).

REFERENCES

Bai, Z.F. and Zhao, Y., 2013, "A Hybrid Contact Force Model of Revolute Joint with Clearance for Planar Mechanical Systems", International Journal of Non-linear Mechanics, Vol. 48, pp. 15-36. doi: 10.1016/j.ijnonlinmec.2012.07.003

Bauchau, O.A. and Rodriguez, J., 2002, "Modeling of Joints with Clearance in Flexible Multibody Systems", International Journal of Solids and Structures, Vol. 39, No. 1, pp. 41-63. doi: 10.1016/S0020-7683(01)00186-X

Big-Alabo, A., Harrison, P. and Cartmell, M.P., 2015, "Contact Model for Elastoplastic Analysis of Half-Space Indentation by a Spherical Impactor", Computers & Structures, Vol. 151, pp. 20-29. doi: 10.1016/j.compstruc.2015.01.005

Boltachev, G.S., Volkov, N.B. and Zubarev, N.M., 2012, "Tangential Interaction of Elastic Spherical Particles in Contact", International Journal of Solids and Structures, Vol. 49, No. 15-16, pp. 2107-2114. doi: 10.1016/j.ijsolstr.2012.04.013

Brown, M.A., 2011, "A Deployable Mast for Solar Sails in the Range of 100-1000 m", Advances in Space Research, Vol. 48, No. 11, pp. 1747-1753. doi: 10.1016/j.asr.2011.01.014

Cannon, J.R. and Howell, L.L., 2005, "A Compliant Contact-Aided Revolute Joint", Mechanism and Machine Theory, Vol. 40, No. 11, pp. 1273-1293. doi: 10.1016/j.mechmachtheory.2005.01.011

Fang, X., Zhang, C.H., Chen, X., Wang, Y.S. and Tan, Y.Y., 2015, "Newly Developed Theoretical Solution and Numerical Model for Conformal Contact Pressure Distribution and Free-Edge Effect in Spherical Plain Bearings", Tribology International, Vol. 84, pp. 48-60. doi: 10.1016/j.triboint.2014.11.020

Gruber, P., Häuplik, S., Imhof, B., Özdemir, K., Waclavicek, R. and Perino, M.A., 2007, "Deployable Structures for a Human Lunar Base", Acta Astronautica, Vol. 61, No. 1-6, pp. 484-495. doi: 10.1016/j.actaastro.2007.01.055

Hattori, G. and Serpa, A.L., 2015, "Contact Stiffness Estimation in ANSYS Using Simplified Models and Artificial Neural Networks", Finite Elements in Analysis and Design, Vol. 97, pp. 43-53. doi: 10.1016/j.finel.2015.01.003

Johnson, K.L., 1985, "Contact Mechanics", Cambridge University Press, Cambridge, UK.

Lankarani, H.M. and Nikravesh, P.E., 1990, "A Contact Force Model with Hysteresis Damping for Impact Analysis of Multibody Systems", Journal of Mechanical Design, Vol. 112, No. 3, pp. 369-376. doi: 10.1115/1.2912617

Li, T.J., Guo, J. and Cao, Y.Y., 2011, "Dynamic Characteristics Analysis of Deployable Space Structures Considering Joint Clearance", Acta Astronautica, Vol. 68, No. 7-8, pp. 974-983. doi: 10.1016/j.actaastro.2010.08.039

Liu, C.S., Zhang, K. and Yang, L., 2006, "Normal Force-Displacement Relationship of Spherical Joints with Clearances", Journal of Computational and Nonlinear Dynamics, Vol. 1, No. 2, pp. 160-167. doi: 10.1115/1.2162872

Liu, C.S., Zhang, K. and Yang, R., 2007, "The FEM Analysis and Approximate Model for Cylindrical Joints with Clearances", Mechanism and Machine Theory, Vol. 42, No. 2, pp. 183-197. doi: 10.1016/j.mechmachtheory.2006.02.006

Liu, R.Q., Zhang, J., Guo, H.W. and Deng, Z.Q., 2013, "Nonlinear Characteristics of Revolute Joints with Clearance", Journal of Aerospace Technology and Management, Vol. 5, No. 3, pp. 341-348. doi: 10.5028/jatm.v5i3.236

Pereira, C., Ramalho, A. and Ambrosio, J., 2014, "Applicability Domain of Internal Cylindrical Contact Force Models", Mechanism and Machine Theory, Vol. 78, pp. 141-157. doi: 10.1016/j.mechmachtheory.2014.03.010

Puig, L., Barton, A. and Rando, N., 2010, "A Review on Large Deployable Structures for Astrophysics Missions", Acta Astronautica, Vol. 67, No. 1-2, pp. 12-26. doi: 10.1016/j.actaastro.2010.02.021

Qi, Z.H., Xu, Y.S., Luo, X.M. and Yao, S.J., 2010, "Recursive Formulations for Multibody Systems with Frictional Joints Based on the Interaction Between Bodies", Multibody System Dynamics, Vol. 24, No. 2, pp. 133-166. doi: 10.1007/s11044-010-9213-z

Rajaei, M. and Ahmadian, H., 2014, "Development of Generalized Iwan Model to Simulate Frictional Contacts with Variable Normal Loads", Applied Mathematical Modelling, Vol. 38, No. 15-16, pp. 4006-4018. doi: 10.1016/j.apm.2014.01.008

Sarkar, S., Venkatraman, K. and Dattaguru, B., 2004, "Dynamics of Flexible Structures with Nonlinear Joints", Journal of Vibration and Acoustics, Vol. 126. No. 1. pp. 92-100. doi: 10.1115/1.1596548

Shin, Y.S., and Lee, I., 2000, "Investigation of Equivalent System Modeling and Dynamic Characteristics Using Reduced Models", AIAA Journal, Vol. 38, No. 1, pp. 102-109. doi: 10.2514/2.928

Tian, Q., Zhang, Y.Q., Chen, L.P. and Flores, P., 2009, "Dynamics of Spatial Flexible Multibody Systems with Clearance and Lubricated Spherical Joints", Computers & Structures, Vol. 87, No. 13-14, pp. 913-929. doi: 10.1016/j.compstruc.2009.03.006

13

Application of Artificial Neural Networks for Fog Forecast

Rosângela de Oliveira Colabone[1], Antonio Luiz Ferrari[1], Francisco Arthur da Silva Vecchia[2], Adriano Rogério Bruno Tech[3]

ABSTRACT: This study examines the development of a system that assists in planning flight activities of the Academia da Força Aérea (AFA) so that meteorological data can be used to predict the occurrence of fog. This system was developed in MATLAB 8.0 by applying multilayer perceptron-type artificial neural networks and using an error correction algorithm called backpropagation. The methodology used to implement the network comprises eight input variables, five neurons in the intermediary layer, and one neuron in the output layer, which corresponds to the presence or absence of fog. The fog phenomenon is very important for the study and definition of flight strategic planning. Data taken from 1989 to 2008 and related to the input variables were used for the training and validation of the proposed network. Consequently, the multilayer perceptron network has a 95% reliability compared with the data collected. This high level of reliability is an exceptional result for the management, planning, and decision making team of the AFA strategic group. Thus, it can be concluded that the proposed system is efficient and will subsidize, with good safety margin, AFA's flight activity planning and could also be applied to other air activities in Brazil.

KEYWORDS: Strategic planning, Operational management, Intelligent systems, Decision support systems.

INTRODUCTION

One of the concerns of the scientific community and certain sectors of society in recent years with respect to meteorological factors and their consequences in relation to all the social, environmental and economic implications. This leads companies and institutions to develop strategies in order to predict future situations to mitigate possible adverse situations.

Low visibility in bad weather conditions (even in fog) may adversely affect society. Fog affects air, sea, railway, and road transport systems, requiring specific safety measures to prevent accidents, delays, and even cancellation in transport operations (Gultepe *et al.*, 2007).

Fog is formed due to condensation of water vapor into liquid droplets or ice crystals as a result of air cooling, humidification, and/or by mixing contrasting air portions. A most common form of fog occurs when its formation on land involves nocturnal radiative cooling under low wind conditions (Roach, 1995) when dissipation usually occurs a few hours after sunrise because of the heat flux on a surface heated by solar radiation.

Fog has significant importance in the airport sector, taking into account the considerable restriction of landing and takeoff, affecting aircraft and passengers' safety. The forecast of weather events is becoming increasingly vital in various sectors of society, such as in agriculture, construction, economics, public health, transportation, especially in aviation, as it is susceptible to weather conditions, which may cause delays and flight cancellations or even air accidents or incidents due to the adverse conditions, indicating the need for confronting possible problems. Although the occurrence of fog is inevitable, airport and road authorities specifically require a more reliable

1.Academia da Força Aérea – Pirassununga/SP – Brazil. 2.Universidade de São Paulo – Escola de Engenharia de São Carlos – São Carlos/SP – Brazil. 3.Universidade de São Paulo – Faculdade de Zootecnia e Engenharia de Alimentos – Pirassununga/SP – Brazil.
Author for correspondence: Rosângela de Oliveira Colabone | Rua 11, N. 3.015 | CEP: 13.504-102 – Rio Claro/SP – Brazil | Email: rocolabone@uol.com.br

forecast. The interest in studying the meteorological variables in airfields, especially the fog phenomenon, is in the need to obtain, in advance, information on weather conditions at the local scale, for changes in the planning of flight activities if necessary.

The concepts of the meteorological variables presented herein can be found in the Manual of Surface Weather Station (MCA 105-2; Brasil, 2013). For the Departamento de Controle do Espaço Aéreo, fog formation is characterized when the prevailing visibility is reduced to below 1,000 m.

The restriction of visibility due to fog formation occurs sharply at Academia da Força Aérea (AFA) aerodrome, in the municipality of Pirassununga, São Paulo State, as shown in Fig. 1. The occurrence of fog causes cancellations of flights, leading to delays and changes in the training schedule.

Several studies have been conducted to establish a pattern of the phenomena occurrence, with the objective of short-term forecasting, which, theoretically, will allow the reduction of operating costs and time (Rezende, 2005). Tarmeño and Altamirano (2004) performed a climate study of fog events in Jorge Chavez Callao International Airport in Lima, Peru. The analysis of a time series of 30 years of climatological data (1968 – 1997) recorded 329 events, with the highest monthly frequency of fog between March and June, and the lowest frequency in November, which allows managers to optimize and perform, in short and medium terms, contingency plans that will enable restructuring and, consequently, a significant reduction of budget funds, as well as a greater respect to target audience or users of services offered by institutions.

Photos: Rosângela de Oliveira Colabone.

Figure 1. Illustrative photos of reduction visibility due to fog occurrence on the AFA airfield – Pirassununga/SP.

Tardif and Rasmussen (2007) investigated fog events in New York City, United States of America, using a 20-year time series data. Surface hourly data were used to identify fog events in 17 locations under the influence of various physiographic characteristics such as the type of land cover (urban, suburban, and rural) and contrasts between water and land.

The events in each site were ranked according to an algorithm based on the fog formation process. The results showed that the probability of fog occurrence is negatively influenced by the presence of New York City urban heat islands, reinforced in places on direct influence of sea air.

In Brazil, França (2008), using the numerical weather forecast model Eta in 4-km resolution, evaluated the fog forecasting methodology and horizontal visibility from the National Centers for Environmental Prediction analysis and Centro de Previsão de Tempo e Estudos Climáticos/Eta model predictions of 40 km. Two methods for estimating horizontal profile were evaluated: Kunkel (1984) and Gultepe *et al.* (2006) methods. Test cases were chosen for the evaluation depending on the season when fog occurs.

Integrated experiments on Afonso Pena airport, in Curitiba, Paraná State; Guarulhos International airport, in Guarulhos, São Paulo State; and Salgado Filho airport, in Porto Alegre, Rio Grande do Sul State, suggested that the 4-km Eta model may be used as a tool in predicting fog occurrences and horizontal visibility or even in regions with probability of fog within 36 to 48 h in advance.

Currently, artificial neural networks (ANNs) have increasing applications as a tool for prediction of events owing to its competence to simulate the human brain's ability to recognize, associate, and generalize patterns (Russel and Norvig, 2013). A neural network can approximate functional relationships, particularly when relationships are not well-defined and/or are not linear, which makes it difficult to use conventional methods to predict future variations of these relationships (Tafner *et al.*, 1996).

The ANNs have, as a main point, the interaction and manipulation of input data in order to find a satisfactory response within a reliability limit for an expected result. The ANNs are networks trained through entries obtained from external or internal scenarios in the system, and these entries are multiplied by randomly assigned weights. The process takes place throughout the network until an expected error within the proposed limit is reached; otherwise, the process continues to perform new calculations through an error correction of input data called backpropagation (Fig. 2) (Tafner *et al.*, 1996).

After the required value is obtained, it is tested with known data; it is common to use 2/3 of data for training and 1/3 for network validation, with reliability verification, which may provide parameters for managers so they are able to make decisions about future information (Tafner *et al.*, 1996). This allows executives and their teams to analyze future data and propose actions to minimize errors and increase the chances of success, as input data are based on historical data, which served as network training so that one can even perform simulations of future events (Rezende, 2005).

Time series are nothing more than a sequence of measurements for a given event, chronologically organized, which, apparently, does not follow any law or trend. However, in time series related to natural phenomena, one may note certain characteristics that are repeated after a certain period of time (seasonality) and others that maintain itself during the considered range (trends), even without obeying linear patterns.

According to Fabbian *et al.* (2007), the occurrence of fog in aerodromes may negatively impact the air transport operations, thus influencing significantly the overhead activities during this period, with respect to flight safety.

The authors also used ANNs to provide predictions about future events that could occur in Canberra International Airport (YSCB) and observed that, contrary to conventional statistical techniques, ANNs are better suited to solve problems that involve nonlinear complex interactions and thus have a great potential for the fog forecast, which characterizes the importance of developing models that can help those responsible for planning the flight instruction activities to be undertaken by learners in the learning phase and training.

According to Chaudhuri *et al.* (2015), the reduced visibility during fog significantly influences the surface and air transport operations. The fog forecast at aerodromes is a complex and difficult task, despite constant improvements in numerical weather prediction models and, to improve the model, the authors used the neural networks of the type Multilayer Perceptron (MLP) and observed that the forecasting error decreases, thereby improving the learning model in predicting possible fog, which corroborates the use of RNAs for future prediction airfields.

This study proposes a method to forecast fog occurrence through ANNs, and temporal series of meteorological variables related to 19 years of data is used for training a multilayer perceptron neural network with a backpropagation error correction method, focusing on fog forecast in the Brazilian AFA.

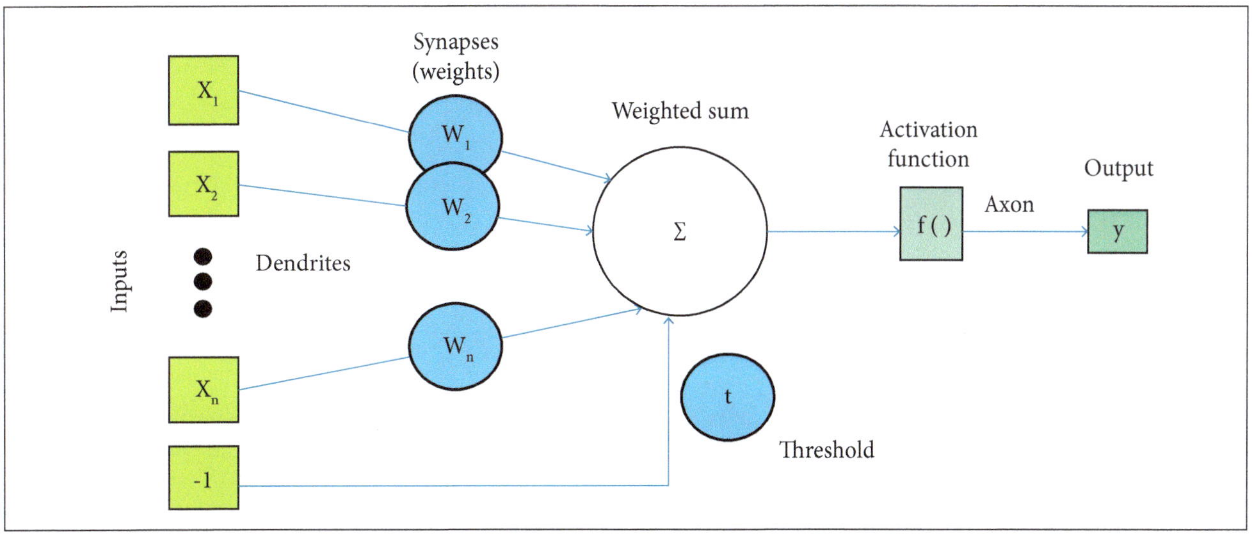

Figure 2. Model of an artificial neuron.

In the AFA, the students of the airmen official board receive, in addition to academic training, flight instruction in two aircrafts, T-25 "Universal", in the basic flight education phase, and T-27 "Tucano", a turboprop type, during the advanced phase to improve and prepare the future official airmen in their end mission. AFA has an airfield called Campo Fontenelle (SBYS encoding) for the operation of its mission. This airfield has two lane sectors, with an average of 160 landings and takeoffs each day.

The process of training or flight instruction, as it is called in AFA, depends on the weather conditions. Thus, severe weather conditions, such as heavy rainfall and incidence of winds and fog formation, are limiting factors, since they may cause human and material risks to the institution.

MATERIAL AND METHODS

The AFA located in the town of Pirassununga, São Paulo State, within the coordinates 21°56'04" to 22°00'29" south latitude and 47°17'16" to 47°22'07" west longitude, is a higher education institution subordinated to the Departamento de Ensino da Aeronáutica, whose mission is to train air force officers.

AFA has a surface weather station on its premises that is responsible for monitoring the (local) weather condition. It is classified as Centro Meteorológico Militar to provide specific meteorological support to military aviation according to the Instituto de Controle do Espaço Aéreo (ICEA 105-2; Brasil, 2013).

The military flight instructions take place, preferably, during the day, starting at 5 a.m. and ending at 11 p.m. By the fact that flight activities are carried out in this period, weather data, including cloud cover, fog, and wind direction, are important variables for the strategic management of the missions (planning).

The data for training and testing of the ANNs were provided by the ICEA, located in São José dos Campos, São Paulo State. Regarding precipitation, there was a need for data collection directly through the forms available at ICEA, in São José dos Campos.

A total of 7,305 forms were assessed to obtain the cumulative sum (24 h) of rainfall, corresponding to what was verified in the AFA airfield. Fog data were similarly evaluated, as records contained only codes that had to be decoded using a table reference.

The fog records were obtained by visual estimates through meteorological observations made by the military, executed at fixed intervals (every hour) and complemented with intermediate observations. In case of significant changes to aviation, depending on the activity, the collection of other variables is considered, such as wind, visibility, visual range, cloud cover, air and dew point temperature, sea level pressure, and weather in the flight track.

Data related to fog, encoded in the range 40 – 49, provided information on fog conditions. The data or codes, as provided in Table 4655 of MCA 105-10 (Brasil, 2011), provided information regarding the fog. The data were also provided by the ICEA and had to be decoded for interpretation.

MATLAB 8.0 was used to develop the system, where the data were processed and the network was built for training, validation, and testing of the proposed model.

The following control variables were chosen as inputs for network training: year, month, day, time, temperature, relative humidity, pressure, and wind speed. These inputs are identified as X_i, and the network output was stipulated as 0 or 1 for fog, where 0 means absence and 1 means presence of fog.

In the network training, the sigmoid function was used, with five neurons in the intermediary layer and one neuron in the output layer, indicating the presence or absence of fog at the AFA airfield. To correct the network error as a function of X_i input and Y_i output, the backpropagation algorithm was used.

According to Russel and Norvig (2013), Eq. 1 represents the input characteristics that will be processed by the network for its training, where ω is the synaptic weights and X_i represents the network entries.

$$\sigma = \omega_0 X_0 + \omega_1 X_1 + \omega_2 X_2 + \omega_3 X_3 + \ldots + \omega_i X_i \tag{1}$$

The sigmoid function is represented by Eq. 2:

$$Y = 1 \frac{1}{(1+e^{-\sigma})} \tag{2}$$

Equation 2 is used as a neuron activation function that will generate an activation value (σ) on the basis of the sum of the inputs X_i multiplied by synaptic weights ω_i (Eq. 1); according to the generated response, this will be used in Eq. 2, where it is possible to obtain the output (Y) that, given by the network, will be compared with the expected response (historical data).

If the results are not the same, it will be reprocessed by error backpropagation depending on Eqs. 3 to 9.

$$O_j = X_j \quad (3)$$

where O and X represent the network entries.

Equation 4 represents the calculation of X:

$$X_j = \sum_i w_{ij} O_i + \theta_j \quad (4)$$

where w denotes the synaptic weights and θ represents the bias.

Equation 5 represents the sigmoid function:

$$O_j = f(X_j) = \frac{1}{1 + e^{-X_j}} \quad (5)$$

where e represents the natural logarithm.

Equation 6 corresponds to error correction of the last or output layer:

$$e_j = O_j (1 - O_j)(Y - O_j) \quad (6)$$

where Y denotes the network output.

Equation 7 corresponds to correction of the hidden or intermediate layers:

$$e_j = O_j (1 - O_j) \sum_k w_{jk} e_k \quad (7)$$

Equations 8 and 9, respectively, denote adjustments of synaptic weights and bias:

$$w_{ij}^{(k)} = w_{ij}^{(k-1)} + n e_j^{(k)} O_j^{(k)} \quad (8)$$

$$\theta_j^{(k)} = \theta_j^{(k-1)} + n e_j^{(k)} \quad (9)$$

where n denotes the neural network learning coefficient.

For the beginning of the training, random synaptic weights were generated and 2/3 of the acquired data were used. After obtaining the smallest or acceptable error, the network was subjected to tests with the remaining 1/3 of the data, corresponding to the period 1989 – 2008, *i.e.* a relatively significant amount of data was used for training and validation of the model.

RESULTS AND DISCUSSION

After the training and the determination of the results, the proposed network achieved a reliability of 95.949% with 2/3 of the data. In other words, the network with input data X_i and output Y_i achieved a satisfactory result by undergoing the network validation test. Using the remaining 1/3 of the data, the network achieved a reliability of 98.107%.

These results demonstrate that the network was capable of learning and the test results were satisfactory, which provides for the executive managers, in this case, the military responsible for the planning of flight activities, an effective tool, enabling them to develop a strategic plan for the following years, with significant and reliable projections on the basis of historical data of approximately 95% reliability. It significantly decreases future changes in plans related to operational cost and time as well as human resources.

Nevertheless, the information provided by the network has the greater purpose of improving planning and designing of campaign activities of cadet airmen, since they also conduct educational activities with other areas of expertise. So there is no loss to the learning process but an advantage to flight and other academic activities.

The result of training can be seen in Fig. 3. The dotted line represents the expected outcome Y_i and the blue line, the results obtained by the network in training, indicating an equation in relation to the obtained results.

Figure 4 shows the network validation after the training was conducted and finalized, with error matrices and synaptic

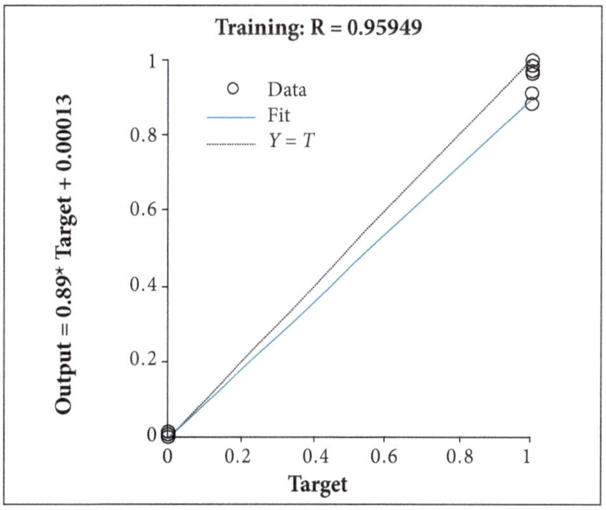

Figure 3. Training of multilayer perceptron network with error correction backpropagation algorithm.

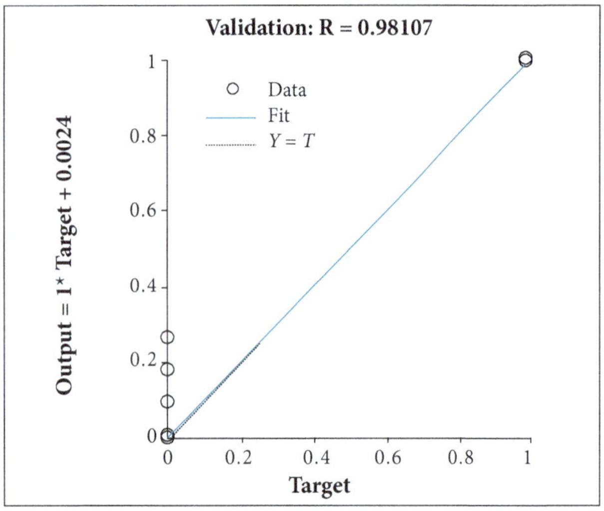

Figure 4. Network validation with 1/3 of the remaining data for the 1989 - 2008 period.

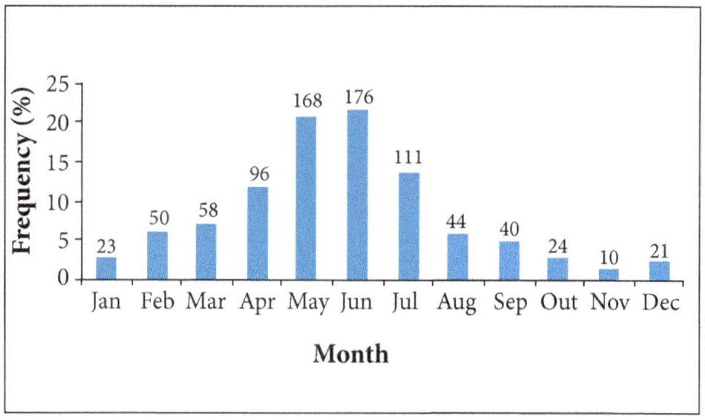

Figure 5. Relative and absolute frequencies of fog occurrence at the AFA airfield in the 1989 - 2008 period.

weights adjusted and ready to infer possible future projections. This will help managers (senior military) to develop a more structured and proper planning. Adjustments could be possible if based on recent data, *i.e.* from the last two or three years.

The data obtained by the network were compared with the statistical data of the same period for comparison and analysis of network performance. The tests performed with the network show values very close to the statistical ones, with the great advantage of agility and speed of the network, since it had already been trained and prepared for this analysis.

The statistical data can be seen in Fig. 5, which are similar to the values obtained by the proposed network. The figure shows the relative and absolute frequencies with respect to fog occurrence at the AFA airfield in the period 1989 - 2008.

From a total of 821 cases of fog occurrence in the whole period, 67% occurred from April to July, with a predominance in May and June. The lower frequencies were observed from October to January, with a total occurrence of 9.5%; and in February, March, August, and September, the occurrence was around 23.4%.

CONCLUSION

The results showed that the use of multilayer perceptron-type ANNs with error correction using the backpropagation algorithm was efficient and fast, providing data to executive managers for better planning, with a reduced error margin, considering that the system was fed with data of 20 years. This favored the network learning process with a 95% percentage of success, which is considered suitable for the use of ANNs.

Thus, the system proposed for the planning and assistance to military officers tends to enhance planning, allowing a more efficient decision-making process, which will greatly benefit the Academia da Força Aérea, in regard to reduction of operational work, reduction of costs in terms of time dedicated to the personnel and possible flight adjustments as well as academic activities related to cadet airmen.

The model was based on past meteorological data. If the data continues in the same pattern, the degree of reliability will remain the same. If there are sharp changes in the input variables, the network provides a new level of reliability, since the previous state has changed.

REFERENCES

Brasil, 2011, Comando da Aeronáutica. Departamento de Controle do Espaço Aéreo, "Classificação dos Órgãos Operacionais de Meteorologia Aeronáutica – ICA 105-2", Comando da Aeronáutica, Rio de Janeiro, Brazil.

Brasil, 2013, Comando da Aeronáutica. Departamento de Controle do Espaço Aéreo, "Manual de Estações Meteorológicas de Superfície – MCA 105-2", Comando da Aeronáutica, Rio de Janeiro, Brazil.

Chaudhuri, S., Sarkar, I., Das, D. and Goswami, S., 2015, "Multilayer Perceptron Model for Nowcasting Visibility from Surface Observations: Results and Sensitivity to Dissimilar Station Altitudes", Pure and Applied Geophysics. doi: 10.1007/s00024-015-1065-2, Retrieved in May 13, 2015, from http://link.springer.com/article/10.1007%2Fs00024-015-1065-2#page-2

Fabbian, D., de Dear, R. and Lellyett, S., 2007, "Application of Artificial Neural Network Forecasts to Predict Fog at Canberra International Airport", Weather and Forecasting, Vol. 22, No. 2, pp. 372-381. doi: 10.1175/WAF980.1

França, V.D.J., 2008, "Avaliação da Metodologia de Previsão de Nevoeiro e Visibilidade Horizontal do Modelo Eta", Master Thesis, Instituto Nacional de Pesquisas Espaciais, São José dos Campos, Brazil.

Gultepe, I., Milbrandt, J. and Belair, S., 2006, "Visibility Parameterization from Microphysical Observations for Warm Fog Conditions and its Application to the Canadian MC2 Model", Proceedings of the Conference on Aviation, Range and Aerospace Meteorology, Atlanta, USA.

Gultepe, I., Tardif, R., Michaelides, S.C., Cermak, J., Bott, A., Bendix, J., Müller, M.D., Pagowski, M., Hansen, B., Ellrod, G., Jacobs, W., Toth, G. and Cober, S.G., 2007, "Fog Research: a Review of Past Achievements and Future Perspectives", Pure and Applied Geophysics, Vol. 164, pp. 1121-1159. doi: 10.1007/s00024-007-0211-x

Kunkel, B., 1984, "Parameterization of Droplet Terminal Velocity and Extinction Coefficient in Fog Models", Journal of Applied Meteorology, Vol. 23, No. 1, pp. 34-41. doi: 10.1175/1520-0450(1984)023<0034:PODTVA>2.0.CO;2

Rezende, D.A., 2005, "Sistemas de Informações Organizacionais: Guia Prático para Projetos em Cursos de Administração, Contabilidade e Informática", Atlas, São Paulo, Brazil.

Roach, W.T., 1995, "Back to Basics: Fog: Part 2 – The Formation and Dissipation of Land Fog", Weather, Vol. 50, No. 1, pp. 7-11. doi: 10.1002/j.1477-8696.1995.tb06053.x

Russel, S. and Norvig, P., 2013, "Inteligência Artificial", Third Edition, Campus, Rio de Janeiro, Brazil. .

Tafner, M.A., Xerez, M. and Rodrigues Filho, W., 1996, "Redes Neurais: Introdução e Princípios de Neurocomputação", FURB, Blumenau, Brazil.

Tardif, R. and Rasmussen, R.M., 2007, "Event-Based Climatology and Typology of Fog in the New York City Region", Journal of Applied Meteorology and Climatology, Vol. 46, No. 8, p. 1141-1168. doi: 10.1175/JAM2516.1

Tarmeño, E.C. and Altamirano, R.A., 2004, "Estudio Climático de las Nieblas en el Aeropuerto Internacional 'Jorge Chavez', Callao, Lima-Peru", Anales Científicos, No. 56, pp. 51-63, Retrieved in April 20, 2014, from http://www.lamolina.edu.pe/Investigacion/web/anales/pdf_anales/Vol.LVI2003_Arch.5-6%20.pdf

Fins Module Conception of the Microsatellite Launch Vehicle Based on Design for Manufacture and Assembly Method

Jéssica Garcia de Azevedo[1], Artur Cristiano Arantes Filho[1], Luis Eduardo Vergueiro Loures da Costa[1]

ABSTRACT: This paper presents the application of the Design for Manufacture and Assembly method (DFMA) during the conception of a space product. This approach enables to capture manufacture and assembly requirements in early phases of development. DFMA guides the manufacture processes and materials that rule the Fins Module interface to the Microsatellite Launch Vehicle (VLM-1) main structure. The results of this process are the requirements captured in the beginning of the project development, allowing avoiding problems that generally occur in the manufacture stage and in assembly, integration and test (AIT) phase. Rework, material loss, wasting time to assembly and integration, over engineering and unnecessary engineering are problems that the method intends to reduce in the physical implementation phase.

KEYWORDS: Design for manufacture, Design for assembly, Concurrent engineering.

INTRODUCTION

The Microsatellite Launch Vehicle (VLM-1) is a project under development at *Instituto de Aeronáutica e Espaço* (IAE) in Brazil and aims at carrying small satellites and small payloads to space. Its development was stated as a cost-driven process which involves the important aspects in anticipation of the requirements to the initial stages of the development (Fulindi, 2011). The VLM-1 is a three-stage solid propelled rocket, which provides the necessary energy to reach low equatorial orbits. This project is developed in cooperation with the National Aeronautics and Space Research Centre of the Federal Republic of Germany (DLR), and the first flight of the VLM-1 will be able to put in low orbit the German payload called Shefex III.

In space projects, different fields of research and engineering, as materials, mechanics, electronics, control and others, are involved. In each conception model of the project, analyses of the functional aspects, with materials to be adopted, mechanical stresses, equipment assembly and preliminary actions are taken in order to integrate such a multifunctional group. This group aims at performing the integrated product team. From the Design for Manufacture and Assembly method (DFMA) analysis, more efficient actions are taken to anticipate requirements for the beginning of the product development. This leads to reduced problems in the assembly, better ideas to prepare the manufacture process, and the decision on materials is taken in the same time when the process is under analysis. Furthermore, all impacts are analyzed in the conception stage of the project. In this work, DFMA is applied aiming at developing the VLM-1 Fins Module.

1. Instituto de Aeronáutica e Espaço – São José dos Campos/SP – Brazil.

Author for correspondence: Artur Cristiano Arantes Filho | Instituto de Aeronáutica e Espaço | Praça Marechal Eduardo Gomes, 50 – Vila das Acácias CEP: 12.220-098 – São José dos Campos/SP – Brazil | E-mail: artur.caf7@gmail.com

Based on DFMA arising from Concurrent Engineering methodology, it was possible to achieve a balanced solution for the VLM-1 Fins Module, which considered organizational aspects, manufacture techniques and assembly strategies. As the main result, one can outstand a Fins Module conceptual design, considering real needs from stakeholders and adapted for the Brazilian technological scenario.

PRODUCT DEVELOPMENT PROCESS AND CONCURRENT ENGINEERING

Product Development Process (PDP) consists of applying a set of activities aiming at reaching the design specifications and its production process. So that, manufacture is able to produce it, from the needs of the market, possibilities and technological constraints, considering the competitive and company's product strategy.

On the other side, as stated by Huang (1996), Concurrent Engineering (CE) improves quality, reduces costs, compresses cycle times, increases flexibility, raises productivity and efficiency, and improves the social image, making an ideal environment for product development.

The practice of CE has applied methodologies during the product development aiming at integrating engineering tools in order to help in decision-making, such as Design for X (DFX), DFMA (Design for Manufacture and Assembly), Computer Aided Design (CAD), Computer Aided Engineering (CAE), Computer Aided Manufacturing (CAM), among others.

Design for X

DFX is one of the most effective approaches to implement CE. The called X can be quality, manufacture, assembly, cost, among other features related to the product life cycle process (Huang, 1996).

DFX has become a powerful tool for project analysis of products, promoting the practice of CE in a prediction error process of subsequent processes. In other words, DFX assists in raising the largest number of project related information, in view of facilitating decisions in the preliminary stages, which generally have less information than in future phases. Since the 1960s, efforts have been made in the study of conceptual and detailed design of products, aiming at manufacture and assembly analysis. Among the various methodologies that compose the DFX, there are some that are already being more strongly applied, having a wider literature and well-structured implementation methods, such as Design for Manufacture and Design for Assembly methodology (Iwaya, 2010).

Design for Manufacture

In Design for Manufacture, or DFM, the product is developed for production, aiming at facilitating and reducing the production cost.

According to Xie (2003), DFM is the method that considers the limitations related to manufacture in the initial phase of the project. Thus, the project engineer can make the selection between different materials and technologies in order to estimate the manufacture time and the product cost quantitatively and to quickly choose between different project plans. The project team compares all types of project plans and technology, and revisions are carried out in the initial phase of the project according to the feedback information. Therefore, the most satisfactory configuration for the product can be determined. There are three objectives in DFM:

- Improving the quality of new products during the development period, including project, technology, manufacture, service and so on.
- Decreasing the cost, including the design cost, technology, manufacture, delivery, technical support, disposal, and so on.
- Shortening the development cycle, including project time, preparation for manufacture, and calculation of repeatability.

Currently, DFM is widely applied with CAD/CAM systems, where the designer can extract the features of a modeled part in order to predict spending and determine manufacture processes. This allows project modifications for the choice of the best product engineering solutions.

Design for Assembly

As stated by Redford (1983), through Design for Assembly (DFA) application in a development early phase, a significantly cost reduction can be achieved during the final integration and assembly stage. The mentioned cost reduction could be attained accomplishing the project of parts that can be easily handled and products that can be easily assembled manually or by machine. In general, improvements can be typically achieved through DFA implementation, resulting in cost reduction. This fact occurs because there are not ideal attention in assembly strategies and time spending to execute this step of development.

According to Boothroyd *et al.* (1994), in order to provide guidance to the designer in reducing the number of parts, the DFA methodology provides three criteria by which each part should be examined as well as how it is added to the product during assembly:

- During the operation of the product, the part moves in relation to all other parts already assembled? Only overall motion should be considered — small movements that can be accommodated by integral elastic elements, for example, are not sufficient for a positive response.
- The part must be of a different material or be isolated from other parts already assembled? Only fundamental reasons related to the material properties are acceptable.
- A piece must be separated from all other parts already assembled because, otherwise, required assembly or disassembly of other separate parts would be impossible.

Design for Manufacture and Assembly

Huang (1996) defines a better alternative in a project by the optimization of the manufacture and assembly process, applying DFMA approach. This approach is a systematic analysis used in the preliminary phases or in the project review, when problems of product manufacture and assembly appear and when a product is losing its market.

This approach is the basis for CE studies, providing guidance to the design team, simplifying the product manufacture and assembly, reducing costs — as a benchmarking tool that aims at studying competitors — and quantifying manufacture and assembly difficulties, as a should-cost tool, helping negotiate suppliers' contracts.

As stated by Boothroyd *et al.* (1994), "to manufacture" refers to the manufacture of the individual component parts of a product or assembly and "to assemble" refers to the addition or joining of parts to form the completed product. Assembly will not be considered a manufacture process in the same sense that machining, molding etc., which are manufacture processes. Hence, the term Design for Assembly (or DFA) means the design for ease of assembly. Thus, Design for Manufacture and Assembly (DFMA) is a combination of DFA and DFM.

The DFMA concerns to anticipate the assembly and integration requirements for the development phase, reducing future problems in the project physical implementation.

METHODOLOGY

The methodology applied in this work is based on references, tools and approaches used in the VLM-1 Project. It is characterized as a qualitative research, according to Silva and Menezes (2001); the interpretation of the phenomena and the attribution of the meanings are basics in the qualitative research process. It does not require the use of methods and statistical techniques. The process and its meaning are the principal focus of the approach.

In order to collect and understand the methods and tools, a line of reasoning considering a framework for the execution of the project can be built, where they are used, starting from the highest to the lowest level. As a result, one has a broader approach as it reaches the final objective (Fig. 1). This approach aims at involving development scenarios to obtain the best design solution.

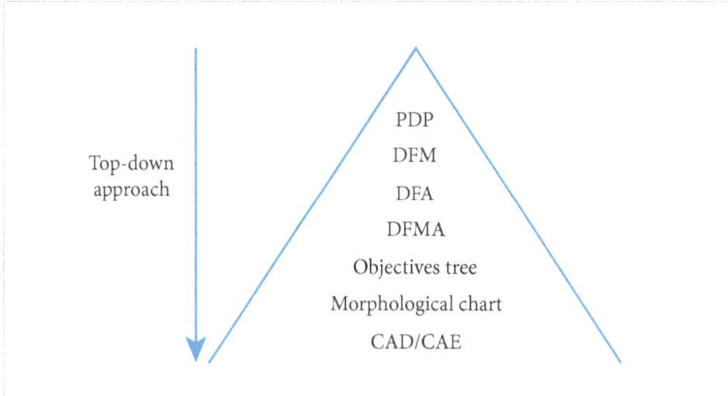

Figure 1. Framework for project of fins module.

The Product Development Process (PDP) model considers, from the beginning, the organization requirements, which, together with the product requirements, guide the process of product development (Fulindi, 2011). Following this model, there is the flow chart shown in Fig. 2. The stages of integrated development are presented, showing the interaction that the DFMA approach provides between the design phases.

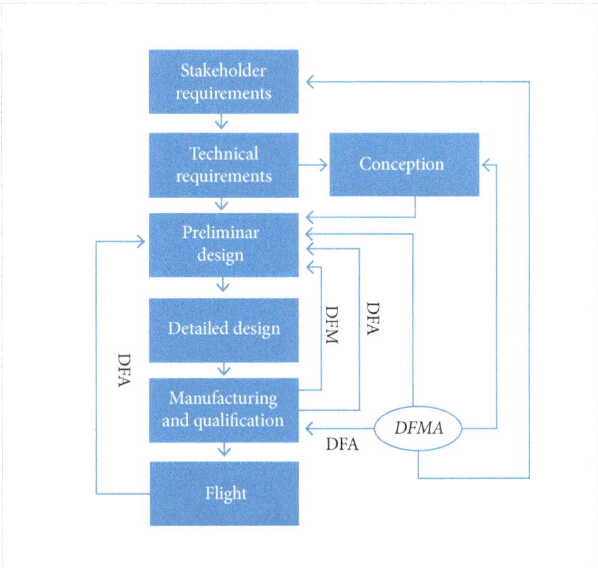

Figure 2. Project flowchart of Design for Manufacture and Assembly method.

From the stakeholders' requirements, the technical requirements that will feed the product conception are defined. The preliminary design is developed based on the requirements anticipation of the manufacture and qualification phase through the DFA and DFM analysis tools, in which the physical elements that will compose the final assembly can be identified as well as how they will be performed. For the processes used in manufacture, materials and types of tools for the visualization and analysis of these elements are used, with CAD and CAE software to validate the concepts generated from the requirements. In this phase, the requirements to the module integration in the vehicle main structure and to aid the launch pad are identified. This analysis looks to the type of the module fixation with the bottom vehicle in the launch pad. In the same time, the question "how to integrate?" the equipment distributed in the vehicle and distributed in the integration tower is analyzed. Though DFA analysis, it is possible to detail and planning all the integration process of the vehicle to the flight phase.

For the VLM-1 Project, a cluster of four fins coupled to a cylindrical membrane structure is required, denominated here Fins Module. The types of manufacture studied attending the project are: carbon, aluminum, fiberglass, and sandwich type in core and aluminum, or core and foam.

The Fins Module also integrates the fins in the vehicle body and aids to protect the equipment inside of the vehicle. Knowing the mechanical and aerodynamics characteristics of the VLM-1, as static margin, the main functions and objectives are identified. These objectives are organized and deployed as an objective tree, which drives to the goals and the metrics that determine the guidelines to the project's physical implementation (Palmerio, 2008).

As stated by Summers (2000), objective trees are used to model the hierarchical nature of the requirements, or objectives, of a design problem. This tool has been used for problem definition and clarification in the early stages of design, though it should be reviewed to ensure that design team is kept on task. Questions about how to measure as well its impact and importance in the design development must be raised, if the objective was achieved. The first question will derive metrics (it means a word as cost, schedule, fuel economy, productivity that encompasses a set of measures). The metrics are deployed to the measures (objectives and quantifiable). Cross (2008) stated that in order to identify the product objectives, the questions "why?", "how?" and "what?" are used to expand and clarify the objectives. Afterwards, the objectives are hierarchically drawn in the tree diagram. In this way, the importance of each objective is analyzed aiming at establishing which one should be achieved first.

Figure 3 shows the objective tree for the VLM Fins Module. The main objectives are deployed towards the lower levels. From these objectives the functions will be attained.

When the objectives unfold, one can identify the features that have become the principal functions of the Fins Module. These functions were represented by physical elements, which, through a morphological chart, allowed building the first physical conceptions.

According to Cross (2008), the morphological chart method exploits all possibilities of combination and encourages the designer to identify novel combinations of elements

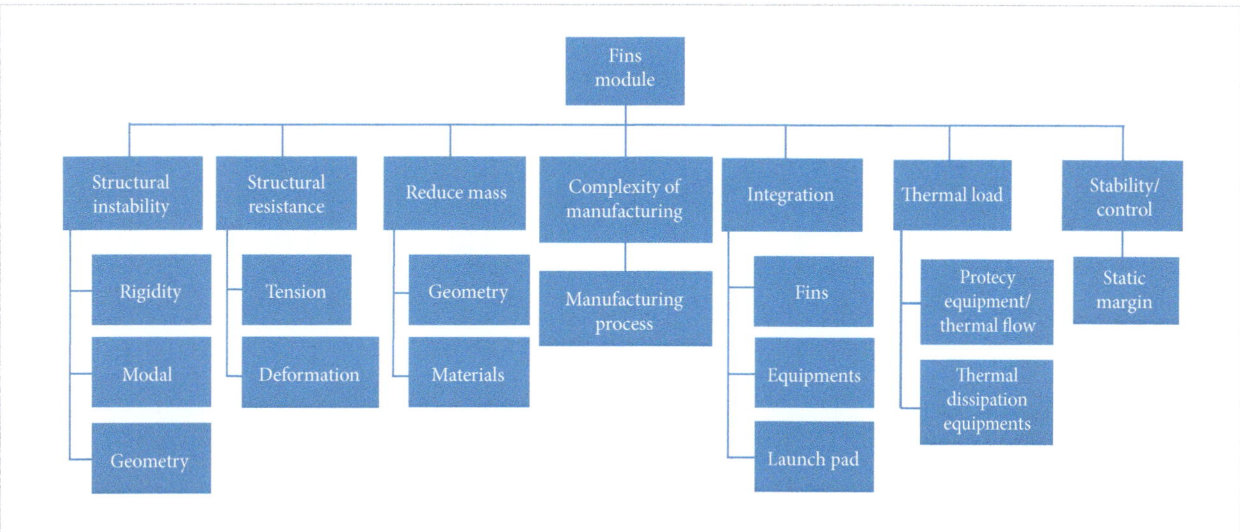

Figure 3. Objective tree of fins module.

or components. The chart sets out the complete range of elements, components or sub-solutions that can be combined together to make a solution. The number of possible combinations is usually very high, and includes not only existing conventional solutions but also a wide range of variations and completely novel solutions.

By analyzing the Fins Module objective tree, the following characteristic were raised: being rigid, resistant, and light, reducing manufacture complexity, enabling integration, allowing thermal dissipation and the vehicle's controllability. The physical elements that meet the Fins Module are known and their dimensions are defined by relevant calculations to other studies.

Table 1 shows the Fins Module Morphological Chart, which can permit a comparison between different solutions for the same requirements arising from the objective tree. The conception defined is about carbon fins and also carbon monolithic membrane based on the principal functions for the system, as mass, stiffness and structural strength.

The carbon was chosen based on the fins material characteristics. Carbon fins are more rigid and resistant, which is one of the most important requirements of the Fins Module. The low carbon density compared to aluminum and the glass fiber make the fins lighter and, as result, it implements the mass reduction requirement. The rigidity and resistance characteristics impacted in the choice of the carbon monolithic membrane. When compared to other materials, it is more adequate for the manufacture, where the performed process reduce the manufacture time.

The choice of the best solution is made based on the manufacture procedures of the module, as well as the integration in the vehicle, taking into consideration the three more relevant scenarios to the system (Fig. 4), chosen from several ones. An analysis with all deployment of the possible scenarios from the life cycle processes is not represented in this work. The scenarios are environments that involve the process of life cycle of the product. In other words, each scenario presents the project elements needed for the development, which are related to the system when subjected to DFMA analysis.

The three scenarios illustrated in Fig. 4 provide the elements that exchange material, energy and information about the system when submitted to the DFMA analysis. These elements for the "Development Organization" are the impacts of the international relationship in this project. Through IAE and DLR interactions, the relevant flows of information to the project conception occur. In the AIT scenario, the resources to implement the assembly, integration and tests are identified. In this scenario, the material types, the processes and the necessary tools to the physical implementation of the project are also identified. Finally, in the Product Operation scenario, the behavior of the system in flight is considered. In this scenario, the geometry and mass of the fins as well as the gravity center of the module are

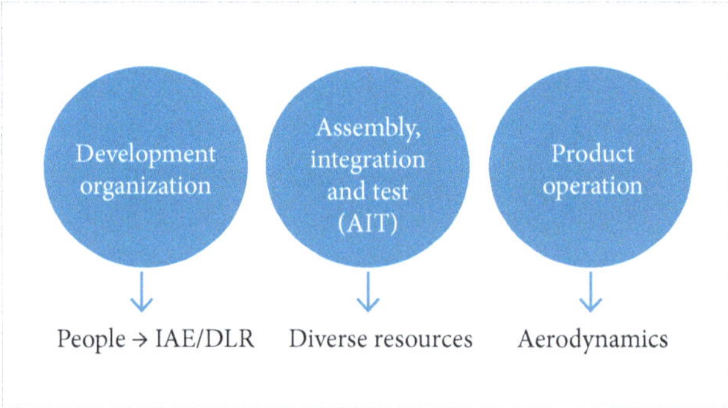

Figure 4. Scenarios from the life cycle process.

Table 1. Morphological chart of fins module.

Function	Physical elements						
	Fin				Membrane		
	Carbon	Aluminium	Fiberglass	Aluminium	Carbon monolithic	Core/Al sandwich	Core/Foam sandwich
Be rigid	x				x		x
Be resistant	x				x		
Be light	x		x			x	x
Reduce manufacturing complexity					x		
Allow integration	x	x	x	x	x	x	x
Allow thermal dissipation		x		x			
Allow controlability	x	x	x	x	x	x	x

identified. All these parameters influence in the aerodynamics and the vehicle's trajectory.

RESULTS

Based on the analysis of the main functions raised and the obtained results using the applied tools, it was possible to describe the first manufacture and assembly for the Fins Module. It is worth to highlight that the number of integration hours and mass limit value for the module derived from calculations belong to other stages of Project VLM-1 development.

The following list presents the first requirements of the Fins Module:

- The Fins Module shall permit the integration of the launch pad.
- The Fins Module shall have mechanical interface to the fins.
- The Fins Module shall have vehicle electrical interface with the launching tower.
- The Fins Module shall be integrated in 3 hours.
- The Fins Module shall have a mass less than 90 kg.
- The Fins Module shall have at most 10 elements for integration.
- The Fins Module shall not allow structural instability.

Table 2 describes each requirement that was allocated for each physical element defined for analysis and studies. Figures 5 and 6 present the first CAD conception of these elements.

Figure 7 shows the physical dimensions of the Fins Module, in a sectional view. These dimensions match with the requirements stated for this part of the vehicle.

Table 2. Allocation of the requirements.

Requirements	Physical element
1	Ring interface of the launch pad
2	Follow-up plans on the cylindrical structure to support the fins' base
3	Umbilical connector
4	Integration plan, environment, and tools
5	Structure and fins in carbon and aluminum rings
6	Fins module + 2 interface rings + fins
7	Monolithic module made of laminated carbon with compatible stiffness

The data of mass, center of gravity and moments of inertia are generated from a finite element model and in the CAD software as well (Table 3). The results show that conception is acceptable based on mass requirements.

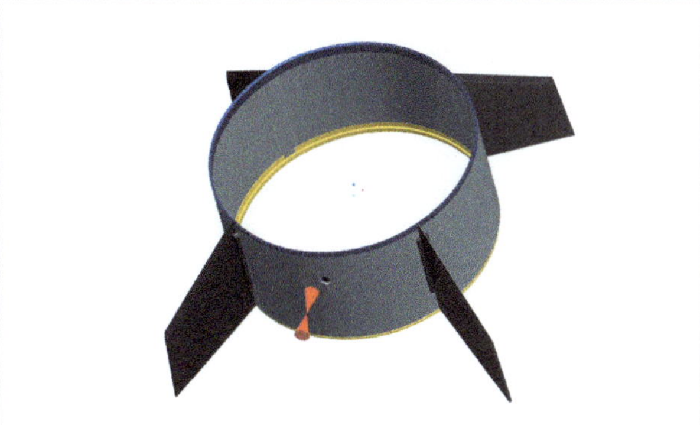

Figure 5. First conception of fins module in CAD.

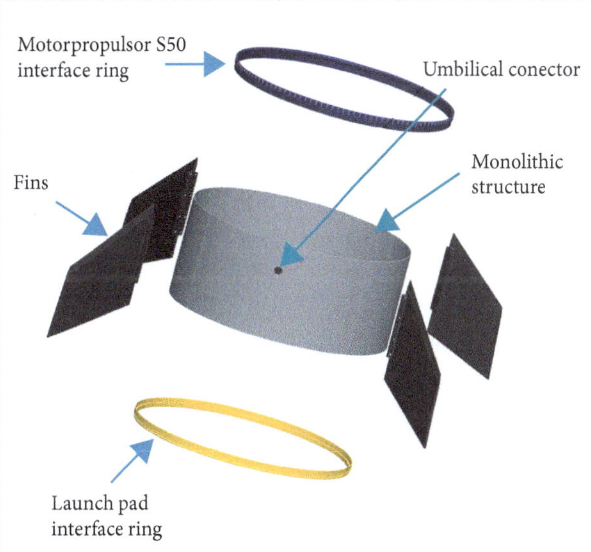

Figure 6. Explode view of fins module in CAD.

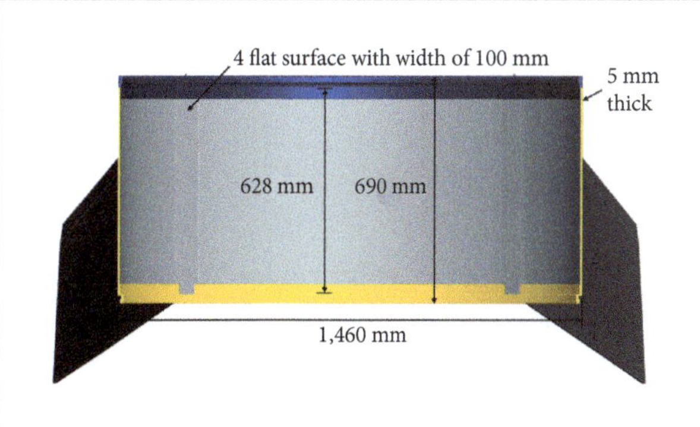

Figure 7. Physical dimensions of fins module.

Table 3. Results of the fins module model analysis.

Volume	3.3557712e-07 mm^3
Surface area	1.3335811e+07 mm^2
Average density	2.3960170e-04 kg/mm^3
Mass	8.0404848e+03 kg
Center of gravity: X	0.0000000e+00 mm
Center of gravity: Y	1.5783229e+00 mm
Center of gravity: Z	8.7176133e+02 mm
Inertia tensor: I_{zx}	0.0000000e+00 kg x mm^2
Inertia tensor: I_{zy}	0.0000000e+00 kg x mm^2
Inertia tensor: I_{zz}	2.0991524e+07 kg x mm^2

A structural CAE static analysis and modal analysis showed that results reached the main objectives proposed, approving the Fins Module conception for use in flight conditions expected for the VLM-1, without problems related to buckling effects and flutter, decoupling frequencies on the fins. Figure 8a shows the finite element method for structural response of the Fins Module, equipped with four fins; on the other side, one notices in Fig. 8b an example of membrane structure subjected to compression load. Therefore, it is concluded that there is a load distribution, which makes the module comply with the requirements of rigidity and resistance.

According to the Fins Module's physical and structural characteristics, and the adopted materials, a manufacture requirements proposal for the Fins Module was raised, following the guidelines of a DFM analysis:

- Manufacture process of carbon structure: type filament winding (winding process of continuous fibers on a rotating mandrel in predetermined pattern).
- Manufacture process of the rings: machining of aluminum plate.

The type filament winding process reduces direct labor in lamination and, consequently, there is a cost reduction of the product. The machining of the rings will follow the profiles drawn in CAD, respecting the DFM guidelines for machining process.

From the conception generated in CAD and their characteristics, it is presented a proposal of components integration of the Fins Module. This was carried out following the guidelines of a DFA analysis, as well as it was taken into account handling the module during transport and integration.

The logical order of the sequence of events was respected:

- Assembly of the rings in the module structure before the launch site transportation.

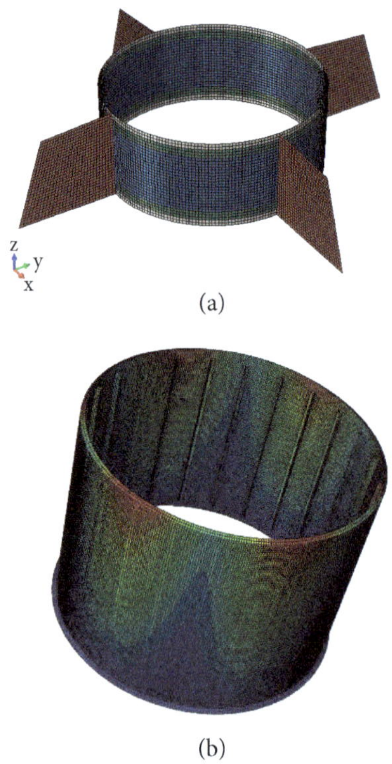

Figure 8. Structural analysis in CAE software with fins (a), and with isolated structure (b).

- Transport of the rings integrated to the module structure with security and protection against impacts.
- Transport of the fins separately to the structure with security and protection against impacts.
- Integration of the Thrust Vector Assembly (TVA) module to the first stage motor case in the horizontal position.
- Integration of the fins to the module structure in the horizontal position.
- Integration of the first stage to the launch pad by the launch pad interface ring.
- Integration of the other modules and stages of the vehicle.
- Integration of the electrical interfaces between the vehicle and the launch pad through the umbilical connector modules present in the vehicle.

From the mentioned integration sequence, it is possible to deploy the Vehicle Integration Plan. In this plan, all steps are detailed, respecting all safety and time requirements. In this work, it will not be showed due to the complexity of the plan and the publishing restrictions. The product conception is carried out simultaneously with the development process in the preliminary phase of the project, considering all the requirements identified using DFMA analysis. Then, the physical implementation of the product was done, with all needed attributes aiming at avoiding rework in future phases of project. This optimizes the time and cost of the manufacture and integration phases.

CONCLUSIONS

The design for manufacture and assembly analysis was presented to the VLM-1 Fins Module conception. A CAD model was developed from the study of the methodology and application of the tools. The DFMA analysis derived the requirements to the physical implementation. From these requirements and the knowledge gained using the DFMA analysis, it was possible to take into account the anticipation of the elements that impact generally in the manufacture, assembly and test phases. This guides to avoid material loss, rework, wasting time to assembly and integration, over engineering and unnecessary engineering throughout the life cycle process.

Thus, describing a framework propose, created from the methodologies analyzed and definitions from VLM-1 project, it was possible to apply the concepts of the DFMA method aiming at defining the characteristics of the Fins Module.

The presented results emerged due to the implemented application, when the guidelines were used aiming at the conception of a CAD model and the survey of the first DFM and DFA requirements for the Fins Module.

Also falls the importance of applying DFMA method throughout the VLM-1 project, as in the case of a low-cost project, where new technologies are approached and used; this method helps in raising the requirements for manufacture and integration present in the development phases in order to minimize the possible errors that may occur in implementation and vehicle's flight.

PROPOSALS FOR FUTURE WORKS

As this work is a part of the current studies on the development of the VLM-1 project, it will serve as the principle for the next related studies, using the presented methods for the development of other subsystems conceptions of the vehicle.

Based on the framework presentation for the product development, this work can be used as a basis for studies of other product development methods as well.

REFERENCES

Boothroyd, G., Dewhrust, P. and Knight, W., 1994, "Product Design for Manufacture and Assembly", Marcel Dekker, Inc., New York.

Cross, N., 2008, "Engineering Design Methods: Strategies for Product Design", Fourth Ed., Wiley, Hoboken, USA,, 230p.

Fulindi, J.B., 2011, "Auxílio Computacional a um Processo de Engenharia Simultânea de Sistemas Espaciais", Instituto Nacional de Pesquisas Espaciais, São José dos Campos, Brazil.

Huang, G.H., 1996, "Design for X: Concurrent Engineering Imperatives", Chapman & Hall, London, UK, 489p.

Iwaya, L.H., 2010, "Estudo sobre a Evolução do DFX", Retrieved in May 20, 2013, from http://leansolutions.com.br/index.php?option=com_phocadownload&view=category&id=1:artigos&Itemid=127

Palmerio, A.F., 2008, "Introdução à Tecnologia de Foguetes", Seventh Ed.

Redford, A.H., 1983, "Design for Assembly", Department of Aeronautical and Mechanical Engineering. University of Salford, UK. Vol 4. No 3.

Silva, E.L. and Menezes, E.M., 2001, "Metodologia da Pesquisa e Elaboração de Dissertação", Third Ed., Laboratório de Ensino a Distância da Universidade Federal de Santa Catarina, Florianópolis, Brazil.

Summers, J.D., 2000, "Design Tool - Objective Tree", Mechanical Engineering Department. Clemson University. Retrieved in January 15, 2015, from http://www.clemson.edu/ces/cedar/images/1/19/05-ObjectiveTree.pdf

Xie, X., 2003, "Design for Manufacture and Assembly", Retrieved in February 18, 2013, from http://home.utah.edu/~u0324774/pdf/DFMA.pdf

The Mapping of Aerospace Meteorology in the Brazilian Space Program: Challenges and Opportunities for Rocket Launch

Amaury Caruzzo[1], Mischel Carmen Neyra Belderrain[1], Gilberto Fisch[2], Daniel Ferreira Manso[3]

ABSTRACT: The meteorological and oceanographic conditions are crucial for the successful launch of aerospace vehicles. However, the decision-making process using environmental information is a complex problem, since it depends on a constant review of current and future weather conditions. To understand this process in the Brazilian Space Program (BSP) context, this paper aims to be the first attempt to map out the systemic view of applied meteorology for the launch missions of aerospace vehicles. Various Brazilian stakeholders were interviewed and their perspectives were analyzed by using a problem structuring method known as Strategic Options Development and Analysis (SODA). With this approach, it was possible to identify different concepts in each group of respondents regarding the current situation of Aerospace Meteorology. One particularly relevant result was identified: weather forecast is not merely a tool to be used to modify the chronology of a mission and to fully provide support in decision-making during the rocket launches in Brazil. Furthermore, the paper shows that the Aerospace Meteorology needs to improve technical processes and to develop a weather decision support system with decision-makers' preferences regarding the uncertainty in weather forecasts. SODA has shown to be a support tool to understand the real situation of meteorology for the launch of aerospace vehicles and appropriate to aid in future planning in the BSP.

KEYWORDS: Spacecraft launching, Decision making, Operational problems, Group behavior, Mapping, Strategy.

INTRODUCTION

The environmental conditions (meteorological and oceanographic) are crucial for the successful launching of aerospace vehicles such as sounding rockets and satellite launch vehicle (Vaughan and Johnson, 2011). Meteorological conditions could also affect other areas of the launch center's infrastructure, therefore it is necessary to take protective action in case of bad weather. However, the decision-making process to protect the facilities of the space center or the exact time of the rocket launch is a complex problem, since it depends on a constant review of current and future weather conditions.

On the other hand, in order to achieve a complete analysis of the decision-making process, it is necessary to evaluate several other factors such as: a) existing operational procedures; b) the limitations of environmental factors; c) current status of the infrastructure of the launching centers, among other factors. In addition to that, the meteorological information should be evaluated by non-expert decision-makers and one should take into consideration all safety requirements during the mission from a systematic approach. That is to say, all weather risks should be evaluated at various stages of the rocket launching procedure, and not just based on the "go" or "no-go" in the launching window.

In the space programs of various countries, many of these procedures have been developed and disseminated in papers and reports (Case *et al.*, 2005; NASA, 2005; Merceret *et al.*, 2006; Kuk *et al.*, 2011; Devyatkin *et al.*, 2012; Dalin *et al.*, 2013). In Brazil, several studies are also being developed in the area of meteorology for rocket launching, for instance, the impact

1.Instituto Tecnológico de Aeronáutica – São José dos Campos/SP – Brazil 2.Instituto de Aeronáutica e Espaço – São José dos Campos/SP – Brazil 3.Instituto de Aplicações Operacionais – São José dos Campos/SP – Brazil

Author for correspondence: Amaury Caruzzo | Praça Marechal Eduardo Gomes, 50, Vila das Acácias | CEP: 12.228-900 – São José dos Campos/SP – Brazil | Email: acaruzzo@gmail.com; acaruzzo@ita.br of current wind for trajectory (Marciotto and Fisch, 2013), rocket exhaust trails/clouds (Moreira *et al.*, 2011), wind tunnel experiments (Avelar *et al.*, 2012) and also weather numerical modeling (Nascimento *et al.*, 2014). However, decision-making in the aerospace and aeronautical sectors, utilizing weather forecasts, is still a scientific and operational challenge, mainly due to the uncertainty of information (Rabelo *et al.*, 2006; Merceret *et al.*, 2013). Furthermore, weather risk perception and preferences of each decision-maker in relation to the probabilistic weather forecast should also be taken into consideration (Joslyn and LeClerc, 2013).

In order to provide a clear understanding of the process from a systemic point of view, this paper aims to map out the applied meteorology for the launching missions of aerospace vehicles within the Brazilian Space Program (BSP). Various BSP's stakeholders were interviewed, and their perceptions were analyzed utilizing a methodology called Strategic Options Development and Analysis (SODA) (Eden and Ackermann, 2001; Georgiou, 2010). Consequently, it was possible to identify different concepts in each group of respondents regarding the current situation of Aerospace Meteorology in Brazil and its relation to the use of weather forecasts on mission launching.

BACKGROUND

BRAZILIAN SPACE PROGRAM

Brazil has been developing its space program since 1965, which begun with the launching of a sounding rocket within Brazilian territory (Brasil, 2008; AEB, 2012). In the late 1970s, the Brazilian government established the "Brazilian Complete Space Mission", which defined the long-term goals of the BSP: that is to say the launching of a Brazilian satellite, with a rocket manufactured and from a launching center in Brazil (Ceballos, 1995).

BSP is currently going through a restructuring process organized by the Brazilian Space Agency, the Ministry of Defense and the Ministry of Science, Technology and Innovation (Brasil, 2012). Furthermore, the BSP has found itself aligned with a new international aerospace industry trend; in other words, it is relying on a greater participation of the private sector (Devezas *et al.*, 2012). In order to make this feasible and to bring a spirit of dynamism to this sector in Brazil, the BSP has created two companies. The first one is the Alcântara Cyclone Space (ACS), a bi-national company (Brazil and Ukraine) to operate the satellite launch vehicle Cyclone-4. The second company is Visiona Corp, a joint venture created from the association between EMBRAER and TELEBRAS, in order to develop the Geostationary Defense and Strategic Communications Satellite (AEB, 2012; Brasil, 2012).

Through the ACS Company, Brazil currently has the possibility of performing satellite operations with liquid propulsion launch vehicles from a Brazilian space center. The Cyclone-4 rocket has the capacity to put into geostationary orbit satellites of up to 1.600 kg (Brasil, 2012; Devezas *et al.*, 2012) and can place Brazil in the private satellite launch market. This context motivated to analyzes the meteorological applications within the BSP.

AEROSPACE METEOROLOGY AND ROCKET LAUNCHING

Weather and climate conditions strongly influence all phases of an aerospace launching mission, to which we highlight:

- Planning: climatological conditions.
- Pre-launching (installation and integration of aerospace vehicle): forecasting and monitoring of weather conditions that can damage the infrastructure of the rocket and the facilities of the launch center.
- Launch (liftoff and tracing the trajectory of the rocket in the atmosphere): observations in real-time and short-term forecasting (nowcasting) of environmental variables (meteorological and oceanographic).
- Post-launch: monitoring and evaluating the impact of weather conditions on the dispersion of the rockets' exhaust gases, payload recovery (for sounding rockets) and the stages of vehicle (for space vehicles), among other activities.

Events in other space programs show that the impact of weather conditions during a mission rocket launch can be critical. For example, during the launch of Apollo XII, in November 1969 (the second manned mission to the moon), the Saturn-V rocket was struck by lightning, which resulted in a computer crash aboard the aerospace vehicle (Uman and Krider, 1989). Despite the fact that the on-board instruments were restored moments later, this event caused significant changes in the operational procedures in the future launches of the United States' space program.

On the other hand, some of the consequences of inclement weather conditions in launch missions may have indirect effects. According to Vaughan (1996), in January 1986, a weather event caused low temperatures at the Kennedy Space Center

(NASA's launching center) — the same week of the Mission of Space Shuttle Challenger. The freezing temperatures, and other factors, caused the O-rings of the rocket fuel tanks to be damaged, originating leaks and, later, the Challenger explosion during launch.

Using probabilistic weather information in decision making process is difficult and requires an analysis of the environmental conditions in an integrated, rational and objective manner. It is remarkable to mention that, for complete success in launching a rocket, various activities are important at different stages of the launch mission. Moreover, the atmospheric conditions are relevant in defining the exact time of the launch window. Thus, all activities related to environmental and atmospheric science that can have an impact on rocket launching missions are classified as Aerospace Meteorology (AM) (Vaughan and Johnson, 2011).

METHODOLOGY

PROBLEM STRUCTURING

One purpose of problem structuring methods is to understand the objectives and perceptions for each of the stakeholders involved in the decision-making situation. For this, several approaches have been intensively used since 1970, when consultants, decision-makers and analysts realized that, by using this method, they would be able to explore, understand and thus make better decisions in order to improve the organizational context under their responsibility (Rosenhead and Mingers, 2001; Georgiou, 2010).

However, there were still some gaps. At the end of the 1980's, a group of researchers devised a new method designated Strategic Options Development and Analysis (SODA), which took place as a primary tool for cognitive mapping, combined with George Kelly's psychological construct theory (Ackermann and Eden, 2001; Georgiou, 2009). SODA is intended to be a method of structuring and problem identification (Eden, 2004; Manso *et al.*, 2015). Through cognitive mapping, and consequently through a hierarchical structure of concepts, individual perceptions of problematic situations are recorded and elicited (Ackermann and Eden, 2001). Individual mappings can be merged, providing a synthesis of the group's perception (Eden, 2004). The final result is a map that provides the analyst with an overview of the investigated context (Rosenhead and Mingers, 2001).

It is necessary to emphasize, however, that the main differences between classical cognitive maps and SODA maps are the theoretical foundations of the SODA, which appropriates part of Kelly's theory of psychological constructs (Georgiou, 2010). Briefly, it can be said that what is called "concept" in cognitive maps is called "construct" in the SODA map. These constructs represent the informal knowledge of the decision-maker and are designed to eliminate ambiguity and subjectivity that may be present in the statements of each client (Ackermann and Eden, 2001; Georgiou, 2009).

In order to make this feasible, we use the so-called "opposite poles" (bipolar design). These are statements about certain actions, situations, or observations separated by three dots and followed by another contrasting statement, able to sufficiently eliminate ambiguity, subjectivity, or even able to clarify the context of what was analyzed (Georgiou, 2010). An example of this approach is shown in Table 1.

Table 1. Example concepts of respondents.

Interviewee	Concepts	Opposite pole
Decision-maker A	Weather forecast is not useful...	Information is accurate
Decision-maker B	Weather forecast is not useful...	Format is suitable

For both decision-makers (A and B), the weather forecast does not appear to be useful. However, by applying the contrast to both of the statements, it is possible to identify the individual (and different) points of view. Therefore, one realizes that the weather forecast may be useful for decision-maker A, if the information presented is accurate. On the other hand, in order to make the weather forecast useful for decision-maker B, the format must be suitable for the intended application.

Henceforth, one may note that the main focus of SODA is to eliminate inherent ambiguities and to provide a clear view of the context under analysis, thus enabling the identification of potential solutions or courses of action emerging from the different perceptions (Georgiou, 2009; Manso *et al.*, 2015). The SODA map is usually employed in the primary stages of approach, when one does not have a clear idea of the context under investigation. However, in some cases, one has verified that the method used to screen more advanced stages serves as decision support, or as a starting point for methods such as system dynamics.

MAPPING OUT AEROSPACE METEOROLOGY

In rocket launch missions, meteorological conditions and weather forecasts are provided, passed on and used by different actors. So, for the systemic mapping of AM in the Brazilian Space Program, several stakeholders, divided into three groups, were interviewed:

- Technical staff (seven interviews): professionals directly related to the provision of weather forecasting and meteorological observation, as meteorologists, engineers and support staff.
- Direct users (ten interviews): actors who use the meteorological information and weather forecasts, directly in their activities during launch missions, and are responsible for the payload, flight safety, team assembly and integration etc.
- Top decision-makers (six interviews): considered to be senior managers in the launch missions and/or the BSP, such as chief operating officer, coordinators and/or directors of the institutions involved.

Thus, a total of 23 professionals in the BSP, spread over 5 different organizations, were interviewed. Through interviews with different stakeholders, it was possible to identify the perceptions of each group in relation to AM and weather forecasts. Surely, these perceptions are related to their various activities in the rocket launch missions, such as infrastructure, personnel etc. According to the SODA method, individual maps of each stakeholder were made. Subsequently, the individual maps were gathered in each group and validated by the highest ranking member (senior) of each respondent group. As a final step, a map aggregate was made of the three group's stakeholders, whom we called the "merged map".

RESULTS AND DISCUSSION

MAPS AND CLUSTER CLASSIFICATION

When all the interviews were finished, the constructs were also classified according to the focus pointed out by the interviewees. In other words, the concepts were grouped into thematic clusters, divided into six different sets:

- Management: constructs related to aspects of legislation, regulation, budget, political support and institutional support for the space program.
- Infrastructure: related to the infrastructure of the launch centers and meteorological activities such as equipment and instrumentation.
- Operational: constructs directly related to the operational activities of the space center, mission launches or decision-makers/managers.
- Staff: related technical staff and other space program professionals.
- Processes: constructs related to the systems and procedures of launch missions.
- Future vision: related to aspects of long-term vision for the future of the BSP.

Table 2 shows the amount of each set of constructs indicated by the stakeholders added and the merged map, separated by clusters.

So, the stakeholders mentioned a total of 118 constructs, which were then merged together and validated by the interviewees, reaching a final total of 74 constructs. The reduction of constructs between the gathered maps from each group and the merged maps is because, quite often, the concepts of each group are similar and can be grouped together in the final version of the SODA map. A complete list of the 74 constructs separated by clusters appears in the Appendix of this paper.

HEAD AND TAILS

Given the properties of the SODA map, it is possible perform a number of detailed analysis of a problematic situation. In the supporting merged map (final version), the "head" may be considered the goal (or goals). That is, through the respondents' perceptions of meteorology, it was possible to define a single objective, which, in this case, is the use of weather as an effective

Table 2. List of constructs by cluster and stakeholder group.

Issue (clusters)	Technical staff	Users	Top decision-maker	Total	Total merged map
Management	4	3	4	11	**9**
Infrastructure	6	5	3	14	**8**
Operational	2	7	9	18	**15**
Staff	5	2	4	11	**7**
Processes	11	13	17	41	**28**
Future vision	4	7	12	23	**7**
Total	**32**	**37**	**49**	**118**	**74**

tool for decision support during rocket launch mission (construct 71, Fig. 1). In other words, the weather forecast (WF) does not modify the activities stipulated in the launch chronology.

As mentioned in the methodology section, through the bipolar design of the SODA map, it is possible to diagnose, to "identify weather risk" and to "establish which procedures in the launch mission" should be followed through, so that the objectives are achieved (construct 71). Another important feature of Fig. 1 is that the constructs with immediate links to a "head" are called "strategic options". That is, these constructs (26, 54, 60, 64 and 70) are possible strategic options available for achieving the objective.

Therefore, in view of the interviewed group, the "strategic options" for the use of weather as a tool for decision support in rocket launch missions in Brazil are presented in Table 3 (in bold).

Grouping by clusters of strategic options, one is in "future vision" (construct 70), three are in "processes" (54, 60 and 64) and one is related to "operational" (26), as demonstrated in the Table 4.

The tails are considered to be the primary cause of the problematic situation. That is to say, for this study, they are the initial concepts that lead to the inappropriate use of weather forecasts in launch missions. In the SODA map, the tails provide a trace of the origin whose state will influence the effectiveness of the goal and its strategic options. In Table 5, the list of 10 identified tails is presented.

It is possible to identify as origin that some concepts are not directly related to AM, for example, the lack of financial resources (construct 8). This can also be seen in the division of concepts into clusters (Table 4): 1 construct in "management" (8), 5 in "infrastructure" (12, 14, 15, 16 and 17), 3 in "operational" (20, 21 and 22) and one related to "future vision" (72). Interviewees perceived the problems of infrastructure and operations as being the initial causes for not using the weather forecast as an integral tool in launch missions. An example of infrastructure highlighted by one interviewee is: "the absence of a comprehensive system of lightning detection at the launch center increases the weather risk" (construct 16, Table 5). Another example is that there are few launch missions in Brazil per year and consequently this also causes a lack of qualified technical personnel (construct 22, Table 5).

IMPLOSIONS, EXPLOSIONS, AND DOMINATING CONSTRUCTS

The relationship between the constructs of the SODA map also allows some other interpretations. In the construct with the largest implosion map (construct 60), *i.e.* the construct that has various constructs leading into it (indegree = 8), it is possible to identify the multiple aspects of the existence of "high risk in decision-making using weather forecasting" (Fig. 2).

In the respondents' point of view, the eight constructs (connected to dominant) are directly related to the risk of making decisions

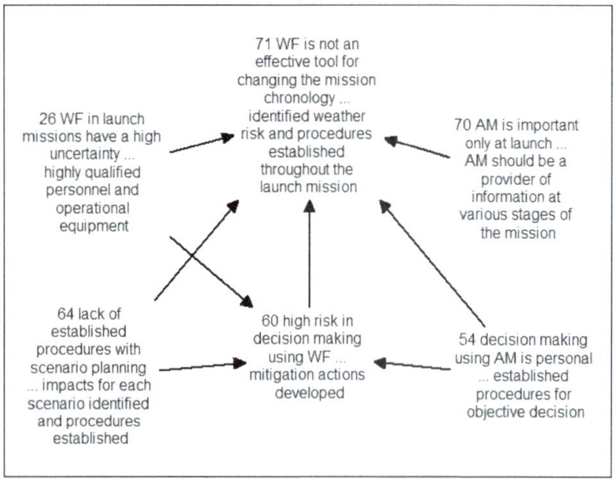

Figure 1. Strategic goal (71) and identified strategic options of the final SODA map.

Table 3. Strategic options and their bipolar design of the construct.

Strategic options
26 WF in launch missions have a high uncertainty... **highly qualified personnel and operational equipment**
54 decision making using AM is personal... **established procedures for objective decision**
60 high risk in decision making using WF... **mitigation actions developed**
64 lack of established procedures with scenario planning... **impacts for each scenario identified and procedures established**
70 AM is important only at launch... **AM should be a provider of information at various stages of the mission**

Table 4. Construct number of tails, heads, strategic options and dominants in the merged map.

Issue (cluster)	Tails	Heads	Strategic options	Dominants (> 4)
Management	1	0	0	4
Infrastructure	5	0	0	1
Operational	3	0	1	3
Staff	0	0	0	1
Processes	0	0	3	15
Future vision	1	1	1	1
Total	**10**	**1**	**5**	**25**

planning utilizing meteorological scenarios (64). The planning scenario related to potential weather events that could negatively impact the launch mission is of great importance for this group of stakeholders.

On the SODA map, the explosion indicates that the construct has a strong influence on the merged map and therefore on the objectives of the problematic situation. There was also a consensus among the three groups of respondents that the lack of operational procedures and the need to identify all the weather-related information demands during the launch mission chronology are an extremely important issue. This feature can

Table 5. Tails in the merged map.

Identified tails
8 low valuation of the space program... sufficient financial resources
12 deficient software of AM (weather systems integration)... software available and operational
14 poor communication data... dedicated link to weather data
15 deficient reception of satellite image... own system available
16 poor lightning and electrical field detection system... financial resources available
17 lack of an observation backup system... multiple observation systems
20 poor training of new staff... replacement staff and continuous training
21 Brazilian rockets are susceptible to weather conditions... rockets with a high protection
22 poor operational team in launching centers... periodic and continuous launch
72 maturation of the importance of "customer"... lack of experience with external payload organizations BSP

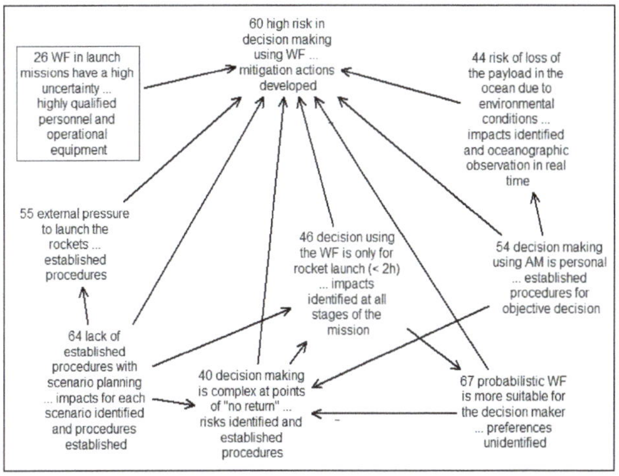

Figure 2. Construct (60) with more implosion on the merged map.

using weather forecasts. These constructs indicate the need for improvements in the identification of meteorological impacts (construct 46), the observation of environmental conditions in real time (44), the definition of the risks and establishment of procedures (40, 54, 55, 64 and 67), as well as keeping qualified technical staff and equipment operating/ready for use (26).

Similarly, the SODA map makes also possible to identify those constructs in which "explosion" occurs, that is, when ideas influence various constructs. In Fig. 3, it is shown that, in construct 56, "few operating procedures are related to AM" and that number has increased connections (total of 8) in the entire model. In other words, there are few sets of procedures that have a direct bearing on the quality and format of weather information (constructs 49, 50, 53, 61, 63 and 66), on the lack of an integrated decision support system (47) and on the lack of

Table 6. Dominant constructs (degree > 7) in the merged map.

Dominant constructs
40 decision making is complex at points of "no return"... **risks identified and established procedures**
46 decision using the WF is only for rocket launch (< 2h)... **impacts identified at all stages of the mission**
47 lack of a weather decision support system... **system designed in accordance with Brazilian demands**
53 WF format is not suitable... **procedures established for each type of mission/rocket**
56 few operating procedures related to AM... **identified demands**
60 high risk in decision making using WF... **mitigation actions developed**
64 lack of established procedures with scenario planning... **impacts for each scenario identified and procedures established**

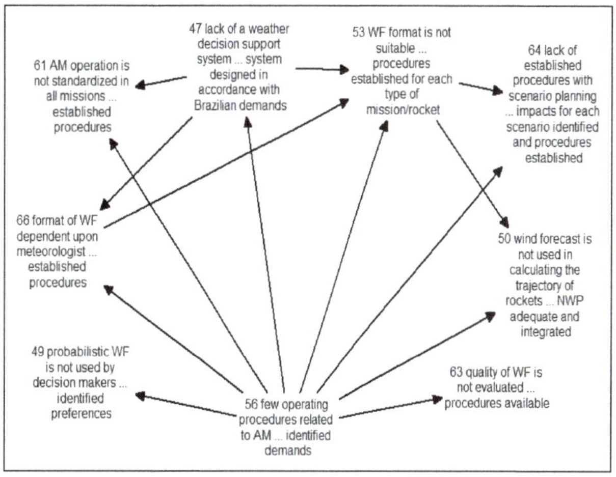

Figure 3. Construct (56) merged with larger explosion in the map (NWP = numerical weather prediction).

also be observed in Fig. 4, which shows the construct 56, that is also the dominant construct in the merged map. In this case, concept 56 has a total of 12 (high total) numbers of constructs connected to it. According to Georgiou (2009), dominants can be interpreted as constructs in a map and indicate focal points in the model where issues or decisions converge to (diverge from) the map.

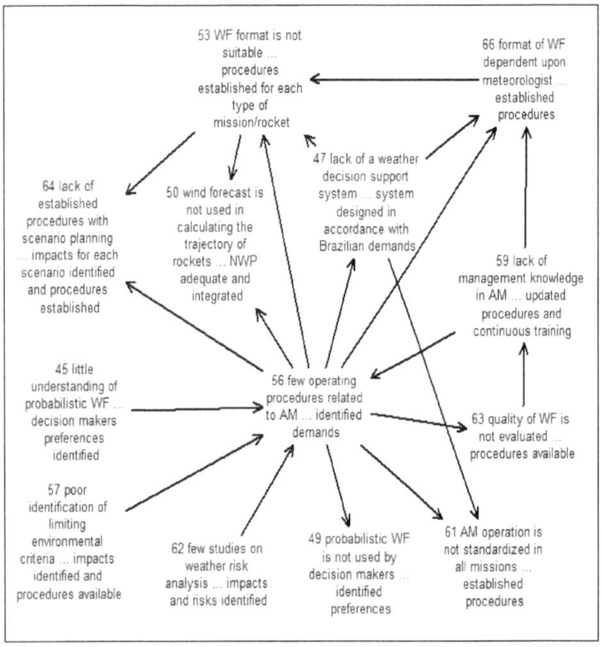

Figure 4. Dominant construct (56) in the merged map.

Naturally, these constructs should be analyzed carefully, as decisions related to the dominant constructs are connected to many other concepts and have a high impact on the model. In Table 6, some dominant constructs (> 7, sum of indegrees and outdegrees) are presented. The actions that should be taken to promote the strategic options are highlighted (in bold).

So, for the most relevant dominant constructs (Table 6), there are three sets of actions that should be taken:

- Identify all demands (construct 56), risks and impacts (40, 46 and 64), as well as mitigation actions related to AM (60).
- Establish operational procedures at various stages of each type of mission and rocket (46 and 53).
- Develop the weather decision support system in accordance with Brazilian demands (47).

FEEDBACK LOOPS

In the SODA map, it is also possible to identify improvement opportunities in a problematic situation (Ackermann and Eden, 2001). This feature is evaluated by feedback loops between the constructs, and can be useful for identifying areas of degenerative or regenerative dynamics. The loops may represent situations of the collapse of a particular decision-makers' concept. In Fig. 5, we have the main feedback loops of the merged map.

The loop constructs 32, 18 and 28 show a degenerative cycle due to the lack of confidence in the weather forecast

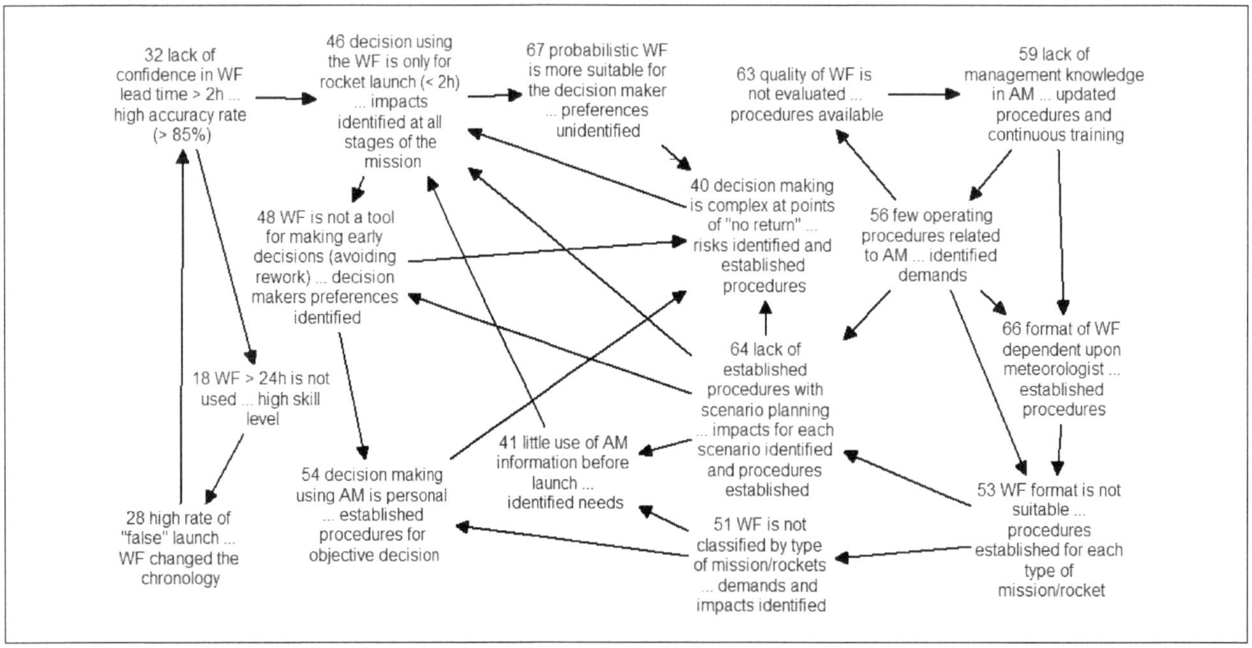

Figure 5. Main feedback loops of the merged map.

because it is not used for periods > 24h, which causes a high rate of false launches. Note that the false entries occur when the countdown is not halted, even with the bad weather forecast. However, when the launch is canceled or postponed around the liftoff window, it causes delays in the space mission and technical staff discontent.

According to some interviewed actors, the false launch can be a big issue, since many sounding rockets need to be integrated several hours before liftoff, and, after numerous false launches, this unnecessary repetition can impair the safety of the mission.

In another loop, constructs 46, 67 and 40 indicate that decision-making using meteorological information only occurs for weather forecasts of up to 2h before the rocket launch. According to respondents, this happens because the impacts, probabilities of accuracy and weather risks were not previously defined. In other words, if the forecast is for bad weather, the decision to cancel the launch is delayed up to a maximum of 2h before the original schedule. This situation is directly related to the false launch, as shown in the previous loop.

For this group of stakeholders, the probabilistic weather forecast (and not deterministic, as it is currently used) would be more appropriate (construct 67). Another negative effect highlighted by respondents of the current model is that decision-making at the point of no return in the launch chronology. This is an extremely complex and challenging moment of the operation because it depends on an individual subjective decision (construct 40) from the Chief Operating Officer responsible for the mission. The point of no return is considered when some rocket system is activated in advance (*e.g.* fuel) and cannot be turned off. That is, even though at the exact time of liftoff there is weather with values above the operating limits, the rocket has to be launched and/or destroyed via remote command.

The loop constructs 40, 46, 48 and 54 show that the preferences and values of decision-makers regarding weather forecasts have not been identified yet, so the decision to use Meteorology information is personal. That is, even if the launches are similar, the decisions could be distinct because they depend on the profile and experience of the Chief Operating Officer.

In Fig. 5, it is also possible to observe the looping of the constructs 56, 63 and 59. This loop indicates that the lack of AM-related procedures negatively impacts the quality of the weather forecast, *i.e.* the hit rate of the forecast is not evaluated. Interviewees perceived this situation as a lack of knowledge between the management and the technical team and also an proper training of new meteorologists.

IMPACTS OF WEATHER INFORMATION FOR ROCKET LAUNCHES

Regarding the constructs identified by stakeholders, some are directly related to the development of weather forecasting and how it is used in the launch mission. Table 7 presents the features of 5 constructs that define technical parameters related to weather forecasting and that can be applied in the parametrization of new decision support systems.

So, in the view of respondents, a weather forecast beyond 24h is not used in decision-making (constructs 18 and 32), due to the low rate of success (< 85%). That is, even if the weather conditions are unfavorable, the decision-makers choose to wait for an updated weather forecast closer to the event, up to 2h before (constructs 32 and 46). In this case, the event can be any chronology activity, where any weather conditions could be a limiting condition (*e.g.* integration rocket on the launch pad).

However, the most important concept regarding weather forecasts is that the values and profiles of decision-makers are not fully identified yet. That is, the interaction with decision-makers showed that it is important to incorporate context in this problem along with the uncertainty of the weather prediction. Therefore, it is mentioned (constructs 47 and 48) that the challenge is to develop a new weather decision support system, as well as to identify and quantify the preferences of the decision-makers, concerning the inherent risk in forecast uncertainty. Nevertheless, in order to incorporate the decision-maker's preferences regarding weather prediction, it would be necessary to develop a new decision support system with a systemic view:

- Identify the demands for weather information throughout the launch mission (construct 46).
- Quantify weather risk to the rocket and the launch center facilities (40).
- Develop a scenario plan (64) for associated weather conditions.
- Develop mitigation actions for each type of scenario, mission and rocket (53 and 60).

CHALLENGES AND OPPORTUNITIES

As a final set of challenges and opportunities for AM in Brazil, we elaborate Tables 8 and 9 based on the concepts of the respondents and our experience in the BSP. Note that these lists are directly related to meteorology. As discussed previously, the improvement opportunities for the problematic situation can be identified by the feedback loop of the SODA map (Fig. 5). In Table 9, the opportunities listed also meet the

Table 7. Constructs directly related to the operation of the weather forecast.

Features of constructs
18 WF > 24h is not used... high skill level
32 lack of confidence in WF lead time > 2h... high accuracy rate (> 85%)
46 decision using the WF is only launch (< 2h)... impacts identified at all stages of the mission
47 lack of a weather decision support system... system designed in accordance with the Brazilian demands
48 WF is not a tool for making early decision (avoid rework)... decision-makers' preferences identified

Table 8. List of key challenges identified.

Challenges	Constructs related
Expand technical staff of Meteorology, maintain qualification and continuous training	4, 19, 20, 33, 34, 35, 36, 38, 39 and 59
Expand and maintain instruments for meteorological and oceanographic observation in full operation	2, 9, 26 and 44
Establish all operational procedures related to AM at the different stages of the chronology of release and classified by type of mission and rocket	25, 40, 52, 53, 54, 55, 56, 57, 58, 59, 61, 63, 66, 70 and 71
Identify and quantify weather risk and mitigation actions in different meteorological scenarios during missions launch	6, 7, 30, 40, 43, 60, 62, 64 and 71
Expand research and development projects related to AM in the Brazilian launch centers, particularly in regional numerical modeling, gas dispersion launches (rocket exhaust clouds) and meteorological hazards	6, 39, 42, 43, 58, 62 and 71

Table 9. List of identified opportunities.

Opportunities	Constructs related
Identify needs for weather information at all stages of the rocket launch mission (not only the liftoff)	43, 46, 47, 48, 51, 54 and 56
Increase interaction between teams of Meteorology, users and decision-makers to develop appropriate procedures in AM demands by type of mission and rocket	28, 40, 52, 53, 54, 55, 56, 59, 61, 63, 64, 66 and 71
Develop products with probabilistic weather forecast (e.g. ensemble)	32, 45, 49 and 67
Develop new projects of research and development in areas related to atmospheric science such as space weather, oceanographic and meteorological instrumentation applied to aerospace	1, 10, 40, 43 and 44
Develop a new weather decision support system, according to the preferences of users and decision-makers from Brazilian Space Program	18, 45, 48, 49, 54 and 67

concepts of the loops of the constructs 18, 32 and 28 (left side of Fig. 5) and the loop 46, 67 and 40 (on the top of Fig. 5). That is, the increasing confidence of users and decision-makers in weather forecasting enables early decision-making (before 24 h) thus avoiding false launches.

CONCLUSIONS

This paper aimed to be the first attempt to map out the systemic view of Aerospace Meteorology at the Brazilian Space Program. The problem structuring methods applied in this study, through interviews with stakeholders, stimulated an overview about the identification and evaluation of weather information for the launch of aerospace vehicles. The SODA map approach increases the decision-maker's knowledge to analyze the problem, since it helps clarify the alternatives and facilitates the understanding of the preferable options needed to apply weather forecasts for the launch mission.

According to the group of Brazilian stakeholders, the constructs related to this problematic situation are divided into 6 clusters (Table 2). By utilizing the SODA method, it was possible to make a set of assessments of decision-makers' values and perceptions by using the constructs of the merged map. As a highlight, we found that the lack of procedures in Aerospace Meteorology (total of 28 constructs) and operational problems at launch mission (total of 15 constructs) constructs were most often remembered by respondents (Table 2). In this context, it has been determined that a weather forecast is not merely a tool to be used to modify the chronology of a mission and to fully provide support in decision-making during the rocket launches in Brazil.

As the strategic objectives for AM in Brazil have been defined by this mapping, weather forecasts should become an effective tool for decision support in launch missions. The initial reasons given by respondents for this strategic objective have not been achieved; the problems are concentrated in infrastructure (constructs 12, 14, 15, 16 and 17 of Table 5) and operational problems during launches missions (20, 21 and 22 of Table 5). On the other hand, there are also some reasons indicated by respondents, which had no direct relationship to AM, for example, the lack of resources for the space program (8) or even changes in the customers' perception of the importance of microgravity research or other payload experiments (72). This also shows that improving Aerospace Meteorology during the launch missions is beyond the activities of the weather forecast.

The strategic options identified by the SODA map (constructs connected to the objective) are clearly concentrated in the absence of processes (constructs 54, 60 and 64 of Fig. 1) during launch mission. As a direct effect, decision-making is personalist (54) and the decision to use the weather forecast is considered highly risky by the decision-maker (60). Furthermore, through the implosions, explosions, dominants and feedback loops, it was possible to identify other highly relevant concepts for the operation of the weather forecast.

We must emphasize this need to develop a weather decision support system, in accordance with the decision-maker's preferences (47 and 56), classified by type of mission and weather scenario planning in rocket launching operations (53 and 64). In other words, forecasts are uncertain, and the resulting risk may be interpreted differently by different decision-makers, depending, in part, on their personal experience and risk perception. In this case, to apply a weather forecast as a reliable tool for decision-making, one needs to incorporate three specific preferences on the demands, procedures, and impacts, which are listed below:

- Value of meteorological variables (wind, rain), according to the boundary conditions of the aerospace vehicle and the launch center facilities (constructs 21, 29 and 53).
- Lead time of the weather forecast (hours, days), *i.e.* the expiration date of the weather prediction presented to the decision-maker (constructs 18, 32 and 46).
- Probability of the weather forecast (%), associated with each variable and lead time (constructs 32, 45, 49 and 67).

Certainly, the challenges related to Aerospace Meteorology have a strong relationship with the challenges of the BSP. For example, to extend the network of meteorological and oceanographic instruments, it is necessary to have consistent funding for the acquisition and maintenance of equipment. In opposition, identifying weather risks and developing mitigation actions throughout the launch mission require greater interaction between the various BSP's organizations and the different technical teams.

Through this mapping, it was possible to identify the main challenges and opportunities (Tables 8 and 9) for improvement within the AM in the BSP. Surely, this is not an exhausted process and requires constant assessment over the years. We must be constantly developing appropriate meteorological products in accordance with the preferences of users and customers. This is a permanent challenge for the meteorologist and technical team. However, in a space program, where risks and costs are much higher, this approach is crucial to the safety of staff and infrastructure.

As a final remark, the SODA map has shown itself to be quite useful for this case, facilitating the understanding of the AM's real situation in the BSP. Furthermore, applying this method in a real case has shown to be a completely appropriate and a reasonable choice for aerospace decision-making problems. In this sense, we can say that the use of problem structuring methods as a first step for decision-making situations is definitely the best option for complex decisions to be made in some topics such as space program planning and/or defense sector.

ACKNOWLEDGEMENTS

Study supported by CNPq (grant nº 142212/2011-3) and CAPES Foundation (grant nº 14552/2013-2). The authors also thank all the Brazilian interviewees (from DCTA, IAE, CLBI, ACS, and INPE) for their contributions, and Maria Pirone, John Madura, and KSC Weather Office Team for comments and suggestions. However, any opinions or conclusions in this paper are those of the authors and do not necessarily reflect the views of the Brazilian Space Program institutions.

REFERENCES

Ackermann, F. and Eden, C., 2001, "SODA - Journey Making and Mapping in Practice", In: Rosenhead, J. and Mingers, J. (Eds.), Rational Analysis for a Problematic World Revisited: Problem Structuring Methods for Complexity, Second Edition, John Wiley & Sons, West Sussex, UK pp. 43-60.

Avelar, A.C., Brasileiro, F.L.C., Marto, A.G., Marciotto, E.R., Fisch, G. and Faria, A.F., 2012, "Wind Tunnel Simulation of the Atmospheric Boundary Layer for Studying the Wind Pattern at Centro de Lançamento de Alcântara", Journal of Aerospace Technology and Management, Vol. 4, No. 4, pp. 463-473. doi: 10.5028/jatm.2012.04044912

AEB, Agência Espacial Brasileira, 2012, "National Program of Space Activities 2012–2021". Retrieved in January 29, 2015, from http:// www.aeb.gov.br/wp-content/uploads/2013/01/PNAE-Ingles.pdf

Brasil, Ministério da Ciência, Tecnologia e Inovação, 2012, "Estratégia Nacional de Ciência, Tecnologia e Inovação 2012–2015", Retrieved in January 31, from http:// www.mct.gov.br/upd_blob/0218/218981.pdf

Brasil, Ministério da Defesa, 2008, "National Strategy of Defense", Retrieved in September 2, 2012, from http://www.defesa.gov.br/ projetosweb/estrategia/arquivos/estrategia_defesa_nacional_ingles.pdf

Case, J.L., Wheeler, M.M., Manobianco, J., Weems, J.W. and Roeder, W.P., 2005, "A 7-Yr Climatological Study of Land Breezes over the Florida Spaceport", Journal of Applied Meteorology, Vol. 44, pp. 340-356. doi: 10.1175/JAM-2202.1

Ceballos, D.C., 1995, "The Brazilian Space Program: a Selective Strategy for Space Development and Business", Space Policy, Vol. 11, No. 3, pp. 202-204. doi: 10.1016/0265-9646(95)92254-B

Dalin, P., Perminov, V., Pertsev, N., Dubietis, A., Zadorozhny, A., Smirnov, A., Mezentsev, A., Frandsen, S., Grønne, J., Hansen, O., Andersen, H., McEachran, I., McEwan, T., Rowlands, J., Meyerdierks, H., Zalcik, M., Connors, M., Schofield, I. and Veselovsky, I., 2013, "Optical Studies of Rocket Exhaust Trails and Artificial Noctilucent Clouds Produced by Soyuz Rocket Launches", Journal of Geophysical Research: Atmospheres, Vol. 118, No. 14, pp. 7850-7863. doi:10.1002/jgrd.50549

Devezas, T., de Melo, F.C.L., Gregori, M.L., Salgado, M.C.V., Ribeiro, J.R. and Devezas, C.B.C., 2012, "The Struggle for Space: Past and Future of the Space Race", Technological Forecasting and Social Change, Vol. 79, No. 5, pp. 963-985. doi:10.1016/j. techfore.2011.12.006

Devyatkin, A.M., Krasnov, V.M., Kuleshov, Y.V., Makov, A.B. and Suvorov, S.S., 2012, "The Use of Information on Atmospheric Conditions for Determining the Impact Areas of Space Rocket Modules", Russian Meteorology and Hydrology, Vol. 37, No. 2, pp. 98-105. doi: 10.3103/ S1068373912020045

Eden, C., 2004, "Analyzing Cognitive Maps to Help Structure Issues or Problems", European Journal of Operational Research, Vol. 159, No. 3, pp. 673-686. doi: 10.1016/S0377-2217(03)00431-4

Eden, C. and Ackermann, F., 2001, "SODA - The Principles", In: Rosenhead, J. and Mingers, J. (Eds.), "Rational Analysis for a Problematic World Revisited: Problem Structuring Methods for Complexity", Second Edition, John Wiley & Sons, West Sussex, UK, pp. 21-41.

Georgiou, I., 2009, "Mapping Railway Development Prospects in Brazil", Transport Reviews, Vol. 29, No. 6, pp. 685-714. doi: 10.1080/01441640902752813

Georgiou, I., 2010, "Cognitive Mapping and Strategic Options Development and Analysis (SODA)", In: Cochran, J.J. (Ed.), Wiley Encyclopedia of Operations Research and Management Science, John Wiley & Sons, West Sussex, UK, pp. 1-10. doi: 10.1002/9780470400531. eorms0974

Joslyn, S. and LeClerc, J., 2013, "Decisions with Uncertainty: the Glass Half Full", Current Directions in Psychological Science, Vol. 22, No. 4, pp. 308-315. doi: 10.1177/0963721413481473

Kuk, B.J., Kim, H.I., Ha, J.S. and Lee, H.K., 2011, "Intercomparison Study of Cloud-to-Ground Lightning Flashes Observed by KARITLDS and KLDN at South Korea", Journal of Applied Meteorology and Climatology, Vol. 50, No. 1, pp. 224-232. doi: 10.1175/2010JAMC2493.1

Manso, D.F., Suterio, R., Belderrain, M.C.N., 2015, "Estruturação do problema de gerenciamento de desastres do Estado de São Paulo por intermédio do método Strategic Options Development and Analysis". Gestão & Produção (article in press).

Marciotto, E.R. and Fisch, G., 2013, "Investigation of Approaching Ocean Flow and its Interaction with Land Internal Boundary Layer", American Journal of Environmental Engineering, Vol. 3, No. 1, pp. 18-23. doi: 10.5923/j.ajee.20130301.04

Merceret, F.J., O'Brien, T.P., Roeder, W.P., Huddleston, L.L., Bauman-III, W.H., Jedlovec, G.J., 2013. "Transitioning Research to Operations: Transforming the 'Valley of Death' Into a 'Valley of Opportunity.'" Space Weather Vol. 11, No. 11, pp. 637–640. doi:10.1002/swe.20099

Merceret, F.J., Short, D.A. and Ward, J.G., 2006, "Radar Evaluation of Optical Cloud Constraints to Space Launch Operations", Journal of Spacecraft and Rockets, Vol. 43, No. 1, pp. 248-251. doi: 10.2514/1.17519

Moreira, D.M., Trindade, L.B., Fisch, G., Moraes, M.R., Dorado, R.M. and Guedes, R.L., 2011, "A Multilayer Model to Simulate Rocket Exhaust Clouds", Journal of Aerospace Technology and Management, Vol. 3, No. 1, pp. 41-52. doi: 10.5028/jatm.2011.03010311

NASA, The Applied Meteorology Unit, 2005, "Severe Weather Forecast Decision Aid (NASA Contractor Report CR-2005-212563)", Retrieved in March 13, 2011, from http://science.ksc.nasa.gov/amu/final-reports/severe-tool-final.pdf

Nascimento, E.G.S., Moreira, D.M., Fisch, G. and Albuquerque, T.T.A., 2014, "Simulation of Rocket Exhaust Clouds at the Centro de Lançamento de Alcântara Using the WRF-CMAQ Modeling System", Journal of Aerospace Technology and Management, Vol. 6, No. 2, pp. 119-128. doi: 10.5028/jatm.v6i2.277

Rabelo, L., Sepulveda, J., Compton, J., Moraga, R. and Turner, R., 2006, "Disaster and Prevention Management for the NASA Shuttle During Lift-Off", Disaster Prevention and Management, Vol. 15, No. 2, pp. 262-274. doi: 10.1108/09653560610659810

Rosenhead, J. and Mingers, J., 2001, "Rational Analysis for a Problematic World Revisited: Problem Structuring Methods for Complexity", Second Edition, John Wiley & Sons, West Sussex, UK.

Uman, M.A. and Krider, E.P., 1989, "Natural and Artificially Initiated Lightning", Science, Vol. 246, No. 4929, pp. 457-464. doi: 10.1126/ science.246.4929.457

Vaughan, D., 1996, "The Challenger Launch Decision: Risky Technology, Culture, and Deviance at NASA", University of Chicago Press, Chicago, USA.

Vaughan, W.W. and Johnson, D.L., 2011, "Aerospace Meteorology: some Lessons Learned from the Development and Application of NASA Terrestrial Environment Design Criteria", Bulletin of the American Meteorological Society, Vol. 92, No. 9, pp. 1149-1157. doi: 10.1175/2011BAMS3133.1

Appendix 1. List of 74 constructs in Merged Map.

Constructs	Cluster	Connection
1 delay in deployment of meteorological instruments... awareness of the importance		+13 +53 +40
2 limited financial resources for equipment and infrastructure for AM... available resources		+1
3 low priority by senior management in the AM activities... awareness of the importance		+37 +2 +4 -7
4 specific and ongoing training deficient in AM... awareness of the importance		+52 -7
5 deficient development of long-term infrastructure plan... awareness by senior management	Management	+13 +1 +2
6 maturation of the importance of AM... low weather risk perception		-63 -59
7 risk perception increased after the VLS accident... low perception of Brazilian decision-makers		+9 +6 -1
8 low valuation of the space program... sufficient financial resources		+35 +27 +2 +3 +5
9 reliability of equipment/data of AM are important... low weather risk perception		-1
10 deficient observation instruments for AM... instruments installed and fully operational		+13
11 NWP inadequate for the mission launch... regional models and short-term WF		+13
12 deficient software of AM (Weather Systems Integration)... software available and operational		+47 +13
13 insufficient general infrastructure for AM... center suitable for all launching operations	Infrastructure	+44
14 poor communication data... dedicated link to weather data		+13 +11
15 deficient reception of satellite image... own system available		+10
16 poor lightning and electrical field detection system... financial resources available		+10
17 lack of an observation backup system... multiple observation systems		+10
18 WF >24h is not used... high skill level		+28 +26
19 limited staff with experience in AM... periodic and continuous launch		+26
20 poor training of new staff... replacement staff and continuous training		+24
21 Brazilian rockets are susceptible to weather conditions... rockets with a high protection		+30 +31 +23 +29
22 poor operational team in launching centers... periodic and continuous launch		+24 +25 +19
23 AM is subject to flight safety team... implementation of WF at different stages of the mission		-31
24 lack of a systemic view of the launch missions... interactions identified and teams prepared		+28 +23
25 low weather risk perception... continuous launches and developed operational culture	Operational	+24 +18 +28
26 WF in launch missions have a high uncertainty... highly qualified personnel and operational equipment		+71 +60
27 lack of interaction between BSP organizations... periodical and continuous launches		+24
28 high rate of "false" launch... WF changed the chronology		+32 +26
29 AM information is relevant at different stages of the mission... low interaction with other sectors		-41 +31
30 accuracy of AM is an important safety factor... activities with low weather risk		+67 -26
31 WF is important throughout the mission chronology... rocket more resistant to environmental conditions		+30
32 lack of confidence in WF lead time > 2h... high accuracy rate (> 85%)		+18 +46 +26
33 hard work during mission launch... no need to move outside staff		+39
34 few forecasters with experience in AM... ongoing training and interaction with more experienced staff		+36
35 high staff turnover in the BSP organizations... adequate career progression and challenging conditions		+27 +37
36 few forecasters and technical teams... adequate operating staff	Staff	+38 +39 +33
37 difficulty in keeping meteorologists and technical staff in the launch center... good infrastructure for housing		+34 +36
38 need to move technical staff for missions launch... local staff available		+39
39 difficulty in conducting research in AM... local and permanent staff available		+42
40 decision making is complex at points of "no return"... risks identified and established procedures		+46 +60
41 little use of AM information before launch... identified needs		+46
42 low interactions between R&D and operation... identified needs		+57 +58 +43 +62
43 nonexistent research in space weather and launching rockets... demands and risks identified		+62
44 risk of loss of the payload in the ocean due to environmental conditions... impacts identified and oceanographic observation in real time		+60
45 little understanding of probabilistic WF... decision-makers' preferences identified		+56
46 decision using the WF is only for rocket launch (< 2h)... impacts identified at all stages of the mission		+67 +48 +60
47 lack of a weather decision support system... system designed in accordance with Brazilian demands		+61 +53 +66 +55 +48
48 WF is not a tool for making early decisions (avoiding rework)... decision-makers' preferences identified		+40 +54
49 probabilistic WF is not used by decision-makers... identified preferences		+48 +40 +54 +46
50 wind forecast is not used in calculating the trajectory of rockets... NWP adequate and integrated		+44 +51
51 WF is not classified by type of mission/rockets... demands and impacts identified		+41 +54 +61
52 poor interaction between AM and other sectors... well-established procedures and hierarchy		+45 +57 +42
53 WF format is not suitable... procedures established for each type of mission/rocket	Process	+44 +64 +50 +51
54 decision making using AM is personal... established procedures for objective decision		+71 +44 +40 +60
55 external pressure to launch the rockets... established procedures		+60
56 few operating procedures related to AM... identified demands		+49 +64 +53 +66 +47 +50 +63 +61
57 poor identification of limiting environmental criteria... impacts identified and procedures available		+56
58 little research on the gas dispersion launches (rocket exhaust clouds)... high interaction between R&D and operation		+62
59 lack of management knowledge in AM... updated procedures and continuous training		+66 +65 +56
60 high risk in decision making using WF... mitigation actions developed		+71
61 AM operation is not standardized in all missions... established procedures		+55
62 few studies on weather risk analysis... impacts and risks identified		+56 +44
63 quality of WF is not evaluated... procedures available		+59
64 lack of established procedures with scenario planning... impacts for each scenario identified and procedures established		+71 +48 +60 +41 +55 +40 +46
65 Brazilian standards based on international standards... appropriate standards for Brazilian characteristics		+47
66 format of WF dependent upon meteorologist... established procedures		+53
67 probabilistic WF is more suitable for the decision-maker... preferences unidentified		-60 -40
68 environmental impact is a concern in commercial launch missions... disinterest in the international market		+69
69 rocket and launch center cannot present problems/delays for commercial missions... interest only in R&D missions		+73 -70
70 AM is important only at launch... AM should be a provider of information at various stages of the mission		+71
71 WF is not an effective tool for changing the mission chronology ... identified weather risk and procedures established throughout the launch mission	Strategic	Head (strategic objective)
72 maturation of the importance of 'customer'... lack of experience with external payload organizations BSP		+74 +69 +68
73 international experience valued AM... low weather risks perception from Brazilian decision-makers		+70
74 quality certification is important for commercial missions... disinterest in the international market		+73

16

Optimal Design of a High-Altitude Solar-Powered Unmanned Airplane

Bento Silva de Mattos[1], Ney Rafael Secco[1], Eduardo Francisco Salles[1]

ABSTRACT: This paper describes a multi-disciplinary design and optimization framework tailored for the conceptual development of high-altitude solar-powered unmanned aerial vehicles. The aircraft baseline configuration that the framework is able to handle is very similar to that of Zephyr, which is developed by the UK based company QinetiQ. The disciplines of aerodynamics, structures, stability, weight, and systems were considered and integrated into a modeFrontier® workflow, capable of providing a relatively simple sizing, but highly realistic airplane.

KEYWORDS: Airplane design, Solar energy, Multi-disciplinary design and optimization, Airplane stability and control.

INTRODUCTION

The present work is concerned with the optimal design of unmanned high-altitude long-endurance (HALE) solar-powered airplanes. In order to accomplish this, a multi-disciplinary design framework was elaborated employing the modeFrontier® (ESTECO, 2011) commercial optimization package. A brief technical retrospective of the development on solar-driven aircraft is provided in this section as well as the reasoning to address the utilization of such kind of airplane.

The first flight of a solar-powered aircraft took place on November 4th, 1974, when the remotely controlled Sunrise I, designed by Robert J. Boucher of AstroFlight, Inc., flew after a catapult launch (Noth, 2008). The flight lasted 20 minutes at an altitude of around 100 m (Noth, 2008).

AeroVironment, Inc. was founded in 1971 by the ultra-light airplane innovator Paul MacCready. Following the AstroFlight's airplane debut, AeroVironment undertook a more ambitious project to design a human-piloted, solar-powered aircraft. The firm initially based its design on the human-powered Gossamer Albatross II and scaled it down to three-quarters of its previous size for solar-powered flights. This was easier to accomplish because, in early 1980, the Gossamer Albatross aircraft had participated in a flight research program at NASA Dryden Research Center. Some of the Albatross flights were performed employing a small electric motor. The scaled-down aircraft was designated Gossamer Penguin (Fig. 1). It had a 71-ft wingspan compared to the 21.64-m (96-ft) span of the Gossamer Albatross (MacCready et al., 1983). Penguin's Empty Weight was just 31 kg, it had a low-power requirement and, therefore, it was an excellent test bed for testing of solar power systems (Noth, 2008).

1. Instituto Tecnológico de Aeronáutica – São José dos Campos/SP – Brazil

Author for correspondence: Bento Silva de Mattos | Instituto Tecnológico de Aeronáutica | Praça Marechal Eduardo Gomes, 50 – Vila das Acácias | CEP 12.228-900 São José dos Campos/SP – Brazil | Email: baviator@gmail.com

Figure 1. The first piloted solar plane, the Gossamer Penguin.

AstroFlight, Inc., of Venice, California, provided the power plant for the Gossamer Penguin, an Astro Cobalt 40 electric motor. This engine was designed to run at between 10,000 and 15,000 rpm, but the 11-ft diameter propeller of Penguin rotates between 120 and 130 rpm. The speed reduction was achieved with a three-stage gearbox with two timing pulley stages providing a reduction ratio of 5/1 each and a final chain reduction stage delivering a 5.12/1 ratio. Robert Boucher, designer of the Sunrise II, worked as a key consultant for both aircraft and for the Solar Challenger. The power source for the initial flights of the Gossamer Penguin consisted of 25 nickel-cadmium batteries and panels of 3,640 solar cells delivering 541W. The battery-powered flights took place at Shafter Airport near Bakersfield, California. Paul MacCready's son, Marshall, was the test pilot. He was 13 years old and weighed approximately 80 lb (36 kg) at that time. The main targets of the test flights were the determination of required power to fly the airplane, the optimization of the airframe/propulsion system, and pilot training. Flights took place on April 7th, 1980, and a short solar-powered one on May 18th. However, the official project pilot was Janice Brown, a Bakersfield school teacher who weighed in at slightly less than 45 kg and was a charter pilot with commercial, instrument, and glider ratings. She checked out in the plane at Shafter and made about 40 solar-powered flights. Wind direction, turbulence, convection, temperature and radiation at Shafter in mid-summer proved to be less than ideal for Gossamer Penguin because of crosswinds at takeoff and decreased solar power output due to higher ambient temperatures. Consequently, the project moved to Dryden in late July, although conditions there also were not ideal. Nevertheless, Janice finished the testing, and on August 7th, 1980, she flew a public demonstration at Dryden in which she was able to cover 3.14 km in 14 minutes and 21 seconds. This was significant as the first sustained flight of an aircraft relying solely on direct solar power rather than batteries. It provided the designers with practical experience for developing a more efficient solar-powered aircraft, taking into account that the Gossamer Penguin was fragile and had limited controllability. The Penguin flew preferably in the morning, when usually wind is minimal and consequently the turbulence level is still low. On the other hand, at this part of the day, the angle of the sun was low, requiring a tilting solar panel to capture properly the solar radiation (Noth, 2008).

From the lessons learned with Gossamer Penguin, AeroVironment engineering staff designed Solar Challenger, a piloted, solar-powered aircraft that was capable of high-altitude and long-endurance flights (Noth, 2008).

The last big step in aircraft powered by solar energy is the first flight of twin-engine Solar Impulse airplane on December 3rd, 2009 (Solar Impulse, 2013). Solar Impulse is a long-range solar powered plane project currently being undertaken at the École Polytechnique Fédérale in Lausanne, Switzerland. The project is promoted by Bertrand Piccard, who piloted the first balloon that was able to circumnavigate the world nonstop. This project is expected to repeat that milestone in aviation history by employing a solar-powered airplane. The first aircraft, officially named HB-SIA, is able to accommodate a single pilot, capable of taking off under its own power, and is intended to remain airborne up to 36 hours. Building on the experience of this prototype, a slightly larger follow-on design (HB-SIB) is planned to make circumnavigation of the globe in 20-25 days. On April 7th, 2010, the HB-SIA underwent an extended 87 minutes test flight (Fig. 2). In contrast to earlier tests, the April flight reached an altitude of 1,200 m (3,937 ft). In February 2012, Solar Impulse pilot Andre Borschberg spent more than 60 hours straight at the controls of the Solar Impulse flight simulator (Fig. 3). Granted, he was able to get some sleep, sometimes napping for a whole 20 minutes at a time (Solar Impulse, 2013). At that time, Borschberg was approaching the end of a 72-hour stint in the simulator, running through a series of tests and challenges to prepare for what lies ahead

Figure 2. Solar Impulse flight on April 7th, 2010.

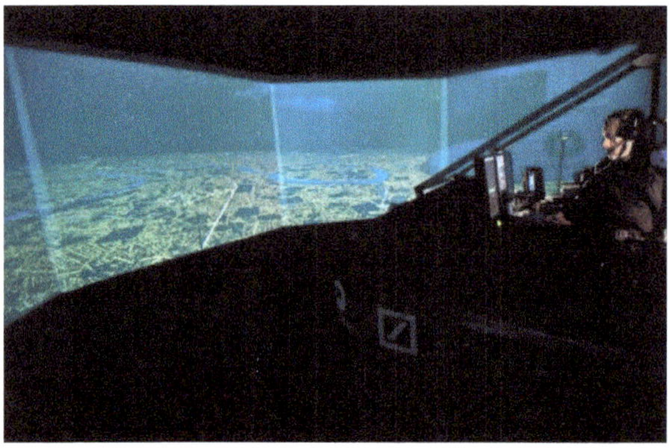

Figure 3. Solar Impulse flight simulator.

when he attempts to fly around the world in a solar airplane in 2014 (Solar Impulse, 2013).

In recent years, European countries are taking a more central role in the development of High Altitude/Long Endurance solar-powered unmanned aerial vehicles (UAVs). Several projects are either ongoing or being planned for advanced solar powered vehicles. In the 2000–2003 period, a team from the Politecnico de Torino, in Italy, together with a team from the University of York in the UK, developed a concept for the Heliplat – a Very-Long Endurance Solar Powered Autonomous Aircraft (VESPAA) (Romeo and Frulla, 2004). Heliplat and other VESPAA UAVs could play the role of a "pseudo satellite", with the advantages of being closer to the ground, more flexible and at a cost lesser than an actual satellite. Heliplat-like HALE flying above a major city will be able to cover a circular area of 1,000 km diameter, and process a predicted 425,000 cell phone conversations simultaneously (Romeo and Frulla, 2004). This means a user community of 8.5 million per unit (although this does not take into account data transmission).

The UAV platform of major interest to the present work is one that is capable to reach stratospheric altitudes and long endurance flights (HALE category) and powered by solar energy. In this context, the Zephyr, developed by the UK based QinetiQ (QinetiQ, 2013), is a configuration that is very similar to that envisaged by the authors. The airplane carries a small payload and has no need for sophisticated assisted takeoff systems or prepared airfields (Fig. 4). The Zephyr is a lightweight airplane

Figure 4. QinetiQ's Zephyr during an assisted take off run (Photo: QinetiQ, 2013).

that uses a combination of a solar array and batteries to stay airborne. The plane weighs just 31 kg and has a wingspan of about 18 m (QinetiQ, 2013). The vehicle can circle over a particular area for extended periods. The military uses the vehicle for reconnaissance and communications platforms. Civilian and scientific programs use it for Earth observation. During the day, Zephyr uses its state-of-the-art solar cells spread across its wings to recharge high-power lithium-sulphur batteries and drive two propellers. At night, the energy stored in the batteries is sufficient to keep Zephyr airborne. The batteries are supplied by Sion. In 2008, the airplane registered a recorded flight, staying airborne for a total of 82 hours and 37 minutes, which is a very impressive number for a solar powered air vehicle (QinetiQ, 2013). The Zephyr went for two straight nights without stopping for recharging and only relying on its solar powered batteries. The airplane is able to climb to altitudes up to 18 km (58,000 ft) (QinetiQ, 2013).

Missions conceived for Zephyr include forest and ocean monitoring, surveillance, and communication relay. Some highlights of Zephyr configuration:

- Zephyr's low-weight structure allows it to travel non-stop over long distances or to remain airborne in a desired station for a considerable long time.
- The unique propeller design provides the aircraft with a higher power-to-weight ratio.
- The aircraft flies day and night powered by solar energy – recharging its batteries along the day.
- No complex launching mechanisms – the aircraft is launched manually by a ground crew of three people.
- Silent flight – neither noise nor pollutant emissions.
- The wing design takes into account thermal air currents to lift the aircraft to higher altitudes.

METHODOLOGY

Before starting with the design of any solar-powered aerial vehicle, it is important to establish how the solar energy will be converted into vehicle motion (propulsion) and how to provide power for the airplane systems. Usually, the wing is an appropriate place to be covered with solar cells, which collect solar radiation and transform it into electrical energy. Batteries, propulsion and systems operate under different conditions and thus a converter package is necessary. An appliance called Maximum Power Point Tracking (MPPT) enables the extraction of maximum power of solar cells and supplies the aircraft systems. During night, there is considerably less energy from solar cells and battery shall fulfill all power needs. Some modern solar cells are able to convert the infrared night radiation into electric power, enabling the batteries continuously to be loaded. This kind of solar cell was not considered in the present work. Figure 5 provides a scheme of the energy management inside the aircraft.

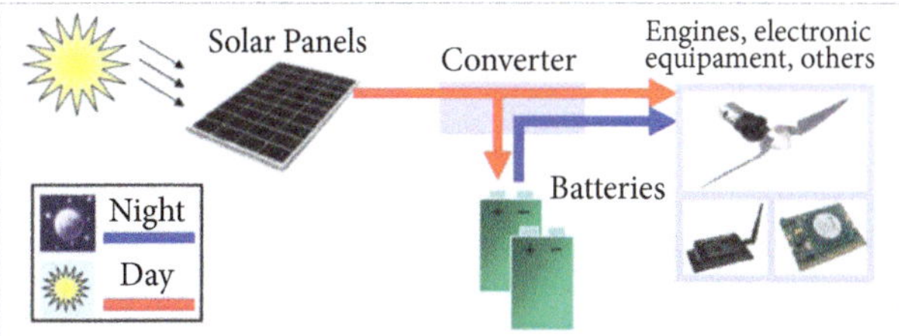

Figure 5. Solar energy conversion system to power engine and other aircraft systems.

SOLAR IRRADIATION MODELS

For a better understanding of the power system, it is necessary to describe the solar radiation model and the systems responsible for its transformation into useful energy.

In order to proceed with the calculation of the solar panel area required on the airplane, it is necessary to estimate the available solar radiation in a given location and date. For this, the relative movement of the Earth around the Sun and of our planet around its axis must be well understood.

The Earth has an elliptical orbit around the Sun, with an average distance of 150 million km. Period of revolution is of 365 days and 4 hours. On January 1st, the nearest distance between our planet and the Sun takes place (perihelion); and on July 1st, the remotest position, the distance is 3.3% farther than that of the closest position (aphelion). Considering that the incident solar radiation is inversely proportional to the square of the distance, the perihelion receives around 7% more radiation than when the Earth is at the aphelion. Another four points also are very important in the astronomy: the autumn and spring equinoxes; and summer and winter solstices (Fig. 6).

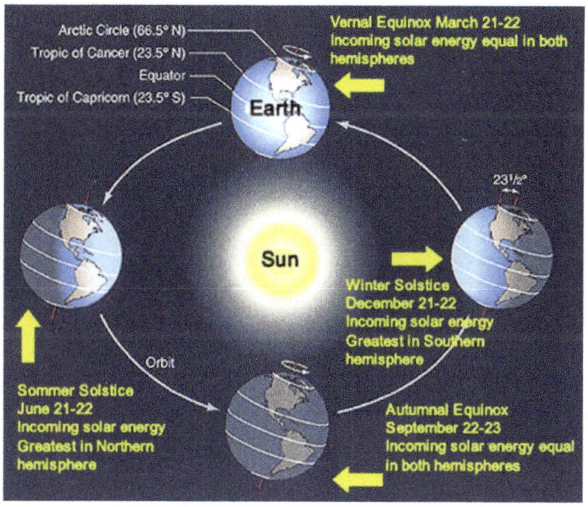

Figure 6. Reference points of Earth's orbit.

The perihelion and the aphelion are the days of most and least intense radiation, respectively. The summer and winter solstices are the days with the longest and shortest time of solar radiation exposure. Being conservative, it is considered that the aircraft should able to fly in any condition and so the point of project is the critical day of the year, in which it receives the least radiation along the day.

The solar irradiance models found in the literature are based on data collected in the Northern hemisphere, mainly in the United States and Europe. For the present work, the model R.SUN was employed (IET, 2013). This sophisticated model was distributed by the Photovoltaic Geographical Information System (PVGIS) of the European Union. PVGIS provides online interactive maps of photovoltaic potentials in Europe and Africa. Figure 7 shows such a map considering the yearly sum of solar electricity generated by 1 kWp system with optimally-inclined modules. R.SUN was developed for the European and Africa environment and it is able to predict the direct and diffuse irradiation and that reflected by the Earth's surface. This last part of the solar irradiation is not taken into account in the present work, since the solar panels are located on the wing upper side of our solar-powered UAV.

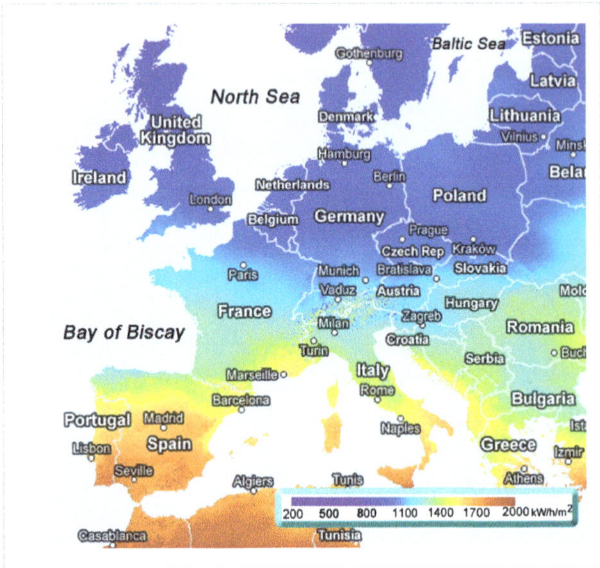

Figure 7. Photovoltaic solar electricity potential in European countries (Source: Photovoltaic Geographical Information System of European Union).

R.SUN is able to estimate the solar irradiation according to latitude, altitude, date and time of the day. In this way, it can be computed the radiation accumulated during an unclouded day. From this estimation, an example of density of energy per area is shown in Fig. 8.

According to the R.SUN model, for an aircraft at 40°N latitude and flying 16 km above sea level, the critical day is December 21st. In this day, the distribution of radiation is shown in Fig. 6. The chosen latitude is the reference for

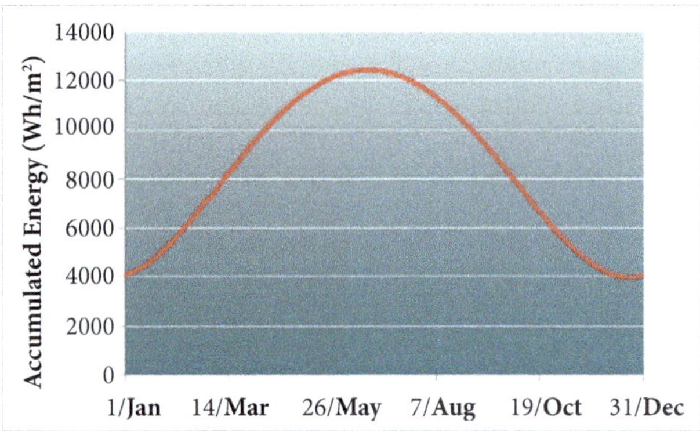

Figure 8. Yearly variation of accumulated daily radiation.

the irradiation model. The power demand should be easily achieved in the inter-tropical region, where Brazil is located, since the incident radiation is larger there.

ELECTRICAL LOAD

In the preceding paragraphs, the available power from solar irradiation was outlined. In this section, the concern is with the electrical power demand for the aircraft.

Power required for the payload package can vary. For this reason, an electrical demand of 100 and 200 W was considered. Power required for the control systems actuators is estimated to be 25.5 W. The engine power can be modeled in the following way (Eq. 1; Anderson, 1989):

$$P = \frac{C_D}{C_L^{1.5}}\sqrt{\frac{(mg)^3}{S}\frac{2}{\rho}} \qquad (1)$$

QinetiQ's Zephyr presents an estimated electrical load of 174 W. This figure occurs close to the stall condition, being necessary to adopt a safety margin. There are some power losses in the power plant system that increase the required power to operate the airplane. Such losses are caused by the propeller, the control box, the engine, and the gearbox, which present efficiencies of 87, 95, 95, and 95%, respectively (Noth, 2008; Roskam, 2000). Taking into account such losses, the required weight-to-power ratio tops 100 kg/hp. Thus, the cruise is not the most critical condition to define the required power. Some critical maneuvers must be addressed for such purpose. The airplane must be able to handle gusts and other severe operating conditions. However, in the present work, the used approach was based on similar weight-to-power ratios at cruise

Table 1. Weight to nominal power ratio for some solar-powered airplanes (Noth, 2008).

Airplane	Weight-to-power ratio (kg/hp)
Solar challenger	60.98
Pathfinder	29.53
Pathfinder plus	23.77
Centurion	27.46
Zephyr (estimated)	22.37

of some solar-powered planes (Table 1). The adopted value was slightly higher than that one of Zephyr: 25 kg/hp.

Typical cruise power for electrical driven airplanes is 56% of the engine nominal power. Thus, for the Zephyr airplane, this represents an increase of 140% and the adopted figure for the present work airplane was of 17.9 kg/hp.

ACTUATORS

Actuators are employed in control surfaces like ailerons, rudder, and elevators. The airplane under development is configured with no high-lift devices and presents no landing gear. German company Volz is the actuator supplier for the Zephyr UAV (Volz Servos, 2013). The temperature of operation for usual actuator varies from 30 to 70°C. However, in the tropopause, the typical temperature is -56.5°C. For this reason, they developed a special actuator able to withstand temperatures below -70°C – model DA 13-05-EXP 11859 (Fig. 9).

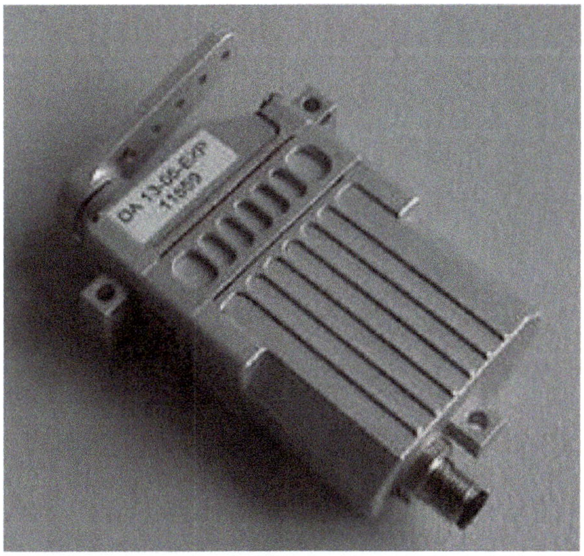

Figure 9. Volz actuator DA 13-05.

All actuators of the DA-13-05 Volz family have the same size (37.4x34.9x13.3 mm), operating voltage (4.8 to 5 V), angle (130°), and casing mass (19 g) but they differ by available output torque, device rotating speed and electrical current (power) (Volz Servos, 2013). Zephyr's actuator is the most advanced available on market in its class. However, there is no public data available and therefore we adopted the specification of the DA 13-05-60 device.

SOLAR PANELS AND BATTERIES

The solar panels are extremely important components, considering that they are the unique source of energy, except the charged batteries in the first day of operation. The impact of solar panels on the project is unquestionable. Their development was recorded to be slow and always has been a critical bottleneck for this kind of airplane. For example, Pathfinder and Pathfinder Plus were developed with the same characteristics, at the same time and using the same technology, and by the same team. However, they differ in service ceiling, which is about 30,000 ft for the Plus version thanks to increased efficiency of solar panels.

Several characteristics of state-of-art solar panels were compared, among them the efficiency and weight per area. Flexible panels (Fig. 10) were then selected. Both specific weights are very similar and they can be considered as ultra-light panels. Efficiency is the factor that was taken into account for panel selection as well. In a conservative way, in this study we consider ISA+15 atmosphere conditions, as the efficiency increases with decreasing temperature. As result, the panel

Figure 10. High-flexible and low-weight solar panels.

with Copper indium gallium selenide (CIGS) technology produced by Swiss company Flisom was selected, providing maximum efficiency of 18.8% (Flisom, 2013).

The most used batteries for portable equipment have lithium in their composition. However, a new generation of batteries based on chemical reaction between lithium and sulfur is available, but the concept is experimental stage. The advantages are weight and volume reduction for the same charge capacity. Lithium-sulfur batteries made by Sion Power are very advanced and some versions have specific energy above 265 Wh/kg (Volz Servos, 2013), value that was adopted in the present study.

Sion Power lithium sulfur (Li-S) batteries were fitted into the Zeyphr airplane in 2010. The batteries played a critical role in the QinetiQ Zephyr smashing the world record for the longest duration unmanned flight. As a result of an intensive joint development effort between Sion Power and QinetiQ, the Zephyr flight exceeded 336 hours (14 days) of continuous flight, significantly surpassing the previous official record of 30 hours 24 minutes set by Northrop Grumman's RQ-4A Global Hawk in March 2001 (Sion Power™, 2013). The Zephyr's world record flight was completed on July 23, 2010, at the US Army's Yuma Proving Grounds in Yuma, Arizona (Sion Power™, 2013).

OPTIMIZATION FRAMEWORK

The simulations for the airplane design under consideration were carried out by using the commercial software package modeFrontier® V4.3 (ESTECO, 2011). The optimization framework modeFRONTIER provides a multi-objective optimization and design environment and it easily enables the coupling of CAD/computer aided engineering (CAE) tools, finite element structural analysis and computational fluid dynamics (CFD) software. It is developed by ESTECO SpA and provides an environment for product engineers and designers (ESTECO, 2011). ModeFRONTIER® is a Graphical User Interface (GUI) driven software written in Java that wraps around the CAE tool, performing the optimization by modifying the value assigned to the input variables, and analyzing the outputs as they can be defined as objectives and/or constraints of the design problem.

In the present work, the airplane configuration that was select for study is very similar to that of the monoplane Qinetiq Zephyr, i.e., featuring a rectangular wing with a conventional tail, which is connected to the wing by a boom. The two engines are located at the inner wing.

The design variables of the present optimization framework are given in Table 2. No geometric wing twist was considered. Two electrical power requirements related to two different payloads were simulated: 100 and 200 W.

Table 2. Design variables.

Variable	Description
AR_w	Wing aspect ratio
AR_{HT}	Horizontal stabilizer aspect ratio
AR_{VT}	Vertical stabilizer aspect ratio
S_w	Wing area
CVH	Horizontal tail volume coefficient
CVT	Vertical tail volume coefficient
y_q	Location of the break station of the wing
L_H	Horizontal tail arm
L_V	Vertical tail arm
Φ_w	Outer wing dihedral angle
T_{Config}	Tail configuration (=0: conventional; =1: "T" tail)

It is reasonable to assume that a lighter airplane will be cheaper. Airplane size is proportional to weight, and if there are two airplanes, one small and one large, with the same payload, the smaller one is structurally more efficient, considering it requires a lighter structure to perform the same mission. Thus, airplane weight is an excellent indicator of configuration efficiency and it was considered the objective to be minimized in the mono-objective simulations. The constraints that were taken into account are: handling qualities, rudder shadowing at stall condition of the horizontal tail, and the area of solar panels, which have to be smaller than the available area on the wing.

Multi-objective simulations were run as well. In this case, weight and energy surplus were taken as objectives and an electrical demand of 200 W for the payload was considered.

DISCIPLINES

The airplane under consideration in the present study must operate at high altitudes and its loiter is performed at very low speeds, which implies a low Reynolds number relative to wing mean aerodynamic chord. Typical Reynolds number lies in the 150,000 – 200,000 range. Due to the low speed on station, high values of lift coefficient (C_l) are experienced by the airplane wing. Taken into account that energy supply is critical, a low-drag configuration is mandatory. Based on Zephyr configuration, a typical C_l of 0.83 at 17 km of altitude was estimated, considering a speed of 60 km/h. In this condition, almost 80% of span operates at $C_l > 0.8$, as shown in Fig. 11. Thus, the biggest aerodynamic issue was to find out a low-drag airfoil for high local lift coefficients. Due to low Reynolds number, a laminar airfoil was considered, with "drag bucket" covering the desired C_l range (Fig. 12). The NACA 6-digit family

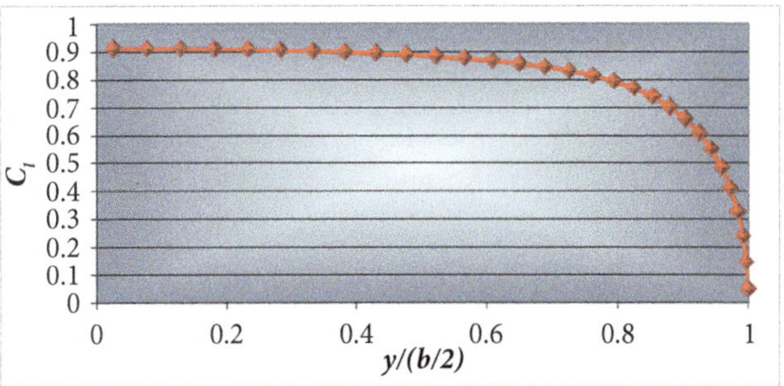

Figure 11. Lift coefficient distribution along wingspan for a Zephyr-like airplane. Calculation performed with Athena Vortex Lattice (Sion Power™, 2013).

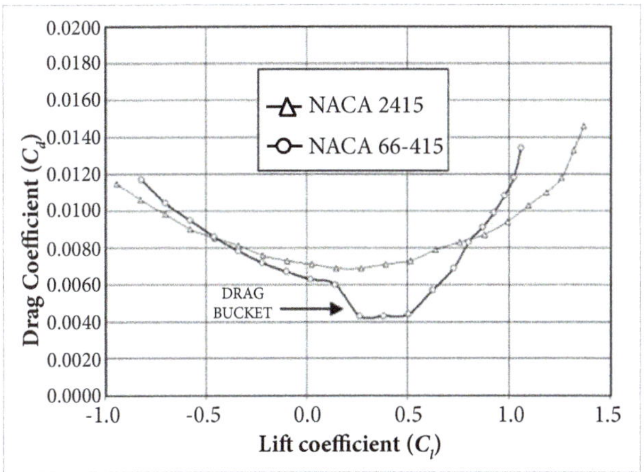

Figure 12. Typical drag polar of laminar airfoils.

offers a variety of laminar airfoils and therefore the NACA 63412 profile was selected to compose the wing. It combines all desired characteristics for the solar-powered UAV under consideration: high maximum lift coefficient (C_{lmax}); moderate to low value for the moment coefficient (C_m); and 12% relative thickness, an adequate one. All the polar curves were computed with the panel code XFOIL (Drela, 2000) and adjusted to fit to experimental results published by Abbott and Doenhoff (Abbott and Doenhoff, 1959) (Fig. 13). Table 3 contains some figures obtained with XFOIL and from experimental data. The correction for the polar curve was obtained by fitting a parabola to the non-laminar portion of the polar curve (Eq. 2):

$$C_d = C_{d,XFOIL} + 0.0006 \quad (2)$$

Where C_d is the drag coefficient.

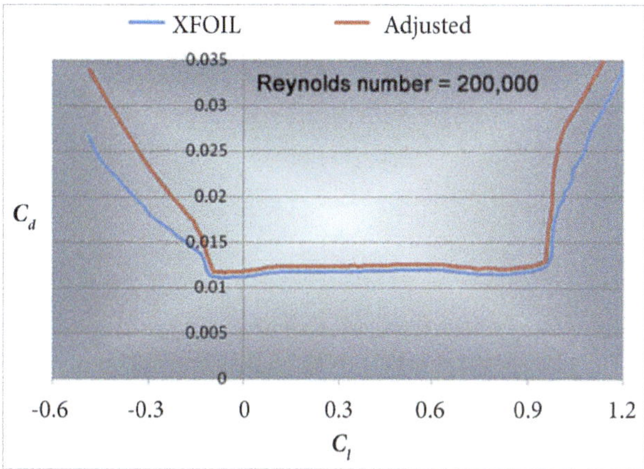

C_d: drag coefficient; C_l: lift coefficient.

Figure 13. Adjusted polar curve for NACA 63412 airfoil according to experimental data.

Table 3. Abbott and XFOIL correlation.

Source	$C_{d,laminar}$	C_{dm}	C_{lm}	k
Abbott	0.0051	0.0063	0.214	0.00732
XFOIL	0.0045	0.0061	0.177	0.00557

The NACA 63010 airfoil was chosen for the horizontal and vertical tailplanes (Fig. 14). Aerodynamic calculations for the complete configuration were performed with the incompressible Athena Vortex Lattice (AVL) vortex-lattice code (Youngren and Drela, 1988). AVL handles lifting surfaces, fuselages, and booms. All required aerodynamic and stability derivative coefficients can be quickly calculated with AVL (Youngren and Drela, 1988).

Concerning load calculation, there is no applicable certification regulation for this kind of aircraft. Estimations that were carried out for the *V-n* diagram elaboration brought no realistic results. In this effort, FAA FAR-23 and JAR-VLA regulations were considered for a Zephyr-like airplane. For this reason, the present work considered a maximum a load factor of 3.8 for the structural sizing (Fig. 15).

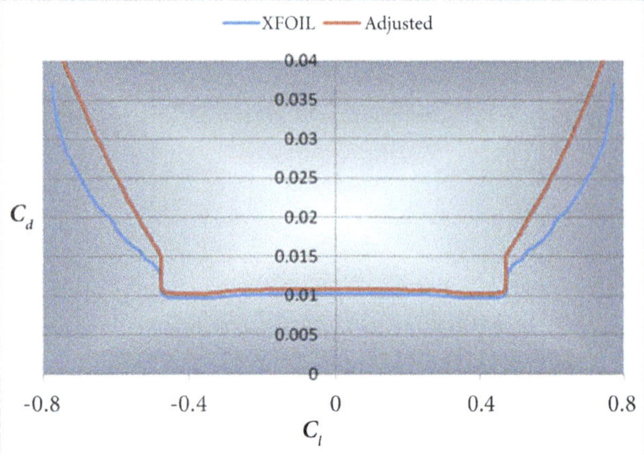

C_d: drag coefficient; C_l: lift coefficient.

Figure 14. Adjusted polar curve for NACA 63010 airfoil according to experimental data.

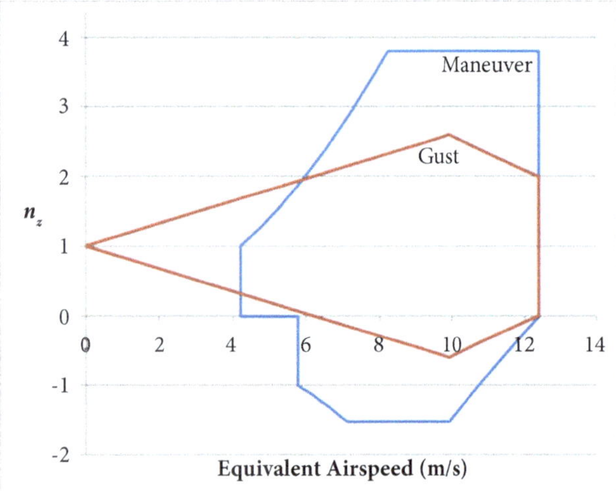

Nz: load factor.

Figure 15. Estimate *V-n* diagram for the Zephyr airplane.

The cruise speed defined by the minimum required engine power for leveled and straight flight. The upper limit of the airplane dive speed was set to be 25% above the cruise speed. The maximum load factor 3.8 was applied for wing sizing; horizontal and vertical tail stabilizers sizing considered the cruise condition, C_{Lmax}, and control surface deflection of 15º; and boom was sized considering simultaneous tails forces. Fuselage and engines were not calculated. The methods of analysis employed in the present study were very simple, as described by Megson (1999). The bending moment is supported by stringers only and torsion loads, by both stringers and skin. The structural layout is shown in Fig. 16, extending from leading edge to 25% of chord, where aerodynamic center is localized. The thickness of the wing skin is the same along the entire span, but stringers may vary for each wing station.

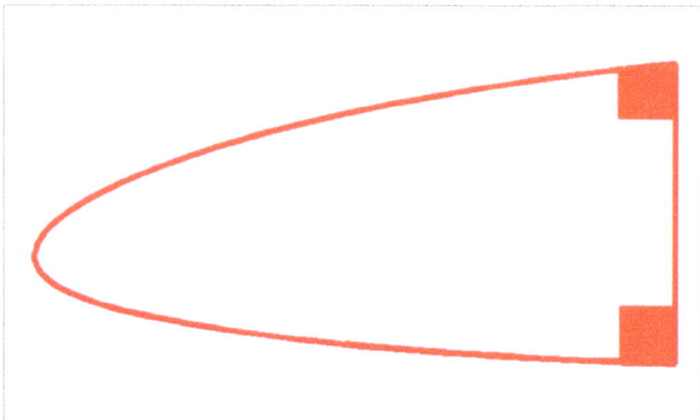

Figure 16. Wing structural layout.

Ribs were not computed but it was assumed putting three of them per meter, with constant thickness.

The boom has a cylindrical form with constant radius and thickness, approximated by thin-walled tube. It can easily withstand the existing loads for this kind of airplane. Its deflection was a constraint and used as sizing criteria. Carbon fiber was the standard material for solar airplanes, except skin, which is built of Mylar. Biaxially-oriented polyethylene terephthalate (BoPET) is a polyester film made from stretched polyethylene terephthalate (PET) and is used for its high tensile strength, chemical and dimensional stability, transparency, reflectivity, gas and aroma barrier properties, and electrical insulation. A variety of companies manufacture BoPET and other polyester films under different brand names. In the UK and the United States, the most well-known trade names are Mylar, Melinex and Hostaphan (Staugaitis and Kobren, 1996).

Densities of the selected materials for the airplane structure are well known and thus the calculation of the mass distribution is an easy task. The fuselage and payload had fixed weight; the weight of batteries depends on the capacity to supply all systems during night; and engine weight was estimated based on methodology described by Noth (2008). Thus, the weight and moment of inertia of each component can be estimated. These calculations are performed through an iterative process until convergence is reached according to some stopping rules. Demand for electric power depends strongly on the drag, which, in turn, depends on the lift coefficient.

Stability and control is a top issue in aircraft design, particularly hard for autonomous airplanes. Unplanned situations may happen and, in this case, there is no pilot to control and stabilize the airplane. Although an on-board computer can control an unstable airplane, this paves the way for accidents. For this reason, flying qualities levels 1 or 2 were imposed, according to MIL-F8785C (Department of Defence of United States of America, 1980), adapted for UAVs (Peters and Andrisano, 1997). This leads to the implementation of lateral-directional analyses in the design framework, addressing short-period, phugoid, roll, divergent spiral and Dutch-roll characteristics.

RESULTS

The modeFrontier® workflow for the task under consideration is displayed in Fig. 17. There are ten design variables in total. They are related to the wing, horizontal tail (HT), and vertical tail (VT). The AVL panel code was employed to calculate the aerodynamic coefficients and stability derivatives. Some MATLAB® routines then evaluated the C_{Lmax} and dynamic behavior of the individuals that arise in the simulation process from data provided by AVL.

The simulations were run on a desktop computer fitted with a single Intel Core2Quad Q6600 2.40 GHz processor, and 8.0 GB of installed Random Access Memory (RAM).

The initial population was generated with the SOBOL algorithm, available in the modeFrontier® framework. SOBOL guarantees that any region of the design space does not become saturated, assuring that the entire domain be fulfilled with a great variety of individuals.

The multi-objective MOGA-II genetic algorithm (ESTECO, 2011) was chosen for running the simulations. MOGA-II is an efficient multi-objective genetic algorithm that uses a smart multi-search elitism. This new elitism operator is able to preserve some excellent solutions without bringing premature convergence to local-optimal frontiers. The efficiency of this algorithm has been orderly proved on six well known test functions for multi-objective optimization.

For simplicity, MOGA-II requires only very few user-provided parameters; several other parameters are internally settled in order to provide robustness and efficiency to the optimizer. The algorithm attempts a total number of evaluations that is equal to the number of points in the Design of Experiment (DOE) table (the initial population) multiplied by the number of generations.

MONO-OBJECTIVE SIMULATION

The first runs revealed that a large number of individuals violated the posed restrictions. It is easy to understand, because in first generations genetic algorithm searches in all

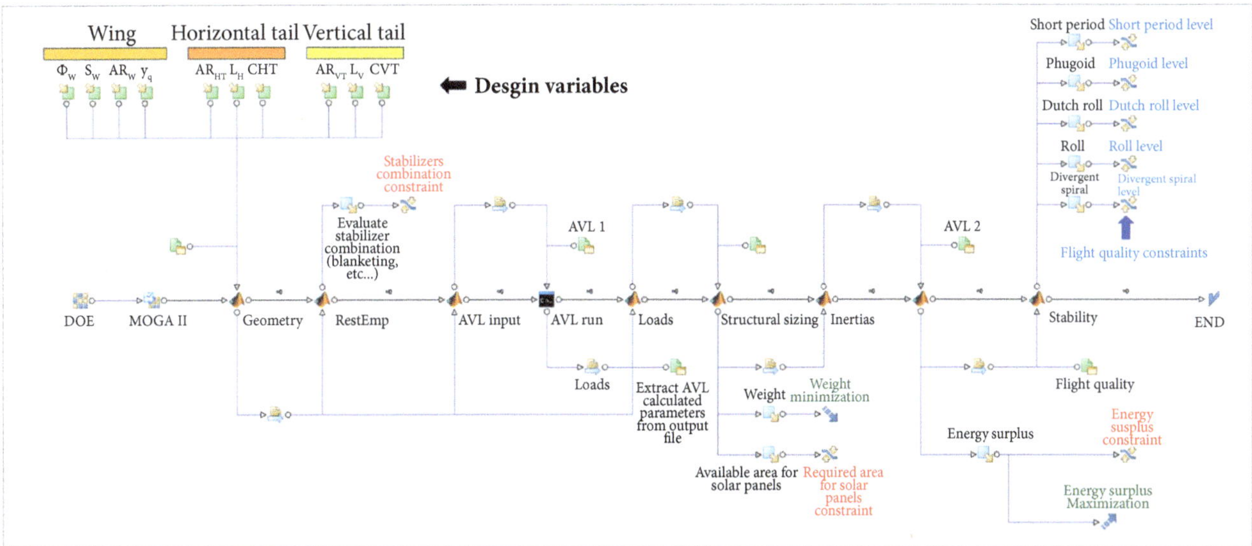

Figure 17. ModeFrontier® workflow for solar-powered unmanned aerial vehicles optimization.

considered domain and more advanced generations are next to optimum. So, algorithm naturally reduced the domain to small range that offers good individuals, except mutations.

For the simulation tasks considering payload electrical demand of 100 W, 20 generations revealed to be enough to provide satisfactory convergence for obtaining an optimal airplane configuration (Table 4). The first runs recorded a large number of individuals that violated the prescribed constraints. Along the convergence process, the optimization framework naturally narrowed the band to generate individuals and most of them fulfilled the imposed constraints (Fig. 18). In other hand, mutation may generate unfeasible individuals along the simulation process.

Table 4. Information for the standard demand simulation.

Variable	Value
Generations	20
Number of individuals	400
Repeated individuals	41
Total processing time	7h51m25s
Average processing time for a single individual	1m19s
Average processing time for four individuals	5m15s

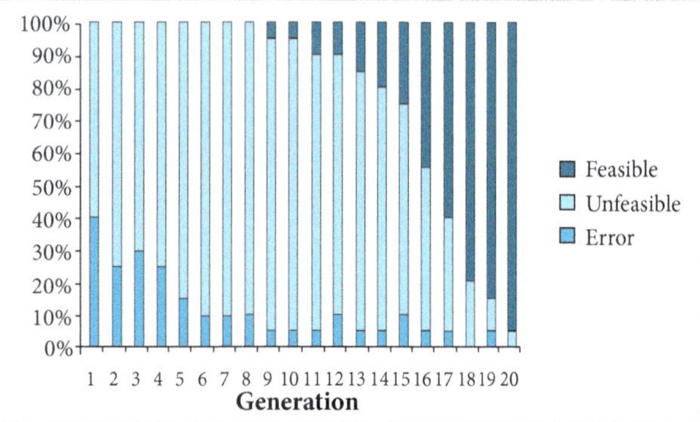

Figure 18. Feasible individual progression along the simulation process.

The violation of power supply constraint is strongly associated with small wing area, while most of other violations are related to tail configuration parameters. Figure 19 provides an overview of the constraint violation by individuals for the entire simulation.

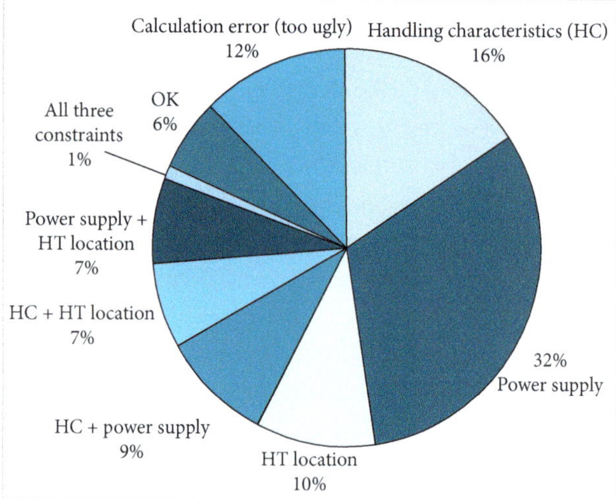

HT: horizontal tail; HC: handing characteristics.

Figure 19. Constraint violators pinpointing.

The area and weight are consistent with data from Zephyr, but its required power supply is unknown. However, the wing loading of the best individual is very close to that of Zephyr.

Considering only the best individuals that did not violate restrictions, it is possible to state that a minimum wing area of 16 m^2 is required for obtaining a feasible configuration. Larger areas can feed systems with greater capacity during the day and must be turned off at night because the battery does not have sufficient storage capacity. For example, a system that requires 350 W can work under daylight conditions and must be turned off at night for airplanes with wing area set to 20 m^2. The day chosen as the project point is very restrictive. If the same airplane flew in the summer, an 1,800 W of power could be generated during the day.

An overview of the best individual from the simulation is provided by Fig. 20. The optimal airplane presents 20 m^2 of wing area and wing aspect ratio of 25 (Table 5). All handling qualities were level 1 on station.

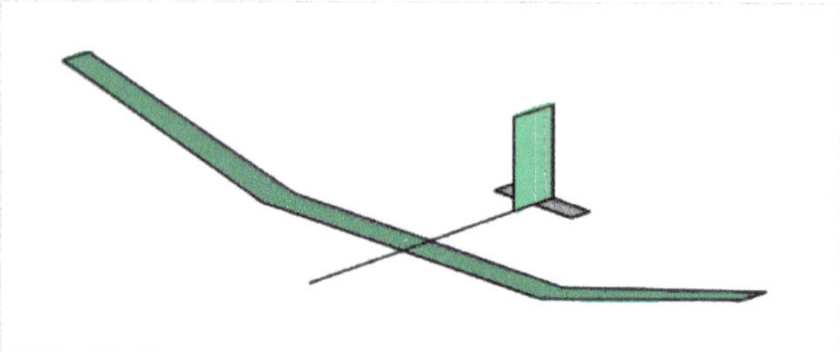

Figure 20. Optimal airplane configuration from the mono-objective simulation.

Table 5. Optimal airplane from mono-objective simulation.

S_w	20 m^2	y_q	0.40
AR_w	25	AR_{HT}	2
Outer wing dihedral angle	14°	AR_V	1.75
L_V	4.25 m	L_H	4.75 m

S_w: wing area; Y_q: break position; AR_w: wing aspect ratio; AR_{HT}: horizontal tail aspect ratio; AR_V: vertical tail aspect ratio; L_V: vertical tail lever; L_H: horizontal tail lever.

Airplane gross weight amounts 30.1 kg. The weights of the airplane components are consistent and the battery weight fraction is huge (Fig. 21), revealing that this equipment presents an issue critical for the design of a solar-powered airplane. Typically, a battery generates 350 Wh/kg (265 Wh/kg considering the wrapper), but further development can increase this figure to 600 Wh/kg, which would have great benefit for the configuration of solar-powered aircraft.

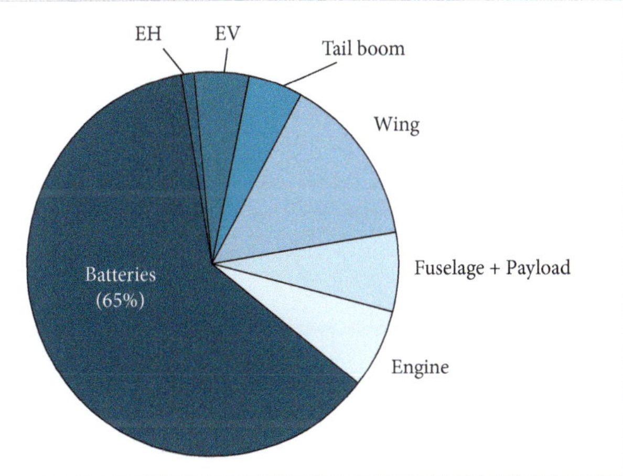

EH: horizontal tail; EV: vertical tail.

Figure 21. The above pie chart provides a weight breakdown of the optimal airplane.

As seen before, batteries are critical for design of solar powered airplanes. If the electrical power requirement for the payload is increased, the structural weight is also increased. This directly affects the weight of the engine and the required thrust to propel the airplane.

Multi-objective optimization

Weight minimization and power surplus maximization were taken as objectives for this task. Wing area was allowed to vary between 30 and 60 m^2. After 40 generations with 30 individuals each, the simulation was stopped. Figure 22 shows the feasible individuals of the present simulation and Fig. 23, its Pareto front.

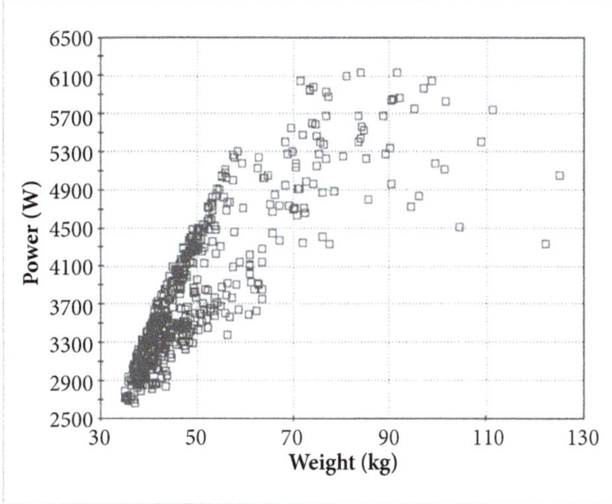

Figure 22. Feasible individuals from the multi-objective simulation.

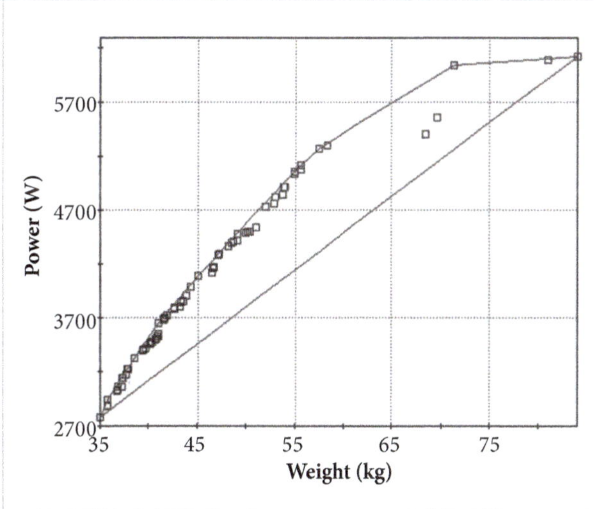

Figure 23. Pareto front of the multi-objective simulation.

Here, the available power from the optimal individuals is considerably higher than that of the optimal airplane from the previous simulation. This can be credited to the lower boundary that was set for the wing area, which was 50% higher than that of the optimal airplane from the mono-objective simulation. The configuration from the Pareto front with the lowest available power weighs just 35.21 kg, slightly over the weight of the optimal airplane from the previous simulation. Table 6 contains some characteristics of this airplane.

Table 6. Characteristics of the optimal lowest power airplane from the multi-objective simulation.

S_w	30 m^2	y_q	0.55
AR_w	21	AR_{HT}	3.25
Outer wing dihedral angle	12°	AR_V	1.50
L_V	4.75 m	L_H	6.50 m

S_w: wing area; Y_q: break position; AR_w: wing aspect ratio; AR_{HT}: horizontal tail aspect ratio; AR_V: vertical Tail aspect ratio; L_V: vertical tail lever; L_H: horizontal tail lever.

CONCLUDING REMARKS

The development of solar-powered airplanes in the past was directly affected by the technology of solar panels. Nowadays, alongside with the solar panel efficiency issue, energy storage became a major concern for the design of solar air vehicles. In addition, considering that there are few airplanes flying in altitudes from 15 to 18 km, there are also few aircraft systems manufacturers able to deliver products capable of working under severe environmental conditions like the ones found in such altitudes.

The application of optimization tools proved to be quite appropriate, as provided in less than one day running a very large variety of airplanes, for which one can identify the strengths, weaknesses, and typical values of the variables for a good design. The case study proposed is an excellent example of this, because there are few semi-empirical mathematical models and built aircraft data, and even then with the application of simple theories, it was possible to design and verify the consistency with the few existing models.

Future work will encompass the wing and tail airfoils as design variables.

ACKNOWLEDGMENTS

The authors thank Fundação de Amparo à Pesquisa do Estado de São Paulo (FAPESP) for its support by the project 2007/00305-5.

REFERENCES

Abbott, I.H. and Doenhoff, A.E., 1959, "Theory of wing sections", Dover Publications, Mineola, NY, USA. 693 p.

Anderson, J.D., 1989, "Introduction to Flight," 3rd Edition, McGraw-Hill, USA.

Department of Defence of United States of America, 1980, "MIL-F-8785C: Military Specification Flying Qualities of Piloted Airplanes", Washington, DC.

Drela, M., 2000, "X-FOIL subsonic development system" [Internet], Massachusetts Institute of Technology (MIT).Available from: http://web.mit.edu/drela/Public/web/xfoil/. Accessed: June, 2006.

ESTECO s.p.a., modeFrontier® 4 User Manual, 2011.

Flisom – Flexible Solar Modules [Internet]. Available from: http://www.flisom.ch. Accessed: April 26, 2013.

Joint Research Centre, Institute for Energy and Transport (IET), "Solar Radiation and GIS" [Internet]. Available from: http://re.jrc.ec.europa.eu/pvgis/solres/solmod3.htm. Accessed: April 26, 2013.

MacCready, P.B., Lissaman, P.B.S., Morgan, W.R., and Burke. J.D., 1983, "Sun-Powered Aircraft Designs", Journal of Aircraft, Vol. 20, No 6, pp. 487-493.

Megson, T.H.G., 1999, "Aircraft structures for engineering students", 3rd Edition, Butterworth Scientific, Oxford.

Noth, A., 2008, "Design of Solar Powered Airplanes for Continuous Flight," Ph.D. Thesis, École Polytechnique Fédérale de Lausanne, Swiss.

Peters, M.E. and Andrisano, D., 1997, "The determination of longitudinal flying qualities requirements for light weight unmanned aircraft," AIAA Guidance, Navigation, and Control Conference, New Orleans, LA, Aug. 11-13.

QinetiQ [Internet]. Available from: http://www.qinetiq.com/. Accessed: April 26, 2013.

Romeo, G. and Frulla, G., 2004, "Heliplat: High Altitude Very-Long Endurance Solar Powered UAV for Telecommunication and Earth Observation Applications", The Aeronautical Journal, Vol. 108, pp. 277-293.

Roskam, J., 2000, "Airplane Design, Part VI: Preliminary Calculation of Aerodynamics, Thrust and Power Characteristics", The University of Kansas, Lawrence. DARcorporation.

Sion Power™ [Internet]. Available from: http://www.sionpower.com/. Accessed: April 26, 2013.

Solar Impulse [Internet]. Available from: http://www.solarimpulse.com/. Accessed; April 26, 2013.

Staugaitis, C. and Kobren, L., 1996, "Mechanical and Physical Properties of the Echo II Metal-Polymer Laminate," NASA TN D-3409, NASA Goddard Space Flight Center.

Volz Servos [Internet]. Available from: http://www.volz-servos.com/en/index.php?m=m. Accessed: April 26, 2013.

Youngren, H. and Drela, M., 1988, "Athena vortex lattice (AVL): an extended vortex-lattice model for aerodynamic analysis, trim calculation, dynamic stability analysis, and aircraft configuration development" [Internet]. Available from: http://web.mit.edu/drela/Public/web/avl/. Accessed: February, 2007.

17

Observational Study of the Surface Layer at an Ocean–Land Transition Region

Luiz Eduardo Medeiros[1], Roberto de Oliveira Magnago[2], Gilberto Fisch[1], Edson Roberto Marciotto[1]

ABSTRACT: High-frequency measurements of wind, and temperature were made during the dry season of 2008 to study the development of an internal boundary layer at the main Brazilian space launching centre, Centro de Lançamento de Alcântara at Alcântara, Maranhão, Brazil. Turbulence measurements taken at the coast, in two different points 227 m apart show different daily cycles of turbulent kinetic energy friction velocity (u_*), and buoyancy flux $\overline{w'T_v'}$. Surface roughness change, surface heating change, and a gap in the natural vegetation seem to be the causes for the variation in these turbulent parameters. The mean wind cycle also shows distinct patterns. It seems that, first, internal boundary layers develop when the oceanic surface layer reaches the continent, and a second when the first internal boundary layer's flow encounters the gap. A direct implication is that turbulence is not horizontally homogeneous and measurements taken at single places are not spatially representative. Knowing how turbulence varies spatially is necessary information to understand the diffusion of pollutants exhausted by rockets near the coast.

KEYWORDS: Coast, Internal boundary layer, Sonic anemometer, Surface layer.

INTRODUCTION

The space vehicle launching centers are usually located at coastal areas due to safety reasons. Examples are the Kennedy Space Centre at Florida, Centre Spatial Guyanais at French Guiana, and Centro de Lançamento de Alcântara (CLA) at Brazil, all of which are located on the coast. CLA, for instance, has its launching pad around 650 m from the seashore. At the coastal regions there are always surface cover changes and in some case changes in topography too. Changes in surface cover create discontinuities in surface roughness and heating, which in conjunction with the topography affect the distribution of winds, temperature, and humidity. It is well known that boundary layers being advected over surface with discontinuities develop internal boundary layers (IBL) (Stull, 1988; Arya, 2001).

A boundary layer is the layer of fluid near a boundary that is affected by friction against that boundary surface, and possibly by transport of heat and other variables across that surface. An IBL is a layer within the atmosphere bounded below by the surface, and above by a more or less sharp discontinuity in some atmospheric property. A surface layer is the same as a surface boundary layer, which is a layer of air of order tens of meters thick adjacent to the ground where mechanical (shear) generation of turbulence exceeds buoyant generation or consumption.

In the case of roughness discontinuities, the mechanically generated IBL might reach equilibrium within or less than a kilometer downwind from discontinuity and its effects are likely to be felt only at the surface layer. In case of surface heating discontinuity, the internal boundary layer might reach equilibrium only at much larger distances (Garratt, 1990;

1.Instituto de Aeronaútica e Espaço – São José dos Campos/SP – Brazil 2.Universidade do Estado do Rio de Janeiro – Resende/RJ – Brazil

Author for correspondence: Roberto O. Magnago | Departamento de Matemática, Fisica e Computação/Faculdade de Tecnologia/Universidade do Estado do Rio de Janeiro | Rodovia Presidente Dutra km 298 – Pólo Industrial | CEP 27.537-000 Resende/RJ – Brazil | Email: ducamobi@gmail.com

Mahrt and Vickers, 2005) and it may affect much deeper layers. Sea and land breezes are examples of thermal IBLs. On this regard, Case *et al.* (2005) analyzed the sea breeze climatology at Kennedy Space Centre using a network of 44 mesonet towers and five wind profiles, while Merceret (2006) studied the rapid temporal variation of the winds using wind profilers. For the case of CLA, Gisler *et al.* (2011) studied the climatology of the surface winds through 5 years of data from a 70 m-high anemometric tower, in order to characterize the wind flow regime in the region. Recently, Moreira *et al.* (2011) described the usage of a software to study the dispersion of pollutants (or toxic gases) released at CLA.

The CLA is located in the northeast (NE) of Brazil, right at the shoreline. The wind regime at its location is influenced by the trade winds and perhaps by a sea/land breeze circulation (Gisler *et al.*, 2011). Besides these influences, there is also influence from a topographic barrier, an escarpment. The barrier is located just where the beach ends, and according to a wind tunnel study/simulation carried by Marinho *et al.* (2009), the flow is supposed to accelerate around the edge of the escarpment and to develop a recirculation a few hundred meters downwind from it. The limitation of this study is that the escarpment was approximated by a step-like barrier with 90 degrees inclination, which is not realistic. The understanding of how the marine boundary layer is affected, and how the surface layer is modified downwind from the escarpment, is extremely important for the evaluation of the stress caused by turbulence and wind on the structure and trajectory of rockets, well as for the dispersion of pollutants exhausted by them.

For the region of CLA, mechanical and thermal IBLs should be developed when the marine boundary layer reaches the coast. Nonetheless, in this work we focus only on the mechanical effects exerted on the near-surface turbulence. The possible deeper effects of the surface heating discontinuity on the local planetary boundary layer are left for a future work.

DATA

Most of the data used in this work are from the Murici II meteorological campaign (Marciotto *et al.*, 2012) conducted at CLA, during September 17–25 of 2008. The measurements were taken right at the coast. The transition zone water-land is composed by an eroded slope about 40 m high, very heterogeneous, which at some points resembles cliffs. The slope is partially vegetated and delimitates the border of a plateau. In Fig. 1, it is shown that the slope with the surface cover change from water to land with small dense bushes typically 3 m tall. In Fig. 2 the local topography with the plateau and the surrounding slope is shown. The topographic surface (Fig. 2) was built using data from the Shuttle Radar Topography Mission provided by NASA Land Processes Distributed Active Archive Center (http://gdex.cr.usgs.gov/gdex/). The spatial resolution is 3 arc-seconds or approximately 92 m for Alcântara region.

Figure 1. The topographic barrier between the ocean and the plateau, the escarpment.

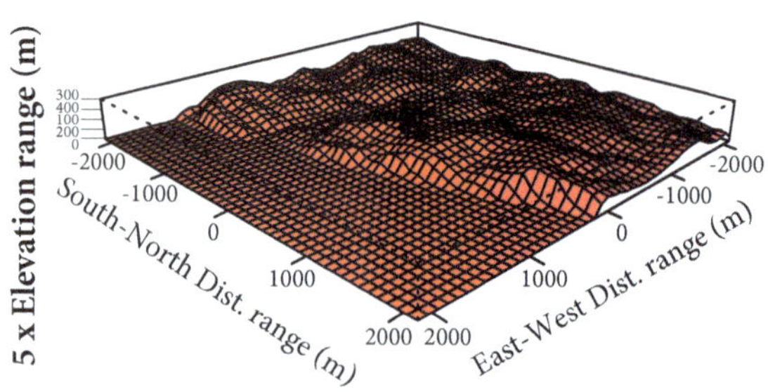

Figure 2. A perspective of the topographic of the region with respect to the prevailing wind direction. The taller vertical segment indicates the position of the tower plus Sonic B, and the shorter segment the mast with Sonic A. The flat part of the surface corresponds to the sea.

Measurements were taken at approximately 675, 700, and 750 m inland at the plateau, as shown in the terrain cross sections (Fig. 3) along the direction of the prevailing wind during the time of the experiment – NE (Fig. 4). The Shuttle Radar Topography Mission data produces very similar cross sections like the ones presented in Fig. 3; however, we preferred to use a local topographic map because there was more data in the region of the slope. The windroses presented in Fig. 4 show the prevailing wind direction in the region. For their determination we used 30 minutes average of zonal (u) and meridional (v) wind components during the period of the experiment. There are 16 directional sectors for each windrose, and every sector corresponds to a range of 22.5 degrees.

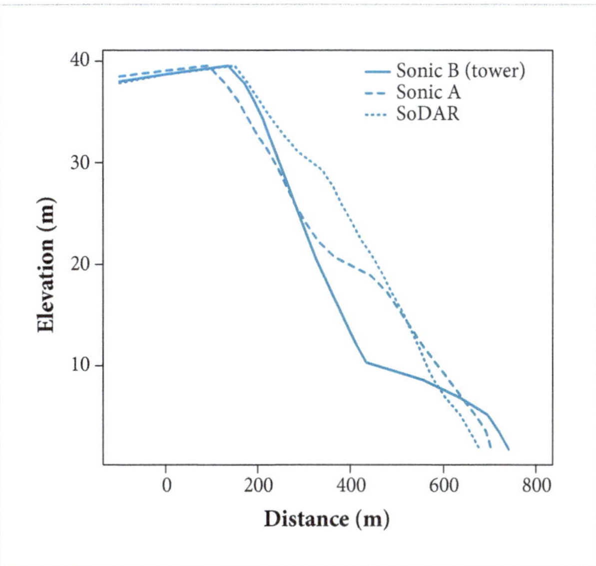

Figure 3. Cross section of the terrain's elevation through Sonic A, Sonic B, and SoDAR, along the prevailing wind direction (NE). Zero in x-axis refers to the exact sensors' position, right x-axis refers to upwind distance, and left x-axis to downwind distance. The elevation data were obtained from a local topographic map.

At 750 m inland there was a 70 m tall tower (taller vertical segment in Fig. 2) with a wind profile composed by six propeller anemometers at the levels 6, 10, 16, 28, 43, and 70 m. A 3D sonic anemometer (Sonic B), from Campbell Scientific, Inc. (model CSAT3), was mounted on the tower at 9.5 m above ground level (AGL). Gisler *et al.* (2011) describes some climatological features of this tower. At 700 m inland, there was a second sonic anemometer (Sonic A – shorter segment in Fig. 2) installed at 9.5 m (AGL) on a mast. This site was located in a clearing with sparse small grass. The clearing, with Sonic A behind, is shown in Fig. 5 upper panel. The prevailing wind direction is from the bushes on the right side of the picture. The upwind natural vegetated fetch of Sonic B is presented in the lower panel of Fig. 5. The sonic boom points towards the direction from where the wind comes from.

The broader view of the details of the landscape of the region, where the tower and Sonic A were located, is presented in an aerial photo of the region (Fig. 6). Inside the clearing, where Sonic A was located with respect to the direction of the

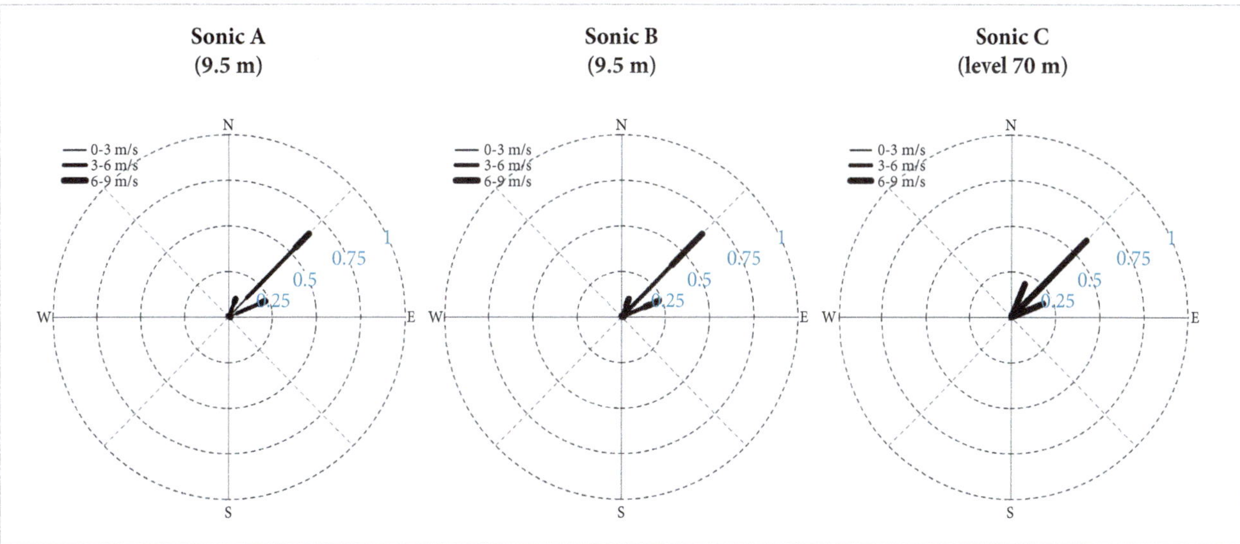

Figure 4. Windroses for Sonic A (700 m inland), Sonic B, and for the anemometer at the top of the tower. Sonic A was 700 m inland, and Sonic B plus tower was 750 m inland.

Figure 5. Sonic A mounted on a mast 9.5 m above ground level at a clearing site with sparse short grass, and lower panel shows Sonic B mounted at the same height but on a tower, with upwind fetch composed by dense bushes of 3 m height.

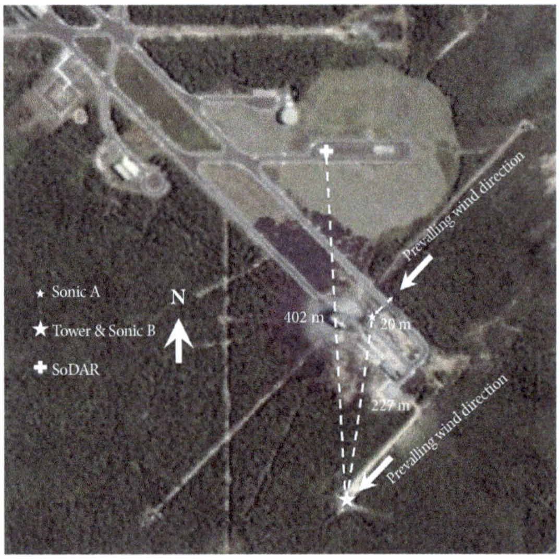

Figure 6. Aerial photo of the region where the tower, the two Sonics (A and B), and the SoDAR were located. This picture was taken in 2003 just after an accident with a rocket. By 2008, the burnt vegetation (black area near Sonic A and building) had recovered and the damaged buildings were removed from the site. The dashed lines indicate the distance between the tower and Sonic B and Sonic A (227 m), between tower and SoDAR (402 m), and the upwind distance between Sonic A and natural vegtation (20 m).

prevailing wind, there is an upwind fetch of 20 m, which is half composed by sparse short grass and half by asphalt. Beyond that, the fetch is composed by same natural vegetation that is present at Sonic B's site. At the tower there is no surface cover change for a fetch of at least 200 m long. Note that these details are also presented in Fig. 5, and we are excluding most part of the slope of the escarpment in the discussion of the fetch here.

The wind profile at the tower is permanent, but the two Sonics were temporary and operated from September 17–25, 2008. This period of two weeks is within the dry season in the region, and according to Fisch *et al.* (2010) is when the easterly winds are the strongest (10–15 m/s). The wind speed and wind direction data from the profile are sampled at 2 seconds but only 10-minute averages are kept. All turbulent fluxes refer to correlation of velocity with velocity, and velocity with temperature during half-hour periods. The Reynolds mean removal was done using 30-minute block averages, and fluctuations were obtained by subtracting the average values from the instantaneous values. Near the observation points the surface was flat and only horizontal rotation was applied to align the x-axis of the reference frame with the direction of the mean wind. Because there was no fast response humidity and pressure measurements available, we approximated the buoyancy flux $\overline{w'\theta_v'}$ by the virtual sonic temperature flux $\overline{w'T_{vs}'}$. Where w', θ_v' and T_{sv}' are, respectively, fluctuations of vertical wind velocity, virtual potential temperature, and derived sonic virtual temperature. We are aware of the error we may introduce when doing such approximation. Fluctuations of virtual temperature T_v' are not exactly the same as T_{sv}'. Nonetheless, fluctuations of sonic virtual temperature are sensitive to fluctuation of humidity as does virtual temperature, but in T_v' the influence of humidity fluctuation is slightly higher. There exists also influence of pressure fluctuation on the buoyancy flux, but we believe it is negligible. In order to simplify our notation in the whole rest of the text, we have dropped the letter "*s*" in the subscript of $\overline{w'T_{sv}'}$.

Wind data from an acoustic Doppler detecting and ranging device - SoDAR (from Atmospheric Systems Corporation with frequency operation range between 4500 and 5500 Hz) were used as complementary data for this study. The wind profiler (SoDAR) operated from October 11 to 14 of 2011, and was located inside a second clearing (Fig. 6) at 6 m east and 402 m north from the tower, approximately 675 m from the ocean (Fig. 3).

RESULTS

INTERNAL BOUNDARY LAYERS

The windroses (Fig. 4) obtained from the tower top level (70 m) aerovane, and from both sonics indicate that the average flow direction is approximately uniform within the height of the tower and within the horizontal extension of the experimental array size. The wind is predominantly from the sector north-north–east (NNE), with ~ 70% of the time with winds coming from NE. This result is somewhat similar to what Gisler *et al.* (2011) found using a longer wind data set. They showed that for dry season, for a period between the years 1996–1999, the predominant winds were 45% from NE and 40% from ENE, and the remaining 15% from NNE and E. In a scale of about 10 km, the costal line is aligned north-north-west (NNW) - south-south-east (SSE), and winds reaching the tower and Sonic A are necessarily from the ocean (Fig. 2). Because in the direction NE to SW the surface drastically changes, the lower part of the marine boundary layer develops an IBL when it reaches the coast (Arya, 2011). Using the IBL formula $z_{IBL} \approx A\, z_{0L}^{0.2} x^{0.8}$ (Elliot, 1958), with A=0.75 (Walmsley, 1989; Garratt, 1990) and a roughness length (z_0) of 0.35 m, and the distance from the steepest part of the escarpment to the tower along the predominant wind direction (200 m), gives an IBL of approximately of 42 m deep at the tower. The 0.35 m roughness length z_0 was obtained by adjusting a neutral logarithmic wind profile to the tower (Fig. 7).

Here, we have to point out that the situation is a bit more complicated than the types of IBLs discussed in the literature (e.g., Stull, 1988; Garratt, 1990; Arya, 2001). Besides the surface roughness changes, there is a topographical barrier. According to Marinho *et al.* (2009) the flow obstruction caused by the barrier forces the flow to accelerate and to develop a recirculation. The recirculation train extends from where the plateau levels off to about 200 m inland. However, the average wind vector field for a layer between 30 and 65 m AGL, obtained from the SoDAR data, shows a horizontal flow with no reversal in the flow direction. The main difference between this study and Marinho *et al.* (2009) is that the analysis presented here is based in the real topography with approximately 8 degrees inclination barrier, while in previous it was assumed step-type barrier with 90 degrees inclination. Because in the real situation

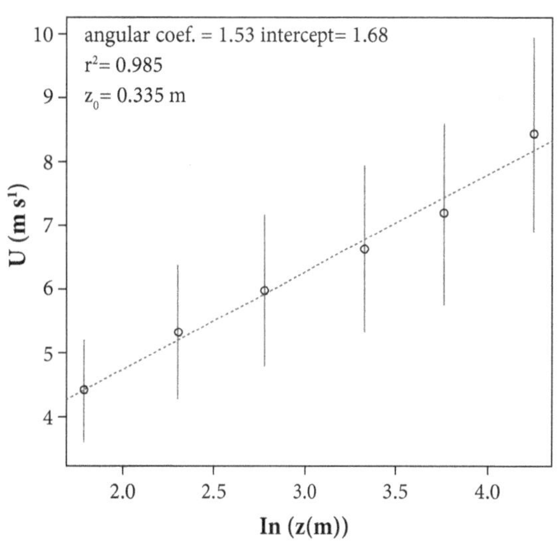

Figure 7. Circles represent tower average wind profile as function of ln(z). The dotted is a neutral logarithmic curve, $U(z) = (u^*/k)\ln(z/z_0)$, adjusted to the tower profile. The angular coefficient, the linear coefficient, r^2, and z_0 refer to the dotted line. The upper and lower ends of the vertical solid lines are the speed standard deviation added and subtracted, respectively, from the average speed.

the slope is significantly more gentle and the measuring points were more than 200 m inland from the steepest part of the escarpment, we believe that the recirculation zone, if existent, will be located upwind and the generated wake will be largely diffused by turbulence by the time the flow reaches the observing points. Therefore, the complicating factors added to the flow due to the barrier should not be the cause for different turbulent regimes at the observing points. In addition, because the IBL height (42 m) is above the height of both sonics (9.5 m), the IBL should be close to equilibrium at the sonics, in sense that the flow is not accelerating or decelerating.

Turbulence and Mean Flow State

The first result, one finds when looking to the average time series of wind, friction velocity, $\overline{w'T_v'}$, and turbulent kinetic energy (*TKE*) for Sonic A and Sonic B (Sonic A is ~ 227 m north of Sonic B), is that the mean wind speed is stronger at Sonic B than at Sonic A (Fig. 8a, b, and c) but u_* and *TKE* are rather weaker. The friction velocity u_* was obtained through $u_* = \sqrt{-\overline{u_r' w'}}$. The u_r and v_r are, respectively, the velocity components along and perpendicular to mean wind direction, obtained through 2D-horizontal rotation, w the vertical velocity component, and primes denote perturbations. The mean wind

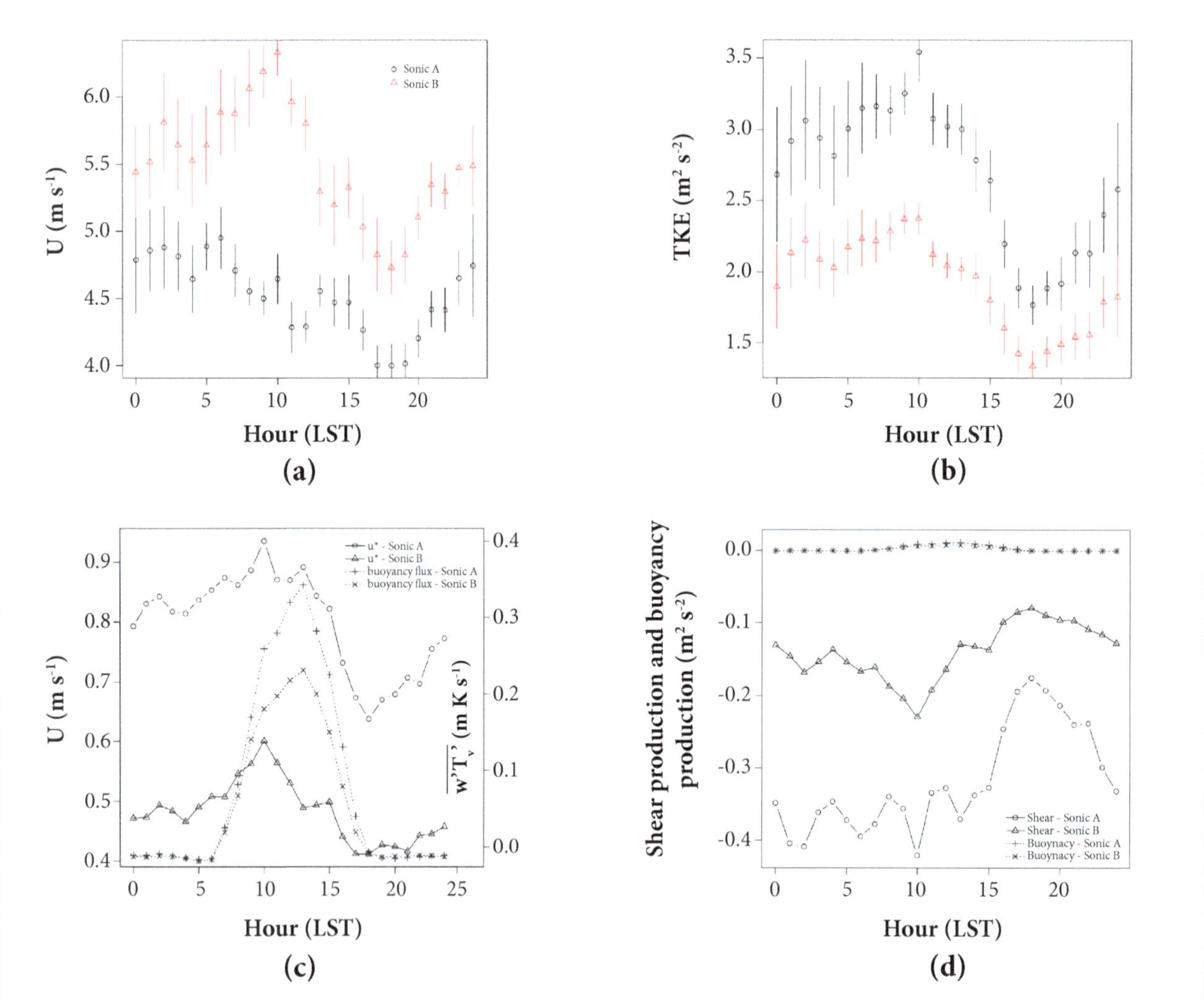

Figure 8. Twenty-four-hour cycles for (a) wind speed for Sonic A (black line) and B (red line), (b) turbulent kinetic energy for Sonics A and B, (c) friction velocity u_* and buoyancy flux $\overline{w' T_v'}$ for sonics A and B, and (d) shear production, and buoyancy production destruction terms for Sonics A and B. The vertical bars in (a) and (b) are the standard errors. Each point refers to 1- hour bin-averages with 18 data points. The error bars include natural variability of the flow and instrumentation errors as well. However, differences between measurements, caused by instrumentation errors, should be minimal because sensors were kept the same.

was obtained through $U = \overline{u_r}$ and $TKE = 1/2\,(\overline{u'^2} + \overline{v'^2} + \overline{w'^2})$. For all these variables, we used the sonic 10 Hz data. The second result is that, shear production term ($SP = \overline{u_r'w'}\,\partial U/\partial z$), and buoyancy production/destruction term ($BPD = (g/T_{ref})\,\overline{T_v'w'}$) are smaller at Sonic B than at Sonic A (Fig. 8d). The vertical gradient of the horizontal wind was determined using the *U*, and the sonic measuring height (9.5 m AGL). Velocity at the surface was set equal to zero. Having in mind that origin of the wind (e.g., pressure gradient and advection of momentum) was the same and that the measurements were taken at the same height AGL at Sonic A and B, these results clearly show the consequence of stronger turbulence at Sonic A. It has to be recalled that Sonic A was located in clearing and Sonic B was not. Wind speed, *TKE*, and *BPD* at the tower (Sonic B) and clearing (Sonic A) come somewhat closer only at night (1700–0500 local standard time (LST)), when the layer is weakly stable. Sonic B shows wider distribution than Sonic A for the stability parameter (*z/L*), indicating that the flow regime at the tower is less neutral than at the clearing as is presented in Fig. 9. Here, *L* is the Obukhov length ($L = u_*^3 / k\,(g/T_{ref})\,\overline{w'T_v'}$, where *k* is the Von Karman constant = 0.4) and *z*=9.5 m, is the height AGL of the sonics. Values of *z/L* at Sonic A resemble more the values found within the surface layer closer to the surface, with wind shearing being the dominant source of turbulence and buoyancy a secondary source. At the tower it resembles more the turbulence found over higher levels in the boundary layer, with somewhat diminished influence of wind shearing on the generation of turbulence.

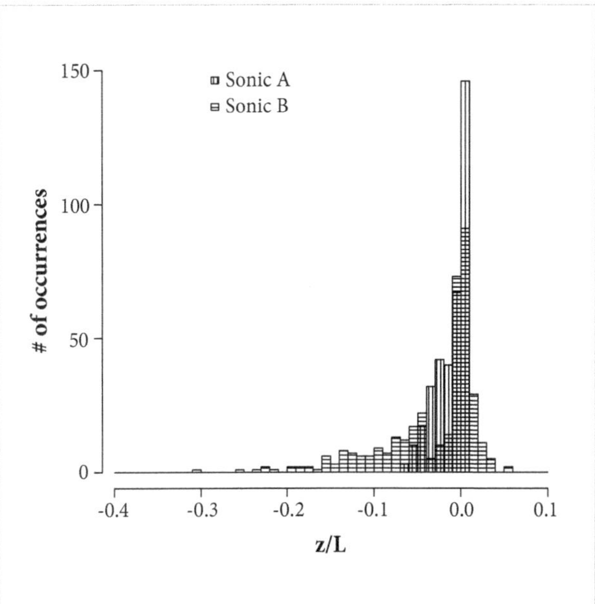

Figure 9. Distribution of *z/L* stability parameter for Sonics A and B. All cases were included.

The wind speed, *TKE*, *SP*, and *BPD* show a clear diurnal cycle, peaking in the morning before noon and being a minimum around 1800 hours of LST. Note though that in *SP* there is a negative peak, with this variable almost completely mirroring wind and *TKE*. In the *TKE* budget equation there is minus sign in front of *SP*, which makes this terms always production term. The objective here is not to close the *TKE* equation, but rather to quantify the sources of turbulence generation, *SP* and *BPD* at the clearing and tower, which makes the turbulence at these places to be distinct. At the clearing, *SP* and *TKE* are in phase with each other but wind is not so. Between 0500 and 1000 LST there is a clear decreasing tendency for the wind but not so for *SP*. Note that just after 0500 LST the wind is maximum but *SP* is not. Shear Production (Fig. 8d) only becomes a maximum at 1000 LST, when the wind is weaker. The lack of clear tendency in *SP* and the apparent tendency for wind to decrease between 0500 and 1000 LST might be a result of averaging a small data record. Nine days of data were necessarily a small sample. In fact, when plotting the entire time, wind series (result not shown), we do see it increasing during this period for a few days. Perhaps, an entire dry season period would be enough to show that *U* increases between 0500 and 1000 LST. Beyond 1000 LST the wind and *SP* become stronger and negatively related (Fig. 8a and d), meanwhile *TKE* seems to be slightly less dependent on *SP*. Just after 1000 LST *TKE* decays not following the increase in decay in *SP*, but further on until 1500 LST it levels off while wind and *SP* vary. During this period the *BPD* is strongest and it probably helped to keep *TKE* near constant. At the tower there is a similar situation. Immediately after 1000 LST, wind and *SP* decrease forcing *TKE* to diminish. Beyond this point until 1500 LST, the approximately constant values of *TKE* are partially kept by the strong *BPD*. As a bottom line, we can conclude that the main difference between the clearing and the tower is that, there is stronger turbulence at the former. The fact that the *U*, *SP*, and *TKE* are not in phase with each other is consequence of sampling a small data record. The observed 24-hour wind cycle maybe a result of effects of variable local horizontal temperature gradient (sea breeze) interacting with the local trade winds. In a recent study conducted at the costal and central regions of Maranhão State, Medeiros and Fisch (2012) observed that 24-hour wind hodograph was the result of the sea breeze interacting with a large-scale southerly flow during the dry season.

Spectra

The normalized spectra $f\,S_\alpha/(u_*^2\ \phi_\varepsilon^{2/3})$, where $\alpha = u_r$, v_r, and w, reveals that at Sonic B the turbulence spectral peak is at low frequencies while at Sonic A the peak is displaced towards high frequencies. This result is clearly seen in Fig. 10, longitudinal velocity u_r (along mean wind direction) and lateral velocity v_r (crosswind – result not shown) but not in the vertical (w) velocity (Fig. 11). Such differences indicate that at Sonic A site turbulence is more fully developed because of higher *SP*. Turbulence at Sonic B is less isotropic than at Sonic A, its w-spectrum shows an inertial, but its u-spectrum has neither subinertial range nor constant slope at high frequencies. The normalized *TKE* dissipation rate, $\phi_\varepsilon = kz\varepsilon/u_*^3$, was determined by adjusting through least-square method $f\,S_u/u_*^2 = 0.3\phi_\varepsilon^{2/3}\,n^{-2/3}$ (Kaimal and Finnigan, 1994) to the inertial subrange of u_r-spectrum, where n is the normalized frequency (fz/U), and f is the frequency in Hertz. The ϕ_ε was calculated for every 30-minute spectrum and later used to obtain the 30-minute normalized spectra of u_r, v_r, and w, which were finally averaged (Figs. 10 and 11). Turbulence mechanically generated (e.g., through *SP*) has spectrum concentrated at higher frequencies interval than do turbulence generated by convection (e.g., through *BPD*). The *SP* supplies energy directly to u_r and *BPD* to w, therefore the shifting in u_r-*spectrum* is a direct consequence of stronger *SP*. For the remaining spectral components (w and v_r), it is an indirect consequence, because they have their energy supplied by correlation pressure term, which is a term that works in a way to make the turbulence more isotropic by removing energy from u_r and putting in w and v_r.

Combining the average wind speed with the inverse of the u_r spectral peak frequency, one finds that typical scale size of horizontal eddies are in the order of 50 m for A and 120 m for B. Consequently the clearing, which is less than 100 m wide, should not support the typical eddy sizes present at B. Its presence not only caused higher $\overline{u_r'w'}\ \partial U/\partial z$ and

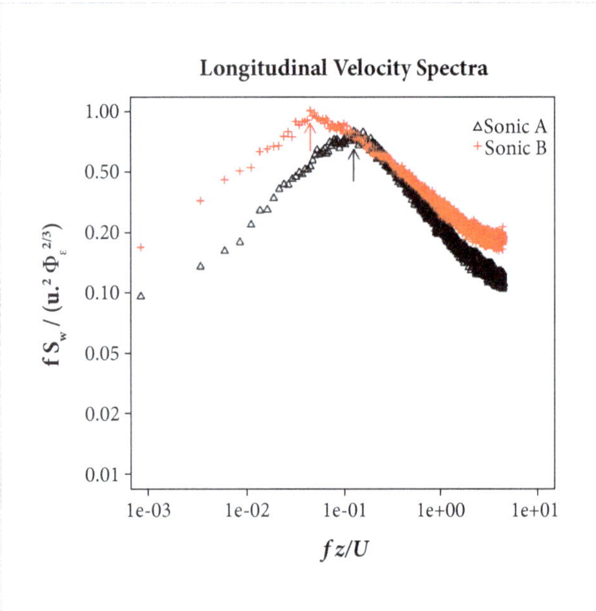

Figure 10. Normalized u-spectral energy density as a function of normalized frequency ($n = fz/U$). Each point represents a binned average of all 30-minute spectra. The bins were defined by constant increments in n. The shifting in the spectral peak of u from Sonic A to Sonic B is 0.044 Hz and in v is 0.061 Hz.

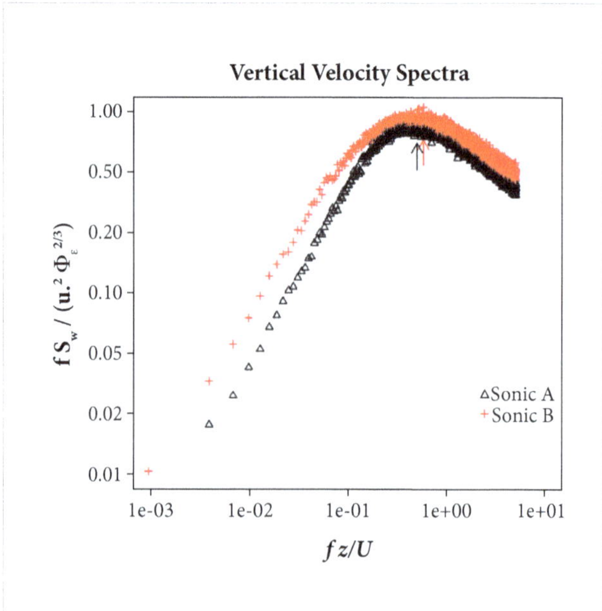

Figure 11. Normalized w-spectral energy density as a function of normalized frequency ($n = fz/U$). Each point represents a binned average of all 30-minute spectra. The bins were defined by constant increments in n. The shifting in the spectral peak of w from Sonic A to Sonic B is 0.021 Hz.

stronger mechanical turbulence, as consequence of a wake behind the trees interacting with the mean wind above, but also a restrictive factor for the size of the horizontal eddies.

WIND SPATIAL REPRESENTATIVENESS

Wind measurements taken 30 and 70 m AGL at the tower and at nearby SoDAR during the dry season 2011 (October 11–14), plus the wind measurements taken by Sonics A and B at 9.5 m AGL during the dry season of 2008 (period of Murici II experiment, September 17–25), were used to define the spatial representativeness of the wind at CLA. The 10-minute wind averages were used for the comparison. The parts of the whole data set came from two different years. However, because surface cover did not change and the wind direction and speed of the prevailing winds, during the measuring period in 2008 and 2011, were very similar, this time period difference should not compromise the analysis. The same level wind measurements were compared through the usage of the statistical indices correlation, normalized mean square error, and fractional shift presented in Table 1 (see caption for details). Although the comparison for the levels 70 and 30 m had the largest distance (402 m) and the measuring techniques were distinct (SoDAR and tower), the 70 m had the best values (scores) for the statistical indices among all. The worst scores obtained for 30 m are partially an experimental artifact caused by different equipments, and the slightly better scores for the 9.5 m are due to identical equipments, and perhaps shorter distance 227 m (Fig. 6).

The results showed that 70 m wind level is more uniform across the region. Surface inhomogeneities, like the clearing where Sonic A was located and the second clearing, where the SoDAR was located (see Fig. 6 for details), have less influence at this level as one would expect. This height is near the top of the calculated IBL, and it is possible that it is above the first real IBL, which forms near the edge of the plateau. The 30-m results indicate that this level is the least spatial homogenous; however, because of the problems mentioned above we still believe that the 30 m level is spatially more homogeneous than the 9.5 m level.

CONCLUSION

Even though the experimental array was limited, two places with turbulence measurements, and the campaign duration was short, less than two weeks, we can still draw some qualitative and quantitative conclusions. For a distance of 200 m (Fig. 3) downwind from the edge of the escarpment, near-surface turbulence is not spatially homogeneous at CLA launching pad site. It is a consequence of the development of other IBL at the clearing, internally to a first one, which develops when

Table 1. For the 70 m and 30 m levels, the tower 70 m and 28 m wind measurement levels were compared with the 30 m and 70 m of the SoDAR. The SoDAR was located 402 m north from the tower (Fig. 6) and operated at that location from October 11 to 14, 2011. For the 9.5m level, we used two sonic anemometers 227 m apart, which operated from September 17 to 25, 2008.

	Spatial representativeness of the wind					
	Zonal wind speed ($\overline{U}$)			Meridional wind speed ($\overline{V}$)		
Measurement height (m)	9.5	30	70	9.5	30	70
Distance between measuring points (m)	227	402	402	227	402	402
Correlation*	0.81	0.75	0.83	0.87	0.76	0.83
Normalized mean square error**	0.04	0.02	0.01	0.03	0.03	0.02
Fractional shift***	0.79	-0.63	-0.32	0.34	0.21	0.08

*The correlation was determined by $\sum_{i=1}^{N} (x_i - \bar{x})(y_i - \bar{y}) / (\sigma_x \sigma_y)$, where x_i and y_i represent, respectively, the i-th measurement from two different sensors, and σ_x and σ_y standard deviations of x and y sensors. Optimum value is one.

** Normalized mean square error was computed by $\sum_{i=1}^{N} (x_i - y_i) / N \bar{x} \bar{y}$. Optimum value is zero.

*** Fractional shift was computed by $2 (\sigma_x - \sigma_y) / (\sigma_x + \sigma_y)$. Optimum value is zero.

the wind first encounters the escarpment. Due to our limited experimental array, we could not determine the extension of the first and second IBLs. *TKE*, shear, and buoyancy are different even within a distance of 227 m (Fig. 6). The point closer to the escarpment (Sonic A), but in a clearing, had stronger u_*, *TKE*, SP (also true for momentum flux), and buoyancy flux (buoyancy production term) but weaker wind. Part of the increment of *TKE* at the tower and clearing during the sunlight hours, was due to stronger buoyancy flux. The stronger buoyancy at the clearing must have been a consequence of the asphalt's higher surface temperature, upwind inside the clearing. The presence of the clearing was also the reason for higher u_* and weaker wind. We suppose that a wake zone is developed behind trees, that delimit the clearing perimeter, and it interacting with the mean wind increased the removal of momentum from it.

As a bottom line of this analysis, the wind, *TKE*, and fluxes measured at the tower are not representative of the conditions at the clearing where the rocket launcher is located. Nonetheless, in the absence of better measurements, the spatial analysis of the wind components ($\overline{U}$,$\overline{V}$) showed that the tower top-level (70 m) wind might be the least surface affected measurement. It can be used as a first approximation to represent the wind for the region at that level, relative to operational decisions concerning rocket launch and diffusion of pollutants exhausted by the rockets.

ACKNOWLEDGMENTS

The authors acknowledge the financial support from the Conselho Nacional de Desenvolvimento Científico e Tecnológico (CNPq) under the grants 510159/2010-9 (Medeiros), PQ 303720/2010-7 (Fisch), 471143/2011-1 (Marciotto), and Fundação de Amparo a Pesquisa do Estado de São Paulo (2010/16510-0). The authors are also thankful to NASA Land Processes Distributed Active Archive Center (LPDAAC); ASTER L1B; USGS/Earth Resources Observation and Science (EROS) Center, Sioux Falls, South Dakota 2001 for providing the topographic data. The authors also acknowledge the valuable suggestions made by reviewers to make the work better.

REFERENCES

Arya, S.P., 2001, Introduction to Micrometeorology. Academic Press, San Diego.

Case, J.L., Wheeler, M.M., Manobianco, J., Weems, J.W. and Roeder, W.P., 2005, A 7-yr climatological study of land breezes over the Florida spaceport. Journal of Applied Meteorology, Vol. 44, No. 3, pp. 340-356.

Elliot, W.P., 1958, The growth of the atmospheric internal boundary layer. Transactions of the American Geophysical Union, Vol. 39, No. 6, pp. 1048-1054.

Fisch, G., Avelar, A.C., Marinho, L.P.B., Gielow, R., Girardi, R.M. and Souza, L.F., 2010, The internal boundary layer at the Alcântara space center: wind measurements, wind tunnel experiments an numeric simulations. In: Proceedings of the Fifth International Symposium on Computational Wind Engineering (CWE2010), Chapel Hill, North Carolina, USA May 23-27, 2010.

Garratt, J.R., 1990, The internal boundary layer – A review. Boundary-Layer Meteorology, Vol. 50, No. 1-4, pp. 171-203.

Gisler, C.A.F., Fisch, G. and Correa, S. C., 2011, Statistical analysis of wind profile in the surface layer at the Alcântara launching center. J Aerospace Management, Vol. 3, No. 2, pp. 193-202.doi: 10.5028/jatm.2011.03022411

Kaimal J.C. and Finnigan, J.J., 1994, Atmospheric boundary layer flows – their structure and measurement. Oxford University Press, New York.

Mahrt, L. and Vickers, D., 2005, Boundary-layer adjustment over small-scale changes of surface heat flux. Boundary-Layer Meteorology, Vol. 116, No. 2, pp. 313-330.

Marciotto, E.R., Fisch, G. and Medeiros, L.E., 2012, Characterization of surface level wind in the Centro de Lançamento de Alcântara for use in rocket structure loading and dispersion studies. J Aerospace Technology and Management, Vol. 4, No. 1, pp. 69-79. doi: 10.5028/jatm.2012.04014911

Marinho, L.M.B., Avelar, A.C., Fisch, G., Roballo, S.T., Souza, L.F., Ralf, G. and Girardi, R.M., 2009, Studies using wind tunnel to simulate the atmospheric boundary layer at the Alcântara Space Center. J Aerospace Technology and Management, Vol. 1, No. 1, pp. 91-98. doi: 10.5028/jatm.200901019198

Medeiros, L.E. and Fisch, G., 2012, Low atmospheric flow at Centro de Lançamento de Alcântara (CLA) and surrounding areas of the north part of the Maranhão State. In: Proceedings of the XVII Congresso Brasileiro de Meteorologia, Gramado, RS, Brazil.

Merceret, F.J., 2006, Rapid temporal changes of boundary layer winds. Journal of Applied Metorology and Climatology, Vol. 45, No. 7, pp. 1016-1020.

Moreira, D.M., Trindade, L.B., Fisch, G., Moraes, M.R., Dorado, R.M. and Guedes, R.L., 2011, A multilayer model to simulate rocket exhaust clouds. J of Aerospace Technology and Management, Vol. 3, No. 1, pp. 41-52. doi: 10.5028/jatm.2011.03010311

Stull, R.B., 1988, An introduction to boundary layer meteorology. Kluwer Academic Publishers, Dordrecht.

Walmsley, J.L., 1989, Internal boundary layer height formulae – a comparison with atmospheric data. Boundary-Layer Meteorology, Vol. 47, No. 1-4, pp. 251-262.

18

Infrared Spectroscopy Applied to Materials Used as Thermal Insulation and Coatings

Natália Beck Sanches[1], Ricardo Pedro[1], Milton Faria Diniz[2], Elizabeth da Costa Mattos[2], Silvana Navarro Cassu[2], Rita de Cássia Lazzarini Dutra[2]

ABSTRACT: It is very important to control the characteristics of a polymer used in rubber compositions and paints in aerospace applications. Hence, the development of simple and fast methodologies that allow the identification of these compositions becomes attractive to researches carried out in this area. This study has evaluated infrared (IR) techniques, such as transmission, universal attenuated total reflection (UATR), and attenuated total reflection (ATR), for the characterization of elastomers and paints. It takes into consideration the characteristics of surface techniques, such as the depth of penetration of the IR beam into the sample. The presence of additives in low concentrations on paints was only detected by UATR after grinding the components. Results show that it is possible to differentiate rubber mixtures with similar IR spectra and to detect small amounts of additives in the surface of coatings.

KEYWORDS: Rubbers, paints, infrared, FT-IR, ATR, UATR.

INTRODUCTION

Polymers are used in many industrial areas, such as the aerospace (Dutra *et al.*, 2002; Oliveira *et al.*, 2011). Their different types, with specific characteristics, are used in propellant (Sciamareli *et al.*, 2012), composites, rubbers (Ferrari *et al.*, 2012; Santos *et al.*, 2013), and coatings (Blackford, 1999).

The resin most often utilized in aerospace applications is hydroxyl terminated polybutadiene (HTPB), which is used in propellants and flexible insulation protection (Lourenço *et al.*, 2006; Crespim *et al.*, 2007). It is known that phenolic resins are employed in composites for rigid thermal insulation, while epoxy are used in adhesive compositions for these same protections and in polymer arrays for the aeronautical sector (Pardini, 2000). Rubbers are used in flexible thermal insulation, and the most cited ones in literature for using in aerospace applications are copolymers of acrylonitrile and butadiene (NBR) and ethylene-propylene-diene rubber (EPDM) (Moraes *et al.*, 2007).

Traditionally, aircraft paints are coatings with specific characteristics and requirements, because in extreme cases the use of a defective one may contribute to injuries in a catastrophic scale (Blackford, 1999). The requirements are associated with the environment in which modern aircrafts are used and how they are painted, among other factors. Thus, there are significant differences in formulations used for the aerospace sector compared to other industrial areas, allowing the use of products that can cause damage to people's health, such as chromate pigments, or to the environment, like large amounts of solvents.

1. Instituto Tecnológico de Aeronáutica – São José dos Campos/SP – Brazil 2. Instituto de Aeronáutica e Espaço – São José dos Campos/SP – Brazil

Author for correspondence: Rita de Cássia Lazzarini Dutra | Praça Marechal Eduardo Gomes, 50 – Vila das Acácias | CEP 12.228-904 São José dos Campos/SP – Brazil | Email: ritarcld@iae.cta.br

Generally, solvents are used to decrease viscosity and facilitate the application of paints. The vehicle or binder is the chemical component that will generate a film on the surface, besides being responsible for keeping the other constituents scattered and clumped. Fillers are materials that provide specific properties to paints, for example, the magnetic nanoparticles that turn aircrafts non-detectable to radars. The additives are compounds used from storage to product application. For example, microbicides preserve the paint during its storage and catalysts reduce the time of film formation during product application (Mello *et al.*, 2012).

With regard to the analysis of polymers used in the aerospace industry, there are studies in the literature (Sanches *et al.*, 2006; ASTM D3677-10, 2010; Romão *et al.*, 2006; Hori *et al.*, 1990) that employ mid infrared (MIR) spectroscopy for identification, characterization, and quantification of materials. In these investigations, analyses of materials using Fourier transform infrared (FT-IR) techniques with variation of the parameters, in different layers of the surface, were not explored.

It is also known that some components present in paints cannot be directly analyzed by using conventional FT-IR techniques, because the characteristic bands overlap with absorptions of other components often seen in high concentrations. Although the study of bands overlapping could be done by using mathematical methods, such as derivatives and deconvolution to quantify components, these ones, in some cases, are complex and depend on the thickness of the analyzed samples. This limitation becomes even more relevant when additives are present in extremely low concentrations. A specific methodology must then be studied.

The FT-IR literature emphasizes that, in the analysis of multiple components systems, there are factors to be considered: influence of sample preparation, total or partial analyses of composition (Allen, 1992; Dutra *et al.*, 1995), and the technique used to obtain the spectrum, single or coupled (Mattos *et al.*, 2004; Almeida *et al.*, 2002; Mateo *et al.*, 2009). In these cases, it is helpful to perform the evaluation of characteristic functional groups, the query to the database of reference spectra (Smith, 1979) using spectra software (Szafarska *et al.*, 2009), and the interpretation of similar studies to highlight the limitations and potential of techniques (Miliani *et al.*, 2002).

Different FT-IR techniques may be suitable for the study of many kinds of materials, according to their characteristics. The most common surface technique is the attenuated total reflection (ATR), which can be used for the analysis of liquid and dried paint films (Mazzeo *et al.*, 2007; Zhang *et al.*, 2009). In the ATR spectroscopy, the radiation passing through the crystal reflects totally on its internal surface (Waltham, 2005). When a radiation absorbing material is placed in contact with a crystal, the infrared (IR) beam penetrates the thin layer of the sample surface and loses energy, causing the attenuated total reflectance. In this technique, good contact between the sample and the crystal is essential (Pandey and Kulshreshtha, 1993).

The penetration depth of the IR beam into the sample depends on the IR radiation angle of incidence, the radiation wavelength, and the refractive indices of the crystal and sample. Ge crystals are able to analyze thinner surfaces (Pandey and Kulshreshtha, 1993). In the ATR technique, different species of the sample surface and interior can be revealed depending on the analysis procedure. The literature reports (Pandey and Kulshreshtha, 1993; PerkinElmer, 2005) and experiences indicate that this aspect must be considered if the study requires selective analysis of the very superficial layer.

Ultimate FT-IR techniques, such as the universal ATR (UATR), have found prominence in the analysis of different materials (PerkinElmer, 2005; Abidi and Hequet, 2005). This technique of internal reflection is able to perform nondestructive analysis of solids, powders, liquids, and gels. In it, the IR beam passes through an ATR element composed of ZnSe (refractive index 2.67) or diamond-KRS-5 (refractive index 2.42), with high refractive index, and reaches the surface of the sample.

One of the prerequisites of the UATR technique is good contact between the crystal and the sample surface. The probe strength can be adjusted to obtain the most suitable contact, since different pressure levels influence directly on the intensities of the obtained spectra. Because the IR beam does not penetrate deeply into the sample, this technique is suitable for analyzing surfaces with thickness of only a few microns (Pandey and Kulshreshtha, 1993).

It was observed in the literature some gaps related to the use of surface FT-IR techniques, varying the degree of sampling depth. Hence, it was concluded that it is essential the development of new characterization methodologies that enable rapid identification of the different materials used in the aerospace industry. The aim of this work was to show the new results obtained by the Chemical Department of the *Instituto de Aeronáutica e Espaço* (AQI/IAE), in the characterization of different types of materials by using FT-IR reflection techniques, such as ATR and UATR. In addition, the transmission FT-IR technique was applied to identify rubber mixtures.

EXPERIMENTAL

The nitrile rubber (NBR) and rubber mixtures used in this study were prepared in AQI/IAE. The paint and additives were provided by Rohm and Haas. The chemical structure of the additives carbendazim (2-methoxycarbonylamino-benzimidazole), Diuron (N'-(3,4-dichlorophenyl)-N, N-dimethylurea)) and OIT (2-N-octyl-4-isothiazolin-3-one) is shown in the text. Portions of the samples, before and after extraction with acetone, were also subjected to Beilstein and acid-resistance tests (Dutra and Diniz, 1993). A treatment with heated ortho-dichlorobenzene was performed as well.

EVALUATION OF TRANSMISSION AND REFLECTION FT-IR TECHNIQUES FOR CHARACTERIZATION OF RUBBERS

NBR was analyzed qualitatively by transmission, preparing the sample by pyrolysis (Smith, 1979) after extraction in methanol for eight hours, using the liquid film technique. For the UATR analysis, the NBR sample was placed in contact with the surface of the crystal and a force of 80 N was applied with the aid of an articulated arm. For ATR (KRS-5 or Ge) analysis, the NBR sample was placed on both sides of the crystal. In this study, KRS-5 (TlBr-crystal mixture TlI) with refractive index 2.4 and Ge with refractive index 4.0 crystals were utilized. The filler was analyzed by transmission, preparing the specimen with the KBr disc technique.

EVALUATION OF TRANSMISSION FT-IR TECHNIQUES FOR CHARACTERIZATION OF RUBBERS AND THEIR MIXTURES

An unknown mixture of rubbers was analyzed qualitatively by transmission, preparing the sample by pyrolysis after extraction for eight hours in acetone, using the liquid film technique (Smith, 1979). All analysis were performed using the spectrometer Spectrum One PerkinElmer (resolution 4 cm^{-1}, gain 1, spectral range 4000–400 cm^{-1}, 40 scans).

UATR ANALYSIS OF PAINTED SURFACE

A paint film was applied on a flat glass substrate. After drying, the film was removed by scraping and then analyzed. The paints additives were analyzed directly, without the need of sample preparation. UATR analysis was performed using the spectrometer Spectrum One, PerkinElmer, DTGS detector (resolution 4 cm^{-1}, spectral range 4000–550 cm^{-1}, gain 1, 20 scans). Spectra were obtained using the accessory UATR and applying a force of 80 N.

RESULTS AND DISCUSSION

This section presents the evaluation of FT-IR techniques for the analysis of elastomers and paints, using methods traditionally applied and developed in AQI/IAE laboratories.

APPLICABILITY OF TRANSMISSION AND REFLECTION FT-IR TECHNIQUES FOR CHARACTERIZATION OF RUBBERS

It is known that conventional FT-IR techniques can be applied to the characterization of rubbers (Matheson *et al.*, 1994; Wake *et al.*, 1983; Williams and Besler, 1995). In this work, different information is obtained by using independently transmission (liquid phase of pyrolyzate) and reflection (ATR and UATR) techniques. These were used for the characterization of the polymer and the filler, respectively.

Figure 1 shows the FT-IR spectra obtained from NBR analysis by using different techniques. It is observed that the analysis of the liquid phase of the pyrolyzate resulted in a spectrum (Fig. 1a) with the base polymer characteristic absorptions. The band observed at 2237 cm^{-1} is assigned to the stretching of the C-N group, and the peaks in the region of 1000–900 cm^{-1} are assigned to the wagging bending vibration of vinyl and trans C=C groups. In the spectra obtained by reflection using different crystals (Figs. 1b, 1c and 1d), the band at 2237 cm^{-1} showed lower intensity and there was a contribution of Si-O groups in the region of 1100–900 cm^{-1}, the latter related to the load used in the formulation (Smith, 1979). Figure 1e shows these load-related bands, analyzed by transmission (KBr pellet), after rubber calcination.

Although the spectra obtained by using reflection techniques did not clearly evidence the absorptions of NBR, they were able to indicate the filler used in the formulation. In order to obtain the same filler related information by using transmission techniques, it is necessary to apply supplementary sample preparation techniques, such as KBr pellet after rubber calcination.

The use of reflection techniques can also be applied to the detection of silica filler in silicone rubbers, instead of using a large number of techniques to achieve the same goal. This is important because applying some sample preparation techniques for transmission analysis, such as calcination, can induce the formation of silicon oxide in this type of rubber.

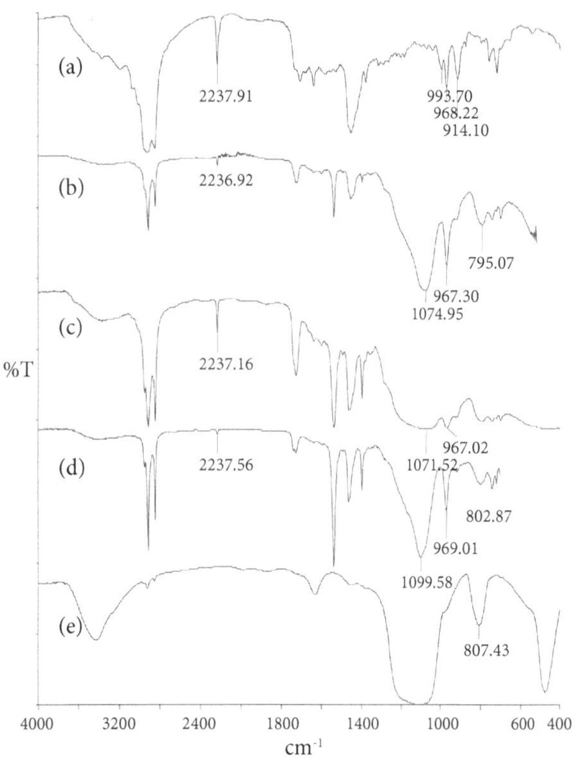

Figure 1. FT-IR spectra obtained from analysis of the nitrile rubber through different techniques: (a) Nitrile rubber transmission pyrolyzed (liquid film), (b) Nitrile rubber – UATR, force 80, (c) Nitrile rubber – ATR/KRS-5, (d) Nitrile rubber – ATR/Ge, and (e) residue obtained after calcination of rubber on Bunsen burner – transmission, KBr pellet.

It might lead to misinterpretation, since not all silicone rubbers have silica as filler.

APPLICABILITY OF FT-IR TRANSMISSION TECHNIQUES FOR THE CHARACTERIZATION OF RUBBERS AND THEIR MIXTURES

As it is known, pyrolysis of rubbers without solvent extraction is performed only to determine the base polymer and the best solvent for its extraction. Because additives bands are also observed, it is necessary to extract the rubber sample in order to obtain a spectrum that shows only characteristic bands of the polymer (Smith, 1979; Wake *et al.*, 1983). The identification of rubber mixtures by using FT-IR is possible, providing that their characteristic bands are different, without the occurrence of overlapping.

Figure 2 illustrates the application of the FT-IR transmission technique (pyrolysis without control of temperature) to an unknown mixture of rubbers, with and without extraction.

The main absorptions observed in the spectrum of the pyrolyzed rubber without extraction (Fig. 2a) and their attributions (Smith, 1979) are: 1376 cm^{-1}, angular bending vibration (δ) of CH_3 groups; 991 and 909 cm^{-1}, wagging bending vibration (ω) of vinyl group; 964 cm^{-1}, wagging bending vibration of trans groups; and 887 cm^{-1}, wagging bending vibration of vinylidene groups. These absorptions positions, shapes, and intensities suggest the presence of EPDM and/or NR (poly cis-isoprene or natural rubber) elastomers, which requires the use of acetone solvent for removal of the plasticizer and other additives (Wake *et al.*, 1983).

The sample was extracted for eight hours in Soxhlet apparatus, using acetone as the solvent. Then, it was pyrolyzed and analyzed as a liquid film (Fig. 2b). The same main bands were observed, as expected, because plasticizers used in EPDM and NR formulations are apolar and their groups do not contribute with different absorptions in the spectrum of the pyrolyzates without extraction. The band at 887 cm^{-1} is characteristic of NR rubber, but it is also present in the EPDM rubber. To further elucidate the chemical structure of the sample, a comparison was made between the IR spectrum of this sample and that of known EPDM and NR rubbers, pyrolyzed after extraction in acetone (Figs. 2b and 2c).

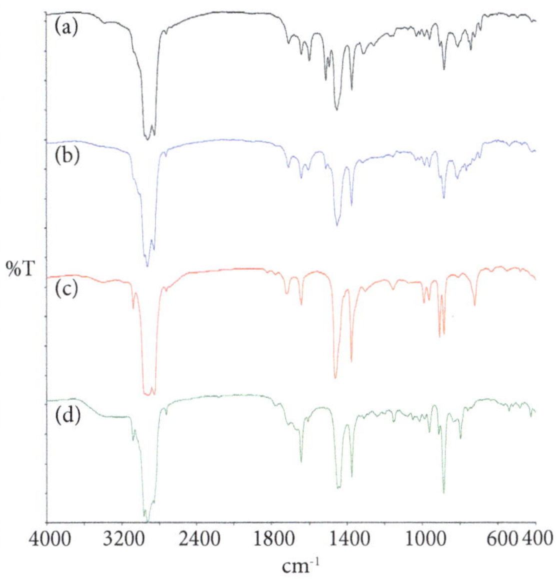

Figure 2. FT-IR spectra of (a) sample pyrolyzed without extraction, (b) sample pyrolyzate after extraction with acetone, (c) spectrum of known EPDM, pyrolyzed after extraction with acetone, and (d) spectrum of known NR, pyrolyzed after extraction with acetone.

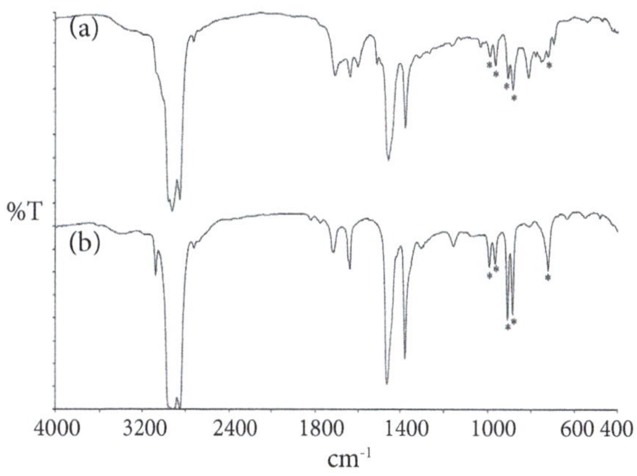

Figure 3. FT-IR spectra of (a) pyrolyzed insoluble portion, after ortho dichloro benzene treatment (previously extracted in acetone for eight hours), and (b) pyrolysed EPDM (reference).

To better identify the EPDM bands, it was also applied a treatment with hot ortho-dichlorobenzene to the sample, which was then pyrolyzed to obtain the IR spectrum. The resultant IR spectrum (Fig. 3a) showed more clearly the presence of EPDM rubber (bands marked with an asterisk), as can be visualized in Fig. 3b.

The Beilstein test was performed before and after extraction in acetone to confirm the absence of halogen compounds, expected because EPDM and NR do not present halogens in their compositions. The negative result corroborates the hypothesis of the presence of these elastomers.

The acid-resistance test, which basically classifies the rubber resistance when treated with an acid mixture at 40°C and 70°C, was also performed in the sample to elucidate the mixture composition. The sample did not show degradation at 40°C in the time established for the test, i.e. 15 minutes (Dutra and Diniz, 1993). At 70°C, it was observed degradation in around three minutes, confirming that the rubber has a certain content of EPDM. If the sample contained only NR, it would resist for only a few seconds, for less than one minute.

Basically, FT-IR analysis of the liquid pyrolysis products, associated with Beilstein and acid resistance tests (Dutra and Diniz, 1993), within the detection limits of the FT-IR technique, indicated that the unknown rubber consists of NR and EPDM.

The utilization of acid-resistance test to characterize rubber mixtures (Dutra and Diniz, 1993), also developed in IAE/AQI, is important especially in EPDM mixtures. This elastomer is considered as a replacement for NBR in the aerospace industry. The addition of NR in a rubber mixture may be intended to lower the cost, but it certainly affects properties. Hence, we can conclude that appropriate methodologies for the quality control of materials, contemplating the application of different tests and analyses, are fundamental to achieve the outlined goals, essentially in the aerospace industry, where materials require specific properties.

APPLICABILITY OF TRANSMISSION AND REFLECTION FT-IR TECHNIQUES FOR CHARACTERIZATION OF PAINTS

As mentioned, aeronautical coatings may contain hazardous products, such as chromate pigments (Blackford, 1999). The identification of chromate anions (CrO_4^{2-}) can be performed by IR spectroscopy in the region of 800 to 900 cm^{-1} (Smith, 1979).

Copper chromite is used in the aerospace industry as a catalyst precursor (Faillace *et al.*, 1999). It was characterized by transmission (KBr disc), reflection, and photoacoustic detection (PAS) FT-IR techniques in an article published by the IAE/AQI group (Campos *et al.*, 2003). It was observed that FT-IR techniques are able to indicate the catalyst method of synthesis. This is possible by using surface analysis techniques, with different degrees of sampling depth, without the need for other characterization methods. Therefore, the techniques used in this study (Campos *et al.*, 2003) have emerged as new alternatives for characterizing other types of catalysts or materials containing these kinds of compounds, such as chromium-based anions.

Certain additives, as OIT, diuron, and carbendazim, are added in small contents to preserve coatings during storage (Mello and Suarez, 2012). In this study, the methodology of chromium-based compounds (Campos *et al.*, 2003) was applied to the analysis of paints, with different degrees of sampling depth.

ANALYSIS OF PAINTS BY UATR

Figure 4 shows the UATR spectra of additives OIT, carbendazim, and diuron, painted with and without preservatives, dried resin, and filler.

The main bands observed in the FT-IR spectrum of OIT (Fig. 4a) are: 2924, 2855 and 722 cm^{-1} (stretching and rocking bending vibration of the CH_2 aliphatic group), 1618 cm^{-1} (stretching of the C=O group), as well as at 1263 and 781 cm^{-1} (vibration of five-membered heterocyclic ring). The spectrum of carbendazim (Fig. 4b) showed bands at: 3320 cm^{-1} (stretching of N-H secondary group), 1710 cm^{-1} (stretching of C=O group), 1627 cm^{-1} (stretching of the C=O group and of the C=N 5-membered heterocyclic ring), 1591 cm^{-1} (stretching of the C-C aromatic and C=N groups), 1266 and 1093 cm^{-1} (stretching of the C-O group), and 726 cm^{-1} (vibration of five-membered heterocyclic ring and of the C-H aromatic substitution group).

The spectrum of the additive Diuron (Fig. 4c) shows bands at: 3280 cm^{-1} (stretching of the N-H group), 1651 cm^{-1} (stretching of C=O and C=N groups), 1584 cm^{-1} (stretching of the C-C aromatic group), 1473, 1298 and 1186 cm^{-1} (stretching of the C-N group), 1132 cm^{-1} (wagging of C-Cl aromatic group), as well as 864 and 813 cm^{-1} (bending vibration of the CH-tri-aromatic substitution group) (Smith, 1979; Silverstein *et al.*, 2005.

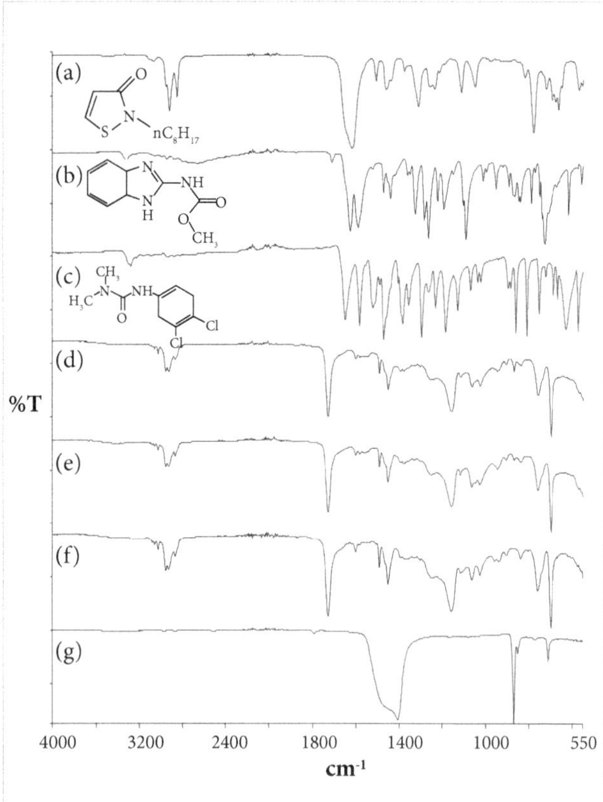

Figure 4. UATR spectra of: (a) OIT, (b) carbendazim, (c) diuron, (d) paint without preservatives, (e) paint with preservatives, (f) dry resin, and (g) calcium carbonate.

The additives OIT, carbendazim and diuron presented intense peaks that could be potentially used as analytical bands. However, the direct analysis of these compounds in the paint, by using UATR, was not possible because spectra of the paint, without (Fig. 4d) and with (Fig. 4e) preservatives are virtually identical. The resin (Fig. 4f) and filler (Fig. 4g) are present in high concentrations, so these ingredients could be the major interferences to the direct analysis. In fact, the main bands observed are associated with the resin (Fig. 4f): 1728 cm^{-1} (stretching of the C=O group), 1159 cm^{-1} (stretching of the C-O group), and 698 cm^{-1} (bending vibration of C-H aromatic substitution group).

Although the characterization of preservatives in the surface of paints by UATR was not possible in the studied conditions, it is known that they are added to the compound to control proliferation of microorganisms (Lindner, 2005). Thus, we can suppose that some content of preservatives is present on the film surface.

It is known that the diffusion phenomenon is a process that depends on the size of the involved particles. The rate of diffusion is inversely proportional to the size of the particle. Hence, a smaller particle will diffuse faster and migrate easier to the surface. Preservatives were grinded to obtain different particle sizes. These samples were studied to investigate the effect of particle size on surface analysis results.

EVALUATION OF THE INFLUENCE OF PARTICLE SIZE IN THE DETECTION OF PRESERVATIVES AT THE PAINTED SURFACE BY UATR

Results of grinded preservatives are presented in Table 1 and show that preservatives reduce their size up to one fifth, when grinded for six minutes. The values represent the average of two determinations. Stabilizers (wetting agents and dispersants) were added to the samples to prevent agglomeration of crystals. Although the resulting particle sizes are very close, there seems to be a tendency of increasing the particle size with time, indicating agglomeration of particles (Welin-Berger and Bergenstahl, 2000). This increasing is negligible compared to the magnitude and extent of the uncertainty in its determination, and therefore it was neglected.

The smaller the particle size, the most effective the preservative should be in the microbiological control. The explanation for this is that with smaller particle sizes, preservatives cover a larger surface area and can be used in lower concentrations. In smaller sizes, these additives also migrate more easily to the surface, becoming more bio-available to control microbes.

UATR ANALYSIS AFTER GRINDING OF PRESERVATIVES

The UATR paint spectra, with and without preservatives and grinding, compared to UATR reference spectra of the additives (OIT, diuron, and carbendazim), are shown in Fig. 5. The band with low intensity around 3320 cm^{-1} (N-H group), detected in the spectra of paints containing grinded preservatives (specially the six-minute grinded sample), and the peaks at 1625 cm^{-1} (C=O group and C=N 5-membered heterocyclic ring), and around 1330 cm^{-1} (C–N group), may indicate the presence of carbendazim. The absorption around 1625 cm^{-1} might have also contributed to the OIT bands around 1620 cm^{-1} (C=O group). The band around 870 cm^{-1} (C–H aromatic substitution group) might be from the additives diuron and/or carbendazim, and the absorption observed at 782 cm^{-1} (vibration of five-membered heterocyclic ring), from the additive OIT.

Table 1. Particle sizes of preservatives.

Sample	Particle size (μm)					
	Without grinding		Grinding for 1 minute		Grinding for 6 minutes	
	Zero time	3 months	Zero time	3 months	Zero time	3 months
1	32.073±0.422	33.148±0.234	14.711±0.102	14.900±0.101	4.963±0.023	5.234±0.036
2	31.834±0.357	35.133±0.289	16.529±0.115	16.987±0.078	6.487±0.109	6.610±0.101
3	32.189±0.309	33.390±0.415	11.870±0.111	13.041±0.099	5.992±0.197	6.123±0.099
4	32.558±0.178	32.098±0.238	4.570± 0.098	15.067±0.121	5.336±0.136	5.897±0.020
5	31.845±0.310	33.022±0.102	15.460±0.201	16.001±0.309	4.196±0.099	4.223±0.021
6	32.012±0.284	32.917±0.287	15.649±0.193	16.122±0.235	5.971±0.099	5.992±0.113

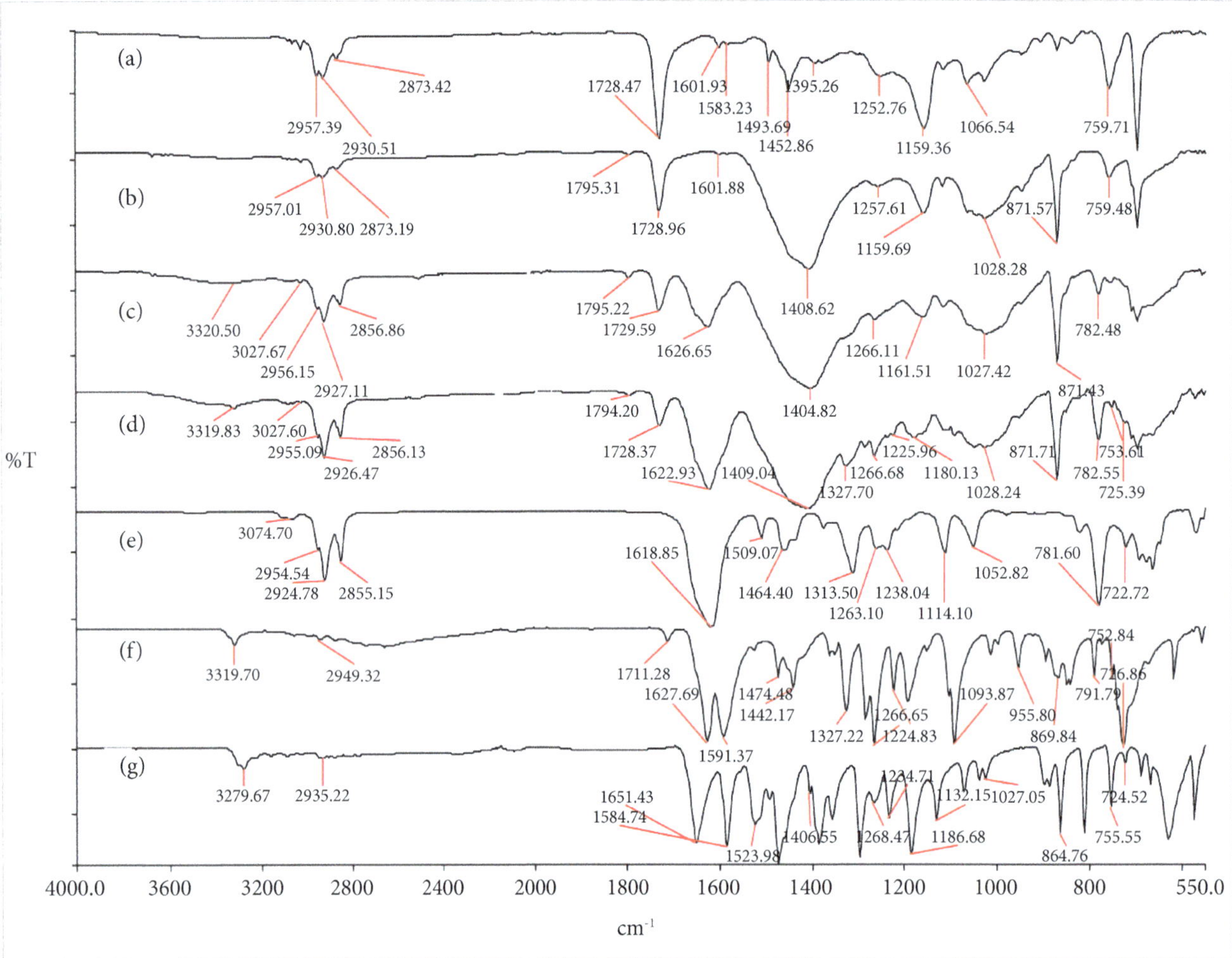

Figure 5. Spectra of UATR – (a) paint without preservatives, (b) paint with preservatives, without grinding, (c) paint with preservatives, grinded during one minute, (d) paint with preservatives, grinded during six minutes, (e) OIT, (f) carbendazim, and (g) diuron.

CONCLUSIONS

The evaluation of IR techniques for the characterization of elastomers and coatings indicated that:

- Transmission and reflection techniques applied to the same rubber sample, NBR in this study, can provide different information, highlighting only the base polymer characteristic absorptions (by pyrolysis/transmission) or the filler ones (by KBr pellet/transmission and ATR or UATR/reflection). This methodology can be applied to other types of rubber, particularly silicone formulations that might or might not contain silica filler. This aspect can be easily verified by using reflection analysis (ATR or UATR).
- The studied EPDM/NR mixture can be identified only by using the pyrolysis technique and after treatment with ortho-dichlorobenzene. This method is suitable only for mixtures whose characteristic bands do not overlap.
- The UATR technique, without grinding of preservatives, did not allow the direct analysis of the painted dry film, for the samples analyzed. However, after the preservatives had been grinded, it was possible to observe bands around 3300 and 1330 cm^{-1} (carbendazim), 1625 cm^{-1} (carbendazim and/or OIT), 870 cm^{-1} (diuron and/or carbendazim), and 782 cm^{-1} (OIT).
- By using different FT-IR techniques, the preservatives carbendazim and OIT allow better visualization of the painted surface after grinding, because of their characteristics bands (medium and intense bands of functional groups and/or higher diffusion).

REFERENCES

Abidi, N. and Hequet, E., 2005, "Fourier Transform Infrared analysis of trehalulose and sticky cotton yarn defects using ZnSe-Diamond UATR", Textile Research Journal, Vol. 75, No. 9, pp. 645-652.

Allen, T.J., 1992, "Paint sample presentation for Fourier transform infrared microscopy", Vibrational Spectroscopy, Vol. 3, No. 3, pp. 217-237.

Almeida, E., Balmayore, M. and Santos, T., 2002, "Some relevant aspects of the use of FTIR associated techniques in the study of surfaces and coatings", Progress in Organic Coatings, Vol. 44, No. 3, pp. 233-242.

American Society for Testing and Materials – ASTM, 2010, "Book of ASTM Standards", Philadelphia, D 3677-10, Standards Test Methods for Rubber-Identification by Infrared Spectrophotometry.

Blackford, R., 1999, "Performance demands on aerospace paints relative to environmental legislation", Pigment & Resin Technology, Vol. 28, No. 6, pp. 331-335.

Campos, E.A., Diniz, M.F., Reis, T.B., Dutra, R.C.L., Rezende, L.C. and Iha, K., 2003, "Aplicação de Técnicas FT-IR na Caracterização de Catalisador Cromito de Cobre. Utilizado na Indústria Aeroespacial", Anais Associação Brasileira de Química, Vol. 52, No. 1, pp. 22-25.

Crespim, H., Azevedo, M.F.P, David, L.H., Cassu, S.N. and Lourenço, V.L., 2007, "Substituição de amianto por silicato de alumínio e grafite expansível em compósito de poliuretano utilizado em motor-foguete", Polímeros: Ciência e Tecnologia, Vol. 17, No. 3, pp. 228-233.

Dutra, J.C.N., Massi, M., Otani, C., Dutra, R.C.L., Diniz, M.F., Urruchi, W.I., Maciel, H. and Bittencourt, E., 2002, "Surface Modification of EPDM Rubber by Reactive Argon-Oxygen Plasma Process", Molecular Crystals and Liquid Crystals, Vol. 374, No. 1, pp. 45-52.

Dutra, R.C.L. and Diniz, M.F., 1993, "Resistência à degradação oxidativa e comportamento aos solventes como indicadores da composição de sistemas elastoméricos vulcanizados mistos", Polímeros: Ciência e Tecnologia, Vol. 3, No. 3, pp. 25-28.

Dutra, R.C.L., Diniz, M.F., Takahashi, M.F.K., 1995, "Importância da Preparação de Amostras em Espectroscopia no Infravermelho com Transformada de Fourier (FTIR) na Investigação de Constituintes em Materiais Compostos", Polímeros: Ciência e Tecnologia, Vol. 5, No. 1, pp. 41-46.

Faillace, J.C., Medeiros, M.E., Filho, A.M.F., 1999, "Comportamento Térmico do Acelerador de Queima de Propelentes Sólidos, Anais do III Encontro Técnico de Materiais e Química, 3th ETMQ, Rio de Janeiro, Brazil.

Ferrari, V.C.G.M., Lourenço, V.L., Dutra, R.C.L., Diniz, M.F., Azevedo, M.F.P., David, L.H., 2012, "Caracterização de um Pré-Impregnado Aeronáutico por FT-IR e Análise Térmica", Polímeros: Ciência e Tecnologia, Vol. 22, No. 4, pp. 369-377.

Hori, K., Iwama, A., Fukuda, T., 1990, "FTIR spectroscopic study on the Interaction between Ammonium Perchlorate and Bonding Agents", Propellants, Explosives, Pyrotechnics, Vol. 15, No. 3, pp. 99-102.

Lindner, W., 2005, "Surface Coatings", In: Directory of Microbicides for the Protection of Materials, Edited by Paulus, W., Springer, New York.

Lourenço, V.L., Kawamoto, A.M., Sciamareli, J., Rezende, L.C., Pires, D.C., Takahashi, M.F.K., Berdugo, A.V., Cruz, S.M., Dutra, R.C.L. and Soares, B.G. 2006, "Determinação da distribuição de funcionalidade de HTPB e verificação de sua influência no comportamento mecânico de poliuretano utilizado em motor-foguete", Polímeros: Ciência e Tecnologia, Vol. 16, No. 1, pp. 66-70.

Mateo, M.P., Ctvrtnickova, T. and Nicolas, G., 2009, "Characterization of pigments used in painting by means of laser-induced plasma and attenuated total reflectance FTIR spectroscopy", Applied Surface Science, Vol. 255, No. 10, pp. 5172-5176.

Matheson, M.J., Wampler, T.P., Simonsick Jr., W.J., 1994, "The effect of carbon-black filling on the pyrolysis behavior of natural and synthetic Rubbers", Journal of Analytical and Applied Pyrolysis, Vol. 29, No. 2, pp. 129-136.

Mattos, E.C., Moreira, E.D., Dutra, R.C.L., Diniz, M.F., Ribeiro, A.P. and Iha, K., 2004, "Determination of the HMX and RDX content in synthesized energetic material by HPLC, FT-MIR and FT-NIR spectroscopies", Química Nova, Vol. 27, No. 4, pp. 540-544.

Mazzeo, R., Joseph, E., Prati, S. and Millemaggi, A., 2007, "Attenuated Total Reflection Fourier Transform Infrared Microspectroscopic Mapping for the Characterization of Paint Cross Sections", Analytica Chimica Acta, Vol. 599, No. 1, pp. 107-117.

Mello, V.M. and Suarez, P.A.Z., "As Formulações de Tintas Expressivas Através da História", Revista Virtual de Química, Vol. 4, No. 1, pp. 2-12, Retrieved on June 11 2013, 2012, from http://www.uff.br/RVQ/index.php/rvq/article/viewFile/248/218

Miliani, C., Ombelli, M., Morresi, A. and Romani, A., 2002, "Spectroscopic study of acrylic resins in solid matrices", Surface and Coatings Technology, Vol. 151-152, pp. 276-280.

Moraes, J.H., Sobrinho, A.S.S., Maciel, H.S., Dutra, J.C.N., Massi, M., Mello, S.A.C. and Schreiner, W.H., 2007, "Surface improvement of EPDM rubber by plasma treatment", Journal of Physics D: Applied Physics, Vol. 40, No. 40 pp. 7747. dx.doi.org/10.1088/0022-3727/40/24/022

Oliveira, J.I.S., Nagamachi, M.Y., Diniz, M.F., Mattos, E.C. and Dutra, R. C.L., 2011, "Assessment of the synthesis routes conditions for obtaining ammonium dinitramide by the FT-IR", Journal of Aerospace Technology and Management, Vol. 3, No. 3, pp. 269-278.

Pandey, G.C. and Kulshreshtha, A.K., 1993, "Fourier transform infrared spectroscopy as a quality control tool", Process Control and Quality, Vol. 4, pp. 109-123.

Pardini, L.C., 2000, "Preformas para Compósitos Estruturais", Polímeros: Ciência e Tecnologia, Vol. 10, No. 2, pp. 100-109.

Perkin Elmer – Technical Note, 2005, Improved Performance with the New Spectrum 100 UATR Accessory, Retrieved on June 11 2013, from www.perkinelmer.com

Romão, B.M.V., Diniz, M.F., Azevedo, M.F.P, Lourenço, V.L., Pardini, L.C., Dutra, R.C.L., Burel, F., 2006, "Characterization of the Curing Agents Used in Epoxy Resins with TG/FT-IR Technique", Polímeros: Ciência e Tecnologia, Vol. 16, No 2, pp. 94-98.

Sanches, N.B., Diniz, M.F., Reis, T.B., Cassu, S.N., Dutra, R.C.L., 2006, "Avaliação do uso de técnicas PIR-G/FT-IR para caracterização de Elastômeros", Polímeros: Ciência e Tecnologia, Vol. 16, No. 3, pp. 211-216.

Santos, R.P., Oliveira Junior, M.S., Mattos, E.C., Diniz, M.F. and Dutra, R.C.L., 2013, "Study by FT-IR Technique and Adhesive Properties of Vulcanized EPDM Modified with Plasma", Journal of Aerospace Technology and Management, Vol. 5, No. 1, pp. 65-74.

Sciamareli, J., Cassu, S.N. and Iha, K., 2012, "Water influence in poly (epichlororydrin) synthesis: An Intermediate to Energetic Propellants", Journal of Aerospace Technology and Management, Vol. 4, No. 1, pp. 41-44. doi:10.5028/jatm.2012.04016011

Silverstein, R.M., Webster, F.X. and Kiemle, D.J., 2005, "Spectrometric identification of organic compounds", John Wiley & Sons, New York.

Smith, A.L., 1979, "Applied Infrared Spectroscopy", John Wiley & Sons, New York, pp. 286.

Szafarska, M., Woźniakiewicz, M., Pilch, M., Zięba-Palus, J. and Kościelniak, P., 2009, "Computer analysis of ATR-FTIR spectra of paint samples for forensic purposes. Journal of Molecular Structure", Vols. 924-926, pp. 504-513.

Wake, W.C., Tidd, B.K. and Loadman, M.J.R., 1983, "Analysis of Rubber and Rubber-like polymer", 3rd edition, Applied Science Publishers, New York, pp. 330.

Waltham, 2005, FT-IR Spectroscopy Attenuated Total Reflectance (ATR), Technical Note, Catalogue PerkinElmer.

Welin-Berger, K. and Bergenstahl, B., 2000, "Inhibition of Ostwald ripening in local anesthetic emulsions by using hydrophobic excipients in the disperse phase". International Journal of Pharmaceutics, Vol. 200, No. 4, pp. 249-260.

Williams, P.T. and Besler, S., 1995, "Pyrolysis-thermogravimetric analysis of tires and tyre components", Fuel, Vol. 74, No. 9, pp. 1277-1283.

Zhang, W.R., Lowe, C. and Smith, R., 2009, "Depth profiling of coil coating using step-scan photoacoustic FTIR", Progress in Organic Coatings, Vol. 65, No. 4, pp. 469-476.

19

Efficiency-Multimission Comprehensiveness Balance for Platform-Based Satellite Family

Otavio Luiz Bogossian[1], Geilson Loureiro[1], Roberto Vieira Fonseca Lopes[1], Edgardo Roggero[2]

ABSTRACT: This paper aims to present the Comprehensiveness Balance for Efficiency (CBfE) method for Platform-Based Satellite Family. The lack of a penalty measurement to assess the performance loss of using a platform could reduce significantly the family performance. The method, taking into account the comprehensiveness of space missions and the platform characteristics defined at the conception phase, assesses the platform inefficiency, in terms of the additional mass required by the platform equipment to cope with the worst environment factors. The method covers the aerodynamic drag and torque, the Earth's magnetic field, the eclipse and Sun energy absorption, the cumulated radiation dose absorbed by the electronic components and the effect on the structure to be prepared for several launchers. Based on this assessment and on an interactive process, the platform designer tunes the comprehensiveness with the suitable level of efficiency. A real case, the Brazilian MultiMission Platform project (PMM), is presented as an example of application. The method covers an existing gap on the platform development process for space applications.

KEYWORDS: Platform, Product family, Satellite family, Development process, Platform efficiency.

INTRODUCTION

The family of products concept became relevant with the transformation of the mass production concept into mass customization aiming to comply with individual client needs (Pine, 1993). The family is a set of similar products obtained from a common platform, given to each product the functionalities required by specific clients (Meyer and Lehnerd, 1997). A product platform is a set of subsystems and interfaces developed to form a common structure (or core) from which a stream of derivative products can be efficiently developed and produced (Meyer and Lehnerd, 1997). Several authors have been working on defining a family of products and the corresponding platform for general applications. The segmentation market grid based on platform was introduced as the way to leverage the family of products across different market niches (Meyer and Lehnerd, 1997). Meyer and Utterback (1993) attempt to map the evolution of a given product family based on platform by means of extensions and upgrades. Bogossian and Loureiro (2011) grouped the family definition in three classes: based on design methods (modularity, platform based, configurational or scalable), based on generation of product variety to target market niches and based on technical aspects for improving the product process, stock reduction and component reutilization promotion.

SPACE CONTEXT

The space context has specific characteristics such as the complexity of the products and the very low production volume. It was remarked that space products are designed to comply with a particular mission (Gonzalez-Zugasti *et al.*, 2000). The space product (the spacecraft) is designed for its particular mission (independent development). This contrasts with products for general applications in which they are designed for a niche market.

1.Instituto Nacional de Pesquisas Espacias – São José dos Campos/SP – Brazil 2.Comisión Nacional de Actividades Espaciales – Buenos Aires – Argentina

Author for correspondence: Otavio Luiz Bogossian | Instituto Nacional de Pesquisas Espaciais | Avenida dos Astronautas, 1.758 – Jardim da Granja | CEP 12.227-010 São José dos Campos/SP – Brazil | Email: otavio.bogossian@inpe.br

A preliminary question stated during the development process is the possibility to define a satellite family and its platform (the common core), taking into account that each mission has specific objective and environment. Based on these specificities, it seems preliminary that the platform will pay a significant penalty.

This paper aimed to present and justify the elements of a method named *Comprehensiveness Balance for Efficiency* (CBfE) to conceive a low Earth orbit satellite family with the scope defined in terms of mission attributes like orbits, wings number and pointing (mission comprehensiveness) and the platform characteristics to cope with the mission attributes and environment like reaction wheel mass and battery mass (efficiency).

This paper is organized as follows: Development Process Section presents a literature review on platforms development approach, mainly for space application; The Penalty and its Unit of Measurement section presents a literature review on penalty and its unit of measurement; The Method section presents the CBfE method and its implementation; Application Case section presents an application case and Conclusions and Further Work section draws some conclusions and sets up some further work.

DEVELOPMENT PROCESS

The satellite platform concept was adopted by some space programs to exploit common aspects of the space missions, requiring satellites of the same category. In general, the space agencies do not have a complete view of the satellite family to be developed before the platform design. However, they aim to increase the reuse of the common part (platform) as much as possible when future missions are defined.

Boas and Crawley (2006) classified the platform development in two approaches, a parallel and a sequential one. The parallel approach corresponds to designing the platform based on a known family of products. The sequential approach corresponds to develop the platform based only on the first product. They stated also that for complex systems, the platform and the first variant (product) are often developed simultaneously, with the follow-on variants developed and deployed in sequence.

Bogossian and Loureiro (2011) concluded that the sequential approach is often applied to the development of multi-mission satellite platforms, demonstrated by missions like Jason 1 with CNES PROTEUS platform (Aerospatiale and Sextant, 1995; Dechezelles and Huttin, 2000; Grivel *et al.*, 2000), like Demeter with CNES Myriade Product Line platform (Bouzat, 2000; Cussac *et al.*, 2004; Alary and Lambert, 2007), previously called as Ligne de Produits Micro-satellite (Buisson *et al.*, 1998), like SkyMed/COSMO with ASI/Alenia PRIMA platform (Galeazzi, 2000) and by SSR with INPE PMM platform (INPE, 2001). During or after the platform design, the space agencies define a certain number of space missions based on platform comprehensiveness in terms of missions, flexibility and constraints (Galeazzi, 2000; INPE, 2001; Dechezelles and Huttin, 2000) such as covered orbits, pointing accuracy, launchers, lifetime, mass and power limits for payloads.

THE PENALTY AND ITS UNIT OF MEASUREMENT

Muffatto (1999) has remarked that adopting the platform concept has several benefits but also some drawbacks; one of them is the open architecture necessary to define new products. It will produce heavier products.

Gonzalez-Zugasti and Otto (2000) mentioned that the main benefits of a platform adoption are the development, manufacturing and operation costs by means of the reuse and scale economy. As a drawback a lower performance or efficiency is obtained when compared with an independent development. They also remarked the need of flexibility to comply with new requirements and also of economical feasibility.

Boas and Crawley (2007) stated that the benefits are tempered by performance penalties and the true benefits and penalties of platform-based product development are difficult to address. According to academic literature, these topics have not been properly addressed in the academic literature and represent an opportunity for improving managerial understanding of platform making.

Gonzalez-Zugasti and Otto (2000) stated that many set of products, such as spacecraft, do not have a single always-increasing desired performance attribute that describes them completely.

THE METHOD

PREMISES AND METHOD SCOPE

To develop the method it was necessary first to determine the development process adopted in the platform design for a low orbit satellite family. Based on real cases as shown on Section Development Process, it was concluded that platform designers usually adopt a sequential development process. The fact of unknowing the family of products tends to increase the platform inefficiency due to the inclusion by the designers, generalities to cover as much as possible, future unknown missions. The absence of penalty measurement does not give to the designers, means to determine the price of these generalities.

Another aspect that was necessary to establish before the method development was the definition of the unit of penalty, described on The Penalty and its Unit of Measurement section.

It was stated as premise that the method should be easily applicable and should provide to the platform designer a quick and simple response. All necessary simulations should be performed previously during the method development. When necessary, the method would interpolate the available data to cover the specific case of the platform under design. As a consequence of these premises, the amount of necessary work to develop the method was significantly increased. For this reason, it was necessary for the first version, to limit the method scope. In order to define the method scope some platforms were considered, reducing, wherever possible, the number of cases and conflicts. The platforms considered were Myriade from Centre National d'Études Spatiales - CNES (Bouzat, 2000; Alary and Lambert, 2007), Proteus from Centre National d'Études Spatiales - CNES (Dechezelles and Huttin, 2000; Grivel *et al.*, 2000), Plattaforma Riconfigurable Italiana Multi-Applicativa - PRIMA from Agenzia Spaziale Italiana - ASI/Alenia Aerospazio (Galeazzi, 2000) and Plataforma Multi-Missão - PMM from the Brazilian National Institute for Space Research - INPE (INPE, 2001).

The method scope is:

- Circular orbits only (low eccentricity).
- Altitudes from 400 to 1,500 km.
- Low inclination orbits from 0° to 25°.
- Sun Synchronous Orbit (SSO) orbits from 400 to 1,500 km with descending node crossing time at 10:00 AM and 12:00 AM.
- Two pointing target, nadir and Sun.
- Satellite configurations with one or two solar wings.
- Cubic platform and parallelepiped shape satellite.

Only significantly affected equipment by the perturbation in the space and launching environments were considered in the present version. The environmental effects and the corresponding affected equipment are the following:

- Drag/orbit decay - tank (kg of fuel).
- Drag/torque - reaction wheels (angular momentum in Nms).
- Magnetic field/unload the reaction wheels - torque rods (Am^2).
- Solar irradiation/energy capture - solar array generator (surface m^2).
- Solar eclipse/continuous energy providing - battery (Ah).
- Radiation belts/Total ionizing dose (TID) - electronic components and equipment radiation capacity (krad).
- Launching/quasi-static and decoupling - structure (kg).

For the present version, the method captures the inefficiency only for these phenomena, dimensioning the equipment with the corresponding unit.

METHOD DESCRIPTION

Each mission requires a specific orbit for the satellite and therefore it will be submitted to a particular environment. In principle, the platform equipment is the same for all missions and the method needs to determine its capacity for the complete set of mission attributes covered by the platform (mission comprehensiveness). From the required capacity for each orbit, the method will determine the best and the worst cases for each equipment.

The method adopted the equipment mass as unit of penalty. It was considered that, for a space mission, the mass is always increasing and is limited by the launcher capacity. The more mass the platform needs the less mass the customized part of the product (payload) will carry. For most of the equipment and, with the same technology, the on-board equipment needs more mass to cope with the increase of environment perturbation. An exception is the radiation effect that increases the cost, reason why it was considered in this work as an indirect mass.

For those equipment and components whose capacity at first is not defined by its mass, it will be necessary to convert them in equipment mass. Table 1 shows an example of the specific capacity used to implement this conversion for the tank. Three commercial tanks could be used (based on the last column only the configuration 3 is being used) and the specific mass corresponds to the rate between the tank mass and the propellant mass. The last column can be used to assign a relative weight to each configuration.

The platform inefficiency will be determined by the platform mass difference. The mass of all covered equipment and required for each mission attribute (comprehensiveness) will be determined. The highest and lowest platform masses will be considered as the worst and best cases respectively. Bogossian and Loureiro (2012) presented a preliminary version of the method.

To determine the platform efficiency for a given comprehensiveness, the method uses four different models according to platform equipment. Only the basic model is presented here (Fig. 1). This model applies to the tank, reaction wheels, torque rods and battery.

According to Fig. 1, the method obtains the worst dimensioning and the best dimensioning cases for three altitudes per inclination. A minimum equipment mass value could be adopted according to what is available in the market. The mass calculation in Fig. 1 corresponds to the conversion of the equipment capacity into equipment mass, using specific mass table as previously explained.

The dimensioning is significantly dependent on the highest and lowest altitudes, reason why the method interpolates the altitude range of the platform being assessed with respect to the three altitudes considered by the method. Considering all inclinations, the inefficiency is the difference between the worst and the best cases multiplied by the number of onboard equipment (by one if no redundancy) and divided by the number of different configurations (scalability concept). The more over dimensioned with respect to the minimum value, the less efficient is the platform.

The scalability has a significant effect on the platform efficiency for a given comprehensiveness. As the required component capacity varies according to the mission parameters, the utilization of different hardware configurations will enable to reduce the inefficiency by choosing the appropriate configuration for each mission.

The method has a specific treatment for some platform equipment. For the solar wings it is always possible to use a symmetric configuration with two identical wings or a free configuration with one or two wings. The panels, as the main wing component, have a standard size (informed as input) and a maximum number of them in the wings.

If the equipment is dimensioned exclusively by one of the environment factors considered by the method (without being based on a budget with several factors) and if it is an off-the-shelf equipment, it is possible to adopt a minimum value to exclude from the inefficiency determination capacity values not available in the market.

The inefficiency related to the TID absorbed by the electronic components of the developed equipment and absorbed by the purchased equipment does not affect directly the mass, but the cost instead. The components with higher TID capacity are more expensive. The method requires the platform designer to establish a level of TID for the components and equipment that is between the best and the worst cases, according to what is available in the market. Based on this level, the method determines the amount of additional shielding (with respect to existing boxes and platform structure) necessary to keep the TID under the maximum supported level for the lifetime.

The method captures the inefficiency from the structure considering the dimensioning for several launchers. It considers only two environmental factors, the quasi-static acceleration

Table 1. Specific mass table.

	Equip. Model	Propelant mass (m^1) (kg)	Tank mass (m^2) (kg)	Specific mass (m^2/m^1)	Relative weight
Conf 1	OST 31/1	136,22	6,4	0,047	0
Conf 2	OST 31/0	80,50	6,4	0,080	0
Conf 3	PSI 80274-1	46,44	6,0	0,129	1
Mean (weighed) specific mass				0,129	1

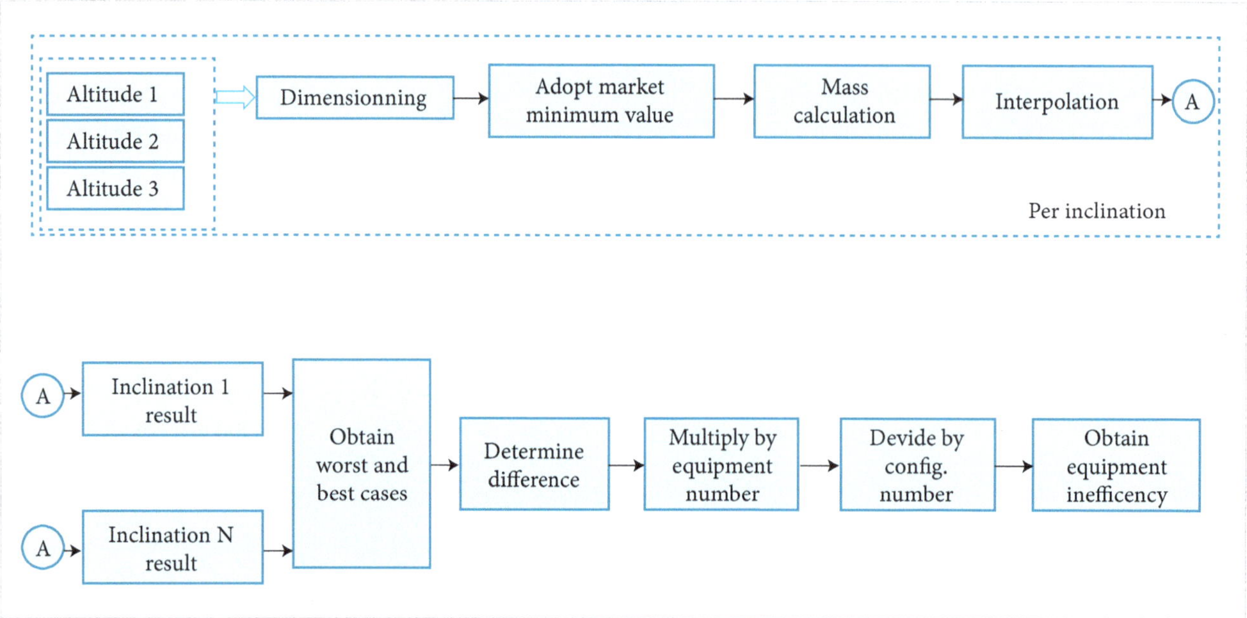

Figure 1. Approach to inefficiency capture.

and the decoupling of the first platform mode longitudinal frequency from the launcher corresponding one. Each launcher has specific values for these two factors and the platform has to consider the worst and the best cases. For the quasi-static effect the method determines the face sheet width of the top, bottom and lateral panels. The width difference between the best and the worst cases determine the amount of additional mass between both cases and corresponds to the inefficiency due to this factor. With respect to the decoupling, the method determines the bottom panel rigidity defining the panel width for the best and the worst cases. From the difference between them it is possible to obtain the inefficiency due to this factor.

The mass resulting from the sum of these two effects corresponds to the structure inefficiency due to the flexibility of the platform to be launched by a set of launchers.

THE METHOD IN THE PLATFORM LIFECYCLE

The method aims to help the platform designer at the conception phase. In space project lifecycle (ECSS, 2009) it will be used at Phase A, as shown in Fig. 2. In this phase, the platform is conceived, the launchers are defined and the mission comprehensiveness established. With these inputs the method assesses the platform and provides consolidated results (level of inefficiency) that could be used by the platform designer to modify the preliminary results.

All required data are available at Phase A such as dimensions, number of wings and wheels, required power and layout.

IMPLEMENTATION

The logic and calculation necessary for the method were implemented by using a set of spreadsheets. Twelve spreadsheets were required, one for each considered environment factor (seven at total), one for the results, two for constants, one for specific capacities and one for the inputs.

For each case, the method shall determine the dimensioning of the equipment to cope with the corresponding environmental factor, based on well-known equations or simulation results.

The method covers satellite orbital altitudes from 400 up to 1,500 km. The highest and lowest altitudes are essential to determine the worst and the best cases. If the altitude range

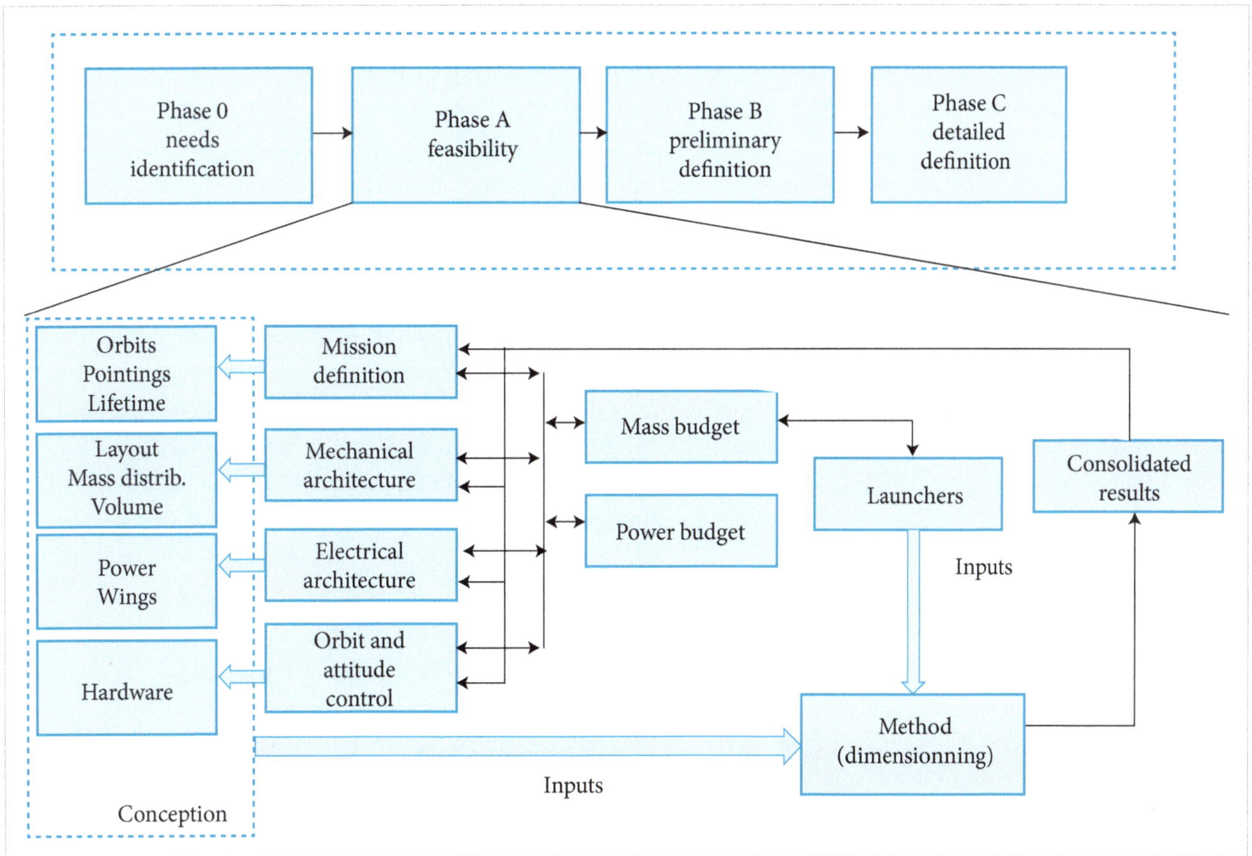

Figure 2. Positioning of the method in the platform lifecycle.

implemented by the method is higher than the platform altitude range under evaluation, the method will interpolate the corresponding values with the best curve.

The next items describes, for each environment factor, the implementation characteristics.

Drag

The atmospheric drag is one of the environmental factors (the others are not considered by the method) that will affect the amount of propellant necessary to maintain the nominal orbit. Based on the drag force, the method determines the amount of propellant needed for each case during a year timeframe, considering the mean orbit attack angle, the air density based on Jacchia model (Jacchia, 1977), the satellite external surface, the orbital speed, the drag coefficient and the accepted orbit error. The minimum mean attack angle was determined by computing simulation (STK, version 8.12).

Aerodynamic Torque

The aerodynamic torque is one of the environmental factors that affect the reaction wheels dimensioning. The angular momentum capacity is dimensioned by the method for each case. The drag force is also used for torque determination, based on the position of the center of pressure with respect to the center of mass.

The reaction wheel mass is strongly dependent on the capacity to store the angular momentum, for a given maximum rotation speed. The angular momentum determination is based on the center of pressure (wings and body), center of mass, number of wings and number of orbits to store the momentum. The required wheel capacity is determined by the torque integration in a specified number of orbits, per axis. All wheels are dimensioned with the same capacity, being considered the highest axis value.

Magnetic Field

The Earth magnetic field defines the torque rods dimensioning necessary to unload the reaction wheels. By simulation (STK, version 8.12) is obtained, for each case and axis, the minimum annual value of the mean orbit magnetic field. The rods must produce a torque that generates a negative angular momentum with respect to the stored momentum. The unload process considers the same number of orbits accounted in the wheels dimensioning. All rods will be dimensioned with the highest axis value.

Sun

For each case considered, the method determines the solar panel surface area necessary to provide the minimum power established by the platform, always using the same surface area per panel. The method considers configurations with always two wings or one or two wings. A fixed mass is considered per wing (e.g. yoke, hold-down, SADA). By simulation (STK, version 8.12) it is obtained the worst orbit case in a year period. All necessary efficiencies were considered in the implementation.

Eclipse

For each case, the annual maximum eclipse duration is determined by simulation (STK, version 8.12), dimensioning the battery capacity. Some inputs required by the method are efficiencies, minimum bus voltage, required power and mean Depth of Discharge (DoD).

Total Ionizing Dose

Based on simulations (SPENVIS, 2011), the method included several TID tables. Based on these tables the method determines the equipment/components absorbed TID, considering only the basic shielding (platform structure and equipment boxes for the components). When the dose exceeds the component/equipment specification, the shielding is increased by including aluminum plates in parallel with the electronic boards, ensuring that the absorbed dose will remain within the specification. The method requires inputs such as lifetime, component/equipment TIDs and quantity and size of the boxes.

Launchers

The implementation for the quasi-static acceleration is based on analytical solution for sandwich plates, supported by a rigid frame. The top panel is in simple square supported (by the frame) configuration with an uniform distributed load on the panel. This implementation is based on the launcher's maximum acceleration and the platform layout that informs in which panel each equipment is located. The relation between the dimensioning between two launchers (best and worst cases) produces the face sheet width difference. From the panel geometry and the adopted material, it is obtained the additional mass between the worst and the best cases. For the lateral panels, the same process is applied taking into account the lateral acceleration.

For the bottom panel, the launcher-satellite interface (circle) is the supported region. The load of the platform upper panels is distributed around the square perimeter. The equipment placed directly on the bottom panel is also considered as presenting an uniformely distributed load. The mass directly placed over the interface is ignored.

For the first mode minimum frequency (decoupling factor), an analytical solution is also considered. The launcher-satellite interface is a rigid supported circle. The dimensioning result is the bottom panel width necessary to place the first longitudinal frequency above a specified value, for a given launcher. Taking into account the worst and the best cases, it is possible to obtain the mass difference between both cases, considering also the informed geometry and adopted material.

METHOD OUTCOMES

The method outcome is the inefficiency mass of each onboard equipment, the total direct and indirect masses and the inefficiency percentage with respect to the platform and the satellite (see example in Application Case section). These percentages could be used as a subjective absolute value or relative value to assess the actions efficacy.

A recommended approach is the variation of several orbits and platform parameters to determine the effect of each one (sensitivity analysis). The parameters that slightly reduce the comprehensiveness and increase significantly the efficiency are the most suitable to be adopted.

Bogossian *et al.* (2011) presented a preliminary version of this method without considering scalability and structure.

APPLICATION CASE

The PMM project began in 2001 at INPE with the objective of providing the necessary means of producing low Earth orbit satellites in a reduced time and cost. The first satellite is the Amazonia-1, shown in Fig. 3, a remote sensing satellite planned to be launched in 2014. The satellite dimensions are 2.35 x 0.95 x 0.95 m. The total mass is 550 kg and the platform mass is around 300 kg. The satellite is pointed to nadir and it has always two wings with a total surface of 6.3 m^2 with a SADA (Solar Array Drive Assembly) to rotate the wings.

Considering that the PMM (INPE, 2001) platform is already designed and it is too late to be balanced by the method, the application case was included in this paper only to exemplify the method outputs.

A sub set of tables is shown in this work to give a real meaning of the method outcomes. Table 2 shows the result of the tank dimensioning process for the considered cases including different orbits (three altitudes and some inclinations), pointing targets and quantity of solar wings. The method determines the mass value that corresponds to the fraction of the tank necessary to store the amount of propellant required to keep the orbit within the established error limit for the lifetime. Table 2 shows data only for the applicable cases (two-wing configuration, $W = 2$). It is remarked in yellow, the interpolation for the PMM orbit range. Considering that the PMM range of altitudes (600–1,200 km) is lower than the method range altitudes (400–1,500 km). Figure 4 shows the interpolation curve to estimate the required tank mass for the PMM altitude range. In the last two lines, Table 2 informs that PMM has on board only one tank, and it informs also that was considered by the project, only one tank size (# Config.).

Table 3 shows the required panel surface area for each case and Table 4 presents the total number of solar panels (considering a 1m^2 panel) and wings mass.

Table 5 shows the consolidated results. The first seven rows show the equipment mass difference between the worst and the best cases, representing the inefficiency of each one (obtained from Tables 1 to 4). Row 8 represents the total direct mass (excluding only the shielding mass) and line 9 includes all equipment. Rows 10 and 11 present the platform and satellite (Amazonia-1) masses

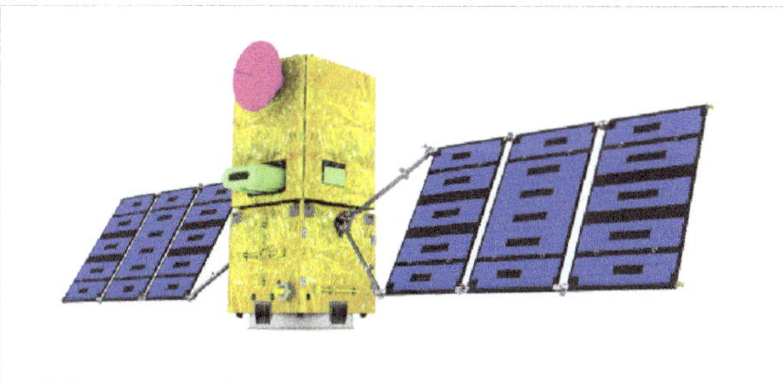

Figure 3. Amazonia -1 satellite.

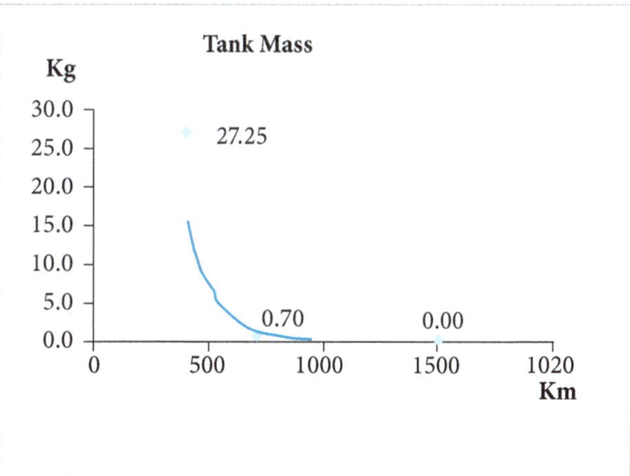

Figure 4. Interpolation curve.

Table 2. Tank dimensioning.

Tank mass								
Orbit		Pointing	# Wings	Orbit altitude			PMM	
Class	Inclination			400 (km)	700 (km)	1500 (km)	600 (km)	1200 (km)
Equator	0	Nadir	1	NA	NA	NA	NA	NA
			2	27.25 km	0.70	0.00	3.47 km	0.09 km
Low inc	12	Nadir	1	NA	NA	NA	NA	NA
			2	27.25 km	0.70	0.00	3.47 km	0.09 km
Low inc	25	Nadir	1	NA	NA	NA	NA	NA
			2	27.25 km	0.70 km	0.00 km	3.47 km	0.09 km
Equator	0	Solar	1		NA	NA	NA	NA
			2		0.88 km	0.01 km	1.39 km	0.04 km
Low inc	12	Solar	1		NA	NA	NA	NA
			2		0.88 km	0.01	1.37 km	0.04 km
Low inc	25	Solar	1		NA	NA	NA	NA
			2		0.86 km	0.01 km	1.36 km	0.04 km
SSO 10H	97/98	Nadir	1	NA	NA	NA	NA	NA
	100		2	27.25 km	0.70 km	0.00 km	3.47 km	0.09 km
SSO 12H	97/98	Nadir	1	NA	NA	NA	NA	NA
	100		2	27.25 km	0.70 km	0.00 km	3.47 km	0.09 km
Max: 3.47		Min: 0.04		Unit: 3.43		Inef: 3.43		
# Config: 1				# tanks: 1				

Table 3. Required surface for solar panels.

Solar panel (surface)							
			Orbit altitude			PMM	
Orbit		Pointing	400 (km)	700 (km)	1500 (km)	600 (km)	1200 (km)
Class	Inclination						
Equator	0	Nadir	3.49 m^2	3.29 m^2	2.98 m^2	3.38 m^2	3.08 m^2
Low inc	12	Nadir	3.78 m^2	3.57 m^2	3.19 m^2	3.66 m^2	3.32 m^2
Low inc	25	Nadir	4.43 m^2	4.00 m^2	3.44 m^2	4.23 m^2	3.62 m^2
Equator	0	Solar		3.01 m^2	2.74 m^2	3.05 m^2	2.84 m^2
Low inc	12	Solar		2.92 m^2	2.61 m^2	2.95 m^2	2.72 m^2
Low inc	25	Solar		2.75 m^2	2.64 m^2	2.76 m^2	2.68 m^2
SSO 10H		Nadir	3.77 m^2	3.43 m^2	3.19 m^2	3.64 m^2	3.25 m^2
SSO 12H		Nadir	3.26 m^2	3.09 m^2	2.83 m^2	3.17 m^2	2.92 m^2

Table 4. Required panels and wings mass.

			Always two wings				
			# Panels (*)		Fx mass	Total kg	
Orbit		Pointing	600	1200	(kg)	600 (km)	1200 (km)
Class	Inclination		(#)	(#)			
Equator	0	Nadir	4	4	9.596	24.69	24.69
Low inc	12	Nadir	4	4	9.596	24.69	24.69
Low inc	25	Nadir	6	4	9.596	32.24	24.69
Equator	0	Solar	4	4	2.856	17.95	17.95
Low inc	12	Solar	4	4	2.856	17.95	17.95
Low inc	25	Solar	4	4	2.856	17.95	17.95
SSO 10H		Nadir	4	4	9.596	24.69	24.69
SSO 12H		Nadir	4	4	9.596	24.69	24.69
			Max: 32.24			Min: 17.95	
				Min. Without Sada 24.69			
						Inef: 7.55	

(*) Always even

Table 5. Consolidated results.

Total		PMM
	Component	Mass (kg)
1	Tank	3.43
2	Wheels	0.74
3	Magnetic torque rods	1.88
4	Solar wings	7.55
5	Battery	1.81
6	Structure	7.29
7	Shielding (indirect mass)	6.67
8	Inefficiency direct mass	22.7
9	Inefficiency total mass	29.4
10	Platform mass	295
11	Satellite mass	557.0
12	% Inef. Direct platform	7.7%
13	% Inef. Direct satellite	4.1%
14	% Inef. Total platform	10.0%
15	% Inef. Total satellite	5.3%
16	Total ineff. Mass lowest orbit	28.68
17	% Tot. Ineff. Lowest orbit - platform	9.7%
18	% Tot. Ineff. Lowest orbit - satellite	5.1%

used as reference. Rows 12 to 15 present the relative inefficiencies in terms of percentage with respect to the platform and satellite masses. Rows 16 to 18 show the best cases for all in the same orbit, corresponding to the orbit with the lowest total equipment mass.

In order to demonstrate how the method is used for balancing the comprehensiveness in terms of altitude with the efficiency, the lowest altitude was reduced from 600 to 550 km as a first case and increased from 600 to 650 km as second case. The result for the first case was an increase in the total inefficiency mass from 29.4 to 31.9 kg (increase of 2.5 kg) that corresponds to an inefficiency increasing from 10.0 to 10.8%. For the second case, the total mass decreased to 27.5 kg (decrease of 1.9 kg) and that corresponds to an inefficiency reduction from 10,0 to 9.3%.

CONCLUSIONS AND FURTHER WORK

The main result of this work is the demonstration that it is possible to obtain an objective platform penalty measurement. It enables, for the products developed by using the sequential approach, to have a feedback from the adopted comprehensiveness with respect to the efficiency. The method itself represents a tool to help space platform designer to balance efficiency with comprehensiveness.

The method scope was reduced for this first version based on the real cases, and it requires further work to augment its scope in terms of space and launching environments as well as platform configuration. It is also recommended that some of the implemented models be upgraded to improve the process of capturing the inefficiencies.

REFERENCES

Aerospatiale and Sextant, 1995, "Aerospatiale/Sextant Avionique brochure", Filière Proteus CNES.

Alary, D. and Lambert, H., 2007, "The Myriade product line, a real success story", ACTA Astronautica, Vol. 61, pp. 223-227.

Boas, R.C. and Crawley, E.F., 2006, "Extending Platforming to the Sequential Development of System Families", INCOSE 2006, 16th Annual International Symposium Proceedings. Orlando, Florida.

Boas, R.C. and Crawley, E., 2007, "Parallel and Sequential Development of Complex Platform-Based Product Family", Engineering Management Conference, IEEE.

Bogossian, O.L. and Loureiro, G., 2011, "Attributes Balance on the Adoption of Platform Based Solutions for Satellites", Concurrent Engineering Proceedings, MIT, Cambridge, MA, USA.

Bogossian, O.L. and Loureiro, G., 2012, "Attributes Balance on the Adoption of Platform Based Solutions for Satellites", Journal of Aerospace Engineering, Sciences and Applications, Vol. 4, No.1. pp. 110 - 115.

Bogossian, O.L., Loureiro, G. and Lopes, R.V.F., 2011, "Architecting Method to Assess Conceptual Design of Platform Based Satellites", International Astronautical Congress, Cape Town, South Africa. IAC-11.D1.6.3.

Bouzat, C., 2000, "CNES Microsatellite Product Line, an approach for innovation", Small Satellites Systems and Services, 5th International Symposium, La Boule France.

Buisson, F., Cussac, T., Lassalle-Balier, G., Laurens, A., Ledu, M., Llorens, J.C. and Chadoutaud, P., 1998, "La ligne de produits Micro-satellite du CNES", Small Satellites Systems and Services, 4th International Symposium, Antibes, San Juan Les Pains, France.

Cussac, T., Buisson, F. and Parrot, M., 2004, "The Demeter Program: Mission and Satellite Description – Early in Flight Results", 55th International Astronautical Congress, 2004 IAC-04-IAA.4.11.2.04. Vancouver, Canada.

Dechezelles, J.J. and Huttin, G., 2000, "PROTEUS: A Multimission Platform for Low Earth Orbits", Air & Space Europe, Vol. 2, No. 1, pp. 77-81.

European Cooperation for Space Standardization (ECSS), 2009, "System Engineering General Requirements", ECSS. [S.1].

Galeazzi, C., 2000, "Prima: A new, competitive small satellite platform", Acta Astronautica, Vol. 46, No. 2-6, pp. 379-388.

Gonzalez-Zugasti, J.P. and Otto K.N., 2000, "Platform-Based Spacecraft Design: A formulation and implementation Procedure", Aerospace Conference Proceeding, IEEE.

Gonzalez-Zugasti, J.p., Otto K.N. and Baker J.D., 2000, "A Method for Architecting Product Platforms", Research in Engineering Design, Vol. 12, No. 2, pp. 61-72.

Grivel, C., Doullet, F., Huiban, T., Sainct, H., Bailion, Y., Terrenoire, P. Schrive, J. and Lazard, B., 2000, "Proteus: European Standard for small satellites", Small Satellites Systems and Services, 5th International Symposium, La Boule France.

National Institute for Space Research (INPE), 2001, "Multimission Platform: Data Package for System Requirement Review", INPE's Internal Document.

Jacchia, L.G., 1977, "Thermospheric Temperature, Density and Composition: New Model", Research in Space Science SAO Special Report No. 375, Smithsonian Institution.

Pine, B.J., 1993, "Mass customization: The new frontier in business competition", Harvard Business School Press, Boston.

Meyer, M. and Lehnerd, A.P., 1997, "The power of product platform – building value and cost leadership", Free Press, New York.

Meyer, M. and Utterback, J., 1993, "The product family and the dynamics of core capability", Sloan Management Review, Vol. 34, No. 3, pp. 29-47.

Muffatto, M., 1999, "Introducing a platform strategy in product development", International Journal of Production Economics, Vol. 60-61, pp. 145-153. doi: 10.1016/S0925-5273(98)00173-X

Software Satellite Took Kit (STK Analytical Graphics, Inc.) version 8.12 Expert Edition.

Space Environmental Information System (SPENVIS), 2011, ESA, Retrieved in January 13, 2014, from http://www.spenvis.oma.be/intro.php.

Analysis of Radar Cross Section Reduction of Fighter Aircraft by Means of Computer Simulation

Luiz Alberto de Andrade[1], Luan Silva Carvalho dos Santos[1], Adriana Medeiros Gama[1]

ABSTRACT: This paper presents a preliminary study of the Radar Cross Section (RCS) reduction on the fighter aircraft. First, it was studied the RCS of the aircraft from computational simulations based on prior knowledge of vulnerable areas of this aircraft to radar threats. Subsequently, the possible applications of Radar Absorbing Materials (RAM) on the surface of the aircraft were evaluated, in order to reduce its RCS. The absorber material used in the simulations was denominated FC70, which has good attenuation in the range of 10 to 12 GHz. The study of this reduction was accomplished by applying RAM in four different scenarios at the frequency of 11.1 GHz, where the material is more sensitive. The RCS simulations of the fighter aircraft and its RCS reduction by RAM application were carried out with the support of the software "Computer Simulation Technology" (CST), version 2012. Such technology makes it possible to simulate the application with an absorber material layer on the surface of the aircraft. For the study of the RCS reduction on the fighter aircraft, it was first necessary to develop a detailed 3D model of the fighter aircraft, and it was developed with the software "Computer Aided Three-Dimensional Interactive Application" (CATIA). In conclusion, it is impossible to make much progress attempting to retrofit stealth onto a conventional aircraft because if the shape is wrong, no amount of absorbing material treatments will reduce the RCS.

KEYWORDS: Radar cross section, Radar absorber material, Computational simulation, Radar cross section reduction.

INTRODUCTION

The goal to reduce a military aircraft's RCS is directly related to the distance at which it can be detected by hostile radars. The radar equation given below (Eq.1) provides a quantitative way to analyze the impact of a target's RCS reduction in its distance for monostatic radars (Knott *et al.*, 1993):

$$R_{max} = [(P_t G^2 \lambda^2 \sigma) / (4\pi)^3 P_{min} L)]^{1/4} \quad (1)$$

Where: R_{max} is the maximum range of the radar detection, P_t is the radar antenna's transmission power, P_{min} is the minimum power detected by the radar, $G = G_r = G_t$ is the radar gain, L are the losses associated to the radar electronics and the environment, and σ is the RCS (Gama and Rezende, 2010; CST, 2012).

When analyzing Eq.1, it is revealed that among the variables of the radar equation, only one possible control by the target aircraft is its RCS, all others are inherent to either the hostile radar system or the environment. Therefore, from the point of view of the aircraft, the radar parameters and the environment can be considered as a constant of the detection system. An examination of Eq.1 shows that the RCS of a target should be decreased by sixteen times, so that the maximum detection distance R falls by its half.

METHODOLOGY

The computational resource used in this study was the CATIA V5 software, responsible for the development of 3-D models of the aircraft which shall have the reduction of its

1. Instituto de Aeronáutica e Espaço – São José dos Campos/SP – Brazil

Author for correspondence: Luiz Alberto de Andrade | Instituto de Aeronáutica e Espaço – Divisão de Materiais | Praça Marechal Eduardo Gomes, 50, Vila das Acácias, São José dos Campos/SP | CEP: 12228-904 – Brazil | Email: andradelaa@iae.cta.br

RCS studied. In order to study the electromagnetic scattering, the package A-Solver (asymptotic solver) of the Computer Simulation Technology software – CST 2012 version – was used (Gama and Rezende, 2005). In the interest of studying the RCS reduction, the RAM FC70 was inserted on the library of the CST (Gama and Rezende, 2010; Gama and Rezende, 2005; Gama *et al.*, 2011),which simulated its application in parts of the surface of the aircraft, called Scenarios 1, 2, 3 and 4.

In order to facilitate the RCS's simulations on the F-5 and to modify the aircraft, because it is still in operation, some of its characteristics were modified (CST, 2013; CATIA, 2010; EMBRAER, 2005). The simplifications were: the setting of a metallic canopy, the removal of external antennas, the withdrawal of navigation and training lights, as well as the removal of "probe" for refueling during flight, weapons and engines. Figure 1 shows, in detail, a three dimensional model split to enable the necessary RAM application according to Scenarios 1, 2, 3 and 4 (Alves *et al.*, 2006).

In Scenario 1, an almost covered aircraft is presented. The RAM were not applied on the canopy because it would compromise the pilot's vision nor on the Radome's surface because it could attenuate the signal emitted and received by its own radar.

In Scenario 2, in addition to the surfaces uncovered in Scenario 1, the turbine palettes and exhaust were not coated either.

In Scenario 3, in addition to the parts not covered in Scenario 2, the air inlet of the aircraft was not coated, and in Scenario 4, only the horizontal and vertical stabilizers were coated, in addition to the uncoated parts in Scenario 3, the wings, fuselage's parts and aircraft's rear.

Figures 2, 3, 4 and 5 illustrate, respectively, Scenarios 1, 2, 3 and 4, where the fighter aircraft is partially coated with RAM, noticing that coated surfaces are yellow and uncoated surfaces are grey.

Figure 1. 3-D perspective model of the fighter aircraft developed with the CATIA software.

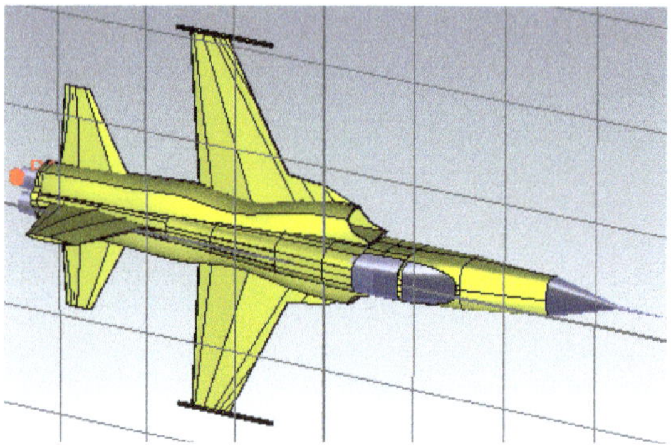

Figure 3. Fighter aircraft partially coated with RAM, Scenario 2.

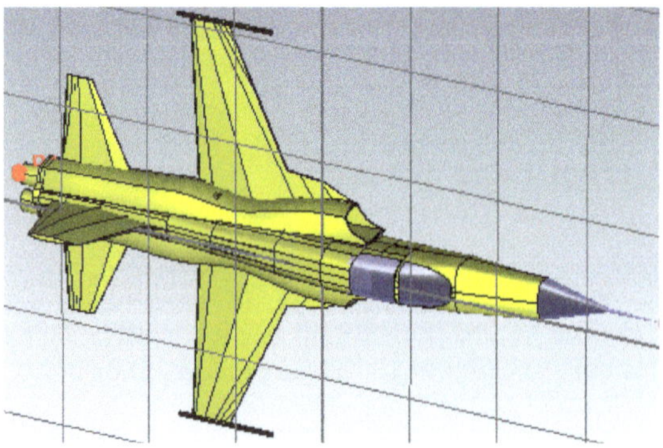

Figure 2. Aircraft fighter aircraft partially coated with RAM, Scenario 1.

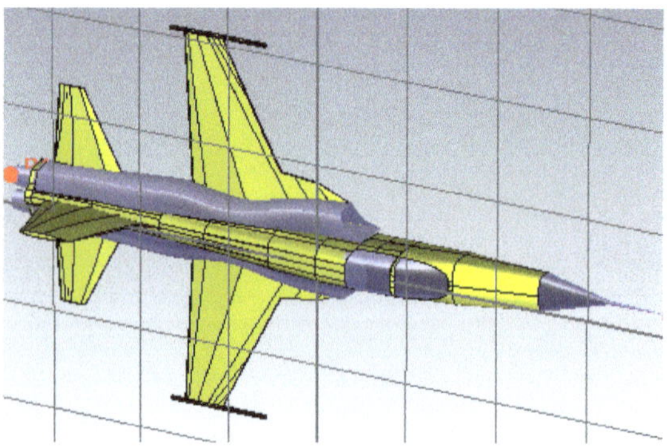

Figure 4. Fighter aircraft partially coated with RAM, Scenario 3.

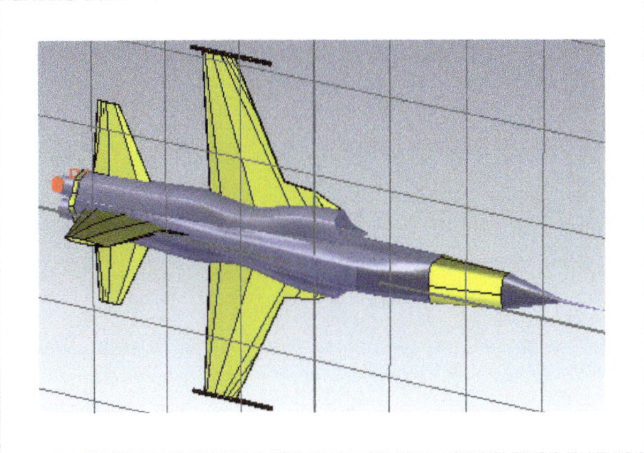

Figure 5. Fighter aircraft partially coated with RAM, Scenario 4.

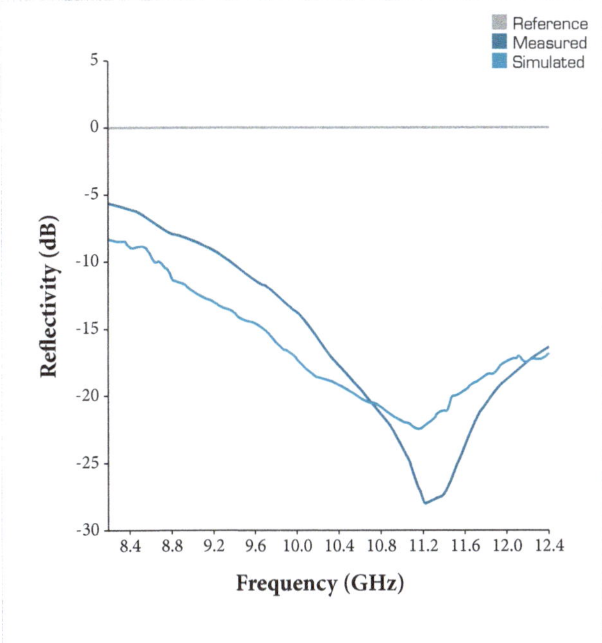

Figure 6. The reflectivity curve of the RAM FC70 used for RCS reduction.

Table 1. Comparison between the measured and the simulated reflectivity of the RAM FC70.

Frequency (GHz)	Reflectivity	
	Measured	Theoretical
8.2	-5.48852	-8.28
11.1	-26.17	-22.60
11.2	-28.06	-22.06
12.4	-16.5625	-16.96

Figure 6 and Table 1 show the comparison between the reflectivity measured and simulated with the CST software, and the RAM FC70 used in this study of RCS reduction.

The difference between the theoretical and the measured curve is due to inhomogeneity of the material and the variation in the thickness of the sample.

RESULTS

The preliminary study of RCS reduction on the fighter aircraft, by managed application of RAM, was performed at the frequency of 11.1 GHz, with aspect angles ranging from 0° to 360° with increments of 1°. All simulations were performed in the same frequency because RAM is more sensitive.

All RCS simulations were made using the package Asymptotic Solver (A-Solver), with a Gaussian excitation, vertical polarization and Triangular Mesh. Around 2,000,000 elements were generated. Figure 7 shows a comparison between RCS of the fighter aircraft at the frequency of 11.1 GHz, partially coated and uncoated with RAM, as shown in Fig. 2 (Scenario 1).

Figure 8 shows a comparison between RCS of the fighter aircraft at 11.1 GHz, partially coated and uncoated with RAM, as shown in Fig. 3 (Scenario 2).

Figure 9 shows a comparison between RCS of the fighter aircraft at 11.1 GHz, partially coated and uncoated with RAM, as shown in Fig. 4 (Scenario 3).

Figure 10 shows a comparison between RCS of the fighter aircraft at frequency 11.1 GHz, partially coated and uncoated with RAM, as shown in Fig. 5 (Scenario 4).

Table 2 shows a comparison between peak values of RCS at 11.1 GHz and RCS in the frontal, lateral, and rear portions of the fighter aircraft.

Tables 3 and 4 show values of RCS at 11.1 GHz in units of decibels square meter (dBsm) and square meter (m^2), respectively. The average was taken in intervals of 10° for RCS in the frontal, lateral, and rear portions of the fighter aircraft.

Typical radars detect aircrafts with a frontal RCS of $5m^2$ within 165 km of distance. Using the Eq. 1 and the results from Table 4, the reach of radars after applications of RAM, according to the scenarios presented, was estimated. Table 5 shows the reach of radars before and after applications of RAM.

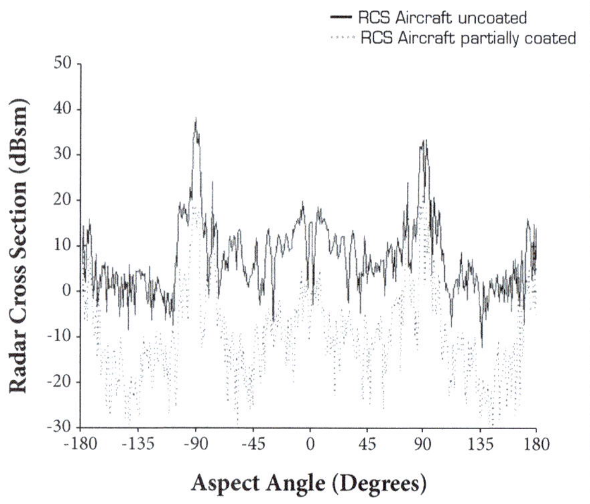

Figure 7. Scenario 1 – Comparison between the RCS of the fighter aircraft at 11.1 GHz, partially coated and uncoated with RAM.

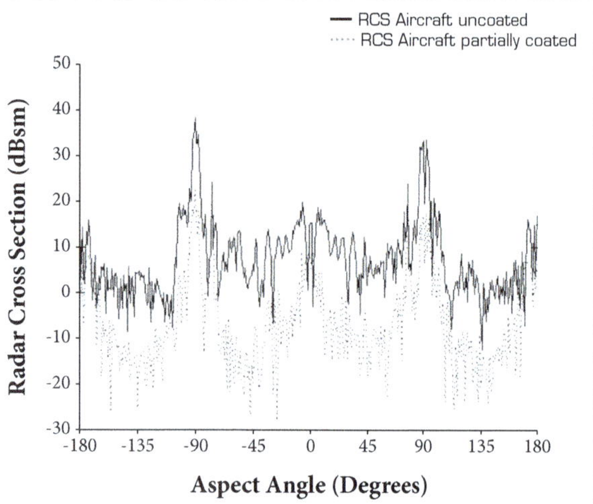

Figure 8. Scenario 2 – Comparison between the RCS of fighter aircraft at 11.1 GHz, partially coated and uncoated with RAM.

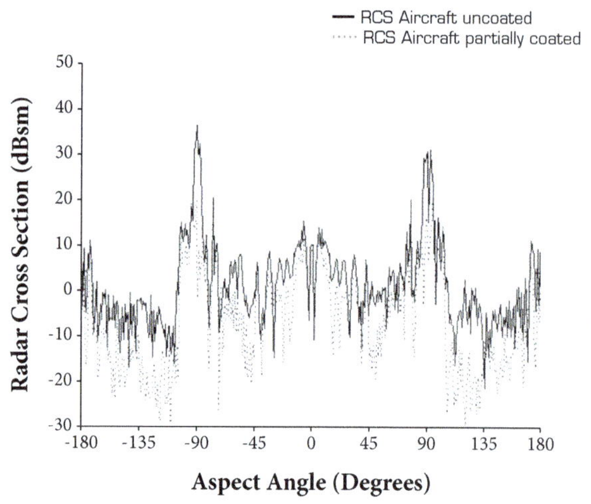

Figure 9. Scenario 3 – Comparison between the RCS of the fighter aircraft at 11.1 GHz, partially coated and uncoated with RAM.

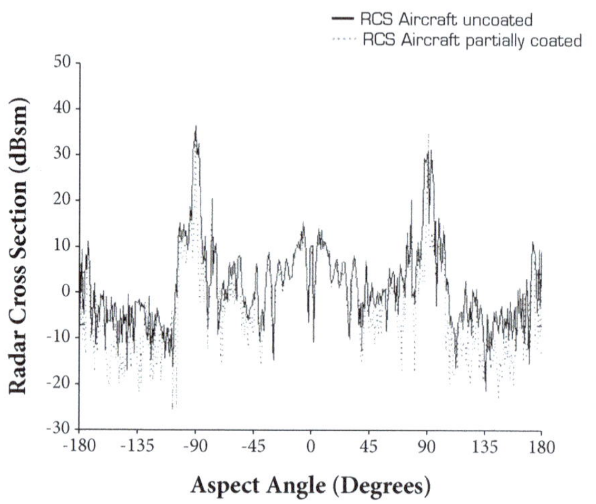

Figure 10. Scenario 4 – Comparison between the RCS of the fighter aircraft at 11.1 GHz, partially coated and uncoated with RAM.

Table 2. Comparison between the RCS of the fighter aircraft at 11.1 GHz frequency, uncoated and partially coated with RAM FC70.

Incidence		Peak RCS in dBsm				
		Uncoated	Scenarios (Coated with RAM)			
			1	2	3	4
Frontal	0°	15.20	-5.42	7.79	13.6	17.2
Lateral	90°	33.10	21.40	17.00	16.6	36.7
Rear	180°	30.50	5.41	11.30	15.3	16.2

Table 3. Comparison between the fighter aircraft RCS average at 11.1 GHz frequency uncoated and partially coated with RAM FC70.

Incidence		Average RCS in dBsm				
		Uncoated	Scenarios (Coated with RAM)			
			1	2	3	4
Frontal	-5° to 5°	10.20	-1.11	5.51	10.80	10.9
Lateral	85° to 95°	26.30	9.70	11.00	16.00	17.50
Rear	-175° to 175°	10.50	0.16	2.75	3.03	3.93

Table 4. Comparison between the fighter aircraft RCS average at 11.1 GHz frequency uncoated and partially coated with RAM FC70.

Incidence		Average RCS in m²				
		Uncoated	Scenarios (Coated with RAM)			
			1	2	3	4
Frontal	-5° to 5°	10.47	0.77	3.55	12.02	12.30
Lateral	85° to 95°	426.58	9.33	12.59	39.81	56.23
Rear	-175° to 175°	11.22	1.04	1.88	2.00	2.47

Table 5. Comparison between the detection distance of the fighter aircraft in the frequency of 11.1 GHz to a typical radar using the average RCS, uncoated and coated with RAM FC70.

Incidence		R_{max} in km				
		Uncoated	Scenarios (Coated with RAM)			
			1	2	3	4
Frontal	-5° to 5°	222.5	116.0	169.8	230.3	231.7
Lateral	85° to 95°	562.2	216.2	233.0	310.7	338.8
Rear	-175° to 175°	226.4	124.8	144.8	147.1	155.1

After the application of RAM with a theoretical attenuation of 22.6 dB at 11.1 GHz, for Scenario 1, the average of frontal RCS presented around 10 dB of attenuation. While the maximum range of the radar detection fell from 222 km to 116 km, i.e. the aircraft fighter could get 48% closer to the target before executing the mission. For Scenario 2, whose differences are only changing by the exhaust turbine and the application of RAM for metal, the average of frontal RCS presented around 7dB of attenuation. While radar reach fell from 222 km to 169.8 km, i.e. the aircraft fighter could get 25% closer to the target before executing the mission. For Scenarios 3 and 4, the application of RAM proved to be inefficient for frontal RCS reduction.

CONCLUSIONS

Nowadays, for an aircraft to be considered stealth, it shouldn't be detected in less than 20 km of distance. However, this value can vary considerably depending on the technological advancement of radar detection systems and stealth technology. It is impossible to make much progress attempting to retrofit stealth onto a conventional aircraft because if the shape is wrong, no amount of material absorber treatments will reduce the RCS. Consideration must be given to any part of an aircraft to which a radar wave can reach to, in order to develop a low observable aircraft. On the other hand, the first critical factor to consider in the design process is the shape of the aircraft. This element has been designed into the aircraft from the beginning.

REFERENCES

Alves, M.A., Andrade, L.A. and Rezende, M.C., 2006, "Influência da discretização de modelos em simulações de seção reta radar" In: Anais do VIII Simpósio de Guerra Eletrônica, São José dos Campos, Br, Vol. 1, pp. 1-4.

CATIA, 2010, Manual do Usuário, Available from: www.catia.com. Acessed on: 20 May 2013.

CST, 2013, Manual do Usuário CST Studio Suite versão 2012. Available from: www.cst.com. Acessed on: 20 May 2013.

EMBRAER. T.O. 1F - 5EM - 1, 2005, São José dos Campos, BR.

Gama, A.M. and Rezende, M.C., 2005, "The relationship between Mn-Zn ferrites with different iron ion contents and the absorption energy in X-band", Proceedings of the IEEE, Vol. 1, pp. 322-325. doi: 10.1109/IMOC.2205.1580009.

Gama, A.M. and Rezende, M.C., 2010, "Complex permeability and permittivity variation of carbonyl iron rubber in the frequency range of 2 to 18 GHz, Journal of Aerospace Technology and Management, Vol. 2, No. 1, pp.59-62. doi: 10.5028/jatm.2010.02015962.

Gama, A.M., Rezende, M.C. and Dantas, C.C., 2011, "Dependence of microwave absorption properties on ferrite volume fraction in MnZn ferrite/rubber radar absorbing materials", Journal of Magnetism and Magnetic Materials, Vol. 323, No. 2, pp. 2782-2785. doi: 10.1016/j.jmmm.2011.05.052.

Knott, E.F., Schaeffer, J.F. and Tuley, M.T., 1993, "Radar Cross Section", 2nd edition, Artech House Inc., USA, Chen, W.K., 1993, "Linear Networks and Systems", Book style, Wadsworth, Belmont, US, pp. 123–135.

Effect of Solid-State Laser Parameters on the Surface's Topography Formation During Texturization of Hard Metal Cutting Tools

Patrícia Helena de Oliveira Teixeira[1,2], Adriano Gonçalves dos Reis[1,2], Andreas Janssen[3]

ABSTRACT: The objective of this paper is to analyze the influence of specific laser system parameters on the microstructure of the ring formation process on the hard metal tool's surface. The parameters analyzed were: working distance, incident beam power and tool surface treatment, flank and rake. The microstructures were analyzed with regard to its diameter and depth by an optical 3-D surface metrology instrument. The analyses were made by comparing the ring's process formation with the dimples' process formation, a well-known process called ablation. The results show that the ring's diameter is increased by increasing incident beam power or the distance of the laser beam's focus for both surfaces. In addition, increasing the working distance decreases the ring's depth for both surfaces.

KEYWORDS: Surface finishing, Pulsed lasers, Ablation, Microstructure.

INTRODUCTION

One of the biggest concerns of industry is the improvement of manufacturing efficiency in order to obtain minimization of production costs, maximization of production rate and maximization of production accuracy, etc.. In aeronautic industry, these are main concerns as well, especially with the prediction that the number of new airplanes will grow by more than 27.000 until 2025 compared to 2005 (Klocke *et al.*, 2012). This prediction leads to an economic pressure on the aviation industry manufacturers, and the search for optimization in the manufacturing processes becomes necessary.

For the aircraft components construction, several manufacturing processes are used, and machining stands between the most applied ones. According to Aurich *et al.* (2011), during machining process, high mechanical and thermal loads occur at the cutting tool, causing tool failure. Due to this, optimization at the cutting tool represents an advantageous possibility. Cutting tools have been optimized by improvements of cutting tool material, coating system and tool macrogeometry. however, cutting tools with higher qualities regarding material removal rates, chip removal of hard-to-machine materials and tool life are each day more requested.

1.Faculdade de Tecnologia de São José dos Campos – São José dos Campos/SP – Brazil 2.Instituto Tecnológico de Aeronáutica – São José dos Campos/SP – Brazil 3.Fraunhofer Institute for Production Technology – Aachen – Germany.

Author for correspondence: Patrícia Helena de Oliveira Teixeira | Faculdade de Tecnologia de São José dos Campos | Avenida Cesare Mansueto Giulio Lattes, s/n, Eugênio de Melo | CEP: 12.247-014 – São José dos Campos/SP – Brazil | Email: patriciahelena.oliveira@yahoo.com.br

LASER-BASED CUTTING TOOL PRODUCTION

Looking on technical and mechanical assemblies, contact surfaces mainly determine the performance of systems in terms of being the interfaces where energy is transferred, converted and dissipated. The preparation of cutting tools using laser radiation is a useful improvement compared to the use of conventional techniques for optimization of these surfaces. Laser processing is a very powerful technique for manufacturing modification of defined cutting tool geometries. It is a non-contacting, wear free technique and not only offers highest flexibility due to the fact that the beam can be moved very fast by optical devices, but also provides excellent control of shape and size of microstructures, and, furthermore, a process relatively clean to the environment (Etsion, 2004).

According to Aurich *et al.* (2011), certain morphology modifications of contacted surfaces on cutting tools can improve tool performance. The so called laser-manufactured "micro-dimples" are microstructures already analyzed in recent research (Scaraggi *et al.*, 2013; Kim *et al.*, 2012). The fundamental formation process is the well-known process of laser ablation, which forms single ablation cavities by single laser pulses or complex superposition of many pulses.

According to Meijer *et al.* (2002), the term "ablation" relates to the effect of vaporizing irradiated material by incident laser beam. According to Poprawe (2004), laser radiation is used as a working tool to structure matter by ablation, defining the lateral precision. For this reason, laser radiation is projected by objectives onto the surface, thereby being formed to a specific spatial intensity distribution. Ablation is a process strongly dependent of pulse duration and wavelength, and it can occur by two main principles: short pulse and ultra-short pulse laser ablation. The type of ablation process determines the finishing quality of the ring (Poprawe, 2004). The characteristic of the ablation process — short or ultra-short laser ablation — is determined by the pulse duration, as shown in Fig. 1.

When the pulse duration is longer than 1 nanosecond (1 ns = 10^{-9} s), the process is called short pulse laser ablation and its major characteristic is the significant amount of material being removed from the irradiated area in liquid phase. At this process, there is a high absorption of laser energy in the vaporized material along the duration of one pulse. When the pulse duration is smaller than 1 ns, the heating mostly takes place after the period of time of pulse duration and the process is called ultra-short pulsed laser. In opposite to pure ablation process, the formation process of rings is still not widely studied. During the rings' formation, there is just a slightly melting of surface material and very few of it is evaporated. According to Cabeza *et al.* (2012), the morphology at the surface can be characterized as solidified melted zone combining fine grain appearance and waves in the radial direction.

Janssen (2012) introduced the possible benefits of such microstructures manufactured by laser ablation on highly stressed parts. Based on this, in the present context, the functionalizing aims of three technological benefits for the cutting process are:

- An optimized surface topography enables the reduction of friction and cutting forces in the regime of direct tool-work piece contact.
- The lubrication shall be improved, leading then, to a reduction of tool's wear and hence to increased tool's life time. Alternatively higher cutting velocities can be applied without decreased life time.
- For several applications, it can be assumed that the current laser process allows supplying a higher surface quality on the treated work piece and thus increases an qualitative process output.

These benefits can be achieved optimizing tool surface by the combination of two phenomena: formation of a beneficial topography and formation of a new material microstructure in a surface near a layer of several hundred nanometers. The target surfaces are the flank and rake surface of cutting tools.

The target surface's texture in the present work is called "rings structure". Figure 2 shows an example of both structures,

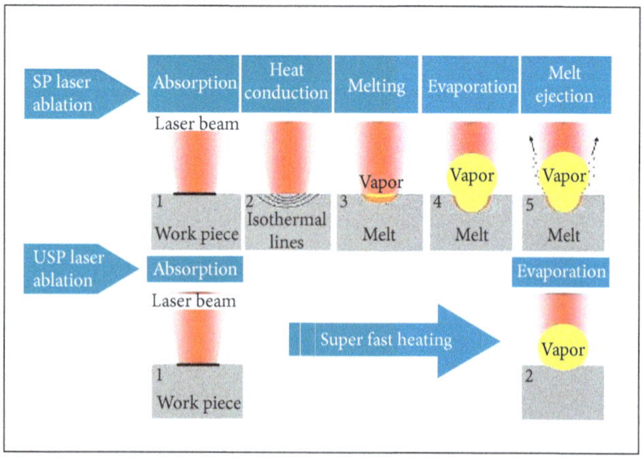

Figure 1. Physical mechanisms in SP and USP laser ablation (Janssen, 2012).

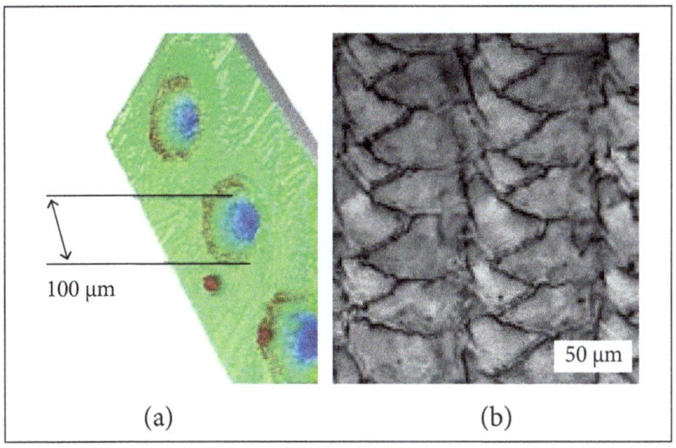

Figure 2. (a) regular dimples, distributed on the surface, and (b) textured composed of overlapping dimples, completely covering the surface (Klocke and Schmitt, 2012).

dimples and rings. The difference between both microstructure is that, unlike dimples, which allow some parts of the original tool material to be in contact with the workpiece, in the ring's microstructure, the tool surface in contact with the workpiece is completely modified by laser texturing. The present work aimed analyzing the formation process of ring's microstructure subjected to the main process parameters.

Due to positive results considering tool life, high material removal rates etc., obtained in previous works, the laser texturing process becomes an important tool in the cutting tool preparation field. Research about the laser functions and operation methods is necessary in order to define a behavior profile of laser parameters in obtaining textured surfaces, especially when these textured surfaces are now starting to be used, as for the case of ring's structures.

METHODOLOGY - STUDY ON RING'S FORMATION

The work was performed at the Fraunhofer Institute for Production Technology (IPT), a research laboratory, localized in the city of Aachen, Germany. For texturing experiments, a custom-designed laser structuring machine has been utilized, which was installed within former research projects.

EXPERIMENTAL SETUP

Based on industrial as well as research demands, the machine has been designed based on a 5-axis milling machine, which was adapted by implementation of an entire nanosecond laser ablation system including laser source, transportation fiber and laser scanner. With this machine, a real 3-D laser structuring process is possible by means of using 8 axes (5 machine, 3 optical, 1 virtual).

The laser source used was the Ytterbium pulsed fiber laser YLPM-1-A4-20-20, from IPG Laser. Regarding to its configurations, the laser presents an extended pulse repetition rate from 1.6 to 1,000 kHz, selectable pulse durations (4, 8, 14, 20, 30, 50, 100 and 200 ns) and maximum peak power of 15 kW.

The tool material used was cemented carbide from the group K (red), a conventional tungsten carbide grade. The material was chosen by the fact that the cemented carbide is one of the most utilized cutting tool materials today and the group derived from tungsten carbide is most readily employed.

The tool utilized was manufactured by SECO, from the model TNMA160404F, 410, 5μ. It was specially developed for specific applications in Fraunhofer IPT. It presents the flank surface specially treated but the rake surface is free of any surface treatment, which means the laser hits a hard metal WC-Co surface.

EXPERIMENTAL APPROACH

It has to be considered that the formation of the desired rings is based on a reduced ablation process. The process parameters are selected for causing a slight melt formation on the surface and subsequent evaporation of a very small amount of heated material. In regular laser ablation, this amount is decreased to the achievable maximum, since it enables high material removal and thus ablation rate. Here, the evaporation is only utilized to evolve a back pressure on the molten material. The molten layer, however, is only several hundred nanometers in thickness, thus the re-solidification takes place very fast, leading to the formation of rings in terms of "frozen liquid structures". For a better understanding, it can be imagined a stone being thrown in the water, and then it immediately solidifies in the first ring's formation.

Among the existing process parameters, which define a laser process, this work focused on the influence of two main process parameters: laser power and working distance. These process parameters control the process. However, they are no real process quantities. In the following, the meaning of these two parameters is described.

The laser power P_L describes the average laser radiation energy emitted by the laser source per time. Pulsed laser emits pulses with single durations much smaller than 1 s. In this case, the size of average output power of the laser is sometimes misleading, since it includes the time between the pulses. In the process of SP and USP laser processing, the pulse pause, which separates each pulse from the following one is commonly several thousand times higher than the pulse duration. Thus, the average power P_L can be several ten Watts only, while the pulse peak power P_{LM}, which is the absolute power within the pulse duration, can reach many thousand Watts.

To meet this fact, it is common in the field of SP and USP laser ablation processing to give not the laser output power but the intensity I or the fluency F, while the intensity gives the energy coupled into a unit area per time (Eq. 1):

$$I = \frac{E_p}{A \cdot t_L} \tag{1}$$

where E_p is the pulse energy; A is the focal spot area and t_L is the pulse duration.

The fluency gives the energy coupled into a unit area without time information, according to Eq. 2.

$$F = \frac{E_p}{A} \tag{2}$$

This is more common in USP laser processing, where effects can be interpreted independently from the pulse duration. In the project around this work, a laser has been utilized not only in the SP regime but it is additionally capable of emitting pulses of variable pulse duration. The pulses can be selected between 4 and 200 ns. Based on preliminary investigations, the pulse duration was kept constant at 100 ns.

In the present work's evaluations, the parameter P_L is given. So, the variables A and E_p could be varied independently.

The working distance is the distance between focusing lens and work piece. To understand the meaning, Fig. 3 shows the beam propagation of a laser beam. By varying the working distance z_W, the spot diameter can be adjusted, offering the possibility to control the focal spot size. This is limited due to the laser beam's propagation profile, which is the intensity distribution within a fixed z-plane.

Having these two control parameters, laser power and working distance, one is able to influence the ring formation in terms of varying the irradiated and the impinged energy per area. The main characteristics which are evaluated concerning the ring's formation are their diameter and depth. The arrangement of rings in a pattern can be easily done by adapting the program which controls the tool path for the laser scanner mirrors and thus is not investigated in this work.

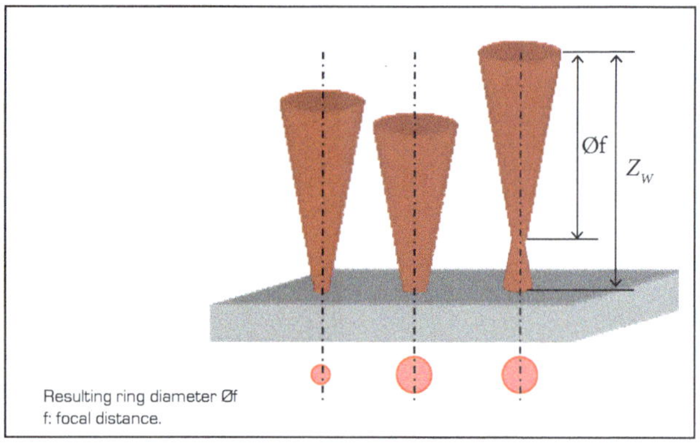

Figure 3. Beam propagation and spot size variation (Janssen, 2012).

Because the rake and the flank surfaces passed by different treatments, each surface has, therefore, different properties. For this reason, both were laser-treated and the laser parameters' influence was also investigated under the circumstances of the two material differences.

The surface measurements — diameter and depth — of the structured tool were made by the surface measurement device Alicona Infinite Focus Real 3-D — an optical 3-D surface metrology instrument. It automatically measures the complete structure of a surface (form, roughness and contour, for example).

For the tests, 20 different laser parameter combinations were defined, as shown in Table 1, in order to achieve different diameters and depth in the rings' structure.

RESULTS AND DISCUSSION

DIAMETER ANALYSIS

The first results were obtained by varying the power percentage. The working distance was kept in -0.8 mm. The graph from Fig. 4 shows the diameter's behavior through the variation of the power percentage. The increment in diameter presents a tendency of growth when the incident beam power is increased. In the flank surface, which received the special surface treatment, the increment in the diameter is constant and it occurs in a smoother behavior. On the other hand, for the rake surface, where there is no treatment, the increasing in the diameter occurs through an irregular manner.

In the second analysis, the power was kept constant at 100%, while the working distance was varied. The graph from Fig. 5 shows a comparison of the diameter's behavior between the two surfaces tested.

Table 1. Laser parameter variation.

Parameter	Working distance (mm)	Power (%)	Tool surface
1	-0.4	100	Rake
2	-0.7	60	Rake
3	-0.7	80	Rake
4	-0.7	100	Rake
5	-0.8	70	Rake
6	-0.8	80	Rake
7	-0.8	90	Rake
8	-0.8	100	Rake
9	-0.8	50	Rake
10	-0.8	60	Rake
11	-0.4	100	Flank
12	-0.7	60	Flank
13	-0.7	80	Flank
14	-0.7	100	Flank
15	-0.8	70	Flank
16	-0.8	80	Flank
17	-0.8	90	Flank
18	-0.8	100	Flank
19	-0.8	50	Flank
20	-0.8	60	Flank

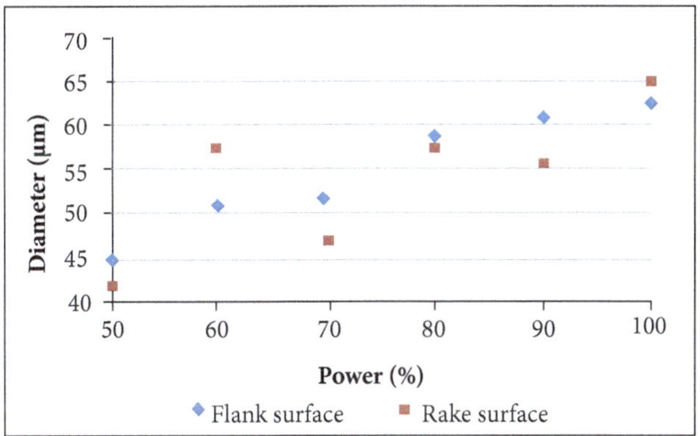

Figure 4. Diameter's behavior through power percentage variation.

Focus position is one of the most critical process parameters and directly affects the remaining geometry after material removal (Bordatchev and Nikumb, 2006).

On both tool surfaces, the decrease in laser focus distance caused an increase in the spot's diameter, which was the result expected according to the situation presented in Fig. 3.

Figures 6 and 7 show the result of the laser surface treatment on both tool surfaces.

Althought increasing the power increases the diameter's values, it also causes other effects on the rings, as it can be seen in Figs. 6 and 7. The images show that the tests where the power percentage was higher presented less precision

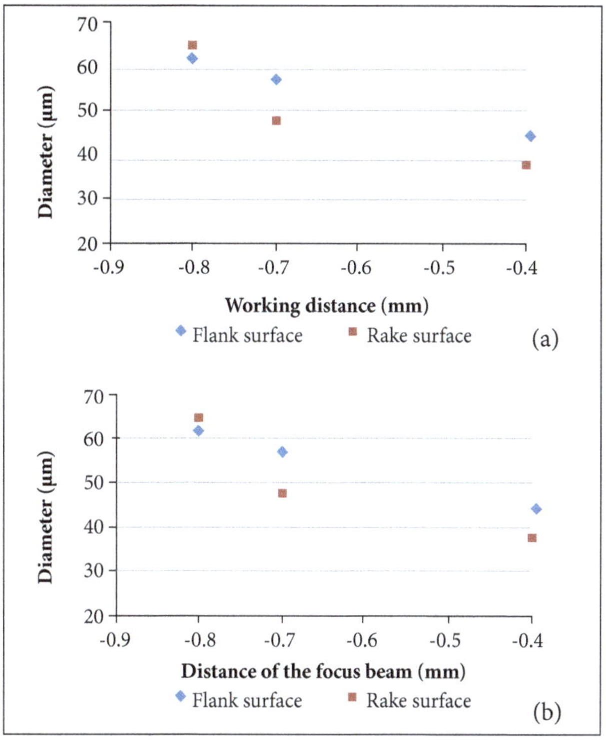

Figure 5. Diameter's behavior through working distance variation.

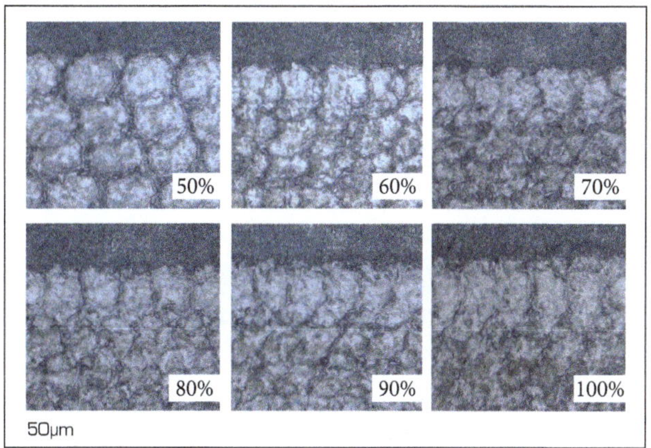

Figure 6. Images of ring's texture for different power percentage values on flank tool surface.

in the diameter's shape. In both surfaces, it is possible to notice that the laser ring's edges become less defined as the power grows.

This phenomenon can be explained by the texturing laser mechanism. Since the utilzed laser pulses where set to 100 ns in duration, the laser matter interaction time is long enough to enable a large amount of material being melted whithin the pulse duration. Evaporating material evolves a huge back pressure on the underlying material, which, in this case, is in melted state and thus can be expelled to the outer region of the irradiation zone.

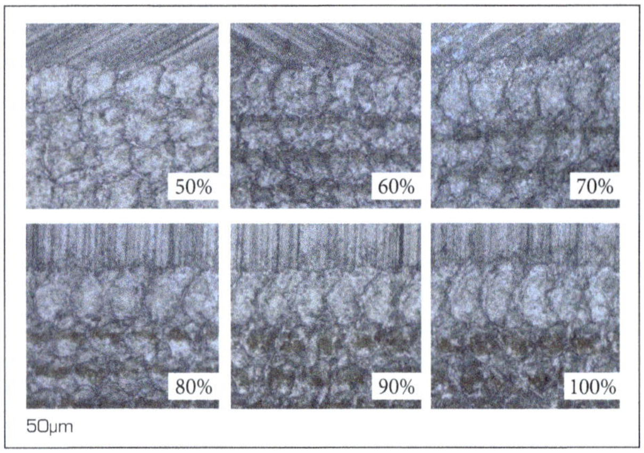

Figure 7. Images of ring's texture for different power percentage values on rank tool surface.

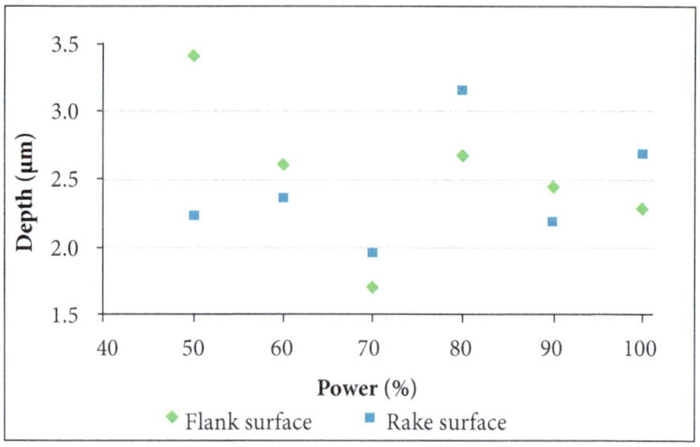

Figure 8. Graph of depth's behavior through power variation on the flank and rake tool surfaces.

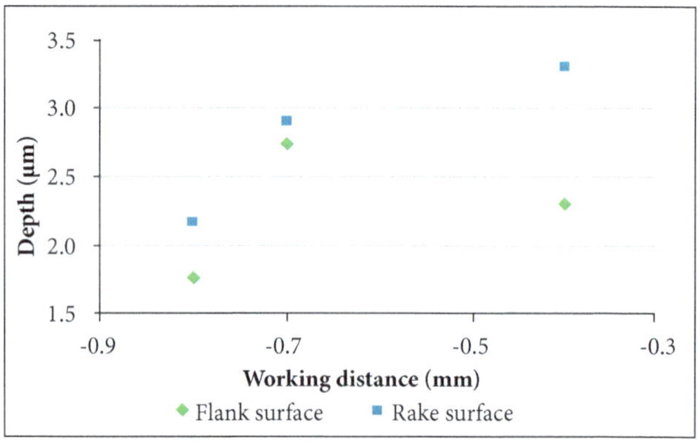

Figure 9. Ablated depth for different fluencies.

DEPTH ANALYSIS

The second microstructure parameter analyzed was the depth of the rings. For this analysis, the same variation of laser parameters used for diameter's analysis was used.

Figure 8 shows the dependency of the resulting ring's depth on laser's power. The working distance was kept constant at 0.8 mm. Figure 8 also shows that, for the flank surface, there is a slight decrease in depth when the power increases. It is also possible to see, for the rake surface, that the depth slightly increases as well, when the power percentage increases.

In previous researches, Dumitru *et al.* (2004) presented a study where one could observe that higher incident fluencies generated larger ablation rate. In other words, it means that with/when increasing the laser power, the depth will also increase.

However, the same behavior is not seen in either the studied cases. Changing the power in the rings' formation does not present the same correlation as in the pure ablation process, when analyzing the depth of the microstructure.

Caused by the formation principle, described before, the resulting surface after the meltiwng is not flat, but with peaks and valleys, created by the solidification of the thermally driven motion of molten material. This surface characteristic makes it difficult to define a right spot to measure the depth. In addition, the depth profile of the rings is just slightly above the order of the original surface's roughness, which also complicates the depth's determination.

For the second analysis of the microstructure's depth, the power was kept constant at 100%, while the working distance was varied. The graph from Fig. 9 shows a comparison of the depth behavior between the two tested surfaces. In the rake surface, it is seen that when increasing the working distance, the depth also decreases. In the flank surface, however, this behavior is not constant, although, in general, there is the decrease of the depth with the increase of the working distance.

It is known that, when the laser beam operates in its exact focus, the incident intensity (and also fluency) in the irradiated surface's area is maximal. For a pure ablation processes, the bigger the incident intensity, the bigger is the ablation rate during the process, which leads to a bigger depth. However, the intensity hitting the surface is decreased when the laser beam works out of its focus. Thus, the manufacturing of the ring's texture needs to be conducted in a defocused range in order to find the optimum intensity. The latter is found when the amount of evaporated material is just sufficient to force the molten material to form a ring exhibiting the desired shape.

CONCLUSION

Based on this investigation, the following conclusions could be drawn:

- Increasing the laser power or the distance of the laser beam focus from the surface led to an increase in the rings' diameter on the flank and rake surface due to enhanced thermal affection of the surface's material. These results are in agreement with the results for the laser ablation process.
- The variation in the incident power beam generated different results in the microstructure's depth of each tested surface. This difference was explained by the process of ring formation and by the superficial melting in the surface's material caused by the laser. The behavior of the ring's structural formation is not the same as the phenomenon of ablation.
- The variation in the working distance caused the same result in depth of both studied surfaces. Increasing the working distance caused the decrease on the ring's depth, which corresponds to the results of the laser ablation process.

Further studies are required to evaluate the tribological performance of such ring's textures on the desired application.

ACKNOWLEDGMENTS

This work was supported by the Deutsche Forschungsgemeinschaft (German Research Foundation) and Coordenação de Aperfeiçoamento de Pessoal de Nível Superior (CAPES), within the research project "Surface conditioning and texturing of cutting tools edges for the optimization of manufacturing processes - ConTexTool".

REFERENCES

Aurich, J.C., Zimmermann, M. and Leitz, L., 2011, "The Preparation of Cutting Edges Using a Marking Laser", Production Engineer, Vol. 5, No.1, pp. 17-24. doi: 10.1007/s11740-010-0275-9

Bordatchev, E. and Nikumb, S.K., 2006, "Effect of Focus Position on Informational Properties of Acoustic Emission Generated by Laser-Material Interactions", Applied Surface Science, Vol. 253, No. 3, pp. 1122-1129. doi: 10.1016/j.apsusc.2006.01.047

Cabeza, M., Castro, G., Merino, P., Pena, G. and Román, M., 2012, "Laser Surface Melting: a Suitable Technique to Repair Damaged Surfaces Made in 14 Ni (200 grade) Maraging Steel", Surface and Coatings Technology, Vol. 212, pp. 159-168. doi:10.1016/j.surfcoat.2012.09.039

Dumitru, G., Romano, V., Weber, H.P., Gerbig, Y., Haefke, H., Bruneau, S., Hermann, J. and Sentis, M., 2004, "Femtosecond Laser Ablation of Cemented Cabides: Properties and Tribological Applications", Applied Physics A – Materials Science & Processing, Vol. 79, No. 3, pp. 629-632. doi: 10.1007/s00339-004-2675-1

Etsion, I., 2004, "Improving Tribological Performance of Mechanical Components by Laser Surface Texturing", Tribology Letters, Vol. 17, No. 4, pp. 733-737. doi: 10.1007/s11249-004-8081-1

Janssen, A., 2012, "Laserstrukturieren senkt Reibung hochbelasteter Serienteile", Maschinenmarkt, Vol. 11.

Kim, B.S., Chung, W.Y., Rhee, M. and Lee, S., 2012, "Studies on the Application of Laser Surface Texturing to Improve the Tribological Performance of AlCrSiN - Coated Surfaces", Metals and Materials International, Vol. 18, No. 6, pp. 1023-1027. doi: 10.1007/s12540-012-0483-2

Klocke, F., Gierlings, S., Adams, O., Aerbach, T., Kamps, S., Veselovac, D., Eckstein, M., Kirchheim, A., Blattner, M., Thiel, R. and Kohler, D., 2012, "New Concepts of force measurement systems for specific machining processes in aeronautic industry", Proceedings of the 5th Conference on High Performance Cutting.

Klocke, F. and Schmitt, R., 2012, "Surface conditioning and texturing of cutting tools for the optmization of machining processes". Proceedings of the 3rd International Meeting of Brazilian-German Collaborative Research Initiative on Production (BRAGECRIM), Aachen, Germany.

Meijer, J., Du, K., Gillner, A., Hoffmannc, D., Kovalenkod, V.S., Masuzawae, T., Ostendorfc, A., Poprawec, R. and Schulzc, W., 2002, "Laser Machining by Short and Ultrashort Pulses, State of Art and New Opportunities in the Age of the Photons", CIRP Annals - Manufacturing Technology, Vol. 51, No. 2, pp. 531-550. doi:10.1016/S0007-8506(07)61699-0

Poprawe, R., 2004, "Lasertechnik für die Fertigung", Springer, Heidelberg, Germany, 526p.

Scaraggi, M., Mezzapesa, F.P., Carbone, G., Ancona, A. and Tricarico, L., 2013, "Friction Properties of Lubricated Laser-Microtextured-Surfaces: An Experimental Study from Boundary- to Hydrodynamic-Lubrication", Tribology Letters, Vol. 49, No. 1, pp. 117-125. doi: 10.1007/s11249-012-0045-2

22

FSM-Based Test Case Generation Methods Applied to Test the Communication Software on Board the ITASAT University Satellite: A Case Study

Arineiza C. Pinheiro[1], Adenilso Simão[1], Ana Maria Ambrosio[2]

ABSTRACT: The software in satellite applications has become increasingly larger, more complex and more integrated, so its verification and validation require exploration of new approaches. In this paper we present a Model-Based Testing (MBT) approach applied to the Communication Module of the ITASAT-1 university satellite. The models are Finite State Machines (FSM) representing the software behavior. In order to manage the difficulties to model the software behavior the approach employs the Conformance and Fault Injection (CoFI) testing methodology associated with the JPlavisFSM tool in the real context of a satellite's critical software. The former advises the modularization of the modelling into different types of behavior into different FSMs, while the latter integrates several FSM-based methods to derive test cases, provides facilities to design and to check properties of the models and computes metrics. The main result of this case study was the evaluation of the drawbacks on the design of the testing models supported by CoFI and JPlavisFSM. The models, test sets, metrics with the application of our approach applied to the Communication Module are presented. The paper discusses the benefits as well as the points requiring new researches.

KEYWORDS: Finite state machine, Model-based testing, Test-case generation methods, Testing methodology.

INTRODUCTION

The increasing development of university satellites has allowed researchers to experiment new approaches in space area. In traditional satellite development, experimental methods are usually avoided to prevent risks and cost increase, whereas university satellites open an opportunity to explore new approaches using a real system (Alencar, 2013).

The software for satellite applications has become increasingly more complex, as it includes more functions and is more integrated. The development of satellite-related software usually requires a series of rigorous tests, along with all the satellite development phases. The development tendency of this kind of software points out the use of formal models for the designing and testing. To adopt formal models, such as the Finite State Machines (FSMs), not only does it benefit the identification of ambiguities and gaps in the requirements (Morais and Ambrosio, 2010; Morais, 2011; Pontes *et al.*, 2012), but it also makes the automatic generation of test cases feasible (Ambrosio *et al.*, 2005; Romero *et al.*, 2012) by applying the vast theory of automatic test case generation.

The specification of test cases for embedded software in satellites should take into account characteristics such as: real-time requirements, integration of technologies and fault tolerance mechanisms. The occurrence of failures in this kind of software may cause large losses, so they should be thoroughly tested using systematic and rigorous approaches, in which

1. Instituto de Ciências Matemáticas e de Computação/USP – São Carlos/SP – Brazil 2. Instituto Nacional de Pesquisas Espaciais – São José dos Campos/SP – Brazil.
Author for correspondence: Ana Maria Ambrosio | Instituto Nacional de Pesquisas Espaciais | Avenida dos Astronautas, 1.758, CEP: 12.227-010 –São José dos Campos/SP | Brazil | Email: ana.ambrosio@inpe.br

various testing techniques are combined to exercise different aspects of the system.

The Model-based Testing (MBT) is an approach that reduces cost with the automation of test cases generation and with the reuse of models created during the software designing and testing activities; besides that, it provides high fault detection and traceability. In MBT, the test designer can specify the test model using different modeling notations, such as FSMs, Labeled Transition Systems (LTS), Unified Modeling Language (UML) state machines, etc. FSM is a formal modeling technique, adopted due to its rigor and simplicity. FSM-based testing has been studied for several decades (Moore, 1956; Chow, 1978) and it has still presented contributions (Simão *et al.*, 2009; Simão and Petrenko, 2010; Yin *et al.*, 2010; Pedrosa and Moura, 2010; Hierons and Ural, 2010; Simão *et al.*, 2005) such as a tool that integrates different FSM-based methods into a single software product named PLAVIS (PLAtform for Software Validation & Integration on Space Systems); similar work is given in Santiago *et al.* (2008).

In MBT, supporting tool plays an essential role, since the cost of building the models for testing only should be balanced by the possibility of generating and executing larger test suites (Utting and Legeard, 2007). However, the difficulties still remain to build the models. In our work, we have applied the Conformance and Fault Injection (CoFI) methodology (Ambrosio, 2005) in order to guide the tester to create the models associated with the JPlavisFSM tool (Pinheiro, 2012). CoFI was firstly proposed to lead a tester to understand the problem well enough while creating a set of FSMs representing normal and abnormal behavior of software on-board a satellite. It defines detailed steps to apply FSM-based testing tools, in a systematic way and it has been applied in several cases (Ambrosio *et al.*, 2007; Pontes *et al.*, 2009; Anjos *et al.*, 2011; Mattiello-Francisco *et al.*, 2013). JPlavisFSM tool comprises not only four methods to automatically generate test cases starting from a given specification in FSM but also the Mutation Testing technique (Fabbri *et al.*, 1994), that allows the adequacy evaluation of a test cases set.

This paper presents a case study in which the CoFI methodology and the JPlavisFSM tool are combined to test the Communication Module of the ITASAT-1 satellite (Sato *et al.*, 2011). All the steps applied to design the models have been illustrated. We employed the JPlavisFSM tool to the test case generation, exploring all the FSM-based methods available. The main result of this case study was the identification of the drawbacks to create the testing models that is the key activity to successfully apply MBT techniques to satellite-related software. On the modelling work, we found that the previous knowledge on the testing methods theory had facilitated the modelling process, which took 16 days only. The length of time to understand the system under test, to do the modeling and to generate the test cases accounted for 33 days. We did not evaluate how the models impacted on the quality of the tests, as the tests were not executed against the implementation of the Communication Module software, because the implementation was not finished before the conclusion of this work. However, we evaluated the quality of the test sets by mutation analysis applied to FSM.

The remainder of this paper presents the background, the CoFI and the JPlavisFSM tool, the case study, and finally, discusses the MBT applicability and the conclusions.

BACKGROUND

In this section, we introduce the main concepts necessary for understanding the results presented in this paper.

FINITE STATE MACHINES

Finite State Machines (FSMs) have been widely used for modeling reactive systems, ranging from simple protocols to complex embedded systems (Lai and Leung, 1995). Among the main advantages of FSMs, there are the solid theoretical background, the expressiveness power and the existence of numerous methods to generate test cases. FSMs are hypothetical machines composed of states and events, which correspond to transitions between the states. A transition is associated to two kinds of events: input and output. When an input event occurs in a given state, the FSM responds with an output event and may move to another state (Gill, 1962).

Formally, a Mealy FSM is defined as a tuple $M = (S, s_0, X, Y, D, \lambda, \delta)$, where: S is the nonempty finite set of states; s_0 is the initial state ($s_0 \in S$); X is the nonempty finite set of input symbols; Y is the nonempty finite set of output symbols; $D \subseteq (S \times X)$ is the specification domain; $\lambda : D \rightarrow Y$ is the output function; and $\delta : D \rightarrow S$ is the transition function.

Figure 1 shows an FSM example with 4 states and 8 transitions, where s_0 is the initial state, $S = (s0, s1, s2, s3)$, $X = (a, b)$ and $Y = (0, 1)$.

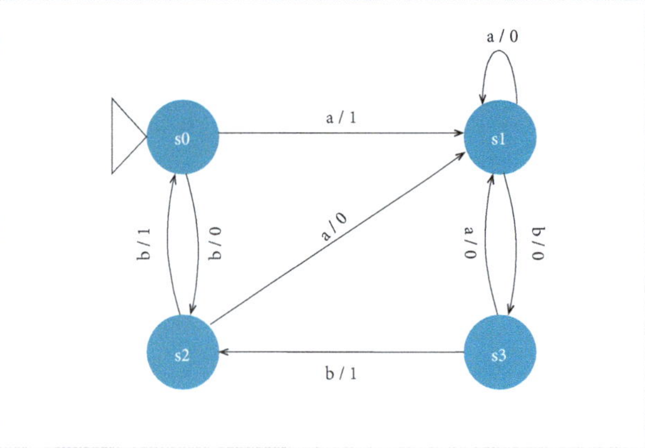

Figure 1. Example of FSM.

Given state s and input x, the transition (s, x) is *defined* if $(s, x) \in D$. An input sequence $\alpha = x_1x_2...x_k$ is *defined* at state s if there exists a sequence of states $s_1, s_2, ..., s_{k+1}$, so that $s = s_1$ and $(s_i, x_i) \in D$, $\delta(s_i, x_i) = s_{i+1}$, for $1 \leq i \leq k$. The set of input sequence defined at state s is denoted $\Omega_M(s_i)$. The output and transition functions are extended to defined input sequences in the usual way; given a state s and the empty sequence ϵ, we have that $\delta(s, \epsilon) = s$ and $\lambda(s, \epsilon) = \epsilon$; given input sequence α and input x, we have that $\delta(s, \alpha x) = \delta(\delta(s, \alpha), x)$ and $\lambda(s, \alpha x) = \lambda(s, \alpha)\lambda(\delta(s, \alpha), x)$. A set of input sequences is initialized if it contains the empty sequence. An input sequence α separates states s and s', if $\lambda(s, \alpha) \neq \lambda(s', \alpha)$. A set of sequences is an *identifier* for state s, denoted $Id(s)$, if for each state $s' \neq s$, there exists a sequence in $Id(s)$ which separates s and s'. A set containing one identifier for each state is *harmonized*, if for each pair of state s and s', there exists an input sequence $\alpha \in Id(s) \cap Id(s')$ which separates them. An input sequence α is a *Unique Input/Output Sequence* for state s, denoted $UIO(s)$, if $\{\alpha\}$ is an identifier for s. A set of input sequence W is a *characterization set* for each pair of state s and s', if there exists an input sequence in W which separates them. A *preamble* for state s is an input sequence α, so that $\delta(s_0, \alpha) = s$, i.e., it takes the FSM from the initial state to s. A *state cover* is a set containing preambles for each state. A *transition cover* is a set P of input sequences, so that, for each defined transition (s, x), there are α and αx in P, where α is a preamble for s.

The FSM M is defined over X. Based on this definition, some structural properties in FSM, which are important for test case generation methods, are listed:

- *Completely specified* – an FSM is *completely specified*, or *complete* if all states (S) have a transition for each input event from the set of input symbols (X), so that $D = (S \times X)$. Otherwise, the FSM is *partially specified*, or *partial*;
- *Strongly connected* – an FSM is *strongly connected* if for every pair of states (s_i, s_j) there is a path that leads s_i to s_j, i.e., there is some input event sequence that performs a path of transitions with source in s_i and destination of s_j. If all other states can be reached from the initial state, the FSM is initially connected;
- *Deterministic* – an FSM is *deterministic* when there is only one transition with a particular input event from any state that allows transition to a next state. Otherwise, the FSM is nondeterministic;
- *Equivalent* – one state s_i is considered *equivalent* to s_j if there is no input event sequence which, when executed from the respective states, generates a different output sequence;
- *Reduced* – an FSM is *reduced* if there is no pair of equivalent states. Otherwise, it is unreduced.

FINITE STATE MACHINE (FSM)-BASED TESTING

In the context of FSM-based testing, a finite sequence of input events is a *test case*, or just a *test*. A *set of test cases* is a finite set of sequences, so that there are no two sequences α and $\beta \in T$, where α is a proper prefix of β. (There are two sequences α and β, α is prefix of β, if $\alpha \leq \beta$, such as $\alpha\omega = \beta$, to any ω. And, α is a proper prefix of β, if $\alpha < \beta$, such as $\alpha\omega = \beta$).

Given two FSMs M and I; two states s from M and t from I are distinguishable if there is an input sequence $\gamma \in \Omega_M(s) \cap \Omega_I(t)$, called *separating sequence*, so that the same input event sequence generates different output sequences for both FSMs, i.e., $\lambda(s, \gamma) \neq \lambda(t, \gamma)$. Two FSMs I and M are *distinguishable* if their respective initial states are distinguishable.

A *fault domain* $\Gamma(X)$ is the set of all possible implementations of M defined over the set of input symbols X. Similarly, $\Gamma_n(X)$ denotes the set of all FSM defined over the set of input symbols X with at most n states. The test set T is said *n-complete*, or just *complete*, for the specification M if for all $I \in \Gamma_n(X)$, so that I is distinguishable from M, there is at least one sequence $\alpha \in T$ that produces different output sequences when applied to M and I in the respective initial states. In other words, a test set is complete when it is possible to distinguish the specified FSM with n states from all other distinguishable FSM with the same set of input symbols and at most n states.

The main objective of the FSM-based testing is to compare a reduced FSM *model M* with *n* states to an *implementation* of the FSM model, which is assumed to be represented by an unknown FSM *I*. When a reduced FSM *M* which represents the correct version of the specified FSM is compared to FSM *I*, the following faults may be revealed (Chow, 1978):

- Output Error: *I* and *M* differ in the output of a transition;
- Transference Error: *I* and *M* differ in the final state. When the final state achieved in both FSM is different after applying a test case;
- Transition Error: it is the general term for output or transference error;
- Missing States: *I* has fewer states than *M;*
- Extra States: *I* has more states than *M.*

Several methods have been proposed for generating set of test cases which guarantee that the implementation does not contain any of the above listed faults, that is, which are complete regarding these faults. The completeness of the generated test case is proved by showing that no faulty implementation FSM would pass the test cases. The main differences among them are the cost to generate the sequences and the effectiveness (the power of the test cases to detect faults). One of the first methods to be proposed was the W (Chow, 1978), which is considered a precursor of the area, since most of the following methods are based on it: UIO (Sabnani and Dahbura, 1988), UIOv (Vuong *et al.*, 1989), Wp (Fujiwara *et al.*, 1991), HSI (Petrenko *et al.*, 1993) and SPY (Simão *et al.*, 2009).

MUTATION TESTING

The Mutants Analysis is one of the most popular criteria of Defects-based testing, which aims at deriving test requirements from knowledge about typical mistakes made by designers or developers. Mutation Testing is a testing technique that can be widely employed as a way of evaluating the test cases generated. The criteria can be used as:

- Black-box testing, when we consider the specification as a test artifact, or
- White-box testing, when the artefact considered is the source code. In this context, the test artifact could be represented formally by a model, as FSM.

The Mutants Analysis consists of assessing the adequacy of a test set *T* for the test artifact *P*. First, you run the test set *T* against P. If a fault occurs, a defect was found. Otherwise, it is assumed that some defects already exist but cannot even be detected.

The next step is to derive one or more products from the original artefact *P*, which gives the new products P_1, P_2, ..., P_n, also called mutants of *P*. Mutants are generated from mutation operators, which are dependent on the programming language and determine what types of syntactic changes can be made in the artefact. The mutation operators' aim:

- Inducing simple syntactic changes based on the typical mistakes made by developers, such as changing the value of a constant, or
- Forcing certain test goals, such as performing each node or arc of the product (Delamaro *et al.*, 2007).

The test set T is executed against each of the mutants generated. The main objective is to "kill" all mutants. It is expected that, due to the changes made, the mutants have different behavior from the original product, featuring a defect.

Occasionally, when the artifact is a source code, a mutant and the original product can still have the same result for any test case in the set. In these cases, the tester has to check the code and determine whether there is equivalence between the products. If so, the mutant is called as equivalent. The equivalence between programs is an undecidable problem and, therefore, there is no automated solution to the problem and the manual intervention of the tester is necessary. However, when the product is a model, there is no equivalency problem, because it could be determined automatically due to the formality of the FSM.

After the mutants' execution and equivalence analysis, the mutation score is calculated. The aim is to determine the adequacy of test cases used in a range from 0 to 1, providing a quantitative measure of how efficient the test set is in revealing the difference between the mutant and the original product. The mutation score *ms*(P, T), to product *P* and the set of test cases *T* is calculated as:

$$ms(P,T)= \frac{DM(P,T)}{M(P)-EM(P)}$$

where: DM (P, T): number of mutants killed by *T*; M (P): total number of mutants generated from *P*, and EM (P): number of generated mutants that are equivalent to *P* (Delamaro *et al.*, 2007).

The mutation score is obtained from the ratio between the number of mutants killed by the set *T* and the number of mutants that could be killed, given by the difference between the total number of mutants generated and the number of mutants classified as equivalent.

THE JPLAVISFSM TOOL AND THE CONFORMANCE AND FAULT INJECTION METHODOLOGY

This section presents the main features of the JPlavisFSM tool and the CoFI methodology which were combined to guide the creation of FSMs, representing the system's behavior.

JPLAVISFSM TOOL

In a joint cooperation project among Brazilian researchers, the test platform named PLAVIS was conceived. This platform integrated existing tools for automatic generation of test cases starting from FSM, which had been developed by PLAVIS-project's members. PLAVIS was a web-based test platform providing different functions to support FSM-based testing (Simão *et al.*, 2005). Years later, Pinheiro (2012) improved usability features in the previous platform, now named JPlavisFSM tool.

The JPlavisFSM is a standalone software, whose graphical user interface (GUI) is imported from the open source tool JFlap (http://www.jflap.org/) to design the FSM. From the GUI, one can easily draw states and transitions, edit and adjust them before automatically generating a set of test cases as well as evaluating the adequacy of a set of test cases.

The JPlavisFSM tool includes a function to analyze FSM structural properties namely *complete, strongly connected, deterministic*, etc. This kind of analysis is very important because the applicability of the test methods depends on presence or absence of some structural properties. For example, the W method requires a FSM with the properties of *complete, strongly connected, deterministic*, and *reduced*. One advantage is to reduce the need of theoretical knowledge, particularly for testers unfamiliar with such theory.

Other functionalities were implemented to assist test case handling, such as test set execution, inclusion/exclusion and enabling/disabling. Test sessions can be created, as shown in Fig. 2. In this window, the user can visualize the FSM on the left and the generated test cases on the right side (each line has one test case). At the bottom, there is an option that helps the tester to create particular test sets by himself.

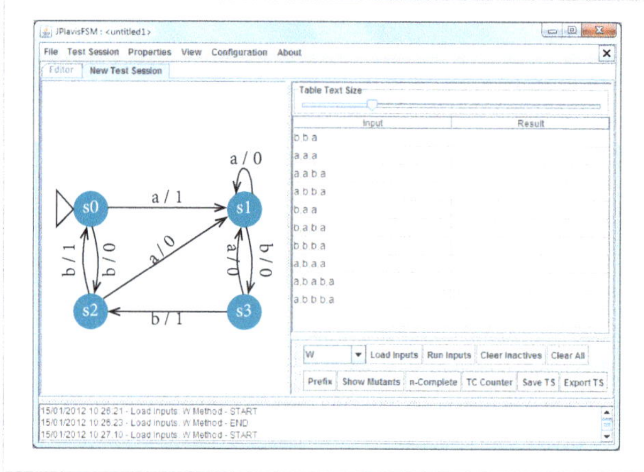

Figure 2. JPlavisFSM Tool.

One of the best features of the JPlavisFSM's test session is to support the construction of a test case set with a high probability of finding defects in the System Under Test (SUT). In a test session, the tester can import (or load) a set of test cases, which was automatically generated through one of the testing methods given by JPlavisFSM or manually generated by the tester. Once imported, the set of test cases can be executed (i.e., applied against the specification) and the mutation score obtained.

The JPlavisFSM releases the following testing methods to generate test cases set: W, UIO, HSI and SPY. The W method generates test cases as follows. First, an initialized transition cover and a characterization set of the specification are determined. Secondly, for each sequence of the transition cover, each sequence of characterization set is appended. The UIO generates test cases as follows. First, an initialized transition cover is determined, as well as UIOs for each state. Secondly, for each sequence of the transition cover, the UIO of the state reached by that sequence is appended. HSI method generalizes the W method, and can be applied to partial FSMs. It first determines a transition cover and a set of harmonized state identifiers. Then, for each sequence of the transition cover, each of the sequence in the identifier for the state reached by that sequence is appended. The SPY method generalizes the HSI method; the main difference is that both the transition cover and the harmonized state identifiers are computed "on-the-fly", i.e., during the execution of the method, trying to minimize the number of test cases which are required to obtain complete test cases. For the HSI and SPY methods two versions are provided: the original one and the auto-completed; the latter auto-completes the FSM before applying the original method. New methods can be incorporated in JPlavisFSM as plugins or be imported at runtime (provided that the compatible input/output formats are used) as well as providing more flexibility to the tool and the possibility to explore all the other features for the test set analysis.

Two different ways of evaluating a test cases set are provided:

- Mutation Analysis; and
- n-Completeness Analysis (Simão and Petrenko, 2010).

The Mutation Analysis, provided by JPlavisFSM, allows a test set to be evaluated and also to be enhanced. This enhancement can be achieved by expanding the test set with test cases targeted at undetected mutations. The mutant operators (Fabbri *et al.*, 1994) were designed according to the classes of faults detected on the FSM-based test models. When the user creates a test session, the mutants are automatically created and the user can access all the mutants to verify their statuses: *alive*, when the ongoing test fails to reveal the defect of the mutant; or *dead*, the other way around. This feature could help the user to upgrade test sets since this criterion guides the user to create tests that will run paths, which may contain defects. The n-Completeness analysis provides another way to verify the quality of a given test set, checking whether the test set is complete or not.

COFI METHODOLOGY

The CoFI methodology (Ambrosio, 2005) guides the functional testing activity. It was proposed considering the needs of space application's validation and therefore defines steps for creating test cases for software embedded on-board of satellites in a systematic way in order to get as much reusability in testing as possible. The main objective is to help the tester to define FSM-based models which represent the behavior of the SUT.

The SUT behavior's decomposition into FSMs starts with the identification of services, which can be a system function from the user's viewpoint. Each service shall be described by a set of FSMs that map different classes of behavior. This classification takes into account different kinds of input events:

- *Normal* (absence of faults);
- *Exceptions* (foreseen faults);
- *Unexpected inputs* (inputs occurring when they are not expected); and
- *Hardware faults.*

In short, the CoFI methodology consists of three steps: identification, modeling and test case generation. Figure 3 shows the three steps of CoFI methodology described hereafter.

(A) Identification

Based on a given specification, the tester has to understand properly the external behavior of the SUT, as in the black-box testing. The information to accomplish this step can be extracted from textual specification documents, requirements, use-cases, sequence-diagrams, and also from interviewing experts in the SUT if necessary. In this step, it shall be identified:

- *Services* that an user can recognize or execute in the SUT;
- *Events* and *actions*: it is necessary to identify all the possible events and actions (or outputs) that can be, respectively, commanded and trigged (or observed). The selected events and actions will be abstracted as inputs and outputs in the FSMs;
- *Control and observation points*: they are addresses or mechanisms to input data and to obtain responses from the SUT;
- *Physical faults*: faults occurring in hardware. They are to be considered whenever the SUT includes embedded software in a hardware and this software must have mechanisms to treat the faults;
- *Facilities and constraints*: the tools and theirs commands that supports the test execution against the SUT and the events that cannot be activated during the test execution.

(B) Modeling

In this step, for each *service*, the tester should develop four classes of behavior:

- *Normal*;
- *Specified exceptions*;
- *Sneak path;* and
- *Fault tolerance.*

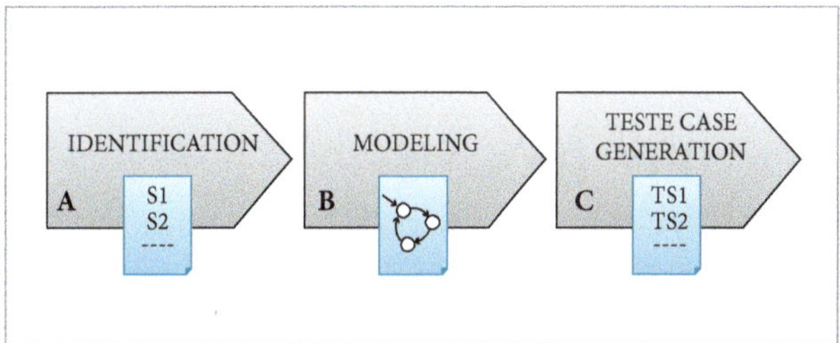

Figure 3. Main steps of the CoFI Methodology.

At least one FSM model shall be created representing the partial system behavior.

The *normal behavior model* defines the sequence of events that the SUT normally expects when it is executed. The tester has to identify the normal mode the SUT will be subordinated to and the events and actions expected for this case. For the *specified exceptions* model the tester has to recognize the exceptions mentioned in the documents, as timeouts and unknown commands and thus define the events expected in this context, called *exception events*. The *sneak path* model deals with the correct inputs arriving in wrong time. The objective is to predict the unexpected behaviors, i.e., to describe what the machine should do if an input occurs in a state where it is not defined. The tester can use the *normal* model as a basis, write it in tabular form (event × state) and fill the blank cells with the corresponding expected behavior. In some cases, there is no possible behavior to model, and then the tester evaluates the best option for the SUT. The transitions trigged by the *unexpected inputs* should be designed based on the normal model. Finally, the *fault tolerance* model shall map the possible physical faults when the SUT implements mechanisms to tolerate hardware faults. For each physical fault, the tester has to adapt the normal model including the fault events. At the end of the modeling step, the tester has a set of models representing the SUT's behavior, from which test cases may be automatically generated.

(C) Test Case Generation

This step can be supported by testing tools that automatically generate a set of test cases (also named test sequences). In the experiences about applying CoFI described in Ambrosio *et al.* (2007), Pontes *et al.* (2009), Anjos *et al.* (2011) and Mattiello-Francisco *et al.* (2013), the Condado tool was used to automate the test cases generation. In our work, the JPlavisFSM tool was used because it provides five different testing methods, including some variations that can be applied in partial FSM. Moreover, it has facilities to analyze the models' properties and the adequacy of test case sets.

CASE STUDY: ITASAT'S COMMUNICATION MODULE

We now discuss the case study we have carried out applying CoFI methodology and JPlavisFSM tool in a MBT approach for a real system. The SUT is the Communication Module's software of the ITASAT-1 satellite. The ITASAT Mission (Sato *et al.*, 2011) is part of a program funded by the Brazilian Space Agency (AEB) in the context of Action 4934 for Development and Launching of Small Technological Satellites developed by the *Instituto Tecnológico de Aeronáutica* (ITA) and other Brazilian universities. ITA is the responsible member for the project implementation and the *Instituto Nacional de Pesquisa Espacial* (INPE) provides technical consulting and laboratory infrastructure. The ITASAT-1 is a university satellite planned to execute experimental payloads namely, Digital Data Collection Transceiver, Heat Pipe Experiment, Micro Electrical Mechanical System for attitude determination and an Inter Satellite Link.

The satellite architecture is composed of five service subsystems: Mechanical Structure (MSS), Thermal Control (TCS), Electric Power (EPS), Attitude Control and Data Handling (ACDH) and Telemetry & Telecommand (TMTC); the last three subsystems are shown in Fig. 4. The functions of the ACDH are commanded by the on-board computer (OBC). The Communication Module (CM), located in the ACDH subsystem, is in charge of the communication between the OBC and the TMTC. Figure 4 highlights the Communication Module and indicates that it is the System under Test (SUT), the focus of this study.

The CM comprises not only two receivers and two transmitters in redundancy, but also the software in charge of receiving commands arriving from the ground stations and transmitting telemetry collected from satellite pieces of equipment to ground stations.

The CM Software performs critical functions so it has to work properly for the space mission success. Its functions are to receive the telecommands from ground stations and to transmit the telemetries collected by the satellite's payloads and sensors, to manage the payloads status and to turn off the pieces of equipment when a critical failure occurs leading the satellite into survival mode.

Considering that the CM is one of the most critical parts of this satellite, it was chosen as case study to experimentally evaluate the proposed approach. The three main steps that guided this study are illustrated in Fig.5.

(A) Identification

The first activity was to study the Requirements Document of ITASAT-1 (Table 1 illustrates some of such requirements)

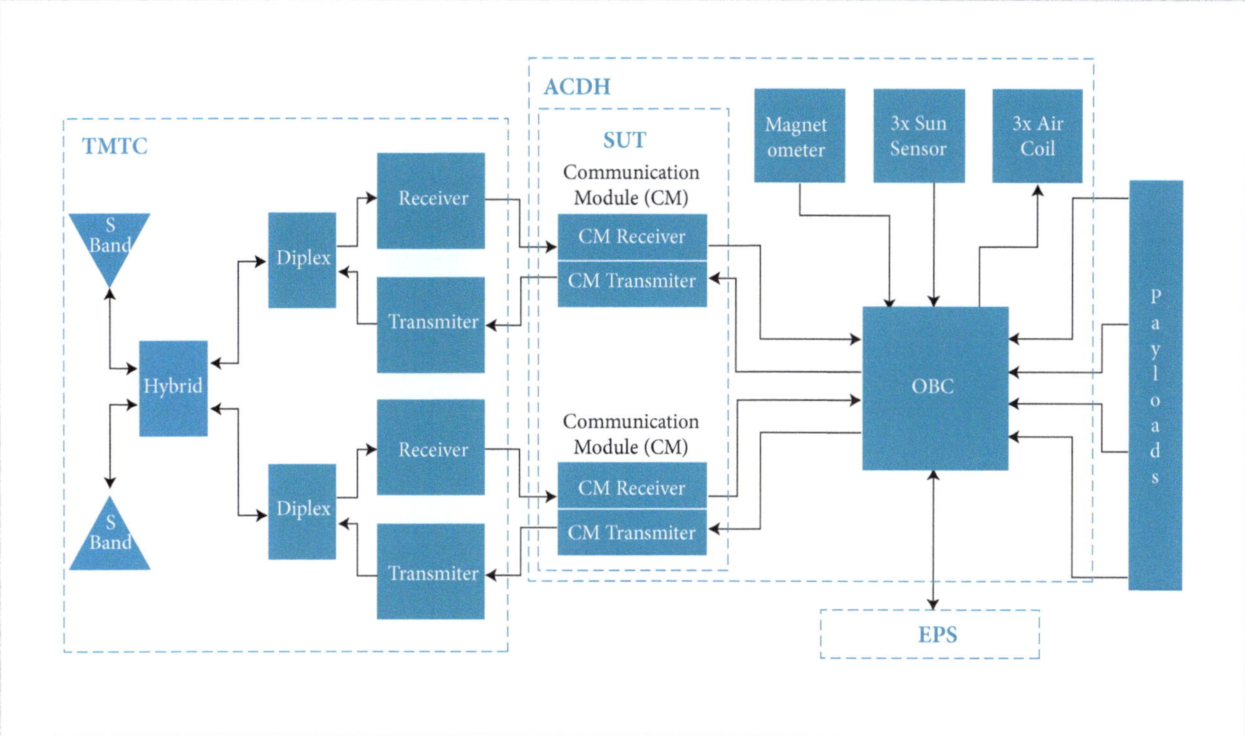

Figure 4. ITASAT-1 satellite architecture.

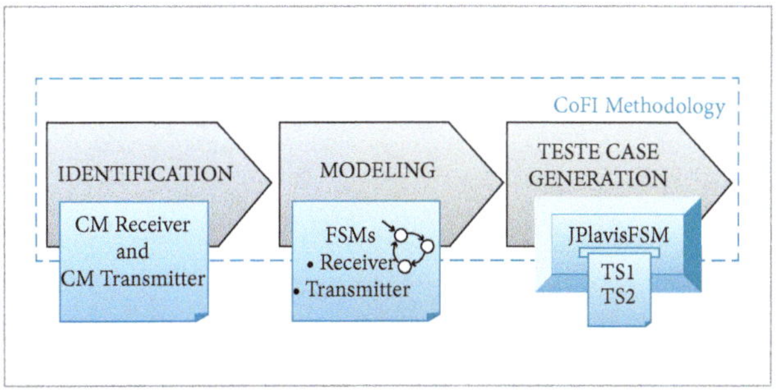

Figure 5. Development of the case study.

in order to understand the context and the behavior of the CM's software. The studies lasted 12 days, five to understand the ITASAT-1 satellite documentation; five to the CoFI guidelines and two to identify the main elements required by CoFI, namely, services, events, actions, faults, and so on.

We identified two *services* corresponding to the two main functions of the Communication Module: (S1) receive commands from ground and (S2) transmit the collected telemetries from on-board equipment. These functions are implemented respectively by the *CM Receiver Software* and by the *CM Transmitter Software*.

The *events* and *actions* extracted from the requirements are those naming the transitions of the FSMs illustrated in Figs. 6 and 7. The events and actions are abstract representations, i.e., they do not indicate a physical signal arriving. One example of a CM Receiver's event is *CmdOBCOK*, that indicates the presence of one command to be executed. One action is *StoreTC1wait*, which indicates that the just-arrived telecommand shall be stored on board. Other examples are *EndTimerB* as event and *StartTimerB* as action to be performed by the CM Transmitter. For the CM Receiver Software, 12 events and 13 outputs were identified, while for the CM Transmitter Software, 19 events and 12 outputs were identified. Specified *exceptions* were identified in seven requirements, which are listed in Table 1. Concerning the *physical faults*, although they can occur, the Communication Module Software is not in charge of dealing with them, therefore, for testing purposes, the *fault tolerance* models were not created.

(B) Modeling

It was observed that on adopting the CoFI methodology's guidelines the state explosion problem was avoided. First, we modelled the *normal* behavior of each service in the CM Receiver FSM (*R_N*) and the CM Transmitter FSM (*T_N*), as shown in Figs. 6 and 7, respectively.

For the FSM models used here, namely the Mealy machine (Mealy, 1955), all the transitions should be represented by a pair of input-output, which means that each event should be associated with an action. Since in this software test some events were not associated with an observed output (or action), the output "*Terminate*" was used in the model to indicate the end of a transition. Another modeling condiction was made up to represent one transition requiring the occurrence of two events or transitions with two outputs. Figure 6 illustrates the *ACKandStartTimerC* output, which, in practice, indicates two actions: "send an *Acknoledge message*" and "start the *Timer C*".

Regarding the *specified exceptions*, the FSMs R_Ex13, R_Ex2, R_Ex4 and R_Ex5 map the exceptions of the CM Receiver, and T_Ex1 and T_Ex2 map those of the CM Transmitter. Table 1 matches the FSMs to the ITASAT-1's requirements they model. For sake of space, not all FSMs are shown in this paper. For more details see Pinheiro and Ambrosio (2013).

The CM Software has to handle timers which trigger timeout events needed for the communication timing. On the first analysis, to handle timers seemed an issue, because classical FSM does not treat time events. Therefore, these events were abstracted away in the model. The CoFI guides the definition of timers as external devices that are started by an action and produce an event to the FSM when the time is expired, indicating a timeout. For the *normal* behavior, the software only signalizes the start of a timer, as represented by *StartTimerC* and *StartTimerA* events. For the *specified exception* models, it was needed to represent events to indicate expired time, so we used the *EndTimerC* and *EndTimerA* events.

The first CM Receiver's FSMs were pointed out as non-deterministic by the JPlavisFSM, because the event *INT2TC* associated to *StoreTC* action occurred in several states. On further analysis of the real software's behavior we defined two abstractions to solve this problem:

- A finite buffer to store at most three TCs, called TC1, TC2 and TC3; and
- Two actions to differentiate TC stored and TC sent.

Then, to create several actions (outputs): *StoreTC1wait, StoreTC1send, StoredTC2wait*, and so on, instead of using only the *StoreTC* action.

Table 1. ITASAT-1 Requirements vs. Specified Exceptions Models.

Specified exceptions			
#	Requirement	Description	FSM
1	FRq11152000-12	"When the OBC does not send a request in TC seconds, the CM receiver finalizes the communication routine"	R_Ex13
2	FRq11152000-13	"The request of the OBC shall be one of the following: the transmission of TC or TC error log from CM receiver to OBC, or the execution of a (correct) command from OBC." Obs.: Terminate execution if the command is not OK.	R_Ex2
3	FRq11152000-16	"If the CPU does not send the acknowledge in TA seconds (value TBD), the CM receiver shall terminate the communication routine."	R_Ex13
4	FRq11152000-27	"The CM receiver shall check the first 16 bits, used for synchronization and packet addressing, and shall discard the TC packet if its address data does not match with the module address."	R_Ex4
5	FRq11152000-29	"If the verification described in Frq11152000-2 (The CM receiver module shall check the mod the repeated data and EDAC field) results in failure, the CM receiver shall store a TC error log (as specified in [AD3]) in the TC buffer"	R_Ex5
6	FRq11152000-52	"When any analog channel verified present unexpected results, the CM transmitter shall generate a pulse in the survival flag pin (which is read by the CM receiver).	T_Ex1
7	FRq11152000-58	"When the OBC does not send a request in TC seconds, the CM transmitter finalizes the communication routine."	T_Ex2

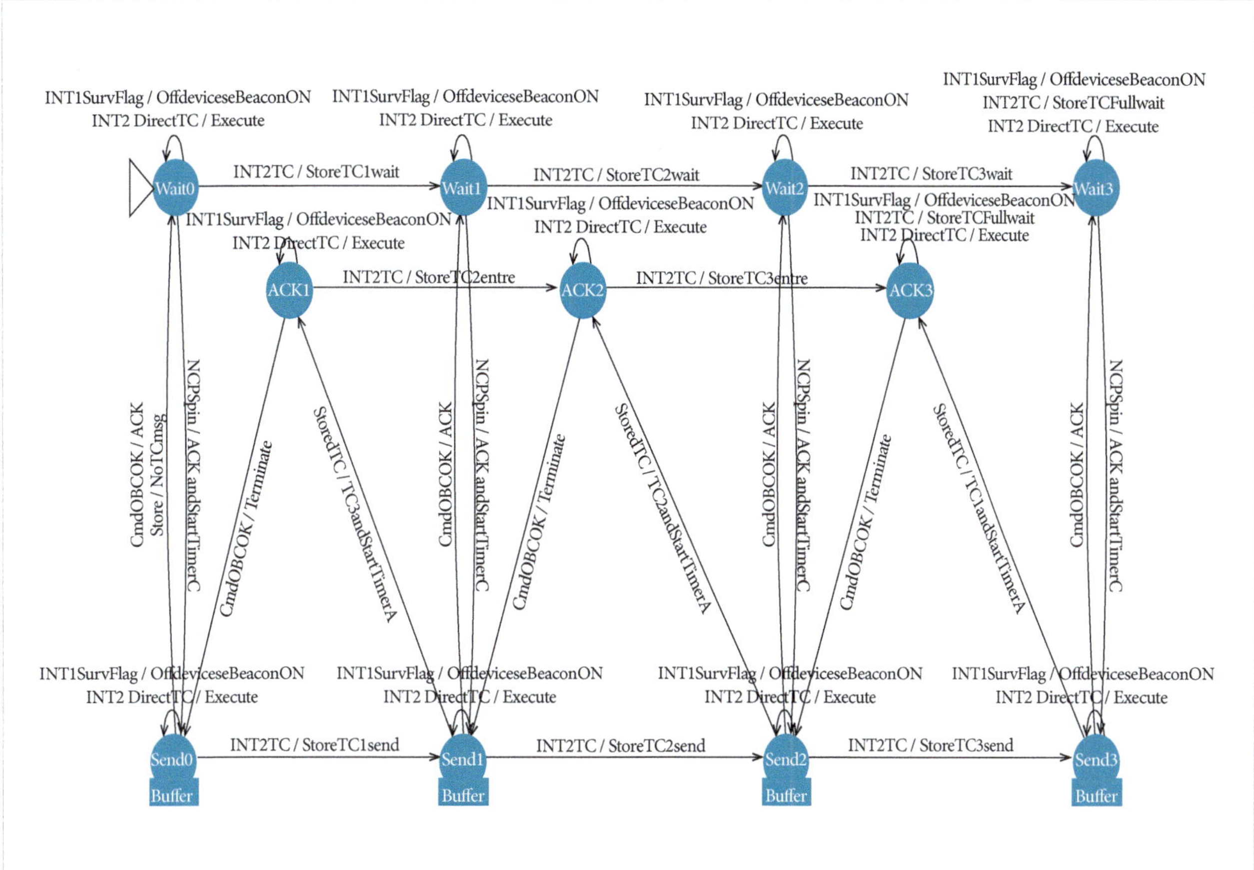

Figure 6. R_N: Finite State Machine representing the normal behavior of the CM Receiver.

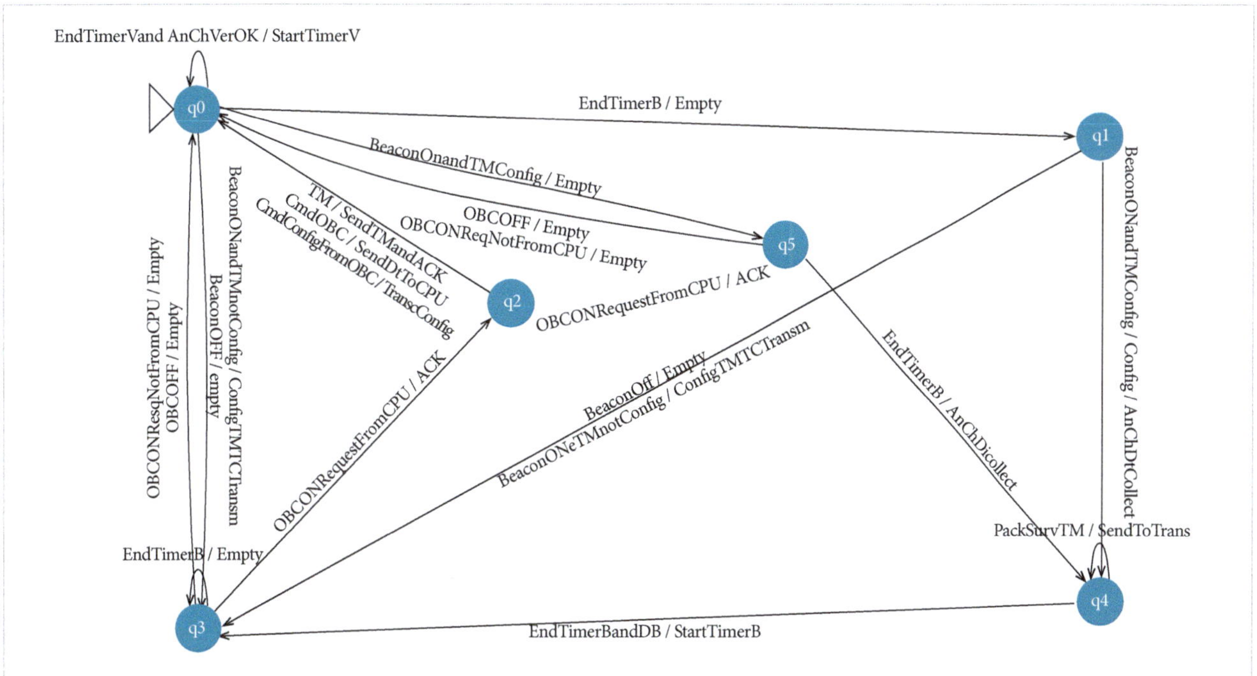

Figure 7. T_N: Finite State Machine representing the normal behavior of the CM Transmitter.

After concluding the normal behavior modeling, we analysed all events presented in the normal model, in order to create the *sneak path* models. However, because of the logic implemented in the hardware design construction, events never occur in an inappropriate state, so sneak path models were not designed. The *fault tolerant* models were not designed as well, because the CM is not in charge of dealing with the hardware faults. In the end of the modeling step, we had eight FSMs, as shown in Table 2.

The JPlavisFSM indicated incompleteness for all FSMs, consequently the traditional W method could not be applied. The completeness is a difficult property to be achieved when modelling the behavior of real systems, that is why the JPlavisFSM tool provides other methods, as HSI and SPY that do not require this propriety.

Summarizing, the modeling steps accounted for 16 days, being 8 days to create the normal and specified exception models including reviews and re-work in order to adequate the models to precisely represent the software behavior; and generate the test cases through the methods. Three adequacies were considered:

- Adequacy to the system – Do models represent the system?
- Adequacy to the FSM concepts and generation methods – How can one abstract the model to achieve the FSM needs?
- Adequacy to the final model – Have the final models achieved the system objectives and the methods needs?

(C) Test Case Generation

We applied all methods for the eight FSMs. The W method's applicability is restricted to complete, deterministic and initially connected FSMs. As the FSMs were not complete we used the JPlavisFSM's function to auto-complete it before using the W method. The auto-complete functionality adds all the missing transitions in a given state as self-loops, i.e., transition that starts and ends in the same state. Althought the auto-complete option leads to the generation of many test cases, which are not significant in practice, it allows the applicability of the W. One test case generated by W starting from the CM Receiver FSM normal behavior (Fig. 6) is given by the following sequence of inputs *INT2TC, StoredTC, NCPSpin,* and its corresponding expected output is *StoreTC1wait,TC1andStartTimerA,ACKandStartTimerC.*

Table 2. Summary of models design.

Services	FSM models by classes of behavior			
	Normal Behavior	Exception Specified	Sneak Path	Fault Tolerance
CM Receiver	1	4	0	0
CM Transmitter	1	2	0	0

The UIO method requires a *strongly connected* FSM, whereas the SPY and the HSI can be applied into a *non-completed* FSM. Thus, these methods have a wider applicability in comparison to the others; moreover, the test set generated by them is smaller than those genereted by W and UIO methods. On the other hand, it is possible to apply the UIO Method because it requires a *separation sequence* defined for each state. One example of a test case generated by UIO method starting from the CM Transmitter FSM is *BeaconONandTMnotConfig, OBCONRequestFromCPU, CmdOBC, EndTimerVAndAnChVerOK.*

The HSI and SPY are based on *separating sequences* by pair of states (see Background section). Thus, each pair of states must have a transition with an event in common with distinct outputs. In the FSMs of the CM Transmitter software no event occurs in more than one distinct state, so these methods were applied neither to T_N nor to T_EX1 and T_EX2. One test case generated by the SPY method is *INT2TC, NCPSpin, StoredTC, INT2DirectTC, INT2TC, INT2DirectTC, INT2TC* which starts with the same input as the example to W method, but explores a valid sequence of events.

In general, all methods of the JPlavisFSM were applied to the created models. Table 3 shows the test sets generated by all methods for each FSM. The '-' symbol indicates the method could not be applied and '*' symbol indicates the use of auto-complete functionality.

Particularlly, the original W method did not generate any test case for the *non-complete* FSM, so there is no column for it. The original W was executed only thanks to the auto-complete feature. This option increases the number of transitions, leading to a large number of test cases.

The UIO method generated smaller sets, but the method does not guarantee the generation of complete sets. The HSI and SPY were successfully applied to CM Receiver FSMs because there was a *separating sequence* that distinguished their states. For CM Transmitter FSMs, it was not possible to use the HSI and SPY methods, therefore using the auto-complete feature they could be applied. The auto-complete feature was executed in CM Receiver FSMs only to show the increase in the number of test cases, caused by the extra transitions artificially included in the FSM to enable the application of the methods.

We have also explored the JPlavisFSM's functionality to calculate the *minimal test set* composed by test cases produced by all the methods. After creating a test set comprising all test cases, all the prefixed test-cases are discarded and the minimal test set is generated by the tool. The test sets shown in the HSI* and SPY* columns were not considered, since the original methods, HIS and SPY, could be applied.

Table 4 illustrates the number of test cases derived from each model (considering the total set and the minimal set), the number of mutants created to each model and the mutation score obtained with both sets of test case. The mutation scores for the minimal and the total sets were the same, since only prefixes are removed from the total set. The number of test cases (total and minimal sets) includes the tests generated by W*, UIO, HIS and SPY, but it does not include HIS* and SPY* to FSMs from CM Receiver.

All the methods (except UIO) generate complete test sets, but it is a theoretical concept which assumes that the implementation can be represented by an FSM with the same number of states of the specification.

In general, to apply FSM-based testing implies handling a very huge number of test cases, so the tester must either be prepared to automate the test execution or select and prioritize the test cases. In our approach, we suggest a test strategy that could be adopted: first is to select the smallest test set generated by original methods; then, if there is still enough time to the testing activity, other sets could be applyed to improve the system's reliability. In this case study, the SPY and UIO methods should be selected for CM Receiver FSMs and CM Transmitter FSMs, because the Mutation Score of these sets were 1.0 for each one, and they are complete sets as well.

The inputs and outputs represented in the models are abstracted to enable the creation of FSMs, this fact led to the generation of *abstract test suites*, so a post-processing of the test sets will be necessary before the test execution. This allows the test cases to be designed before defining some details of implementation and before the implementation is ready. Moreover, in the FSMs of the Comunication Module some events shall be replaced by a set of others, as the case of the *INT2DirectTC*, which corresponds to 23 different telecommands.

DISCUSSIONS AND RELATED WORKS

Various methods have been published to optimize the number of test cases to minimize the efforts required to obtain an acceptable level of software quality (Lai, 2002). However, these methods and test tools are yet rarely used in the software industry. Issues, such as 'What are the hindrances

Table 3. Number of tests generated by method for each Finite State Machine.

FSM	W*	UIO	HSI	HSI*	SPY	SPY*
R_N	234	48	38	78	22	40
R_Ex13	356	44	34	89	26	42
R_Ex2	201	37	27	67	21	40
R_Ex4	201	37	27	67	21	40
R_Ex5	201	37	27	67	19	37
T_N	316	21	-	184	-	72
T_Ex1	316	21	-	184	-	72
T_Ex2	335	19	-	158	-	61

Table 4. Results of the generated Test Sets.

FSM	Number of test cases		Mutation Analisys	
	Total Set	Minimal Set	# Mutants	Mutation Score
R_N	342	285	1852	1.0
R_Ex1	460	408	1720	1.0
R_Ex2	286	264	1320	1.0
R_Ex4	286	263	1357	1.0
R_Ex5	284	265	1357	1.0
T_N	593	432	775	1.0
T_Ex1	593	432	775	1.0
T_Ex2	573	454	591	1.0

to the adoption of the generation methods based on FSM in the industry?' or 'What is the gap between academic and industrial testing methods?' have been investigated in order to ease the difficulties of employing FSMs in the real context. In this section we discuss these questions in general speaking and concerning our approach.

The fault domain, introduced in Background section, is defined for general purpose FSM. That fault domain provides all possible faults that can happen in the model. However, it may be interesting to adapt the model to the real scenario that should be tested. In Srivastava and Singh (2009) a fault model based on FSM to embedded systems was proposed. The fault model provides three kinds of errors that could occur among the connections between the hardware and software components: *unconnected outputs*, when the implementation should activate an output but does not activate, in other words, there is no output after a transition happens; *unconnected inputs*, when an input event has no behavior effect, there is no transition to another state as expected; and *redirected inputs*, which occur when an input is redirected to an incorrect state and generates a wrong behavior. The requirements for the construction of the test set were defined according to the new fault domain. The main focus of Srivastava and Singh is on the definition of the domain applied to embedded systems; and a testing generation method is not applied.

In Santiago *et al.* (2008) a test environment that supports the generation of test sequences based on Statecharts and FSMs is presented. The simplicity of the model is considered the main advantage of employing FSMs as a technique for modeling reactive systems. Methods such as W and UIO were applied in the context of embedded systems.

Although the applicability of MBT has been widely investigated, there are some reasons for reluctance in adopting academic methods for FSM-based testing. We have evaluated some of the reasons presented in Lai and Leung (1995) and Lai (2002), such as *feasibility*, that concerns the use of the testing methods in real cases; *extreme formalism*, that gives the impression of being too academic; *need for training* or education in this area; and *resistance to changing*, because it is not necessary the use of formal methods to do tests, according to the current point of view of many industries acting in the satellite-based software sector.

The difficulty in using MBT starts with the modeling phase. Identifying the system inputs and outputs, as well as the system behaviors, is not a trivial task. A broad knowledge about the system and about FSM is necessary to understand how a non-formal specification could be transformed into a formal one. In our approach, CoFI methodology helped to deal with these difficulties. The tester is guided, step by step, to model the system behavior into several simple FSMs. Moreover, the JPlavisFSM tool facilitated fitting the models to the testing methods, automating the test cases generations, producing test sets' metrics allowing the tester to choose the best testing set.

The tester must yet interpret the generated test sets due to the abstraction of the model, though the practicality of MBT was improved by CoFI and JPlavisFSM. The formalism of the FSM-based testing methods is transparent to the JPlavisFSM tool user. The analysis of FSM properties is automatically generated.

Some validations with the use of FSM-based testing techniques and CoFI methodology have been conducted (Anjos *et al.*, 2011; Pontes *et al.*, 2012; Mattiello-Francisco *et al.*, 2013). In such cases the Condado (or ConData) (Martins *et al.*, 1999) was used as a tool to automatically generate test cases. Condado tool does not have neither the facilities to FSM structural analysis nor different methods to be chosen, however the single FSM property it requires is the FSM be connected. In the approach presented here, the FSMs have more elaborated structures, so the facilities provided by the JPlavisFSM tool allowed improving the application of the CoFI. Besides that, it was applied to real embedded software of satellite application.

Concerning the training aspect, JPlavisFSM tool can be used to teach how the generation methods work. The tool has the option to import extra methods. The user may implement his/her own method and apply the n-Complete tool to analyze if the method is correct. The JPlavisFSM may be used to support the modeling phase too, since the GUI is simple, clean and it provides functionalities for analyzing FSMs. The usability improvements provided by JPlavisFSM and the guidelines of CoFI methodology were tentative of reducing the gap between the good results obtained in academia with testing methods and its use in practical cases of industry as discussed in Lai and Leung (1995).

CONCLUSION

The gap between academic research and industrial practice is still large. Thus, it is important to analyze the performance of testing activity supported by FSMs in real systems, once FSM-based methods are richly explored in the academic context.

This paper has reported on the applicability of the FSM-based testing in the real, developed Communication Module software of the ITASAT university satellite. As testing is a costly activity, automating the generation of test cases can help reduce the costs. In this sense, MBT is an approach that derives test cases from a formal model built to support the testing activity. On the other hand, real systems like ITASAT software should be tested taking into account requirements of timing and reactions to events. In this context, the systematic approach of CoFI methodology supported by the formalism implemented by JPlavisFSM could be an alternative to test the system. We have observed restrictions and benefits on applying this approach. Some observed limitations, which should be issues for new investigations, are:

- The modeling phase is not trivial and some knowledge about FSM is required to start the process. The tester has to learn the basic concepts and structural properties of FSM;
- The applicability of the generation methods is highly dependent on structural properties of FSM generated in modeling phase. The tester has to know the basic theory about the generation methods;
- The abstraction of the system is required to build models, so the test set generated is an abstract set of inputs and outputs, which needs to be interpreted before being executed. So, the post-processing of test sets could aggregate an extra cost to the test execution phase.

The observed benefits were:

- The CoFI methodology guides the modeling phase, which helps the inexperienced testers start using the FSM technique;
- The FSM-based test generation was automated leading to cost reduction of the test activity;
- The JPlavisFSM tool eliminates the need of profound theoretical knowledge about FSM structural properties and testing generation methods.

The MBT still has some practical limitation, but the initiatives that were made in academia helps to reduce them. The definition of new generation methods which do not require properties, such as FSM minimality, has to be explored. At the same time, the industry needs to step forward on the use of formal methods. These initiatives will improve the efficiency and effectiveness of testing activity.

As future work, we envision the following. The generated test sets must be post-processed and executed against the Communication Module implementation of ITASAT and the complexity of these activities should be analyzed against the obtained results.

ACKNOWLEDGMENTS

The authors would like to thank professors Emília Villani, David Fernandez and Wilson Yamagutti for the opportunity of applying the proposed approach in the ITASAT project. The authors would also like to thank the financial support of FAPESP, CNPq and CAPES. The authors are very thankful to reviewers for their useful comments.

REFERENCES

Alencar, W.A.F., 2013, "Model checking aplicado a software embarcado crítico do satélite universitário ITASAT", Dissertação de mestrado, ITA, São José dos Campos, Brazil.

Anjos, J.S., Gripp, J., Pontes, R. and Villani, E., 2011, "Applying the CoFI testing methodology to a multifunctional robot end-effector", São José dos Campos, SP, Brazil.

Ambrosio, A.M., Martins, E., Vijaykumar, N.L. and Carvalho, S.V., 2005, "Systematic generation of test and fault cases for space application validation", In Proceedings of the 9th ESA Data System in Aerospace (DASIA), Edinburgh, Scotland. Noordwijk: ESA Publications.

Ambrosio, A.M., 2005, "CoFI: uma abordagem combinando teste de conformidade e injeção de falhas para validação de software em Aplicações Espaciais", Tese de doutorado, INPE-13264-TDI/1031. 206p.

Ambrosio, A.M., Mattiello-Francisco, F., Santiago, V.A., Silva, W.P. and Martins, E., 2007, "Designing fault injection experiments using state-based model to test a space software", Third Latin-American Symposium on Dependable Computing (LADC), Morelia, México. Lecture Notes in Computer Science (LNCS) series. Springer Editors: Bondavali, A.; Brasileiro, F.; Rajsbaum, S. Springer, Berlin. pp. 170-178. doi: 10.1007/978-3-540-75294-3_13.

Chow, T.S., 1978, "Testing software design modeled by finite-state machines," IEEE Transactions on Software Engineering, Vol. 4, No. 3, pp. 178–187. doi: 10.1109/TSE.1978.231496.

Delamaro, M.E., Maldonado, J.C. and Jino, M., 2007, "Introdução ao Teste de Software", Elsevier.

Fabbri, S.C.P.F., Maldonado, J.C., Masiero, P.C. and Delamaro, M.E., 1994, "Mutation analysis testing for finite state machines", In Fifth International Symposium on Software Reliability Engineering, Monterey, California, USA, pp. 220-229.

Fujiwara, S., Bochmann, G.Von, Khendek, F., Amalou, M. and Ghedamsi, A., 1991, "Test selection based on finite state models", IEEE Transactions on Software Engineering, Vol. 17, No. 6, pp. 591-603. doi: 10.1109/32.87284.

Gill, A., 1962, "Introduction to the theory of finite-state machines", McGraw-Hill, New York.

Hierons, R.M. and Ural, H., 2010, "Generating a checking sequence with a minimum number of reset transitions", Automated Software Engineering, Vol. 17, No. 3, pp. 217-250. doi: 10.1007/s10515-009-0061-0.

Lai, R. and Leung, W., 1995, "Industrial and academic protocol testing: the gap and the means of convergence", Computer Network ISDN System, Vol. 27, No. 4, pp. 537–547. doi: 10.1016/0169-7552(93)E0110-Z.

Lai, R., 2002, "A survey of communication protocol testing", The Journal of Systems and Software, No. 62, pp. 21-46.

Martins, E., Sabião, S.B. and Ambrosio, A.M., 1999, "ConData: a Tool for Automating Specification-based Test Case Generation for Communication Systems", Software Quality Journal, Vol. 8, No.4, pp. 303-319.

Mattiello-Francisco, M.F., Ambrosio, A.M., Villani, E., Martins, E., Dutra, T. and Coelho, B., 2013, "An experience of the technology transfer of CoFI methodology to automotive domain", LADC, Industrial track. BDBComp Biblioteca Digital.

Mealy, G.H., 1955, "A Method for Synthesizing Sequential Circuits". Bell System Technical Journal, Vol. 34, No. 5, pp. 1045-1079.

Moore, E.F., 1956, "Gedanken–experiments on sequential machines", in Automata Studies. Princeton University Press, pp. 129-153.

Morais, M.H.E and Ambrosio, A.M., 2010, "A new model-based approach for analysis and refinement of requirement specification to space operations", In Proceedings of the 10th Conference on Space Operations. Huntsville, Alabama, USA. SpaceOps 2010. American Institute of Aeronautics and Astronautics (AIAA). doi: 10.2514/6.2010-2231.

Morais, M.H.E., 2011, "Uma abordagem para a melhoria da qualidade de requisitos baseada em modelos de estados", Dissertação de mestrado. INPE, 2011. São José dos Campos.

Pedrosa, L. and Moura, A., 2010, "Generalized partial test case generation method", In 4th International Conference on Secure Software Integration and Reliability Improvement Companion (SSIRI-C 2010). Washington, DC, USA: IEEE Computer Society, pp. 70-77.

Petrenko, A., Yevtushenko, N., Lebedev, A. and Das, A., 1993, "Nondeterministic state machines in protocol conformance testing", in Proceedings of the 6th International Workshop on Protocol Test systems VI (IFIP TC6/WG6.1). Amsterdam, The Netherlands, The Netherlands: North-Holland Publishing Co., pp. 363-378.

Pinheiro, A.C. and Ambrosio, A.M., 2013, "Modelagem do Módulo de Comunicação do Satélite ITASAT – Segundo a metodologia CoFI", Relatório de Pesquisa, sid.inpe.br/mtc-m19/2013/10.04.19.58-NTE. INPE, São José dos Campos. Retrieved in March, 19th 2014, from http://urlib.net/8JMKD3MGP7W/3EUF338

Pinheiro, A.C., 2012, "Subsídios para a aplicação de métodos de geração de casos de testes baseados em máquinas de estados", Dissertação de mestrado. ICMC. USP. São Carlos, Brazil.

Pontes, R.P., Morais, M.H.E., Véras, P.C., Ambrosio, A.M. and Villani, E., 2009, "A comparative analysis of two verification techniques for DEDS: Model checking versus model-based testing", In 4th IFAC Workshop on Discrete Event System Design (DEDS), Valencia, Spain, pp. 70-75.

Pontes, P.R., Véras, P.C., Ambrosio, A.M. and Villani, E., 2012, "Contributions of model checking and CoFI methodology to the development of space embedded software", Empirical Software Engineering – An International Journal, Vol. 10, No. 2, Springer. Online FirstTM, doi: 10.1007/s10664-012-9215-y - ISSN 1382-3256.

Romero, A.G., Ambrosio, A.M. and Souza, M.L.O, 2012, "Finite state-machine verification applied to hybrid systems", 21st SAE BRASIL International Congress and Exhibition, Expo-Center Norte, São Paulo, Brazil. doi:10.4271/2012-36-0429.

Sabnani, K. and Dahbura, A., 1988, "A protocol test generation procedure", Computer Networks and ISDN Systems, Vol. 15, No. 4, pp. 285-297. doi: 10.1016/0169-7552(88)90064-5.

Santiago, V., Vijaykumar, N.L., Guimarães, D., Amaral, A.S. and Ferreira, E., 2008, "An environment for automated test case generation from statechart-based and finite state machine-based behavioral models", In Proceedings of the 2008 IEEE International Conference on Software Testing Verification and Validation Workshop. Washington, DC, USA: IEEE Computer Society, pp. 63-72.

Sato, L.S., Saotome, O., Timm, C., Fernades, D. and Yamaguti, W., 2011, "Itasat-1: Brazilian university microsatellite for payload test and validation in low earth orbit", In Proceeding of the 8th Symposium on Small satellites for Eath Observation, Berlin, Germany.

Simão, A.S., Ambrosio, A.M., Fabbri, S.C.P.F., Amaral, A.S.M.S., Martins, E. and Maldonado, J. C., 2005, "Plavis/fsm: an environment to integrate FSM-based testing tools". In Simpósio Brasileiro de Engenharia de Software - Sessão de Ferramentas. Universidade Federal de Uberlândia. Uberlândia, Brazil.

Simão, A.S., Petrenko, A. and Yevtushenko, N., 2009, "Generating reduced tests for fsms with extra states," In Proceedings of the 21st IFIP WG 6.1 International Conference on Testing of Software and Communication Systems and 9th International FATES Workshop, ser. TESTCOM '09/ FATES '09. Berlin, Heidelberg: Springer-Verlag, pp. 129-145.

Simão, A.S. and Petrenko, A., 2010, "Checking completeness of tests for finite state machines", IEEE Transactions on Computers, Vol. 59, No 8, pp. 1023-1032. doi:10.1109/TC.2010.17.

Srivastava, S. and Singh, A., 2009, "Testing of embedded system using fault modeling", pp. 177-180. doi:10.1109/ ELECTRO.2009.5441142.

Utting, M. and Legeard, B., 2007, "Practical model-based testing: A tools approach", 1st ed. Morgan Kaufmann.

Yin, Y., Li, Z. and Liu, B., 2010, "Real-time embedded software test case generation based on time-extended EFSM: A case study", In Proceedings of 2nd WASE International Conference on Information Engineering (ICIE'10), pp. 272-275.

Vuong, S., Chan, W. and Ito, M., 1989, "The UIOv-Method for protocol test sequence generation", In Proceedings of the 2nd International Workshop Protocol Test Systems, pp. 161-175.

Location Issues of Thin Shell Parts in the Reconfigurable Fixture for Trimming Operation

Hu Fuwen[1]

ABSTRACT: The location of thin shell parts is always a knotty problem during machining, welding, forming, assembling and inspection operations. This paper mainly focuses on the special location issues in the digital and flexible trimming process for aircraft skins fastened by reconfigurable fixture. Firstly, in view of the dynamic change of effective locators, the "X-2-1" location principle was proposed with reference to the "3-2-1" and "N-2-1" location schemes. Secondly, the standard procedure for solving location parameters was summarized in consideration of location admissibility, holding posture, locator layout and so on. Thirdly, a locating experiment was conducted to investigate the positional accuracy of the reconfigurable fixture and the calculation accuracy of location parameters solution. Fourthly, a quantitative evaluation method to evaluate the dynamic stiffness of the fixturing system was put forward. Moreover, the effects of location layout on the dynamic stiffness were analyzed using the finite element simulation system for the trimming process. A noticeable appearance had been observed that the cliff effect of the dynamic stiffness of the flexible fixturing system may be induced due to the dynamic change of effective locators. Finally, some conclusions and discussion on future works were given.

KEYWORDS: Thin shell part, Reconfigurable fixture, Location principle, Dynamic stiffness, Trimming.

INTRODUCTION

The location of a component is one of the most important aspects to achieve the required accuracy during a machining process. Generally, the proper design of clamping and locating arrangements is a difficult task, considering its workpiece geometry, tool paths, set-up time, production scale, resistance to cutting forces, costs, adaptability and so forth. For a long time, towards fixture design becoming a science rather than an art, and towards achieving automation and intelligence of fixture design activities, many theories, tools and approaches have been introduced by many researchers (Wang *et al.*, 2010; Cecil, 2001; Boyle *et al.*, 2011).

However, most part of this research has concentrated on the fixture design for workpieces which can be regarded as rigid bodies. Actually there are also some low rigidity or flexible components in industrial fields, e.g. aircraft skin panels, automotive cover panels, rocket shells and other thin-wall structures. They cannot be seen as rigid bodies due to their inherent flexibility and inevitable deformation caused by external forces or their own weights. Thereby, their location issues have always been a particularly knotty problem. To overcome it, the single purpose and dedicated fixtures are often employed in practical production. Although this approach would be economical for large scale mass manufacturing (e.g. automobile manufacturing), it would lead to long lead time, a large quantity of tooling, and high production costs under small batch manufacturing environments such as the fabrication of aircrafts and rockets. Therefore, the flexible tooling solutions like reconfigurable fixtures or modular fixtures have been increasingly adopted. These flexible tooling can be changed, reconfigured or reused to suit dimension and shape changes of

1.North China University of Technology – Beijing – China.
Author for correspondence: Hu Fuwen | College of Mechanical and Electrical Engineering | North China University of Technology | 100144 | Beijing–China | E-mail: hfw@ncut.edu.cn

compliant parts. In the past decade or so, a great deal of attention has been increasingly directed toward the design, analysis and synthesis of flexible fixturing systems. There are a number of strategies in order to achieve flexibility (Shirinzadeh, 2002), but this paper mainly focused on the reconfigurable fixtures designed for machining operations of thin shell parts.

According to available documents, the earliest idea of a reconfigurable fixture for thin shell parts trimming process may date back to 1987 (Douglas and Ozer, 1987). About ten years later, the reconfigurable fixture for sheet metal machining was reduced to practice (Proctor, 1998). Afterwards, several commercially available reconfigurable tooling systems were designed and built, for example, the Pogo® flexible tooling system by CNA manufacturing systems (U.S.A.), and the TORRESTOOL® universal holding fixture by the M. Torres group (Spain). Application results from Northrop Grumman indicated that the Pogo® flexible tooling system can cut down setup time by about two-thirds on trimming and hole drilling operations on more than 100 different fuselage skin parts (Koelsch, 1998). But these commercially available reconfigurable tooling systems for holding compliant parts are extremely expensive (Walczyk and Longtin, 1999). Therefore, many researchers and manufacturers are inclined to develop the reconfigurable fixturing devices independently. A reconfigurable fixturing system was specifically designed for the drilling of sheet metal parts, which included a T-slot base plate, vertical supports and locating pins (Youcef-Toumi *et al.*, 1987). A novel reconfigurable modular system was presented for the fixturing of thin-walled flexible objects, which consists of a base plate, and height-adjustable locators and clamps (Sela *et al.*, 1997). A computer-controlled reconfigurable fixturing device for compliant parts, based on a matrix of individually-stoppable pins lowered by a single rigid platen, has been developed (Walczyk and Longtin, 2000). Zhou *et al.* (2008) designed a flexible tooling system for large-scale aircraft skins by means of the manipulator's drivability. For aircraft skins trimming based on the combination of the reconfigurable fixture and 5-axis gantry type milling machine tool, the author and co-workers have been studying the process planning method, the trimming process simulation, as well as the equipment development (Hu and Li, 2011; Hu *et al.*, 2012a; Hu *et al.*, 2012b). Besides, to evaluate the milling forces in the trimming process of aircraft skins, the author presented an integrated approach based on the combination of the mechanistic model of milling forces, 3D finite elements simulation of milling operations, and the support vector regression methodology (Hu and Li, 2012). Along with the promotion of research, a better systematic understanding of the issue of thin shell components locating in the flexible fixture has been recognized. These would be described as follows.

In the next section, the author would introduce the flexible manufacturing technology for aircraft skins based on the reconfigurable discrete dies and the reconfigurable fixture. And the third section mainly focuses on the distinctive location principle that appears in the trimming process of aircraft skins supported by the reconfigurable fixture. Then the fourth section illustrates the location parameters solution method, considering location allowance, holding posture, locator layout, etc. Then a location experiment was presented to investigate the location accuracy.Subsequently a quantitative evaluation method for the dynamic stiffness of the fixturing system was put forward. Finally, some conclusions and discussion on future works are given.

FLEXIBLE MANUFACTURING MODEL OF AIRCRAFT SKINS

Aircraft skin panels with the aerodynamic profile are the key components to fabricate fuselage, wings, empennage, and so forth. Generally, skin panels are made of aluminum alloys, however, in the design of new aircrafts, composite panels increasingly tend to be adopted. The accuracy of their shape and dimension and the integrity of their surfaces and edges are extremely stringent. For instance, laser cutting of high-strength aluminum alloys for aerospace applications has been limited, due to its inevitable heat-affected zone (Riveiro *et al.*, 2008). Therefore, the periphery cutting of aircraft skins mainly employs the machining method or the water jet cutting method. In addition to that, skin panels always have low structural rigidity, and complex internal structure and outer shape. These factors make their forming and machining process very difficult. Thereby, the fabrication technology of aircraft skin panels is always a key technology of aircraft manufacturing.

TRADITIONAL FABRICATION PROCEDURE OF AIRCRAFT SKINS

Traditionally, a typical fabrication process of metal skins is illustrated in Fig. 1. The entire process for any skin part needs

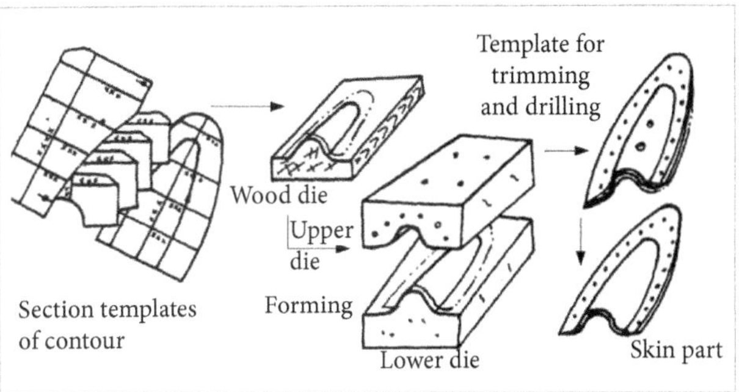

Figure 1. A traditional fabrication procedure of aircraft skins.

a set of dies, templates and other tooling. The setup time of tooling fabrication accounts for approximately 60 to 80 per cent of the cycle time of one part. Furthermore, due to the analog data transfer through the whole process, the accumulative error is often relatively large. Additionally, the whole process has to be executed sequentially rather than concurrently. Once a fault (e.g. springback) occurs on any tooling, repair and rework would be necessary.

FLEXIBLE MANUFACTURING MODEL OF AIRCRAFT SKINS

In order to improve the fabrication process of metal skins, a flexible, digital and tooling-less manufacturing model was presented, namely, to replace the dedicated dies by discrete multi-point dies and to replace the dedicated trimming template by the reconfigurable fixture (Papazian *et al.*, 2002; Li *et al.*, 2009). As illustrated in Fig. 2, the entire process was accomplished jointly with the digital design, digital tooling, numerical control equipments (e.g. stretch forming machine and CNC machine tool) and digital measure technology. Obviously, the good flexibility of this new manufacturing model can adapt to the dimension and shape changes of aircraft skins, and radically change the traditional fabrication model. As proved in an application experiment performed by Papazian *et al.* (2002), in this new "tool-less" sheet metal fabrication environment, tooling fabrication time would be reduced by a factor of 8 and labor hours would be reduced to 1/3 of the traditional values.

In the aspect of the flexible stretch forming process, based on the discrete-die concept, undoubtedly, it was David Hardt at MIT and his research group that had done the pioneering work, including the prototype reconfigurable multi-points tooling development and the closed-loop shape control technology (Hardt *et al.*, 1981; Walczyk and Hardt, 1998; Munro and Walczyk, 2007). On the basis of the original MIT work, since 1996 to 2002, a reconfigurable tooling for flexible fabrication project was sponsored by the Defense Advanced Research Projects Agency. And the project participators included MIT, Northrop Grumman Corporation, Cyril Bath Company, etc. As a result of this project, a commercially available reconfigurable multi-point tooling was designed and built by the Cyril Bath Company in 2002. It was the first production reconfigurable tooling, which had a working volume of $1.06 \times 1.83 \times 0.30$ m^3 and consists of 2,688 movable pins (Munro and Walczyk, 2007). In China , approximately since 2003, the scholarly research upon the reconfigurable tooling for the stretch forming of aluminum sheet began, and it has been performed chiefly by the Li Mingzhe's research group of Jilin University (Cui *et al.*, 2008; Zhou *et al.*, 2005; Cai *et al.*, 2009) and Li Dongsheng's research team of Beihang University (Chen, 2006; Huang *et al.*, 2008; Yu *et al.*, 2011). At 2007, a larger-size ($1.216 \times 1.824 \times 0.305$ m^3) production reconfigurable tooling was designed and built by the Beijing Aeronautical Manufacturing Technology Research Institute (Zhou, 2007). In short, a persistent and concerted effort by many industry and academic researchers has brought the concept of reconfigurable "discrete-die" tooling to the production environment.

The natures of both the reconfigurable multi-point tooling and the reconfigurable fixture are to use discrete points to approximate a surface. However, as another aspect of the flexible manufacturing model of aircraft skins, the flexible trimming process, based on the reconfigurable fixture, seems to have a reverse development history, namely, from reduction-to-practice to academic research. Perhaps there is no such difficult problem as the shape control that remedies the forming defects (e.g. dimpling) which occur in the flexible stretch forming process. Nevertheless, as the chief issue in the development and application of the expensive flexible fixture, the location issue has not been quite clear in academic research. The following factors suggest the particularity and complexity of the workpiece location issues in the flexible trimming process.

- The end effectors are both locators and positioners. Namely, they not only clamp the thin shell part through the vacuum absorption, but also locate the part consistently to prevent the possible deformation. Moreover, the locating and clamping are synchronous.
- When the "flexibility" of multi-point supporting would encounter the "elasticity" of the low-rigidity thin shell part, the location stability would be difficult to assure. Additionally,

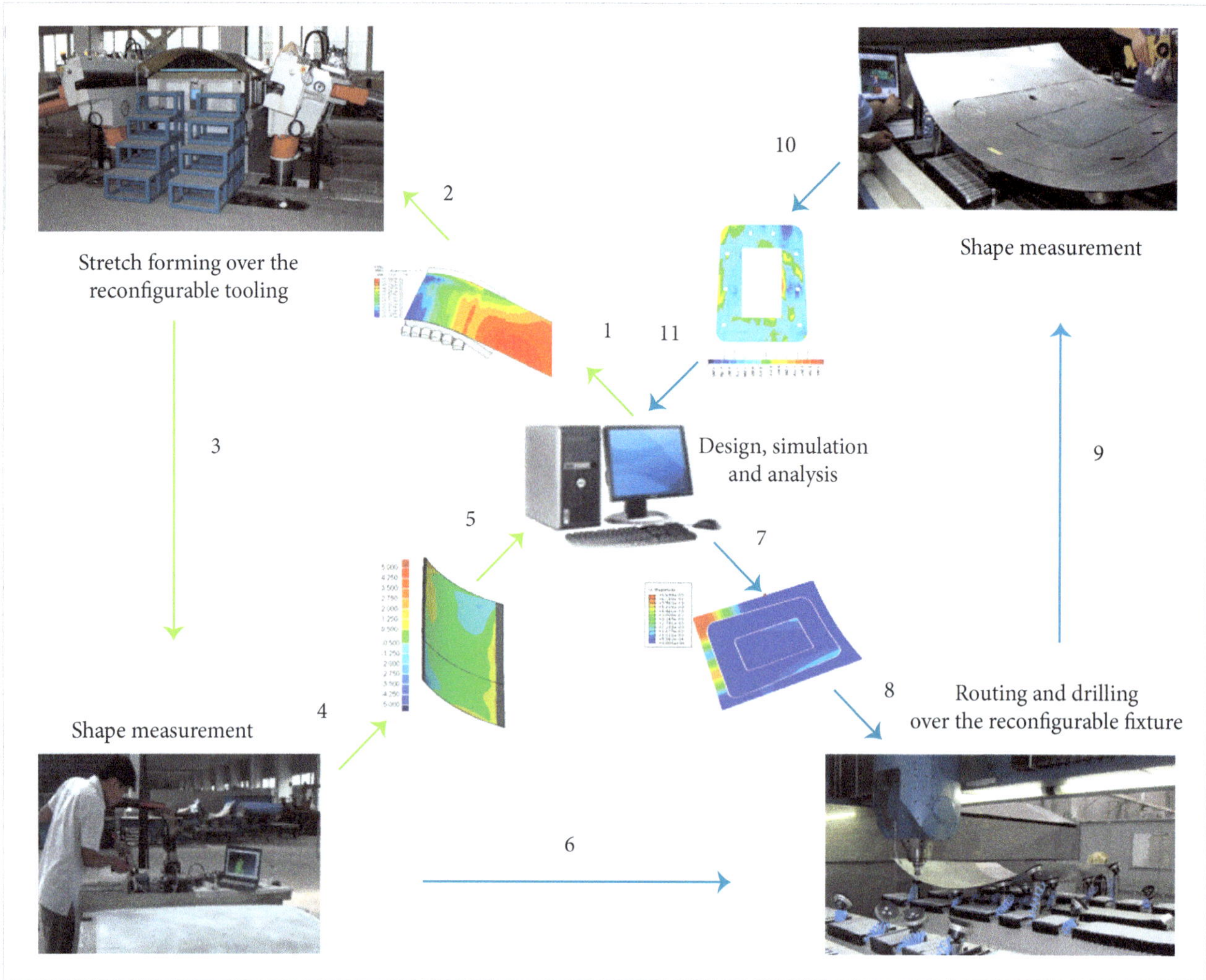

Main contents of step [1] include pin height calculation, stretching trajectory calculation and springback compensation; In step[2] the sheet metal would be formed over the reconfigurable multi-point tooling; Then, the geometrical precision would be measured via laser scanning; In step[4] the error distribution can be obtained after data processing; If the formation precision cannot meet the requirements, step [1]-[4] begin the cycle again; Otherwise, the surface part would be located in the reconfigurable fixture for machining operations; Certainly, before loading the surface part, the location parameter of the reconfigurable fixture would be solved; In step [8] the surface part would be machined over the reconfigurable multi-point tooling; Then, the machining precision would be measured via laser scanning; In step [10] the error distribution can be obtained after data processing; If the machining precision cannot meet the requirements, step [7]-[10] begin the cycle again.

Figure 2. Flexible manufacturing model of aircraft skins.

with the gradual separation of the thin shell part, and with the loading of the periodical milling forces, the process stability would also be difficult to assure. The whole process stability turns out to be a particularly knotty problem.

- The main task of location layout is to determine the positions and numbers of vacuum end effectors. Two types of restriction should be taken into account: one is geometric constraint, which ensures there is no interference between any two adjacent actuator rods, between any two adjacent lateral frames, and between the machining cutter and the vacuum end effectors; another one is physical constraint, which mainly considers the clamping and processing stability. The method for planning feasible location layout should be studied.

LOCATION PRINCIPLES

ADEQUATE LOCATION

As it is known, a rigid body has six movement degrees of freedom (3 linear and 3 rotary ones) in a three-dimensional space. If a workpiece can be regarded as a rigid body, it is necessary to arrest the DOF (degrees of freedom) (≤ 6), which affects the machining accuracy. However, due to low rigidity and easy deformation, the thin shell part cannot be considered as a rigid body. As illustrated in Fig. 3, if the elastic surface part is supposed to consiste of n mass points, it would be essential to limit all DOF of n mass points to avoid machining errors. Theoretically, a mass point has three DOF (3 linear), and n tend to be the infinity of integers. Therefore, the only way to limit all 3n DOF entirely is "surface location" provided by the dedicated hard surface fixture (Fig. 4). Moreover, for the cutter penetration, trimming paths should be slotted on the surface of the rigid fixture. Also the trimming or drilling operations can be done by manual cutters, if the plate thickness is less than 2mm, otherwise by the numerical control machine tool.

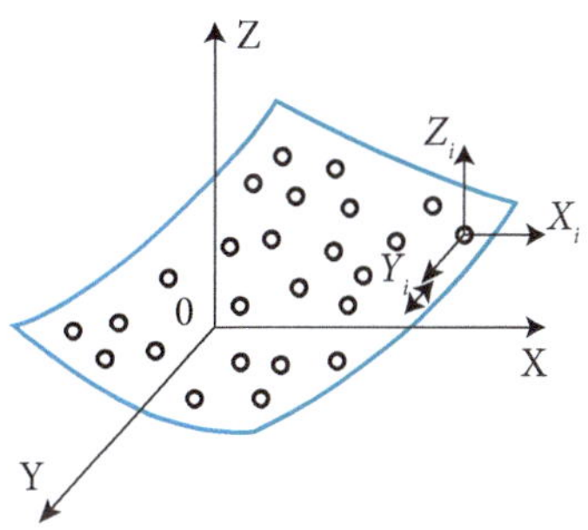

Figure 3. Discrete points model of a thin shell part.

Figure 4. Manual trimming of aircraft skins on the dedicated tooling.

Apparently, when using the reconfigurable fixture to locate the thin shell part (Fig. 2), only the DOF of the mass points close to the vacuum end effectors can be partly restricted. Due to the loaded milling forces and inherent low rigidity, especially in the normal direction, vibration would be unavoidable. In this case, if the required machining accuracy would not be affected, the location status also can be called adequate location. Evidently, adequate location is one of the essential requirements for guaranteeing the machining accuracy.

LOCATION SCHEMES

Generally, for a rigid part, the basic location concept is to arrest the motion that may affect the machining accuracy. Specifically, for a prismatic part without an existing hole, the "3-2-1" location principle can arise proper arresting of all the motions. As illustrated in Fig. 5, "3" refers to 3 locators (passive fixture elements) on the primary datum surface; and "2" refers to 2 locators on the secondary datum surface; then "1" refers to 1 locator on the tertiary datum surface. However, the 3-2-1 principles can only be applied for prismatic-workpiece fixturing, and the three perpendicular datum planes and their corresponding features must be well defined. For a prismatic part with an existing hole, the most efficient way of locating the workpiece is to apply three supports and a single internal locator to restrict nine movements at once. And for nonprismatic parts, the fixturing techniques are often dependent on the workpiece shape. Moreover, due to the complex nature of workpiece geometry, there are no generalized fixture-design principles for nonprismatic parts.

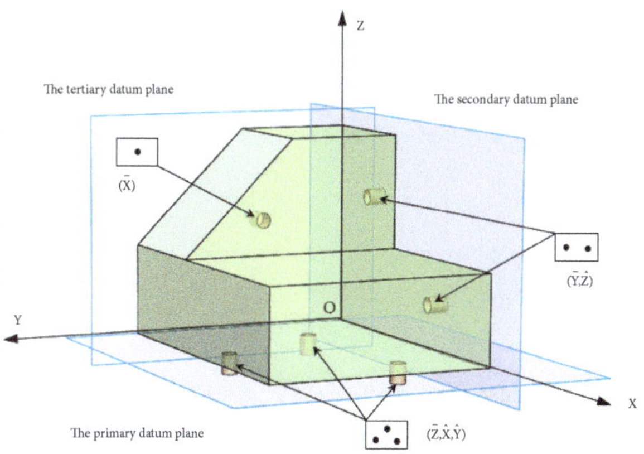

Figure 5. "3-2-1" location principle for a prismatic part without an existing hole.

The basic "3-2-1" location scheme for rigid parts may not be suitable for thin shell and compliant objects, due to their inherent flexibility and low rigidity. In such case, the "N-2-1" locating scheme (Fig. 6) was formally introduced for deformable sheet metal parts, as compared to the widely accepted "3-2-1" principle for rigid bodies, where N supports are required in order to support the metal sheet part in the primary datum direction to avoid excessive deflection (Cai *et al.*, 1996). It must be pointed out that the "N-2-1" locating scheme was presented considering that of when drilling forces or resistance spot welding forces were induced. Therefore, the "N-2-1" location principle is appropriate when the reconfigurable fixture are used for inspection, laser scribing and for the drilling process.

Nevertheless, in case of trimming process, the "N-2-1" locating scheme can not reveal the appearance that some locators gradually lose their effectiveness. Both the researches of Walczyk and Longtin (1999) and the initial research of the author (Hu and Li, 2011) neglected this singular detail. As illustrated in Fig. 7, at the beginning of trimming, there are X locators which take action; while with the separation of the desired shape from the blank, there are Xs locators that gradually lose their effectiveness. Until the end, only Xw locators continue to work. In short, the locators are always changing due to the workpiece separation. The author called this particular location scheme "X-2-1" location principle. "X" refers to the locators of dynamic changes. In order to guarantee the stability of the whole trimming process, designing the locators' initial layout should take the dynamic change into account. Otherwise, the trimming operation may failure due to machining errors or other accidents.

LOCATION PARAMETERS SOLUTION

Location parameters refer to the desired positions that vacuum end effectors and reference positioners should reach, according to the dimension and shape of the surface part. The nature of location parameters solution is to transfer the aircraft skins digital model into the numerical control data, which drive the reconfigurable fixture into quick and accurate reconstruction. Two different solution methods have been presented by the author: one is based on the geometric structures of locators and positioners (Hu *et al.*, 2012a); another is based on the virtual locating and positioning (Hu *et al.*, 2012b). In this paper, the standard procedure of location parameters solution is summarized as shown in Fig. 8. In a previous research by the authors (Hu *et al.*, 2012a; Hu *et al.*, 2012b), the procedures of holding posture design, locators layout design and location parameters solution were discussed in details. Whereas the location admissibility was omitted in previous researches, it would be described emphatically as follows.

Location admissibility indicates whether the thin shell part can be located and supported properly according to the configuration of the reconfigurable fixture and the part model. Specifically, the listed aspects should be taken into account:

- The dimensions of the thin shell part are beyond the working stroke of the reconfigurable fixture?

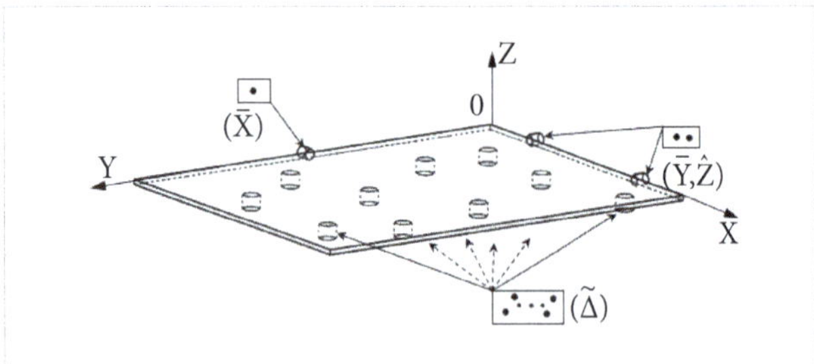

Figure 6. "N-2-1" location principle for deformable metal sheet parts in drilling and welding.

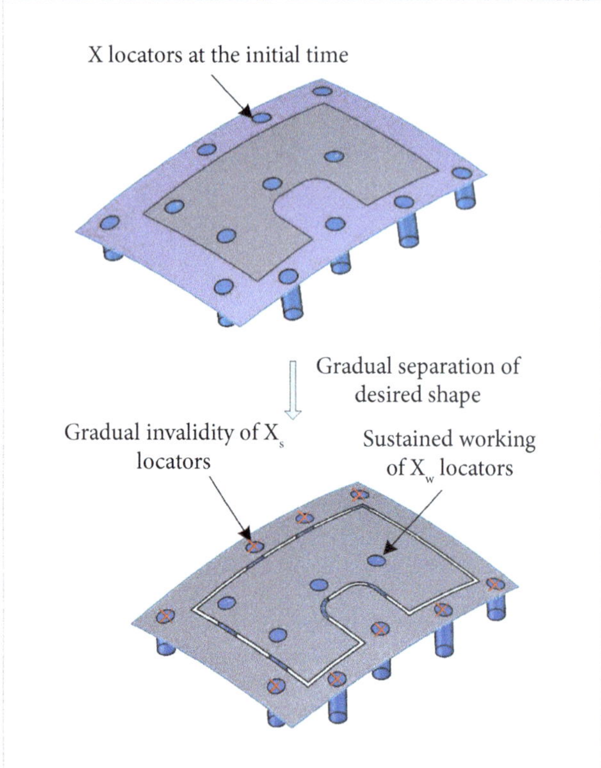

Figure 7. "X-2-1" location principle for thin shell parts during flexible trimming process.

- Due to the part dimensions and the cutting trajectory, the locators that can be arranged are too few?
- The local curvature of the thin shell part is beyond the maximum allowable swing angle of end effectors?

The first two aspects are comparatively easy to judge and address, according to the part model and the features of the reconfigurable fixture. As for the third point, when the part surface is adsorbed onto the vacuum suction of end effectors, the centre line of the end effector should be aligned to the normal direction n_i of the contact center point P_i on the locating surface. Evidently, the center line would always be in the maximum conical angle θ_{max}, when the end effector swings freely to suit the local curvature of the thin shell part. After the holding posture is defined, as illustrated in Fig. 9, transfer the normal vector n_i of P_i point in the workpiece coordinate system into the normal vector n_i^L in the end effector coordinate system. Then figure out the angle θ between n_i^L and the axis Z_L If

$$\theta \leq \theta_{max} \quad (1)$$

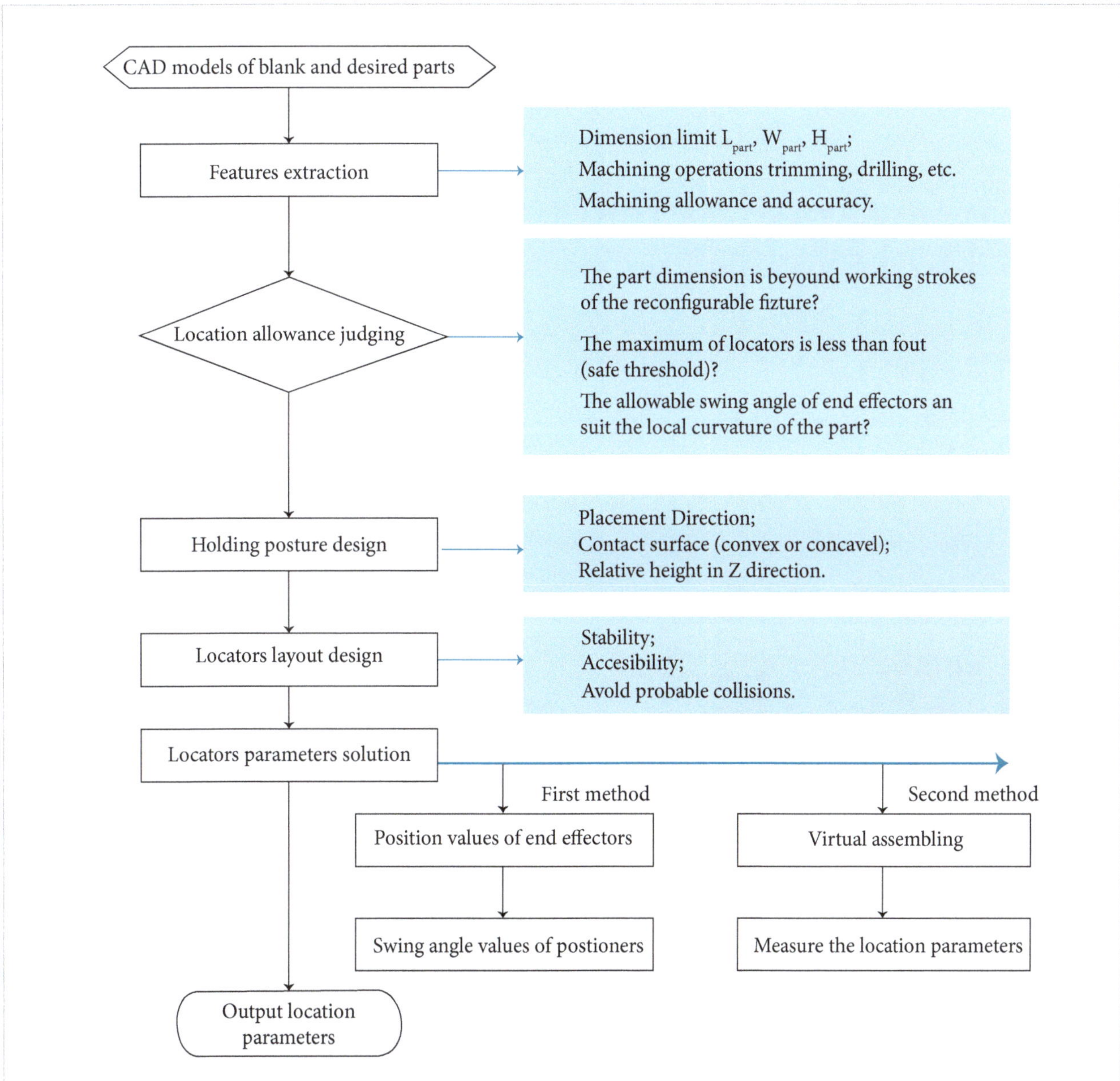

Figure 8. Location parameters solution procedure.

the end effectors can be located at this point. Otherwise, the end effectors cannot suit the local curvature of this point. For instance, the curvature of the leading-edge skin components is often great; thereby the free swing holding cup cannot be available. Therefore, the commercially available reconfigurable tooling system would generally provide a quick-change style adapter such as extension arms (Fig. 10).

For the automation of location parameters solution, a special location parameters solution software system (Flextrimming system) has been improved and seamlessly integrated into the CATIA V5 software via the component application architecture (CAA) development kit, a component object model-like interface. The Flextrimming system can, in sequence, perform the location parameters solution procedure interactively: features extraction, locating allowance judging, holding postures design, locators layout and location parameters solution.

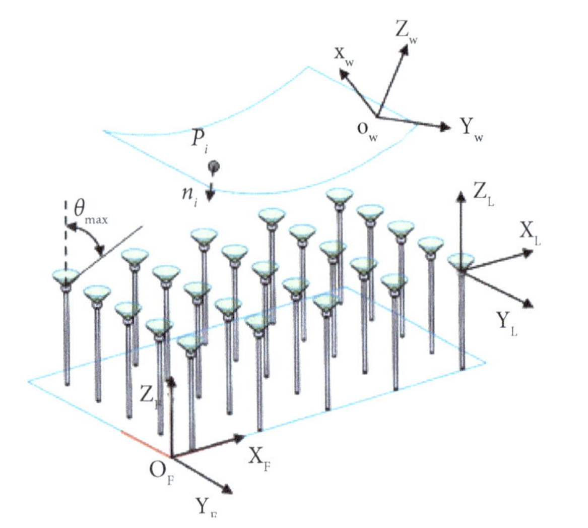

Figure 9. Local curvature and the maximum conical angle.

Figure 10. Quick-change style adapter.

LOCATION ACCURACY

The geometrical accuracy of a machined feature on a workpiece depends on, partially, the fixture's ability to precisely locate the workpiece, which is in fact related to the locator's configurations and positional accuracy of each locator. The reconfigurable flexible, developed independently by our research group, maintains a positioning accuracy of 0.1 mm and a repeatability of 0.05 mm. On the other hand, the solution accuracy of location parameters can reach 0.1μm. Namely, the final location accuracy depends on both the hardware and the software systems, and in order to validate the final location accuracy, an aluminum alloy skin part with a thickness of 1.27 mm was selected and located on the reconfigurable flexible fixture (Fig. 11). Then the located workpiece was measured using the API Tracker3 laser tracker (Fig. 12).

Comparative analysis between the digital model and the measurement data were performed, and the obtained error contour is shown in Fig. 13. The final errors may be caused by the multiple effects including measurement error, forming errors and location errors. Thereby errors distribution was in the reasonable scope and it can be acceptable in practical productions. Meanwhile it may be observed that there is little deformation caused by the vacuum absorption of end effectors. Though the aircraft skin is very thin and poorly rigid, its local rigidity is enough to make the end effectors swing freely to automatically adjust to the local curvatures change of the thin shell part.

DYNAMIC STIFFNESS OF THE PROCESS SYSTEM

EVALUATION OF DYNAMIC STIFFNESS

Dynamic stiffness is a measure of the machining system's ability to dampen the vibrations from input forces, which would directly affect machining accuracy and stability. Theoretically, under the action of machining forces, the whole deformation of the process system is caused by each individual component including the deformations of machine tool, cutter, fixture and workpiece. In fact, the thin shell part has relatively poorer stiffness, thus the deformation of thin shell part is the largest part of the whole deformation of the process system. Therefore, other deformations are neglected and only the workpiece deformation is considered.

Figure 11. Locating set-up.

Figure 12. Measurement via laser tracker.

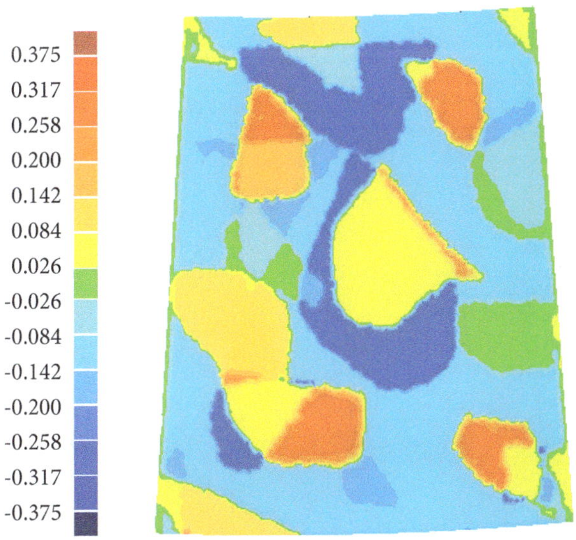

Figure 13. Error distribution contour.

As previously mentioned, the surface part can be regarded as the constitution of discrete points. Then the points close to end effectors could be fully restricted and the displacements of other points would vibrate along with the trimming process. Obviously, the closer the point is to the vacuum adsorption area, the less its vibration will be. Conversely, the farther the point is to the cutting path, the greater its vibration will be. In order to evaluate the dynamic stiffness of the whole fixturing system, n points near to the cutting paths was selected evenly from nodes on the elements as reference points (Fig. 14) and the dynamic stiffness is defined as follows,

$$K_t = \frac{F_t}{\Delta(F_t, t)} = \frac{1}{1/k_{1t} + 1/k_{2t} + \cdots + 1/k_{nt}} \tag{2}$$

$$\delta_i(F_t, t) = \frac{F_t}{k_{it}} \tag{3}$$

$$\Delta(F_t, t) = \delta_1(F,t) + \delta_2(F,t) + \cdots + \delta_n(F,t) \tag{4}$$

Where F_t denotes the external loads, and the subscript t is the trimming time; $\delta_i(F_t, t)$ is the instantaneous displacement of the i-th reference point; and $\Delta(F_t, t)$ is the instantaneous displacement of the whole process system.

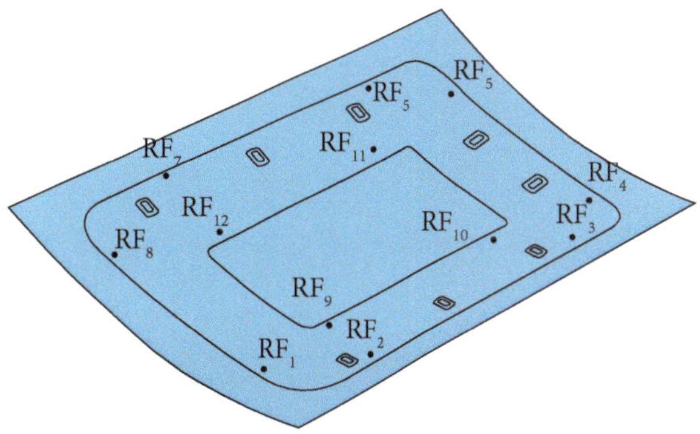

Figure 14. Reference points to evaluate the dynamic stiffness.

EFFECTS OF LOCATION LAYOUT ON DYNAMIC STIFFNESS

In order to analyze the effects of location layout on the dynamic stiffness, the finite element simulation was performed using the specially developed system for the flexible trimming process (Hu *et al.*, 2012b).The basic principle is to sequentially "kill" the elements on the trimming paths in order to simulate the separation process of the desired part. Meanwhile, the milling forces are loaded on the elements that would be "killed" in the next step. As regards the milling forces magnitude, the proposed method (Hu and Li, 2012) was employed. Hereby, the displacements of the reference points can be solved dynamically. The whole procedure is illustrated in Fig. 15.

For the part as shown in Fig. 14, to follow the proposed steps (Fig. 8), five different location layouts were designed and the location parameters were listed in Table 1. The reference coordinate system was given in the Fig. 16. Therefore, the relative distance can be distinguished easily by Table 1 and Fig. 16.

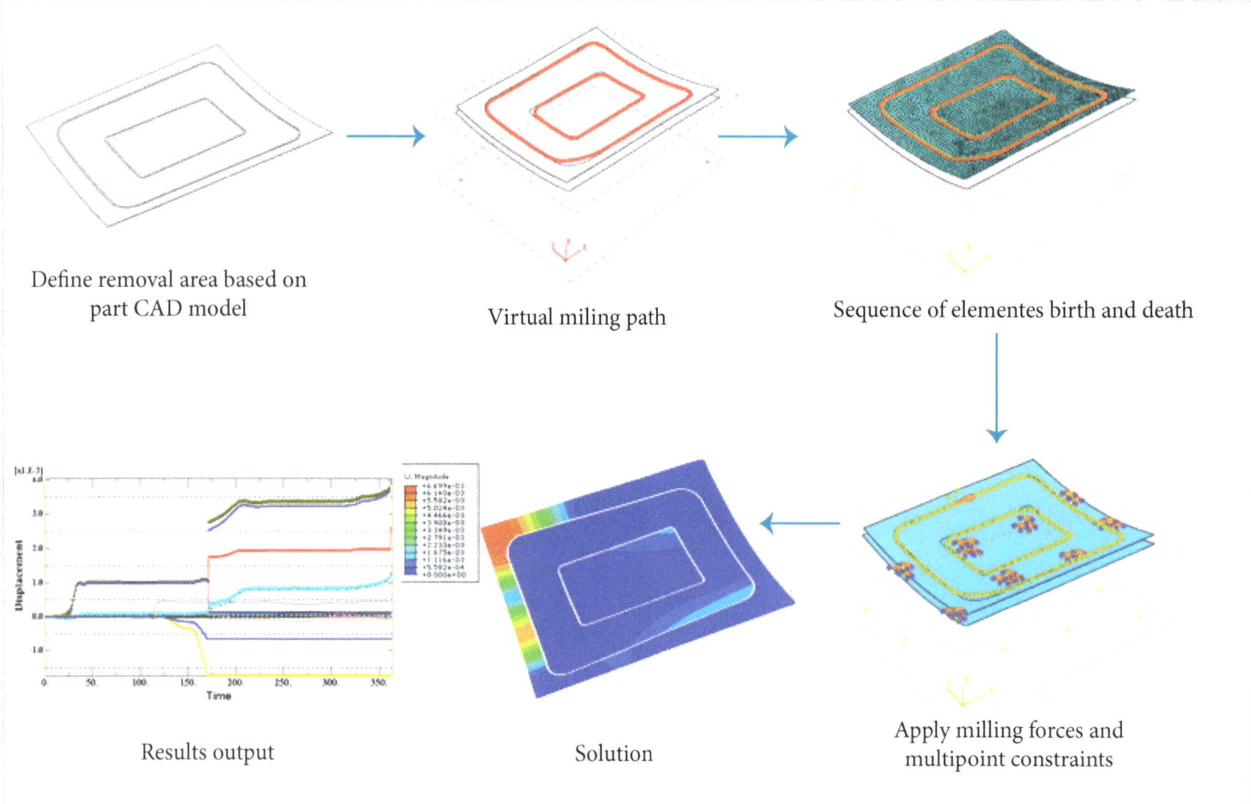

Figure 15. Simulation procedure for trimming process.

Table 1. Location layout mm.

Layout No.	x_1	x_2	x_3	y_{11}	y_{12}	y_{13}	y_{21}	y_{22}	y_{23}	y_{31}	y_{32}	y_{33}
1#	169	579	1112	39	481	855	175	583	991	39	481	923
2#	46	579	1030	141	617	991	39	447	991	175	549	991
3#	46	456	1153	107	515	923	141	515	957	39	481	923
4#	39	481	923	175	549	923	39	481	991	39	447	991
5#	128	661	1112	141	583	991	73	617	991	39	549	957

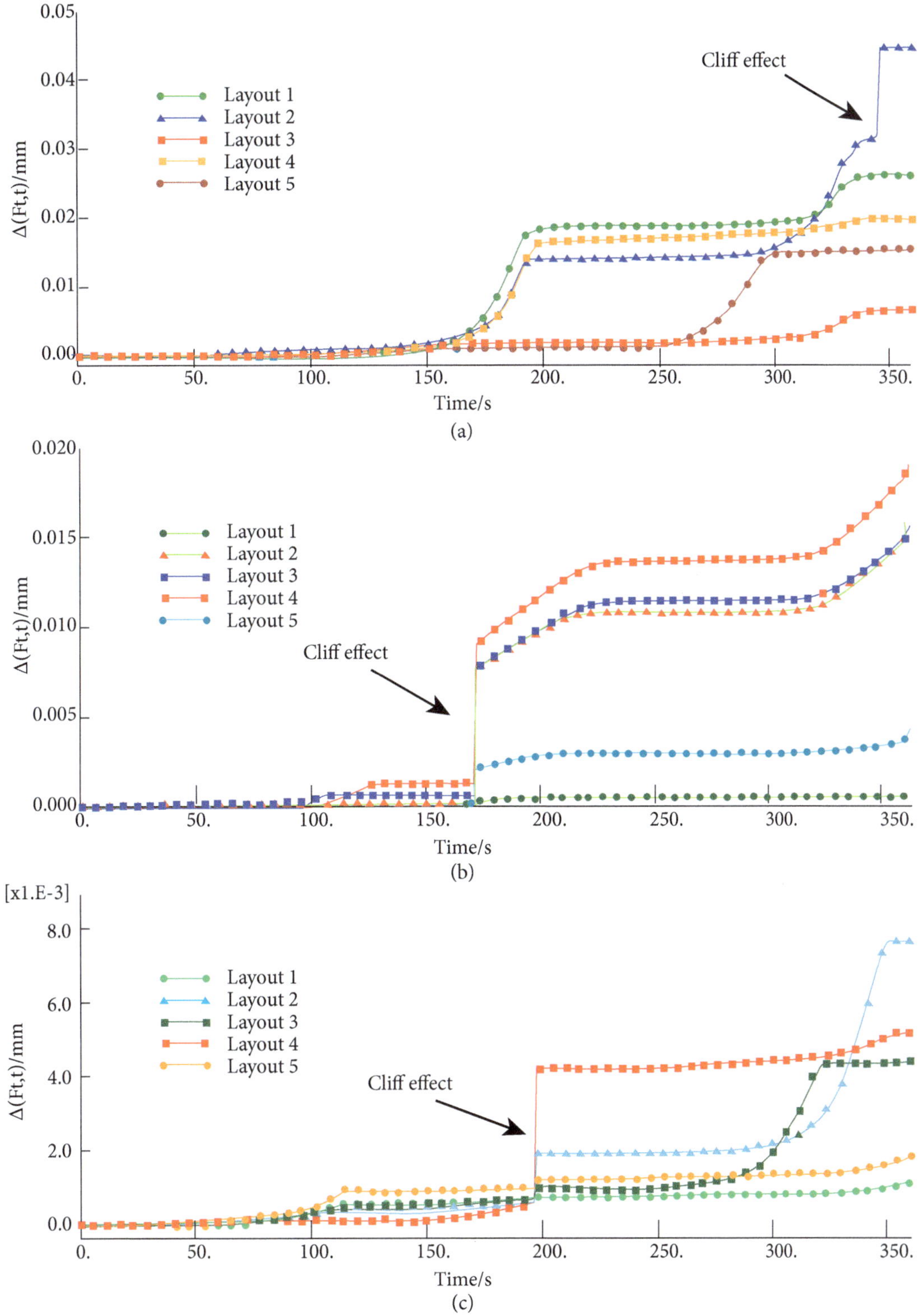

Figure 16. Simulation procedure for trimming process. Cutting parameters: (a) n = 10000 r/min, aP= 1.27 mm, fZ = 0.1 mm; (b) n = 10000 r/min, aP= 2.54 mm, fZ = 0.1 mm; (c) n = 10000 r/min, aP= 3.5 mm, fZ = 0.1 mm.

From the finite element situation results (Fig. 16), it was quite easy to tell that the first and fifth layouts were the best arrangements when compared to the other three layouts. The symbols n , a_p and f_Z, respectively, denote the speed of spindle, the part thickness and the feed per tooth. The good location layout could maintain the dynamic stiffness along with the trimming time. Especially, with the decreasing of the part thickness, the location layout has an increasing effect on the dynamic stiffness of the whole process system.

Another noticeable appearance, as marked in the Fig. 16, should not be ignored, i.e., the cliff effect of the instantaneous displacement of the reference nodes. In the case, the dynamic stiffness of the fixturing system would drop suddenly, due to some locator, becoming valid. At this time, dramatic vibrations may occur and the machining error would be induced. With the decrease of the part thickness, the effect of layout on the dynamic stiffness would increase. However, the effect of layout on the time of the cliff effect occurrence may not be obvious. If the location layout was designed properly, the cliff effect could be restrained within narrow limits, such as the first and fifth layouts.

Therefore, when planning the location layout, the cliff effect case should be avoided.

CONCLUSIONS

The numerical control trimming process for aircraft skins in the reconfigurable fixture is a digital and flexible fabrication technology. However, due to the multi-point location, low rigidity of thin shell part and the ongoing separation of desired contours, this promising technology becomes a knotty operation to guarantee good workpiece-holding stability and machining accuracy. This paper mainly contributes to the following issues:

- The unique location schema, namely the "X-2-1" location principle, was revealed in view of the gradual reduction of effective locators of the reconfigurable fixture during trimming process. This work may perfect the location principles of low rigidity parts. Meanwhile, the "discrete points" model of thin shell parts was introduced; thereby it would help to figure out the nature of thin shell part location (i.e., to restrict the degrees of freedom of the discrete mass points) and the quantitative evaluation for the dynamic stiffness of the fixturing system (i.e., to calculate the displatments of the referred nodes).
- The location parameters solution method was further summarized. Especially, the location admissibility was mainly presented; this work improved the location design procedure. In addition, the location accuracy was discussed through a locating experiment in order to validate the location parameters solution accuracy as well as the positional accuracy of the reconfigurable fixture.
- A quantitative evaluation method of the dynamic stiffness of the fixturing system was put forward. This work provided a criterion for the quantitative analysis of location layout. Moreover, the effects of location layout on the dynamic stiffness of the flexible fixturing system were analyzed using the finite element simulation system for the trimming process. Noticeably, the cliff effect of the dynamic stiffness can be observed from the simulation results. This case should be alleviated or avoided through the optimization of location layout.

ACKNOWLEDGEMENTS

Hu Fuwen would like to thank the Doctor Foundation Project of North China University of Technology and the Science and Technology Innovation Platform Project of Beijing Municipality (PXM2014_014212_000016) for their financial support, and Li Dongsheng and Xu Honghai for their constructive opinions.

REFERENCES

Boyle, I., Rong, Y. and Brown, D.C., 2011, "A Review and Analysis of Current Computer-Aided Fixture Design Approaches", Robotics and Computer-Integrated Manufacturing, Vol. 27, No. 1, pp. 1-12. doi: 10.1016/j.rcim.2010.05.008.

Cai, W., Hu, S.J. and Yuan, J.X., 1996, "Deformable Sheet Metal Fixturing: Principles, Algorithms, and Simulations", Journal of Mechanical Design, Vol. 118, No. 3, pp. 318-324. doi: 10.1115/1.2831031.

Cai, Z.Y., Wang, S.H., Xu, X.D. and Li, M.Z., 2009, "Numerical Simulation for the Multi-Point Stretch Forming Process of Sheet Metal", Journal of Materials Processing Technology, Vol. 209, No. 1, pp. 396-407. doi: 10.1016/j.jmatprotec.2008.02.010.

Cecil J., 2001, "Computer-Aided Fixture Design – a Review and Future Trends", The International Journal of Advanced Manufacturing Technology, Vol. 18, No. 11, pp. 790-793. doi: 10.1007/s001700170004.

Chen, L., 2006, "Research and Key Techniques on the Establishment of Reconfigurable Flexible Stretch-Formed Die", Beihang University, Beijing, China.

Cui, X., Xu, X., Li, G., Li, M. and Cai, Z., 2008, "Development and Application of the Multi-Point Stretch-Forming Equipment for Aircraft Skins", China Metal forming Equipment & Manufacturing Technology, Vol. 43, No. 3, pp. 35-37. doi: 10.3969/j.issn.1672-0121.2008.03.011.

Douglas, W.A. and Ozer, T., 1987, "Universal Holding Fixture", U.S.A patents 4684113.

Hardt, D.E., Olson, B.A., Allison, B.T. and Pasch, K., 1981, "Sheet Metal Forming With Discrete Die Surfaces", Ninth North American Manufacturing Research Conference Proceedings, pp. 140–144.

Hu, F. and Li D., 2011, "Process Planning and Simulation Strategies for Perimeter Milling of Thin-Walled Flexible Parts Held by Reconfigurable Fixturing System", IEEE 2011 Third International Conference on Measuring Technology and Mechatronics Automation, Vol. 2, pp. 922-926. doi: 10.1109/ICMTMA.2011.513.

Hu, F. and Li D., 2012, "Modelling and Simulation of Milling Forces Using an Arbitrary Lagrangian–Eulerian Finite Element Method and Support Vector Regression", Journal of Optimization Theory and Applications, Vol. 153, No. 2, pp. 461-484. doi: 10.1007/s10957-011-9927-y.

Hu, F., Li, D., Li, X. and Zhu, M., 2012a, "Locating Simulation for Aircraft Skins NC Trimming Based on Flexible Holding Fixture", Computer Integrated Manufacturing Systems, Vol. 18, pp. 993-998.

Hu, F., Li, D., Li, X. and Zhu, M., 2012b, "Process Planning of Aircraft Skins NC Trimming Based on Reconfigurable Fixture", Journal of Beijing University of Aeronautics and Astronautics, Vol. 38, pp. 675-680.

Huang, L., Li, D. and Luo, H., 2008, "Research on Closed-Loop Shape Control System of Stretch Forming Over Reconfigurable Tooling", Journal of Plasticity Engineering, Vol. 15, No. 6, pp. 38-42.

Koelsch, J.R., 1998, "Hold It", Machine Shop Guide, Vol. 3, pp. 26-33.

Li, D., Luo, H., Wang, L. and Li, X., 2009, "Numerical Forming Technology of the Aircraft Skin", Journal of Plasticity Engineering, Vol. 16, pp. 61-65.

Munro, C. and Walczyk, D., 2007, "Reconfigurable Pin-Type Tooling: a Survey of Prior Art and Reduction to Practice", Journal of Manufacturing Science and Engineering, Vol. 129, No. 3, pp. 551-565. doi: 10.1115/1.2714577.

Papazian, J.M., Anagnostou, E.L., Christ, R.J., Hoitsma, Jr.D., Ogilvie, P. and Schwarz, R.C., 2002, "Tooling for rapid sheet metal parts production", 6th Joint FAA/DoD/NASA Conference on Aging Aircraft, San Francisco, CA,USA.

Proctor, P., 1998, "Pogo Fixtures Enhance Tooling Flexibility", Aviation Week & Space Technology, Vol. 149, No. 8, pp. 50-51.

Riveiro, A., Quintero, F., Lusquiños, F., Pou, J. and Pérez-Amor, M., 2008, "Laser Cutting of 2024-T3 Aeronautic Aluminum Alloy", Journal of Laser Applications, Vol. 20, pp. 225-230. doi: 10.2351/1.2995769.

Sela, M.N., Gaudry, O., Dombre, E. and Benhabib, B., 1997, "A Reconfigurable Modular Fixturing System for Thin-Walled Flexible Objects", The International Journal of Advanced Manufacturing Technology, Vo1. 13, No. 9, pp. 611-617. doi: 10.1007/BF01350819.

Shirinzadeh, B., 2002, "Flexible Fixturing for Workpiece Positioning and Constraining", Assembly Automation,Vol. 22, No. 2, pp. 112-120. doi: 10.1108/01445150210423143.

Walczyk, D.F. and Hardt, D.E., 1998, "Design and Analysis of Reconfigurable Discrete Dies for Sheet Metal Forming", Journal of Manufacturing Systems, Vol. 17, No. 6, pp. 436-454. doi: 10.1016/S0278-6125(99)80003-X.

Walczyk, D.F. and Longtin, R.S., 1999, "Fixturing of Compliant Parts Using a Matrix of Reconfigurable Pins", Journal of Manufacturing Science and Engineering-transactions of The Asme, Vol. 122, No. 4, pp. 766-772. doi: 10.1115/1.1314599.

Wang, H., Rong, Y.K., Li, H. and Shaun, P., 2010, "Computer Aided Fixture Design: Recent Research And Trends", Computer-Aided Design, Vol. 42, No. 12, pp. 1085-1094. doi: 10.1016/j.cad.2010.07.003.

Youcef-Toumi, K., Liu, W.S. and Asada H., 1987, "A Computer Integration of Reconfigurable Fixtures and Drilling of Sheet Metal Parts", Robotics and Factories of the Future '87, Proceedings of the 2nd International Conference San Diego, California, pp. 751-759. doi: 10.1007/978-3-642-73890-6_90.

Yu, C., Li, D. and Li, X., 2011, "Application of the Preprocessing System of Stretch Forming Over Multi-Point Tooling in Real Factory Environment", Advanced Science Letters, Vol. 4, No. 6-7, pp. 2396-2399. doi: 10.1166/asl.2011.1457.

Zhou, C., Cai, Z. and Li, M., 2005, "Stretching Process Based on Multi-Point Die and Its Numerical Simulation", Journal of Jilin University of Technology (Natural Science Edition), Vol. 35, pp. 287-291.

Zhou, F., 2007, "A New Generation Multi-Point Flexible Tooling for Aircraft Skin Stretch Forming", Aeronautical Manufacturing Technology, Vol. 11, pp. 30-33.

Zhou, K., Qian, Q. and Men, Y., 2008, "A Robotized Reconfigurable Fixture", China patents 101269466.

Thermal Control Design Conception of the Amazonia-1 Satellite

Douglas Felipe da Silva[1], Issamu Muraoka[1], Ezio Castejon Garcia[2]

ABSTRACT: Amazonia-1 is a Brazilian remote sensing satellite providing mainly images, in order to observe and monitor deforestation, especially in the Amazon region. This paper describes the thermal control design, which uses passive and active concepts. The active thermal control is based on heaters regulated by software via thermistors. The passive thermal control consists of multi-layer insulation blankets and radiators, paints, surface finishes to maintain temperature level of the overall carrier components within an acceptable value. The thermal control design is supported by thermal analysis using thermal mathematical model. The temperatures and heater power are predicted for critical cases.

KEYWORDS: Amazonia-1 satellite, Satellite thermal control, Thermal mathematical model.

INTRODUCTION

The Amazonia-1 Thermal Control Subsystem (TCS) consists of active and passive thermal control elements in order to maintain the spacecraft's components and structure within a controlled range of temperature throughout the mission of the spacecraft, from the Beginning of Life (BOL) to the End of Life (EOL). The payload equipment, the propulsion subsystem components and batteries, along with their operational requirements, are the main drive to develop Amazonia-1 thermal control design and analysis.

The Amazonia-1 mission's main goal is to provide image data, improving the deforestation monitoring capability in the Amazon region. The use of a new instrument called Advanced Wide Field Imager (AWFI) with a 40-meter spatial resolution on board will improve the capability of the system (Scaduto *et al.*, 2010). The current Wide Field Imager (WFI) camera acquires images in two spectral bands, while in the AWFI the images will be acquired in four bands. Figure 1 presents the satellite.

The remote sensing Amazonia-1 satellite is the first mission to use the Multi-Mission Platform (MMP) concept developed by the Instituto Nacional de Pesquisas Espaciais (INPE), coordinated

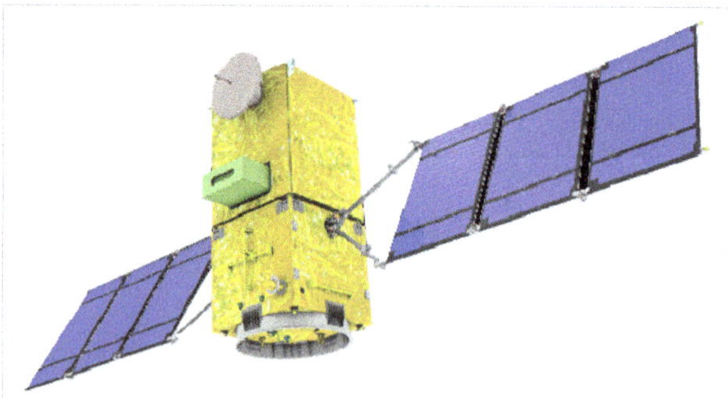

Figure 1. Amazonia-1 satellite.

1. Instituto Nacional de Pesquisas Espaciais – São José dos Campos/ SP – Brasil 2. Instituto Tecnológico de Aeronáutica – São José dos Campos/ SP – Brazil

Author for correspondence: Douglas Felipe da Silva | Instituto Nacional de Pesquisas Espaciais – Av. dos Astronautas, 1.758, São José dos Campos/SP | CEP: 12.227-010 | Brazil | Email: douglas.silva@inpe.br by Brazilian Space Agency (AEB). The MMP platform concept provides capability to support a variety of low Earth orbit missions using the same basic three-axis stabilized platform, with different payload instruments (Santana *et al.*, 2012). The MMP main design goal is modularity, allowing separated integration from the satellite payload module. This provides independent design, construction and test of each module before payload-to-MMP integration and final compatibility tests.

There is an extended amount of works available in the literature about satellite thermal control and, for instance, the handbook of Gilmore (2002), as well as the text books of Karam (1998) and Messeguer *et al.* (2012), are some of the best work presented.

FEATURES OF THE AMAZONIA-1 SATELLITE

A few salient features of Amazonia-1 satellite are as follows:

- Overall dimensions of the main body: 2,200 x 950 x 950 mm;
- Mass: About 500.0 kg;
- Orbit: Sun-synchronous, 752.4 km high, 98.405° inclination and passage time 10:30 a.m.;
- Structure: Aluminum honeycomb structure;
- Power: 420 W (average), InGAP/InGaAs/Ge solar panels, lithium ion battery;
- Stabilization: 3 axis stabilized;
- Mission life: 4 years.

The exploded view of the MMP is presented in Fig. 2, and Fig. 3 presents the payload.

During nominal operation mode, the -Y face will be always pointed to Earth. In emergency mode, the satellite attitude control will point -Z, facing the Sun, in order to warm up the propulsion subsystem elements, and two rotations per orbit will be imposed around the Z axis (in negative direction), in order to distribute external heat loads equally on the lateral panels.

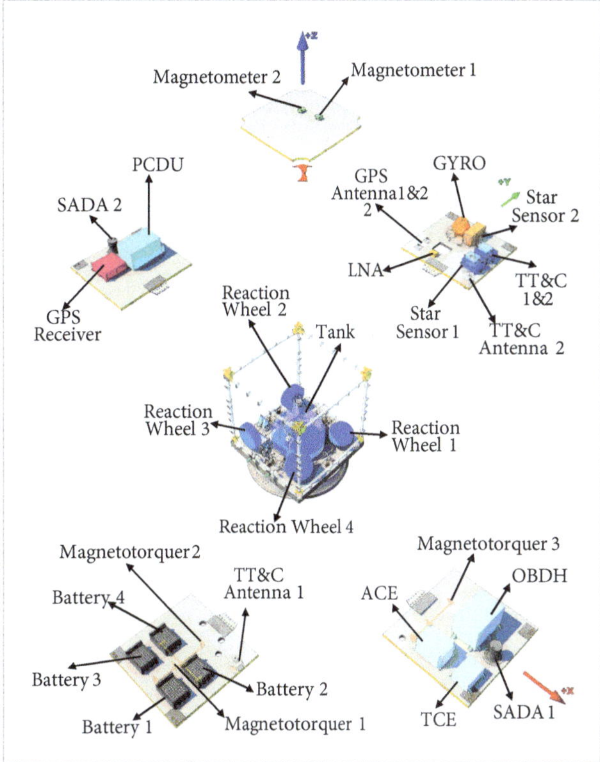

Figure 2. Exploded view of the MMP.

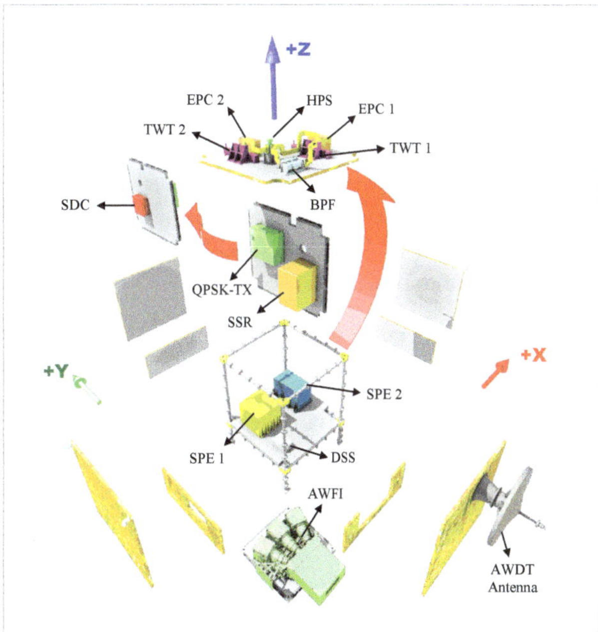

Figure 3. Exploded view of the payload.

CONSTRAINTS AND REQUIREMENTS TO TCS

The Amazonia-1 TCS shall provide a thermal environment which ensures reliable performance of all components during all mission phases, minimizing the thermal control mass and heater power budgets. The main functions of the Amazonia-1 thermal control are:

- To provide temperature distribution in order to have all onboard equipment operate within their designed operational temperature range, under all possible attitudes experienced during all mission phases;
- To allow the dissipation of the excess energy generated with no detrimental effect.

The Amazonia-1 thermal design conception is based on the balance of the following criteria:

- Insulate, as well as possible, the satellite of external environment. Making use of this approach, temperatures are less sensitive to the external radiation and the satellite can perform better during all mission stages. Radiators are used in order to reject internal heat dissipation;
- Minimization of thermal gradients in the satellite by using high emittance coatings on the majority of the internal surfaces (equipment and structural parts) and improving the conductive coupling between equipment and mounting panels, through thermal interface filler materials;
- For AWFI, batteries and propulsion components, which have operational temperature limits quite different from the remaining equipment in the satellite, the paragraph above is not applicable. These components shall be thermally insulated — by radiation and conduction — from the satellite, having their temperatures controlled individually by heaters;
- Insulate the equipment placed in direct view to space from the space environment. This is necessary in order to avoid large temperature fluctuation on these equipment, through illuminated and eclipse periods;
- Insulate MMP from payload module, considering the premise that the MMP is a multi-mission platform and its thermal design shall be qualified for any payload;
- The electrical power allowance for heaters shall not exceed 30 W during mission nominal modes and 60 W during emergency mode;
- The temperature limits (operational and non-operational) and the heat dissipation for all equipment are disposed in Table 1 (payload) and Table 2 (MMP). The cases in these tables will be described further on in "Cases under evaluation".

THERMAL CONTROL SYSTEM USED IN AMAZONIA-1

As explained before, in order to minimize temperature gradients, all internal surfaces of the satellite are coated with

Table 1. Payload equipment temperature limits and heat dissipation.

Equipment	Temperature Range (°C)	Heat Dissipation (W)		
		Case A	Case B	Case C
AWDT Antenna	-100 ~ 80	0	0	0
BPF	-30 ~ 60	0	0	10.0[(1)]
DSS	-10 ~ 45 (-20 ~ 55)[(2)]	0	3.0	3.0
EPC (x2)	-20 ~ 60 (-30 ~ 75)[(2)]	0	0.1	1.5[(3)]
HPS	-20 ~ 50 (-30 ~ 60)[(2)]	0	0	3.2[(1)]
QPSK-TX	-10 ~ 45 (-20 ~ 50)[(2)]	0	0	12.0[(1)]
SDC	-10 ~ 45 (-20 ~ 55)[(2)]	0	0	4.0[(1)]
SPE1	-10 ~ 45	6.0	6.0	90.9[(4)]
SPE2	-10 ~ 45	22.1	22.1	101.6[(4)]
SSR	-10 ~ 45 (-20 ~ 55)[(2)]	0	5.0	10.0
TWT (x2)	-20 ~ 60 (-30 ~ 75)[(2)]	0	0	40[(3)]

[(1)]15 min of operation in illuminated period; [(2)]non-operational temperature range; [(3)]15 min of operation in illuminated period (only main unit; the redundant one is on stand-by); [(4)]25 min of operation in illuminated period.

high emittance black paint. All external surfaces are wrapped with 20 layers of Multilayer Insulation (MLI) blankets with areas available for positioning the radiators. In order to reduce heater power consumption during emergency mode, the internal face of the adapter cylinder, positioned at the MMP -Z panel, is kept in bare aluminum, as well as the areas of the -Z panel which are out of the cylinder circle.

The propulsion components were dealt with different approach since they shall be kept at a temperature range from +10 °C to +50 °C. The lower limit of +10 °C is so to avoid hydrazine freezing. Considering that the lower temperature limit of the propulsion subsystem is considerably higher than those of the remainder MMP components, the thermal design was set according to the following baseline:

- Radiative insulation of all components (except thrusters) wrapping them with 15 layers of MLI blanket;
- Conductive insulation of all attachment points with the mounting panel using fiber glass-epoxy composite washer;
- Electrical heating using skin and spiral heaters on all components.

A total of seven heater circuits (four main and three redundant ones) are installed on the propulsion components, using correspondently 7 switched heater lines. All lines are controlled by ON/OFF commands from the On Board Data Handling (OBDH), which uses the thermistor telemetry as input data. The thermistors which will be used are space qualified, interchangeable, NTC type with a resistance of 10 kΩ at 25 °C, and manufactured by Measurement Specialties™.

The batteries panel also received specific design since their range, 0 to 20 °C, is narrower than most of the electronic equipment in the satellite (-10 °C ~ 45 °C). The concept is to couple the battery packs to the mounting panel and to use

Table 2. MMP equipment temperature limits and heat dissipation.

Equipment	Temperature Range (°C)	Heat Dissipation (W)		
		Case A	Case B	Case C
ACE	-10 ~ 45	18.0	14.0	20.0
Battery (x4)	-10 ~ 20(1)	2.3	2.3	2.8
GPS Antenna (x2)	-20 ~ 55 (-20 ~ 60)(2)	0	0	0
GYRO Elect. Module	-20 ~ 60 (-40 ~ 75)(2)	0	21.6	30.0
GYRO ICU	-20 ~ 60 (-40 ~ 75)(2)	0	0.1	0.2
LNA	-10 ~ 45 (-20 ~ 50)(2)	0	0.6	0.6
Magnetometer (x2)	-20 ~ 50	0	0	0
Magnetotorquer (x3)	-10 ~ 50	2.7	0	2.7
OBDH	-10 ~ 45	47	47	47
PCDU	-10 ~ 45	6.3	6.3	31.8(3)
Propulsion Elements	+10 ~ 50	0	0	0
Reaction Wheels (x4)	-15 ~ 55	6.7	6.7	40.4(4)
SADA (x2)	-20 ~ 60	4.6	5.1	6.5
Star Sensor (x2)	-20 ~ 50 (-40 ~ 70)(2)	0	0.2	13.5
TCE	-10 ~ 45 (-25 ~ 45)(2)	0	0	0.8
TT&C (x2)	-20 ~ 50	6.2	6.2	14.6(5)
TT&C Antenna (x2)	-80 ~ 55	0	0	0

(1)At mounting panel; (2)Non-operational temperature range; (3)Operation in illuminated period; (4)6 min of operation in illuminated period; (5)15 min of operation in illuminated period (only main unit; the redundant one is on stand-by).

this panel as a radiator. The mounting panel is insulated, conductively (through fiber glass-epoxy washers) and radiatively (15 layers of MLI blanket), from the other parts of the satellite.

In order to avoid the temperature at battery panels below -10 °C, heaters are placed at the panel, near the batteries' feet. The temperatures at the panel are monitored by four thermistors, placed near the heaters. The heater circuits will be activated by ON/OFF commands from the OBDH. The main circuit will turn on at -8 °C and turn off at -6°C, while the redundant ones will turn on at -10 °C and turn off at -8 °C. The temperature refers to the lowest value among the four thermistors.

The AWFI has autonomous thermal control, and it is wrapped with a MLI blanket and conductively insulated from the mounting panel through washers. Also, it has its own radiator and dedicated heaters. Since its thermal control is independent from the satellite (Scaduto *et al.*, 2010), in this global analysis, its temperature was considered to be between 15 °C and 20 °C.

The thermal optical properties of coatings used in the satellite are presented in Table 3.

VERIFICATION OF DESIGN BY THERMAL MATHEMATICAL MODELING

The thermal modeling is based on a nodal or lumped parameter method. In this method, the satellite is divided in a number of regions, assumed isothermal, which is called nodes. These nodes exchange heat among each other by conduction and radiation, and with space by radiation. Also, they can receive heat loads from external sources or from electronic components. The temperature of each node is the result of these interactions. The governing equation is the energy equation which consists of transient, conduction, and radiation terms, plus boundary conditions (solar radiation, albedo and Earth radiation), as a source term. This equation can be written as presented in Eq. (1) (Karam, 1998):

$$(mcp)_i \frac{dT_i}{dt} = Q_i^D + (\alpha Q^S + \alpha Q^A + \varepsilon Q^E)_i - \sum_{j=1}^{n} K_{ij}(T_i - T_j) - \sum_{j=1}^{n} \sigma R_{ij}(T_i^4 - T_j^4) \quad (1)$$

Where the thermal mass is mcp [J/K]. The external loads are Q^S, Q^A, and Q^E [W], while internal heat dissipation is Q^D [W]. The conduction and radiation exchange factors are represented by K_{ij} [W/K] and R_{ij} [m^{-2}], respectively. The Stefan-Boltzmann constant is σ [W/m^2/K4], and T_i and T_j [K] are the temperatures of nodes *i* and *j*, respectively.

The temperatures have been calculated in transient regime, considering variations on external heat loads and equipment heat dissipation profiles. The model was built using the C&R Technologies ThermalDesktop/Radcad - SINDA/FLUINT thermal software package, and details of this package can be obtained in Panczak *et al.* (1998).

The MMP is represented by 2,764 diffusion nodes and 1,698 arithmetic ones, while the payload is represented by 4,161 diffusion, 1,444 arithmetic and 2 boundary nodes. Thus, the whole model is composed by 6,925 diffusion nodes, 3,142 arithmetic nodes and 2 boundary ones (total of 10,069 nodes).

The arithmetic nodes are used in order to represent the MLI blankets, which have small mass, since these elements have very low thermal capacitances and respond almost instantaneously to the thermal changes of environment.

The boundary nodes are used in order to represent a component with constant temperature. In this case, the equipment is the AWFI, since it has its own active thermal control and it shall operate between +15 °C and +20 °C.

Table 3. Coatings' thermal optical properties.

Coating	Emissivity (ε)	Absorptivity (α)		Effective Emissivity (εeff)
		BOL	EOL	
Bare Aluminum	0.05	0.15	0.15	-
Black Paint	0.88	0.95	0.95	-
MLI	0.80	0.41	0.51	0.02
White Paint	0.94 (BOL) 0.95 (EOL)	0.20	0.33	-

CASES UNDER EVALUATION

The design philosophy is to find the critical cases and analyze them supposing that all other cases result in intermediate temperatures. Three critical cases are evaluated: cold, hot and emergency cases.

- Case A: Cold case represents the nominal operational mode in which minimum external heat loads (albedo: 0.34; Earth radiation: 208 W/m²; solar radiation: 1326 W/m²), minimum internal heat dissipation and BOL optical properties are considered. The orbit parameters are adopted for winter solstice, and the satellite -Y face will be pointing towards Earth;
- Case B: Hot case represents the nominal operational mode in which maximum external heat loads (albedo: 0.42; Earth radiation: 233 W/m²; solar radiation: 1418 W/m²), maximum internal heat dissipation and EOL optical properties are considered. The orbit parameters are adopted for summer solstice, and the satellite -Y face will be pointing towards Earth;
- Case C: Emergency case represents the satellite with attitude different from the nominal mode. In this case, the satellite -Z face is permanently oriented towards the Sun and the solar panel is oriented parallel to the -Z panel. The spin rate is two revolutions per orbit around the Z axis, in negative direction. The power consumption is minimized in this mode. Some equipment are turned off during this case and the temperature limit is the non-operation limit. The external heat loads, orbit parameters and optical properties are the same considered in case A (nominal cold case).

TEMPERATURES AND HEATERS POWER CONSUMPTION PREDICTIONS

The temperatures have been calculated in transient condition, and the stabilization criterion was considered when the temperature variation between two consecutive orbits is less than 0.1 °C in the same orbit point, for hot case. In cold and emergency cases, it is not possible to accomplish this criterion, since heaters are required and their ON/OFF dynamic does not allow the same conditions at the same orbit point. In this case, a minimum of 15 orbits are simulated.

The extreme predicted temperatures for payload (minimum for emergency case and maximum for hot case) are shown in Fig. 4. By analyzing Fig. 4, it should be noted that the predicted range of the equipment are the extremes of the central (black solid) bar. The empty space between this bar and the other two ones (minimum limit to the left and maximum to the right) is the temperature margin of the equipment.

The extreme predicted temperatures for MMP are shown in Fig. 5.

All maximum temperatures were predicted with at least 5 °C of margin, and the cold ones are controlled by heaters. It should be noted that the predicted temperatures of the solar array generator exceeded the acceptable range (-80 °C to 80 °C), reaching -84.6 °C in the emergency case and 82.1 °C in the hot case. Although its temperature does not affect the satellite thermal behavior considerably, a detailed thermal analysis will be accomplished in order to satisfy its temperature requirement.

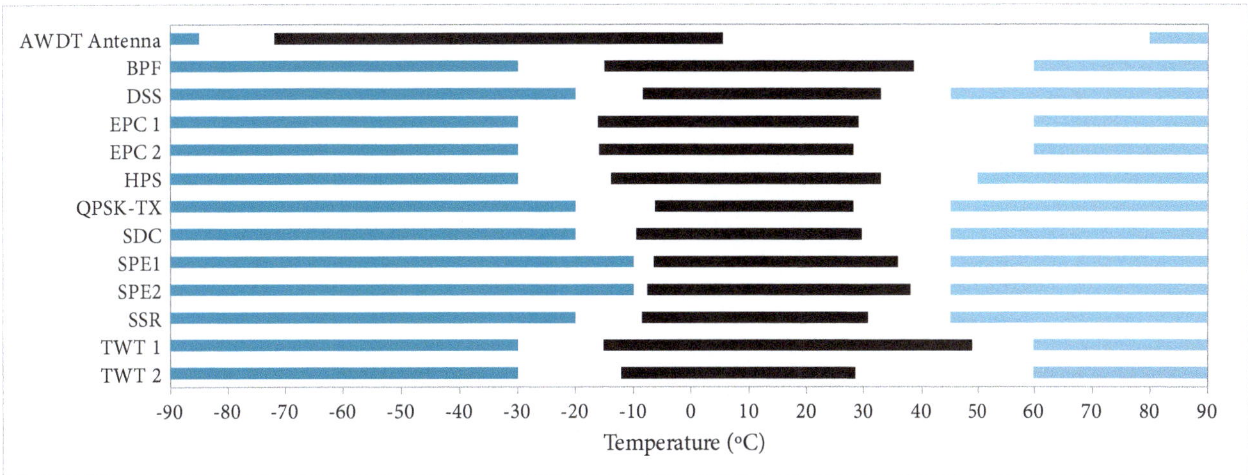

Figure 4. Payload predicted temperatures.

To achieve the temperatures presented in Figs. 4 and 5, it was necessary to determine radiators areas. The process to find these areas, presented in Table 4, is iterative.

The average power consumption for all heaters, during cold and emergency cases, is presented in Table 5.

For the nominal cold case (Case A), the heater power consumption is within the requirement (30 W) with 25% margin, according to U.S. Department of Defense (1982). For emergency cases, the predicted power does not comply with the requirement (60 W). This value is under review, since in emergency cases the overall power consumption is reduced and more power is available. The emergency case is thermally critical due to its distinct attitude, in which the payload module remains in the Sun shadow, and almost all equipment are turned off.

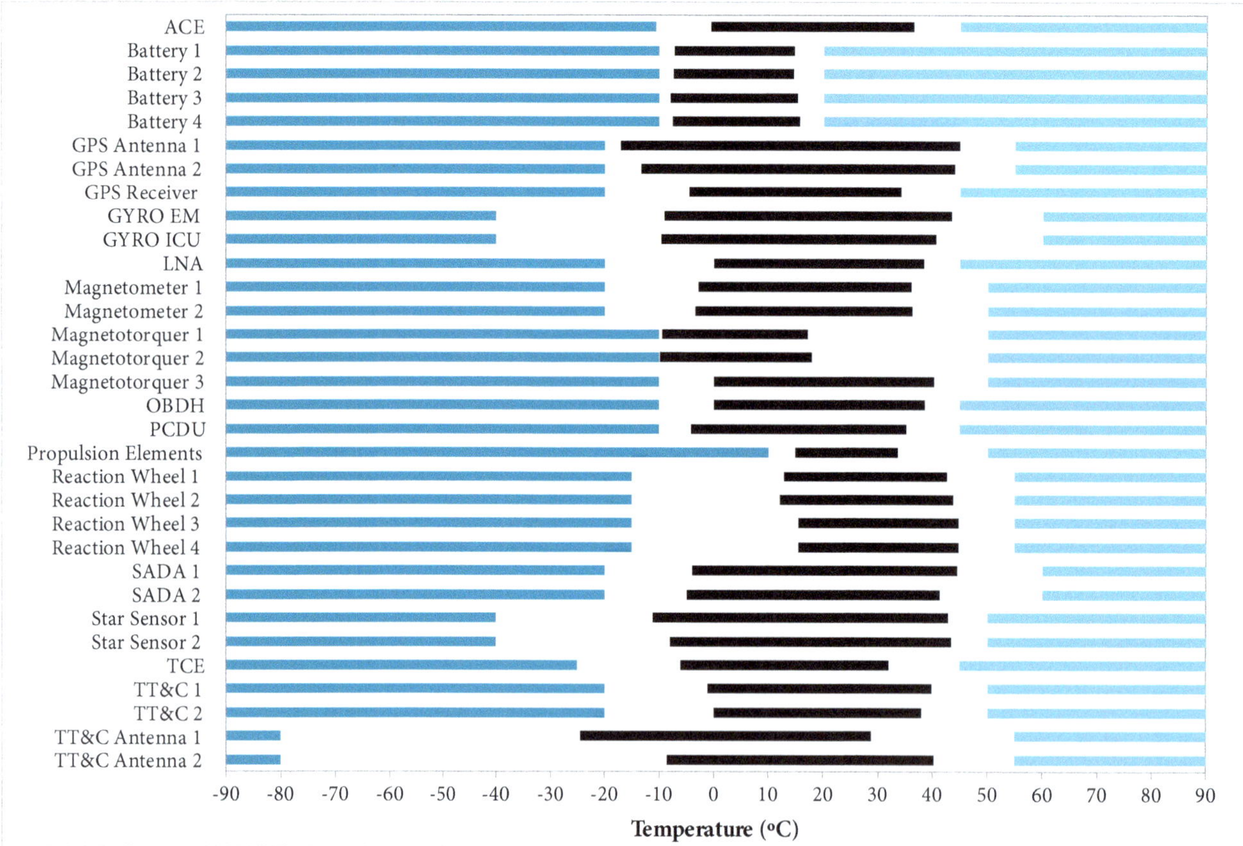

Figure 5. MMP predicted temperatures.

Table 4. Radiators areas.

Radiator Location	Area (m²)	
	MMP	Payload
+X Face	0.3624	0.2356
-X Face	0.1035	0.015
+Y Face	0.24	0.1596
-Y Face	0.432	0
Total	1.1379	0.4102

Table 5. Heaters power consumption.

Heater	Power Consumption (W)	
	Case A	Case C
Batteries Panel	1.44	32.42
Payload Module	12.45	44.37
Propulsion Elements	3.69	2.85
Total	17.58	79.64

CONCLUSIONS AND FUTURE ACTIONS

The Amazonia-1 thermal control design and analysis were presented. This design has been verified through the use of the SINDA/FLUINT thermal analyzer, and established by using passive means and heaters. All subsystems have been in the specified temperature ranges, for nominal and emergency cases. The heater power predicted for nominal operational cases complies with the specification, however, in an emergency case, the predicted value is beyond specification. This requirement is under review and may possibly be increased.

A thermal model based in this design will be submitted to the Thermal Balance Test, in order to qualify the thermal design. Afterwards, the results will be correlated, and the thermal mathematical model validated.

REFERENCES

Gilmore, D.G., 2002, "Spacecraft Thermal Control Handbook", 2nd Edition, The Aerospace Corporation Press, El Segundo, CA, Vol. I.

Karam, R.D., 1998, "Satellite Thermal Control for System Engineers", Progress in Astronautics and Aeronautics, AIAA, Cambridge, Vol. 181, pp. 274.

Meseguer, J., Pérez-Grande, I. and Sanz-Adréz, A., 2012, "Spacecraft Thermal Control", Woodhead Publishing, Cambridge.

Panczak, T., Ring, S. and Welch, M., 1997, "A CAD-based Tool for FDM and FEM Radiation and Conduction Modeling", Cullimore and Ring Technologies, 28th ICES Conference, SAE Paper 981577.

Santana, A.C., Martins-Filho, L.S., Duarte, R.O., Arantes Jr., G. and Casella, I.S., 2012, "Attitude Control of a Satellite by Using Digital Signal Processing", Journal of Aerospace Technology and Management, Vol. 4, No. 1, pp. 15-24. doi: 10.5028/jatm.2012.04014611.

Scaduto, L.C.N., Carvalho, E.G., Santos, A.R., Soares, A.L., Castellar, A., Azeka, L.A., Souza, W.M., Evangelista, S., Santos, A.G., Malavolta, A.T., Vales, L.F., Segoria, D., Santos, F.S., Candeloro, L., Almeida, M.R., Yasuoka, F.M.M., Modugno, R.G., Cartolano, R., Barbalho, S., Santos, P.A., Rebolho, D., Otoboni, J.A., Correa, Y., Stefani, M.A. and Castro Neto, J.C., 2010, "The Advanced Wide Field Imaging Camera (AWFI) for the Amazonia 1 Brazilian Satellite", International Conference on Space Optics (ICSO), Rhodes, Greece.

U.S. Department of Defense, 1982, "Test Requirements for Space Vehicles - MIL-STD-1540B", U.S. Air Force, Washington, D.C., USA, pp.16-17.

Evaluation of Chemical Equilibrium and Non-Equilibrium Properties for LOX/LH_2 Reaction Schemes

Carlos Henrique Marchi[1], Luciano Kiyoshi Araki[1]

ABSTRACT: Nine chemical reaction models for equilibrium schemes and six chemical models for non-equilibrium ones are studied, considering different conditions found in real liquid oxygen/liquid hydrogen rocket engines. Comparisons between two eight-species models have shown that the most complex is the best one. Besides, it was also verified that the most complex model has been the fastest, among six- and eight-species models. Both combustion temperature and thermochemical/transport properties depend only on the chemical species considered by the used model. Comparisons among results from the implemented code (Gibbs 1.3), Chemical Equilibrium with Applications and Thermochemical Information and Equilibrium Calculations, these last two codes from NASA, have shown that Gibbs 1.3 evaluates correctly both combustion temperature and thermochemical properties. Furthermore, analyses have shown that mass generation rates are very dependent on third body reaction equations and forward reaction constants.

KEYWORDS: Chemical reaction models, Chemical equilibrium, Combustion temperature, Non-equilibrium.

INTRODUCTION

The hydrogen/oxygen (H_2/O_2) combustion systems have many attractive features, including high specific impulse for rocket engine applications, clean-burning feature for achieving environmentally friendly combustion products, reliable ignition and high combustion efficiency. Even though there are only two elements involved, the chemical reaction mechanisms of H_2/O_2 systems are quite complex, involving various steps of chain reactions. This system is also important as a subsystem in the oxidation process of hydrocarbons and moist carbon monoxide (Kuo, 2005; Turns, 2000).

Due to the fact that complex mechanisms are evolutionary products of the chemists' thoughts and experiments (Turns, 2000), several studies have been published recently about the mathematical models for hydrogen ignition, the reaction kinetics and/or the flame propagation. For example, Bedarev and Fedorov (2006) compared three mathematical models of hydrogen ignition, for the reactive flow behind the front of an initiating shock wave. Smith *et al.* (2007) studied the combustion zone near the injection of a liquid rocket engine thrust chamber in order to evaluate the influence of reduced pressure on the flame formation and combustion efficiency. Fernández-Galisteo *et al.* (2009) proposed and studied a short mechanism consisting of seven elementary reactions for hydrogen-air lean-flame burning velocities.

The knowledge of species formed by liquid oxygen/liquid hydrogen (LOX/LH_2) combustion is the first step for a complete study involving combustion mixture gases flow, heat transfer and coolant pressure drop in a real cryogenic H_2O_2 rocket engine, such as Space Shuttle or Vulcain Engine of Ariane Launcher.

1. Universidade Federal do Paraná – Curitiba/PR – Brazil.

Author for correspondence: Luciano Kiyoshi Araki | Universidade Federal do Paraná | Departamento de Engenharia Mecânica | Caixa Postal 19040 | CEP: 81.531-980 Curitiba/PR – Brazil | Email: lucaraki@ufpr.br

At high temperatures, such as those found in rocket thrust chambers, effects of dissociation become important and new species must be considered (Anderson Jr., 2003). In such situations, it is necessary the study of chemical equilibrium and non-chemical equilibrium.

For both equilibrium and non-equilibrium reaction schemes, thermodynamic properties of species are required to define properly the mixture of gases' composition. Such properties can be easily taken from thermodynamic tables and/or graphics. In this work, however, it is used the method presented by McBride *et al.* (1993), in which libraries for thermodynamic and transport properties for pure species can be found. Nine different reaction models are used to obtain equilibrium composition. In the simplest one, only 3 chemical species are considered: H_2, O_2 and H_2O, without chemical dissociation reactions. In the most complete scheme, 8 species (H_2, O_2, H_2O, H, O, OH, H_2O_2 and HO_2) and 18 chemical reactions are studied. For non-equilibrium studies, 6 different models are used, involving only 6 or 8 species. These models are essentially the same ones used at equilibrium schemes, but taking into account third body species (and also, third body reactions).

Two different kinds of comparisons are presented in this work. Firstly, five couples of temperature and pressure values are chosen, intending to cover all range of values found in a real rocket thrust chamber. The results were compared with those obtained from Chemical Equilibrium with Applications (CEA; Glenn Research Center, 2005) and Thermochemical Information and Equilibrium Calculations (Teqworks; Gordon and McBride, 1971), both codes from NASA. Secondly, it is studied the combustion temperature of LOX/LH_2 reactions at different oxidant/fuel mass ratios and combustion pressures. The obtained results were compared with CEA and Teqworks codes and, for some cases, with other results from literature. Furthermore, results for species generation rates for non-equilibrium models are presented.

The importance of this study is to compare both equilibrium and non-equilibrium models, specially in relation to the importance of the chemical intermediate reactions, which are the base of non-equilibrium flows. The numerical code, whose results are presented in this study, is available from: ftp://ftp.demec.ufpr.br/CFD/projetos/cfd5/relatorio_tecnico_1.

CHEMICAL EQUILIBRIUM

The applied methodology in this work to obtain chemical equilibrium parameters of a mixture of gases can be found in Kuo (2005) and is briefly explained in this section. Firstly, it is considered a chemical reaction in the form:

$$\sum_{i=1}^{N} \nu'_{ij} A_i \underset{kb_j}{\overset{kf_j}{\rightleftarrows}} \sum_{i=1}^{N} \nu''_{ij} A_i \qquad (j = 1, 2, \dots L) \tag{1}$$

where A represents each chemical species' symbol, i corresponds to the number of each chemical species, N is the total number of chemical species, j corresponds to a specific chemical reaction, L is the total number of chemical reactions, ν'_{ij} and ν''_{ij} are, in this order, the stoichiometric coefficients of chemical species i in reaction j existent in reagents and products and kf and kb represent, respectively, forward and backward reaction constants.

At chemical equilibrium conditions, for each single chemical reaction j, there is a specific equilibrium constant (K_j), based on partial pressures (p_i). This constant shows how the forward and the backward ways of a reaction j are related at a specific temperature and can be calculated by:

$$K_j = \prod_{i=1}^{N} p_i^{(\nu''_{ij} - \nu'_{ij})} \tag{2}$$

The K_j can also be obtained from the variation of the Gibbs 1.3 free energy (ΔG_j), which can be evaluated by:

$$\Delta G_j = \sum_{i=1}^{N} \nu''_{ij} g_i - \sum_{i=1}^{N} \nu'_{ij} g_i \tag{3}$$

where g_i corresponds to the Gibbs 1.3 free energy of species i, that can be calculated with the methodology presented by McBride *et al.* (1993), in which values of constant-pressure specific heat (c), enthalpy (h) and Gibbs 1.3 free energy (g) can be estimated by polynomial equations:

$$c_i = Ru\left(a_1 + a_2 T + a_3 T^2 + a_4 T^3 + a_5 T^4\right)_i \tag{4}$$

$$h_i = Ru\, T_i \left(a_1 + a_2 \frac{T}{2} + a_3 \frac{T^2}{3} + a_4 \frac{T^3}{4} + a_5 \frac{T^4}{5} + \frac{a_6}{T}\right)_i \tag{5}$$

$$g_i = Ru\, T_i \left(a_1(1 - \ln T) - a_2 \frac{T}{2} - a_3 \frac{T^2}{6} - a_4 \frac{T^3}{12} - a_5 \frac{T^4}{20} + \frac{a_6}{T} - a_7\right)_i \tag{6}$$

where coefficients for thermodynamic properties estimative (a_k) have particular values for every single chemical species i,

Ru is the universal constant for perfect gases (8.314510 J/molK) and *T* corresponds to the absolute temperature (in K) at which the property is been evaluated.

Therefore, K_j can also be determined by:

$$K_j = \exp\left(-\frac{\Delta G_j}{RuT}\right) \quad (7)$$

Similarly, it is possible to evaluate transport properties, such as viscosity (μ) and heat conductivity (k) for a single species i, using the following relations, adapted from McBride *et al.* (1993) to SI units:

$$\mu_i = 10^{-7}\exp\left[b_1\ln(T) + \frac{b_2}{T} + \frac{b_3}{T^2} + b_4\right]_i \quad (8)$$

and

$$k_i = 10^{-4}\exp\left[d_1\ln(T) + \frac{d_2}{T} + \frac{d_3}{T^2} + d_4\right]_i \quad (9)$$

where b_k e d_k (k = 1 to 4) are the particular coefficients for each chemical species i.

REACTION MODELS FOR CHEMICAL EQUILIBRIUM

Nine different chemical reaction models were studied in this work and implemented at Gibbs 1.3 code. These models have 3, 4, 6 or 8 chemical species and from 0 to 18 dissociation chemical reactions. Table 1 summarizes the adopted models.

It is observed that some models have the same species and number of chemical reactions, like models 3 and 4. However, they differ from one dissociation equation at least, reason for considering them independent models.

Model 0

Model 0 is the simplest model implemented at Gibbs 1.3. Dissociation effects are not taken into account by this model, presenting the worst results at high temperatures. The importance of this model, however, is on its equations, that will be used in other models for further calculations. Its global reaction equation is given by:

$$H_2 + \alpha O_2 \rightarrow n_1H_2O + n_2O_2 + n_3H_2 \quad (10)$$

where n is the number of moles of a given chemical species and a is a function of informed value for oxidant/fuel mass ratio (*OF*), obtained by:

$$a = OF\frac{M_3}{M_2} \quad (11)$$

where, M_3 and M_2 are, respectively, the molecular weight of molecular hydrogen (H_2) and the molecular weight of molecular oxygen (O_2).

Model 9

Model 9 is the most complex model implemented at Gibbs 1.3, since it includes 8 species and 18 different reaction equations. This is the same model used by Kee *et al.* (1990), at Chemkin code. The global reaction equation is given by:

$$H_2 + \alpha O_2 \rightarrow n_1H_2O + n_2O_2 + n_3H_2 + n_4OH + n_5O + n_6H + n_7HO + n_8H_2O_2 \quad (12)$$

Table 1. Chemical equilibrium reaction models implemented in Gibbs 1.3.

Model	*L*	*N*	Species	Observations
0	0	3	H_2O, O_2, H_2	Ideal model
1	1	3	H_2O, O_2, H_2	–
2	2	4	H_2O, O_2, H_2, OH	–
3	4	6	H_2O, O_2, H_2, OH, O, H	4 reactions with 3rd body – Barros *et al.* (1990) and Smith *et al.* (1987)
4	4	6	H_2O, O_2, H_2, OH, O, H	4 reactions – Svehla (1964)
5	8	6	H_2O, O_2, H_2, OH, O, H	8 reactions (4 with 3rd body) – Barros *et al.* (1990)
7	8	6	H_2O, O_2, H_2, OH, O, H	8 reactions (4 with 3rd body) – Smith *et al.* (1987)
10	6	8	H_2O, O_2, H_2, OH, O, H, HO_2, H_2O_2	4 reactions from model 3 and 2 from Kee *et al.* (1990) – all reactions including 3rd body
9	18	8	H_2O, O_2, H_2, OH, O, H, HO_2, H_2O_2	18 reactions (5 with 3rd body) – Kee *et al.* (1990)

Table 2 presents the dissociation equations related to model 9.

The resulting system of equations, which must be solved in order to obtain the composition at chemical equilibrium, is presented in Table 3 (where p is the total pressure).

Table 2. Chemical reactions taken into account by model 9.

1	$H + OH + D_3 \leftrightarrow H_2O + D_3$
2	$H_2 + D_3 \leftrightarrow 2H + D_3$
3	$O_2 + D_3 \leftrightarrow 2O + D_3$
4	$H + O_2 + D_3 \leftrightarrow HO_2 + D_3$
5	$H_2O_2 + D_3 \leftrightarrow 2OH + D_3$
6	$H_2 + O_2 \leftrightarrow 2OH$
7	$OH + H_2 \leftrightarrow H_2O + H$
8	$H + O_2 \leftrightarrow OH + O$
9	$O + H_2 \leftrightarrow OH + H$
10	$H + 2O_2 \leftrightarrow HO_2 + O_2$
11	$OH + HO_2 \leftrightarrow H_2O + O_2$
12	$H + HO_2 \leftrightarrow 2OH$
13	$O + HO_2 \leftrightarrow O_2 + OH$
14	$2OH \leftrightarrow O + H_2O$
15	$H + HO_2 \leftrightarrow H_2 + O_2$
16	$2HO_2 \leftrightarrow H_2O_2 + O_2$
17	$H_2O_2 + H \leftrightarrow HO_2 + H_2$
18	$H_2O_2 + OH \leftrightarrow H_2O + HO_2$

D_3: 3rd body.

Table 3. Non-linear equations system of model 9.

1	$K_1 p n_4 n_6 = n_1 n$
2	$K_2 n_3 n = n_6^1 p$
3	$K_3 n_2 n = n_5^2 p$
4	$K_4 p n_2 n_6 = n_7 n$
5	$K_5 n_8 n = n_4^2 p$
6	$K_6 n_2 n_3 = n_4^2$
7	$K_7 n_3 n_4 = n_1 n_6$
8	$K_8 n_2 n_6 = n_4 n_5$
9	$K_9 n_3 n_5 = n_4 n_6$
10	$K_{10} p n_2 n_6 = n_7 n$
11	$K_{11} n_4 n_7 = n_1 n_2$
12	$K_{12} n_6 n_7 = n_4^2$
13	$K_{13} n_5 n_7 = n_2 n_4$
14	$K_{14} n_4^2 = n_1 n_5$
15	$K_{15} n_6 n_7 = n_2 n_3$
16	$K_{16} n_7^2 = n_2 n_8$
17	$K_{17} n_6 n_8 = n_3 n_7$
18	$K_{18} n_4 n_8 = n_1 n_7$

ALGORITHM FOR CHEMICAL EQUILIBRIUM

This section presents the Gibbs 1.3 chemical equilibrium algorithm. It consists in ten steps:

- Definition of data: reactive model number, T, p, OF and numeric parameters (number of iterations and tolerances).
- Reading of thermodynamic coefficients (a_{ki}) for constant-pressure specific heat, enthalpy and Gibbs 1.3 free energy (Eq. 4 to 6).
- Coefficients' evaluation of Eq. (10): a is based on Eq. (11) and n_1, n_2 and n_3 are obtained from a simple chemical equilibrium calculus.
- Initialization of reaction degrees (ε_j) for all dissociation reactions, with null-value.
- Evaluation of initial number of moles for every single reactive model species (n_i) and total number of moles (n), based on variables from items Coefficients' evaluation and Initialization of reaction degrees.
- Evaluation of g_i, Gibbs 1.3 free energy change and equilibrium constant based on partial pressure for reaction j by Eq. (7).
- Iterative evaluation, for the first dissociation reaction (ε_1), by Newton-Raphson method (Turns, 2000), with calculations made until the maximum number of iterations or up to the tolerance is achieved. This is made for the second dissociation reaction and so on, until the last reaction (L).
- Actualization of the number of moles for n_i and n, with obtained results for ε_j calculated in item "Iterative evaluation".
- Evaluation of total number of moles variation (Δn). Return to item "Iterative evaluation" until the maximum number of iterations is achieved or while Δn is greater than the chosen tolerance.
- Evaluation of mixture properties, such as mean molecular weight, molar and mass fractions, gas mixture density, constant-pressure specific heat, ratio of specific heats, gases mixture constant, viscosity and thermal conductivity.

COMBUSTION TEMPERATURE

Combustion temperature determination (also known as adiabatic flame temperature) is basically the same problem of chemical equilibrium; however, in this case, only the temperatures (or enthalpies) of reagents are known, as well as the chamber pressure and the *OF*. The problem is basically to equalize the enthalpy of reagents with the enthalpy of products. However, as the composition of gas mixture is dependent on the temperature, an iterative method for obtaining the combustion temperature is needed. In this work, it was used a methodology based on bisection method (Chapra and Canale, 1994), with chemical equilibrium parameters calculated by equations presented by Kuo (2005). The same chemical reaction models presented in Table 1 are valid for the implemented combustion temperature determination code.

Table 3 presents the non-linear equation system of model 9.

ALGORITHM FOR ADIABATIC COMBUSTION

The implemented algorithm for adiabatic combustion temperature's determination is presented in this section. It consists of the following steps:

- Definition of data: reactive model number, temperature or enthalpy of reagents, *p*, *OF* and numeric parameters (iteration numbers and tolerances).
- Idem steps "Reading of thermodynamic coefficients" to "Evaluation of initial number of moles for every single reactive model species and total number of moles" of the section "Algorithm for chemical equilibrium".
- Evaluation of total enthalpy of reagents by:

$$H_R = h_{fuel} + ah_{oxidant} \tag{13}$$

  where h_{fuel} and $h_{oxidant}$ indicate, respectively, the fuel enthalpy (in the studied case, hydrogen) and the oxidant enthalpy (oxygen).
- First guess: the combustion temperature is considered to be 3,150 K (mean value between 300 and 6,000 K, which corresponds to the range of applicability for a_{ki}).
- Idem steps "Evaluation of Gibbs 1.3 free energy for each single species" to "Evaluation of total number of moles variation" of the section "Algorithm for chemical equilibrium".
- Evaluation of total enthalpy of products by Eq. (14):

$$H_P = \sum_{i=1}^{N} n_i h_i \tag{14}$$

- New estimative for the combustion temperature with bisection method (Chapra and Canale, 1994).
- Evaluation of temperature variation (ΔT) between old and new temperature. Return to item "Idem steps 'Reading of thermodynamic coefficients' to 'Evaluation of initial number of moles for every single reactive model species and total number of moles' of the section 'Algorithm for chemical equilibrium'" until the maximum number of iterations is achieved or while ΔT is greater than the chosen tolerance.
- Evaluation of mixture's properties, such as mean molecular weight, molar and mass fractions, gas mixture's density, constant-pressure specific heat, ratio of specific heats and gases mixture's constant.

CHEMICAL NON-EQUILIBRIUM

Chemical and vibrational processes take place because of molecular collisions and/or radiation. Third body species have an important role on these processes, providing the energy required to other species by collisions, in order to allow chemical reactions occur. These interactions, however, demand a period of time, something that may not come true in some practical situations, such as the gas flow in a rocket nozzle engine. In such cases, equilibrium conditions are not reached, once it demands infinitely large reaction rates, *i.e.* instantaneous adjustments to the local temperature and pressure of gases. Therefore, regions where non-equilibrium phenomena take place will appear through the flow and, for these cases, an alternative methodology must be used, in order to consider the reaction time over the composition of the gas mixture (Anderson Jr., 2003).

To evaluate the mass generation rate of chemical species, some variables must be presented and defined. The first one is the progression rate of reaction *j* (γ_j), which is concerned about the difference between forward and backward reaction rates and is given by:

$$\gamma_j = \left\{ kf_j \prod_{i=1}^{N} C_i^{\nu_{ij}} - kb_j \prod_{i=1}^{N} C_i^{\nu_{ij}} \right\}, \tag{15}$$

where C_i is the molar concentration of species *i* in mixture. The species generation rate (θ_j) is obtained from the product between Eq. (15) and the third body effective concentration and can be estimated by:

$$\Theta_j = \lambda_j \cdot \gamma_j \tag{16}$$

where λ_j is defined as:

$$\lambda_j = \sum_{i=1}^{N} \alpha_{ij} \cdot C_i \tag{17}$$

in which α_{ij} is the efficiency of chemical species *i* at a dissociation reaction *j*.

Unfortunately, there is a lack of information about the efficiency of the existent molecules in a dissociation reaction model. Consequently, it is commonly adopted that the efficiencies of all chemical species are equal to unit and, in this case, the concentration of third body species corresponds to the whole number of existent species (Barros *et al.*, 1990).

The determination of mass generation rates is obtained by the product of Eq. (16) and molecular weight (M_i) of the species, resulting in:

$$\dot{\omega}_i = M_i \sum_{j=1}^{L} \left(\Delta \nu_{ij} \cdot \theta_j \right) \tag{18}$$

where $\Delta \nu_{ij}$ is calculated by $\Delta \nu_{ij} = \nu''_{ij} - \nu'_{ij}$ and represents the difference between formed and consumed number of moles for a reaction *j*.

REACTION MODELS FOR NON-EQUILIBRIUM

Six different non-equilibrium reaction models are studied in this work and implemented at code Gibbs 1.3. These models have 6 or 8 chemical species, corresponding to models 3, 5, 7, 9 and 10 from equilibrium schemes, according to Table 1. It is noted that model 3 was split into two non-equilibrium models. Despite the fact that each model has the same species, two different equation systems for forward reaction constants (*kf*) were considered from works of Barros *et al.* (1990) and Smith *et al.* (1987), which justifies to treat them as two independent models in this section.

MODEL 3.1

Model 3 from equilibrium schemes was split into two different models in non-equilibrium schemes, named model 3.1 and model 3.2. Both models include six different species and the chemical reactions presented in Table 4.

However, for each chemical reaction of model 3.1, one should take into account the forward reaction constants from Barros *et al.* (1990), which are obtained by the relations presented in Table 5.

Table 4. Chemical reactions by models 3.1 and 3.2.

1	$H + OH + D_3 \leftrightarrow H_2O + D_3$
2	$2H + D_3 \leftrightarrow H_2 + D_3$
3	$2O + D_3 \leftrightarrow O_2 + D_3$
4	$O + H + D_3 \leftrightarrow OH + D_3$

D_3: 3rd body.

Table 5. Forward reaction constants by model 3.1.

1	$kf_1 = 3.626 \times 10^{19}/T$
2	$kf_2 = 7.5 \times 10^{17}/T$
3	$kf_3 = 3.626 \times 10^{18}/T$
4	$kf_4 = 3.626 \times 10^{18}/T$

Based on the forward reaction constants, the progression rates for all reactions can be evaluated by Eq. (15). Taking the efficiencies of all species equal to unity, the parameter λ corresponds to the total concentration (*C*), and the reaction parameters (θ_j) can be evaluated as $\theta_j = C \times \gamma_j$, according to Eq. (16). The $\dot{\omega}_j$ can be evaluated by the equations presented in Table 6.

Table 6. Mass generation rates of model 3.1.

1	$[H_2O]: \dot{\omega}_1 = M_1 . \theta_1$
2	$[O_2]: \dot{\omega}_2 = M_2 . \theta_3$
3	$[H_2]: \dot{\omega}_3 = M_3 . \theta_2$
4	$[OH]: \dot{\omega}_4 = M_4 . (\theta_4 - \theta_1)$
5	$[O]: \dot{\omega}_5 = -M_5 . (2\theta_3 + \theta_4)$
6	$[H]: \dot{\omega}_6 = -M_6 . (\theta_1 + 2\theta_2 + \theta_4)$

ALGORITHM FOR NON-EQUILIBRIUM

The implemented algorithm for non-equilibrium is presented in this section. It consists of the eight following steps:

- Definition of data: reactive model number, *T*, *p* and initial mass fractions (Y_i).
- Reading of a_{ki} for constant-pressure specific heat, enthalpy and Gibbs 1.3 free energy — Eq. (4) to (6).
- Evaluation of C_i for each species *i* and *C*.
- Evaluation of: g_i; ΔG_j; and equilibrium constant based on partial pressures (K_j), for each reaction *j*.
- Evaluation of forward reaction constants (kf_j), for each reaction model.
- Evaluation of backward reaction constants (kb_j), for each reaction model, by the equation:

$$kb_j = \frac{kf_j}{Kc_j} \quad (19)$$

where Kcj are the equilibrium constants on molar basis, which are related to equilibrium constants on partial pressure basis by the relation:

$$Kc_j = \frac{K_j}{(Ru \cdot T)^{\Delta\nu_j}} \quad (20)$$

- Evaluation of parameters γ_j, λ_j and θ_j by Eq. (15) to (17).
- Determination of $\dot{\omega}_j$ for each chemical species by Eq. (18).

NUMERICAL RESULTS

Three different kinds of problems were studied: firstly, chemical equilibrium calculation, in which both combustion chamber temperature and pressure are known; secondly, combustion temperature determination, in which only the enthalpies of reagents and the chamber pressure were known, and temperature of reactions, as well as chemical equilibrium composition, is one of the expected results; thirdly, non-equilibrium mass generation rate evaluation, in which several chemical reaction schemes are compared.

CHEMICAL EQUILIBRIUM

Five different problems were studied, intending to cover all range of values found in a rocket thrust chamber: temperatures vary between 600 and 4,000 K and pressures, between 0.002 and 20 MPa. *OF* chosen was the stoichiometric one (7.936682739), because this is the mass ratio which offers the greatest difficulties for numerical convergence. The chosen tolerance to solve the reaction degrees of dissociation equations was 10^{-12}, the same value for the chosen tolerance to solve the total number of moles; and the maximum number of iterations to evaluate the total number of moles was fixed on 50,000, the same value for maximum number of iterations to evaluate the dissociation rate for each reaction degree. For each problem, the 9 chemical reaction models were used and the results were compared with those obtained from CEA and Teqworks, both codes from NASA, which include the 9 species: the 8 ones considered in models 9 and 10 from code Gibbs 1.3 (proposed in the current work) and ozone (O_3).

Table 7 presents the thermochemical and transport properties obtained for chemical equilibrium from codes Gibbs 1.3, CEA and Teqworks for 4,000 K and 20 MPa, being ρ the density, frozen c the constant-pressure specific heat, γ the ratio between specific heats and R, the constant of gases. All these parameters are evaluated for the mixture of gases.

It can be seen, from Table 7, that models considering chemical dissociation, with the same number of species (such as models 3, 4, 5 and 7), present the same results for thermochemical parameters. A briefly view on results presented in Table 8 helps to understand this fact: these models have exactly the same mass fraction configuration. It does not matter the chemical scheme assumed, but only the species taken into account by the model.

Considering the necessary computational time, however, it was verified that model 9 is much faster than model 10, in spite of the greater number of chemical reactions. For a temperature of 2,000 K and a pressure of 0.2 MPa, the necessary time for model 9 to achieve the solution was 0.12 s, while model 10 requires about 13 times more (1.60 s). Even comparing model 9 with 6-species models, model 9 was the fastest. Models 3, 4, 5 and 7 require 0.41, 0.16, 0.15 and 0.23 s, respectively, to achieve the solution. For all cases, it was verified that model 9 was the fastest one (or the second one, only 0.01 s slower than the fastest) among 6- and 8-species models.

Table 7. Thermochemical and transport properties obtained for chemical equilibrium from Gibbs 1.3, CEA and Teqworks codes for 4,000 K and 20 MPa.

Model	*M* [kg/kmol]	ρ [kg/m³]	Frozen *c* [J/kgK]	γ [adim.]	*R* [J/kgK]	μ [Pas]	*k* [W/mK]
0	18.015	10.8336	3,295.5	1.1629	461.53	1.1737E-04	5.6004E-01
1	16.865	10.1421	3,300.0	1.1756	493.00	1.1636E-04	6.2587E-01
2	16.196	9.7395	3,288.8	1.1850	513.37	1.1538E-04	6.2006E-01
3, 4, 5, 7	15.536	9.3425	3,293.5	1.1940	535.19	1.1534E-04	6.3917E-01
10, 9	15.537	9.3433	3,293.6	1.1940	535.14	1.1526E-04	6.3904E-01
CEA	15.516	9.3309	3,290.8	-	-	1.2123E-04	5.5509E-01
Teqworks	15.503	9.3230	-	-	-	-	-

The percent error, shown at some of the following tables and presented in the sequence, is evaluated according to the following expression:

$$E = 100 \; \frac{(reference - value)}{reference} \tag{21}$$

in which *reference* corresponds to CEA's results and *value* refers to results from other sources (such as Gibbs 1.3 and Teqworks).

Table 8 presents the chemical mass fractions for a mixture of gases at 4,000 K and 20 MPa.

Results from model 9 of Gibbs 1.3 have the same accuracy of the results obtained by Teqworks, when both codes are compared with results of CEA (confuse). The maximum errors for thermochemical parameters are 0.15% for Gibbs 1.3 (model 9) and 0.085% for Teqworks (according to Table 7). About mass fractions, the absolute errors are 6.4 x 10^{-3} and 3.6 x 10^{-3}, for Gibbs 1.3 and Teqworks, respectively (based on Table 8). For the other models, the results presented also good accuracy.

It is observed that the model 5, which has 6 species, presents results very close to the ones from model 9, which agrees with the other studied problems, not shown in this work (Marchi and Araki, 2005). Otherwise, a model with only 4 species (such as model 2) does not predict properly the chemical composition and the thermochemical properties for the gases mixture at high temperatures, since, for these conditions, this model presents significant errors. Large errors are obtained when transport properties result from Gibbs 1.3 (models with six and eight species) and CEA are compared. For viscosity, relative errors are lower than 5% and, for thermal conductivity, the errors achieve 15%. The relative errors decrease when temperatures and pressures are reduced. A possible reason for these differences lies in the fact that codes do not have the same polynomial coefficient sources: McBride *et al.* (1993) for Gibbs 1.3 and Svehla (1995) for CEA. However, even these ranges of errors are acceptable based on results presented by Reid *et al.* (1987), who obtained errors for viscosity in binary systems about ± 12% and, for a particular binary system, the error for thermal conductivity achieved 16%, being mostly between 5 and 7%.

COMBUSTION TEMPERATURE

For combustion temperature evaluation, 17 different problems were studied. Only the temperature of oxidant (LOX) and fuel (LH_2) were kept constant, with values of 90.17 K (corresponding to a enthalpy of -12.979 J/mol) and 20.27 K (-9.12 J/mol), respectively.

The results obtained from CEA and Teqworks codes were compared and, when it was possible, with other literature sources. The chosen tolerance to solve reaction degrees was 10^{-12}, the same value for the tolerance to solve the total number of moles; the maximum number of iterations to evaluate the total number of moles was fixed on 5,000, which is the same value for maximum number of iterations to evaluate the dissociation rate for each reaction degree. The maximum number of iterations for solving the combustion temperature (*Tc*) was 500.

Table 9 presents the combustion temperatures for a chamber pressure of 20 MPa and stoichiometric reactants' composition.

Table 8. Mass fraction's results for 4,000 K and 20 MPa.

Model	H_2O	O_2	H_2	OH	O	H	HO_2	H_2O_2	O_3
0	1.0000E-0	0	0	---	---	---	---	---	---
1	8.6362E-01	1.2112E-01	1.5260E-02	---	---	---	---	---	---
2	7.7532E-01	7.7639E-02	1.7462E-02	1.2958E-01	---	---	---	---	---
3, 4, 5, 7	7.5268E-01	7.7291E-02	1.7347E-02	1.2886E-01	2.1134E-02	2.6914E-03	---	---	---
10, 9	7.5214E-01	7.6915E-02	1.7376E-02	1.2865E-01	2.1082E-02	2.6935E-03	9.2804E-04	2.1200E-04	---
CEA	7.4839E-01	7.4654E-02	1.7424E-02	1.3508E-01	2.0636E-02	2.6850E-03	9.2359E-04	2.0703E-04	2.6050E-06
Teqworks	7.478E-01	7.8259E-02	1.7690E-02	1.318E-01	2.1167E-02	2.7045E-03	5.6768E-04	5.534E-013	1.3402E-06

Table 9. Combustion temperatures. Chamber pressure of 20 MPa, stoichiometric reactants composition (OF = 7.936682739).

Model	Tc Gibbs 1.3 [K]	Tc CEA [K]	Error Gibbs 1.3* [%]	Tc Teqworks [K]	Error Teqworks* [%]
0	4,674.85		-25		
1	4,060.30		-8.6		
2	3,838.08	3,737.73	-2.7	3,748.86	-0.30
3, 4, 5, 7	3,742.51		-0.13		
10, 9	3,741.97		-0.11		

*CEA's results are taken as reference.

It is observed that the model 0, which involves 3 species and no dissociation reaction, overestimates the combustion temperature in 25% for a pressure of 20 MPa. In case of pressure value equal to 2 MPa, the evaluated error is 37%. Model 1, which counts 3 species and only 1 dissociation reaction, presents better results, with errors of about 8.6 and 11% for pressures of 20 and 2 MPa, respectively. Model 2, which presents 4 species and 2 dissociation reactions, presents, respectively, errors of 2.7 and 4.1% for pressures of 20 and 2 MPa. The best models, however, are those which have six and eight species. The maximum errors for the models with 6 species were 0.13%, for 20 MPa, and 0.11%, for 2 MPa, while the maximum errors for the models with 8 species were 0.11% for 2 and 20 MPa. The results obtained in these cases are *more* accurate than those obtained with Teqworks code, which presents errors equal to 0.30 and 0.19% for 20 and 2 MPa, respectively.

At chemical equilibrium condition, combustion temperature depends on only the number of chemical species. Therefore, models with same species, such as models 3, 4, 5 and 7 or models 9 and 10, present equal results. Comparing six-species models with eight-species models, it is verified that these models present very close results, like those observed at chemical equilibrium. It must also be noted that model 9, which has the major complexity, is not the slowest: its convergence was faster than 2 models with 6 species and 1 with 8 species. While model 9 requires 0.39 s, models 3, 4 and 10 require 0.56, 0.50 and 0.92 s, respectively, and models 5 and 7 require 0.23 and 0.35 s. Taking into account the necessary computational time, model 5 was the fastest among the 6-species models and model 9 was the best among the 8-species ones.

Table 10 provides a series of data, in which Gibbs 1.3 code's results are compared with those of CEA and other sources from literature. It can be noted that Gibbs 1.3's results are very close to those provided by CEA code. While Gibbs 1.3's errors vary from 0.032 to 0.20%, the results from literature present errors in the range between 0.47 and 11%. Based on this, it can be concluded that Gibbs 1.3 provides good estimates for combustion temperature problems.

CHEMICAL NON-EQUILIBRIUM

Six different physical situations were analyzed for the non-equilibrium mass generation rate evaluation, the same conditions studied for chemical equilibrium determination and a new case corresponding to 3,000 K and 20 MPa. In all cases, the stoichiometric oxidant/fuel ratio was considered and the mass fraction's results, from chemical equilibrium module, were used as input data. Only 6- and 8-species models were analyzed, which can be seen at Table 11. The model 3 from chemical equilibrium analyses was split up into two different models (models 3.1 and 3.2).

Differently from results obtained for chemical equilibrium, at chemical non-equilibrium condition, the same number of species does not mean equal mass generation rates, as it can be observed at Table 11. Even models with the same chemical reaction scheme, such as models 3.1 and 3.2, have different results. Therefore, neither only the number of species nor only the

Table 10. Comparison of combustion temperatures obtained with Gibbs 1.3 (model 9) and CEA codes and other sources from literature.

Total pressure [MPa]	OF	Tc CEA [K]	Tc Gibbs 1.3* [K]	Error Gibbs 1.3** [%]	Tc another source [K]	Error** [%]
20	2	1,797.78	1,796.65	0.063	[Tw] 1,798.71	-0.052
20	4	2,974.69	2,976.10	-0.047	[Tw] 2,986.92	-0.41
20	6	3,595.43	3,599.98	-0.13	[Tw] 3,610.55	-0.42
20	10	3,644.31	3,649.47	-0.14	[Tw] 3,658.22	-0.38
20	12	3,507.10	3,513.33	-0.17	[Tw] 3,523.28	-0.46
20	14	3,368.28	3,374.95	-0.20	[Tw] 3,385.28	-0.50
20	16	3,234.72	3,241.35	-0.20	[Tw] 3,251.62	-0.52
20.241	6.00	3,596.61	3,601.17	-0.13	[Wang and Chen, 1993] 3,639.0	-1.2
0.51676	8	3,237.61	3,240.86	-0.10	[Kim and Vanoverbeke, 1991] 3,300.0	-1.9
0.51676	16	2,964.90	2,970.91	-0.20	[Kim and Vanoverbeke, 1991] 3,073.0	-3.6
6.8948	4.13	2,998.45	3,000.31	-0.062	[Huzel and Huang, 1992] 3,013.0	-0.49
6.8948	4.83	3,235.70	3,238.85	-0.097	[Huzel and Huang, 1992] 3,251.0	-0.47
6.8948	3.40	2,668.70	2,669.55	-0.032	[Sutton and Biblarz, 2010] 2,959.0	-11
6.8948	4.02	2,954.33	2,956.01	-0.057	[Sutton and Biblarz, 2010] 2,999.0	-1.5
6.8948	4.00	2,946.10	2,947.75	-0.056	[Sarner, 1966] 2,977.0	-1.0

*Model 9 only; **CEA's results are taken as reference.
Tw: Teqworks.

number of chemical reaction equations but also the third body reactions and the forward reaction constants are fundamental for mass generation rate evaluation.

Large discrepancies for chemical species generation were verified; the mass generation rate for atomic oxygen obtained from models 3.2 and 7 is about 10^8 kg/m³s and, for model 9, it is about 10^4, *i.e.* four orders of magnitude smaller, reinforcing that the 3rd body reaction equations and forward reaction constants are the basis for the correct mass generation rate determination. Unfortunately, no appropriated data was found in literature to compare results with those obtained by Gibbs 1.3 code. Therefore, there is no way to define which of the listed chemical schemes is the best in physical aspects.

As all models require less than 0.001 s for solving the first problem (presented in Table 9), another methodology was proposed to evaluate the computational CPU time execution: in this case, all models were executed 10^7 times to evaluate the average value. Comparing the required times, it was verified that the slowest model was model 9, which requires 63.4 s. Models 5, 7, 3.2 and 3.1 require 50, 38, 20 and 18%, respectively, of the period time required by model 9. In this case, even model 10 (with 8 species and 6 reaction equations) requires about 26% of the time required by model 9, being faster than models 5 and 7, which are 6-species models that consider 8 reaction equations. Therefore, the computational time is directly associated with the number of reaction equations adopted by the chemical model.

The mass generation rates are important to evaluate the mass fraction rates of a gas mixture in a chemical reactive flow. Although these rates cannot be used directly to evaluate the chemical composition from equilibrium flow conditions, in Fig. 1, numerical results of a simple exercise are presented to show how the chemical composition of a gas mixture can change if the mass generation rates are taken into account. It must be observed, however, that these results are only illustrative; actually, to evaluate the chemical composition, a system of partial differential equations given by the mass, momentum and energy conservation laws, besides a state relation, such as the perfect gas equation, must be solved — which is beyond the scope of this work.

In Fig. 1, equilibrium conditions are related to results shown for time intervals smaller than 10^{-12} s. Once mass reaction rates are given in kg/m³s, a time interval must be considered to obtain a mass fraction distribution. This time interval, for example, can be related to the flow velocity and a given control volume length, for which the chemical mass fraction is supposed to be evaluated. Variations on the chemical composition are easily observed for larger time intervals (specially 10^{-6} s) and the results for both models are not the same due to the different efficiency values of chemical species associated to each model. However, it must

Table 11. Mass generation rates for 4,000 K and 20 MPa [kg/m³s].

Model	H_2O	O_2	H_2	OH	O	H	HO_2	H_2O_2
3.1	-1.538E+09	-2.356E+07	-1.255E+06	1.427E+09	4.738E+07	8.881E+07	-	-
3.2	-1.263E+09	-4.066E+06	-1.239E+07	1.032E+09	1.550E+08	9.254E+07	-	-
5	-1.538E+09	-2.356E+07	-1.255E+06	1.427E+09	4.738E+07	8.881E+07	-	-
7	-1.263E+09	-4.066E+06	-1.239E+07	1.032E+09	1.550E+08	9.254E+07	-	-
10	-1.537E+09	2.679E+07	-1.257E+06	1.879E+09	4.723E+07	9.034E+07	-5.183E+07	-4.530E+08
9	-7.306E+08	8.255E+10	-4.880E+05	1.143E+09	2.712E+04	2.642E+09	-8.515E+10	-4.530E+08

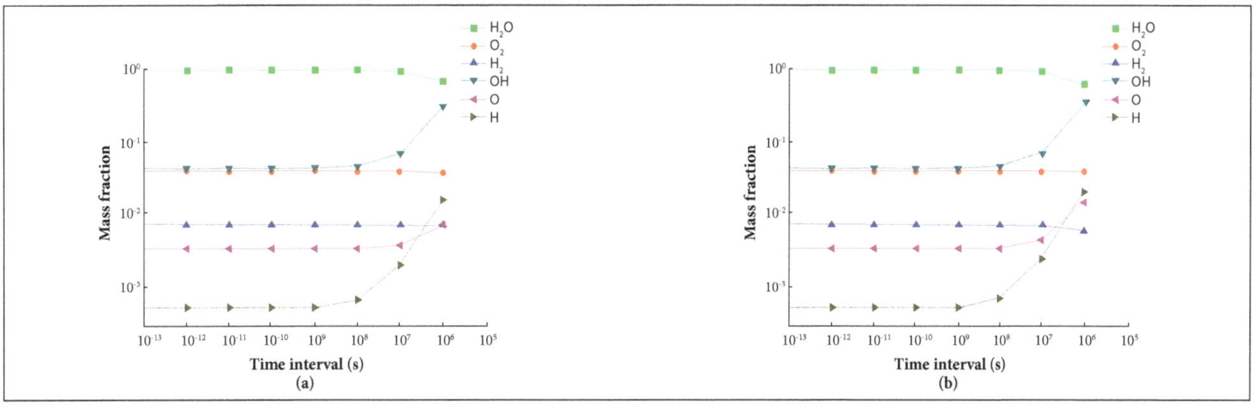

Figure 1. Mass fractions of chemical species after a given time interval for: (a) model 3.1 and (b) model 3.2. Values obtained for mass generation rates for 3,000 K and 20 MPa.

be emphasized, again, that the results shown in Fig. 1 refer only to an exercise, once the system of partial differential equations which models the real flow phenomenon is not solved.

It also must be noted that equilibrium conditions are not achieved in real flows, once it implies that chemical reaction rates must to be infinitely large (Anderson Jr., 2003). Because of this, chemical non-equilibrium models are preferable to evaluate mass fraction rates in a reactive flow, since they take into account both reaction rates and flow velocities.

CONCLUSION

A code for prediction of thermochemical and transport properties at chemical equilibrium and mass generation rates for chemical non-equilibrium was implemented using equilibrium constant methodology. The code, called Gibbs 1.3, evaluates also combustion temperature for a mixture of hydrogen/oxygen. Chemical reaction schemes implemented for equilibrium schemes varied from a model with only 3 species, without chemical dissociation reactions (called model 0) and a model with 8 species and 18 chemical dissociation reactions (model 9), while 6 chemical reaction models were considered for non-equilibrium.

The results for chemical equilibrium property evaluation were validated by comparison with results obtained from Gibbs 1.3 code and other sources, such as CEA and Teqworks. For thermochemical properties, the 6-species and the 8-species models presented global errors of about 0.085%, presenting the same magnitude order of Teqworks when results were compared with those from CEA. Most significant discrepancies in numerical results were observed for the thermal conductivity (about 15%) and viscosity (about 5%). These errors can be attributed to polynomial coefficient sources: McBride *et al.* (1993), for Gibbs 1.3, and Svehla (1995), for CEA. It must be noted that such magnitude of errors are acceptable, as it can be verified in Reid *et al.* (1987).

It was verified that global variables, such as total density or total specific heat, depend only on the species considered by the model, inasmuch as mass fractions obtained are equivalent. The computational time required for convergence is not equal for models with same species; against the common sense, models with more chemical reactions were, in general, faster than the ones with fewer reactions. Therefore, for chemical equilibrium and combustion temperature evaluation, the 6-species model recommended is model 5, while model 9 is the recommended 8-species one. Models with less than 6 species (such as models 0, 1 and 2) are not recommended since they present higher errors.

Analyses for non-equilibrium mass generation rate show that third body reaction equations and *kf* are fundamental for results, even if two models have the same number of species and the same third body reaction equations; if there are different values for *kf*, the results will be different. Otherwise, computational time required is directly associated with the number of chemical reactions considered by the chemical reaction scheme. Despite of model 9 being one of the fastest models in problems involving chemical equilibrium and combustion, this model was the slowest in non-equilibrium problems.

ACKNOWLEDGEMENTS

The authors acknowledge Universidade Federal do Paraná (UFPR), Coordenação de Aperfeiçoamento de Pessoal de Nível Superior (CAPES), Conselho Nacional de Desenvolvimento Científico e Tecnológico (CNPq) and the "UNIESPAÇO Program" of the Brazilian Space Agency (Agência Espacial Brasileira - AEB) by physical and financial support given for this work. The first author is also supported by a scholarship from CNPq.

REFERENCES

Anderson Jr., J.D., 2003, "Modern Compressible Flow", Third Edition, McGraw-Hill, New York, USA.

Barros, J.E.M., Alvin Filho, G.F. and Paglione, P., 1990, "Estudo de Escoamento Reativo em Desequilíbrio Químico através de Bocais Convergente-Divergente", Proceedings of the Encontro Nacional de Ciências Térmicas, Itapema, Brazil.

Bedarev, I.A. and Fedorov, A.V., 2006, "Comparative Analysis of Three Mathematical Models of Hydrogen Ignition", Combustion, Explosion, and Shock Waves, Vol. 42, pp. 19-26. doi: 10.1007/s10573-006-0002-1

Chapra, S.C. and Canale, R.P., 1994, "Introduction to Computing for Engineers", Second Edition, McGraw-Hill, New York.

Fernández-Galisteo, D., Sánchez, A.L., Liñán, A. and Williams, F.A., 2009, "One-step reduced kinetics for lean hydrogen-air deflagration", Combustion and Flame, Vol. 156, No. 5, pp. 985-996. doi: 10.1016/j.combustflame.2008.10.009

Glenn Research Center, 2005, "CEA - Chemical Equilibrium with Applications", Retrieved in February 16, 2005, from http://www.grc.nasa.gov/WWW/CEA Web/ceaHome.htm

Gordon, S. and McBride, B.J., 1971, "Computer Program for Calculation of Complex Chemical Equilibrium Compositions, Rocket Performance, Incident and Reflect Shocks, and Chapman-Jouguet Detonations", NASA, Cleveland. SP-273.

Huzel, D.K. and Huang, D.H., 1992, "Design of Liquid Propellant Rocket Engines", AIAA, Washington, DC, USA.

Kee, R.J., Grcar, J.F., Smooke, M.D. and Miller, J.A., 1990, "A Fortran Program for Modeling Steady Laminar One-Dimensional Premixed Flames", Sandia National Laboratories, Albuquerque, USA.

Kim, S.C. and Vanoverbeke, T.J., 1991, „Performance and Flow Calculations for a Gaseous H2/O2 Thruster", Journal of Spacecraft and Rockets, Vol. 18, pp. 433-438.

Kuo, K.K., 2005, "Principles of Combustion", Second Edition, John Wiley & Sons, Hoboken, USA, 732p.

Marchi, C.H. and Araki, L.K., 2005, "Relatório Técnico 1: Programa Gibbs 1.3, Technical Report, Federal University of Paraná, Brazil", Retrieved in September 01, 2014, from ftp://ftp.demec.ufpr.br/CFD/projetos/cfd5

McBride, B.J., Gordon, S. and Reno, M.A., 1993, "Coefficients for Calculating Thermodynamic and Transport Properties of Individual Species", NASA Technical Memorandum 4513, NASA Lewis Research Center, Cleveland, USA.

Reid, R.C., Prausnitz, J.M. and Poling, B.E., 1987, "The Properties of Gases & Liquids", Fourth Edition, McGraw-Hill, Boston, USA.

Sarner, S.F., 1966, "Propellant Chemistry", Reinhold Publishing, New York, USA.

Smith, J.J., Schneider, G., Suslov, D., Oschwald, M. and Haidn, O., 2007, "Steady-State High Pressure LOX/H2 Rocket Engine Combustion", Aerospace Science and Technology, Vol. 11, No. 1, pp. 39-47. doi:10.1016/j.ast.2006.08.007

Smith, T.A., Pavli, A.J. and Kacynski, K.J., 1987, "Comparison of Theoretical and Experimental Thrust Performance of a 1030:1 Area Ratio Rocket Nozzle at a Chamber Pressure of 2413 Kn/M2 (350 psia)", NASA Technical Paper 2725, NASA Lewis Research Center, Cleveland, USA.

Sutton, G.P. and Biblarz, O., 2010, "Rocket Propulsion Elements", Eighth Edition, John Wiley & Sons, Hoboken, USA.

Svehla, R.A., 1964, "Thermodynamic and Transport Properties for the Hydrogen-Oxygen System", NASA SP-3011, NASA Lewis Research Center, Cleveland, USA.

Svehla, R.A., 1995, "Transport Coefficients for the NASA Lewis Chemical Equilibrium Program", NASA TM–4647.

Turns, S.R., 2000, "An Introduction to Combustion: Concepts and Applications", Second Edition, McGraw-Hill Higher Education, Boston, USA, 676p.

Wang, T.S. and Chen, Y.S., 1993, "Unified Navier-Stokes Flowfield and Performance Analysis of Liquid Rocket Engines", Journal of Propulsion and Power, Vol. 9, pp. 678-685.

26

Experimental Investigation of a Monopropellant Thruster Using Nitrous Oxide

Luciano Hennemann[1], José Carlos de Andrade[1], Fernando de Souza Costa[1]

ABSTRACT: Nitrous oxide is a monopropellant with potential for use in propulsion systems for orbit correction, positioning and attitude control of satellites. This paper presents experimental results with a 2N thruster prototype employing nitrous oxide gas decomposed by a rhodium oxide catalyst supported on alumina. Initially, the thruster design, the test bench and test procedures are described. Then the main experimental results, including thrust curves, mass flow rates, pressures and temperatures along the thruster are presented for pulsed operation. At the end specific impulses, characteristic velocities and thruster efficiencies are determined, thus indicating a catalytic decomposition efficiency of about 88%.

KEYWORDS: Satellite propulsion, Nitrous oxide, Rhodium oxide catalyst, Propulsive efficiency.

INTRODUCTION

Several factors, such as atmospheric drag, the gravitational attractions of the Moon and the Sun, the non-uniformity of the Earth gravitational field and solar radiation contribute to change the attitude and move satellites from their original orbits (Sutton, 2001). For example, in geo-stationary satellites, Luni-solar perturbations cause a north-south drift that requires corrections of 50 m/s/year and the variation of the Earth gravitational field can generate an east-west drift of 1 degree in 60 days (Stark and Swinerd, 2003). Therefore, the use of an orbit correction and attitude control system is necessary for continuity of the normal service conditions of a satellite.

At first, the control system of the rocket launcher provides input for the upper propulsion stages, for initial stabilization in orbit, alignment and firing of small thrusters (actuators). These are responsible for positioning and attitude control of the satellite in its initial orbit, having sensors, algorithms and actuators to aid in these tasks. The sensors determine the satellite's instantaneous position in orbit and provide reference to an algorithm that controls the thrusters to reposition the satellite (Plumlee and Steciak, 2004; Sanscrainte, 1961).

Liquid propellant thrusters can generate high specific impulses and thrusts, and allow precise control of thrust and exhaustion kinetic energy. In general, these systems use pressurizing tanks with an inert gas (nitrogen or helium). In bipropellant systems, the fuel and oxidant are injected, mixed and burned in the combustion chamber to form gaseous products at elevated temperatures that are ejected through a nozzle to generate thrust. Monopropellant thrusters use a single propellant decomposed by a catalyst, or by heating, in order to form hot gases to be ejected at high speeds through the nozzle.

1. Instituto Nacional de Pesquisas Espaciais – Cachoeira Paulista/SP – Brazil.

Author for correspondence: Luciano Hennemann | Laboratório Associado de Combustão e Propulsão – Instituto Nacional de Pesquisas Espaciais | Caixa Postal 1, Cachoeira Paulista/SP | CEP: 12.630-000 – Brazil | Email: hennemann@lcp.inpe.br

Both systems can operate in pulsed mode for attitude control and orbit correction and stay idle long periods of time. However, liquid monopropellant systems can provide smaller and more precise pulsed firings and have a lower number of components compared to bipropellant systems. These can be expensive and difficult to be applied in some cases, and require care in the design to operate with zero gravity (Brown, 1995; Humble *et al.*, 1995).

The use of monopropellant or bipropellant systems depends on the desired application. Comparatively, bipropellant systems have specific impulses between 300 s to 450 s, producing a thrust up to 1.2×10^7 N. Monopropellant systems operate with specific impulses of 130 s to 280 s providing 0.5 N to 500 N of thrust (Makled and Belal, 2009). Nevertheless, monopropellant systems are more commonly used than bipropellant systems for orbit correction and attitude control of satellites due to the low thrust levels required, lower mass and greater simplicity in construction.

Hydrazine (N_2H_4) is the most used liquid monopropellant for space applications; however, in the last decade, there has been a significant interest for using green monopropellants, such as nitrous oxide (N_2O) and hydrogen peroxide (H_2O_2). Nitrous oxide can be used as an energy source and for generation of oxygen on board spacecrafts or launch rockets, in cold gas propulsion systems for attitude control, in monopropellant or bipropellant thrusters, in gas generator systems, turbine operation and resistojets (Zakirov *et al.*, 2000).

A research team from the University of Surrey, in England, has tested several catalysts for decomposition of nitrous oxide. The tests were performed without choking the nozzle, at ambient pressure. The reported temperature for initiation of decomposition was 250°C and the final decomposition temperature reached 1500°C. It was concluded that the catalytic decomposition process depends on the mass flow rate and also that the start of the catalytic decomposition can be faster with higher preheat energy supply (Zakirov *et al.*, 2001).

At Tsinghua University, in China, hot tests were made with nozzle chocking. Based on the research from the University of Surrey, the catalytic bed dimensions were reduced and the transient preheat time decreased from 15 min to 52 s (Zakirov and Li, 2004). Also in Tsinghua, a computational model of the operation of a nitrous oxide monopropellant thruster was developed (Zakirov and Zhang, 2008).

In Surrey and Tsinghua Universities, the focus of research was in the decomposition of nitrous oxide. In contrast, researchers at Stanford University evaluated the characteristic velocity efficiency and predicted the thrust of a nitrous oxide thruster, although without providing information on the thruster design. A gas generator was developed to operate at high mass flow rates of nitrous oxide ranging from 0.9 g/s to 2.3 g/s, thus obtaining characteristic velocities with high efficiency in steady state. The mixing of CH_4 and N_2O was also studied for initiation of the reaction (Lohner *et al.*, 2007; Scherson and Lohner, 2009).

In the University of Beihang, in China, the effect of the catalytic decomposition of nitrous oxide on specific impulse was studied in a 140 mN monopropellant thruster. Based on preliminary studies with a numerical model considering effects of turbulence, the thruster was designed and tested in vacuum conditions. The catalyst used was alumina impregnated with iridium; yielding decomposition starting around 523 K. The experiments have shown that the bed length influences the nitrous oxide catalytic decomposition with longer beds, favoring complete decomposition. Therefore, it was found that to achieve high specific impulse, the optimization of the catalyst bed is required to assure that the products reach a high temperature at the nozzle inlet. On the other hand, effects such as heat losses to the environment cooperate negatively to reaction temperature. A model for the consumption of nitrous oxide and its self-pressurizing conditions in the tank was also developed. It follows from experimental and simulated data that the tank pressure decreases during the feeding process of self-pressurized nitrous oxide. Large pressure drops in the final stages of the feeding process can interfere with the required mass flow (Cai *et al.*, 2011).

Therefore, this work describes an experimental study of a 2 N monopropellant thruster using nitrous oxide as propellant. The propulsive parameters are determined experimentally and the decomposition performance is evaluated. A glow-plug heater system and a rhodium oxide catalyst supported on alumina were used to heat and decompose the gaseous propellant.

NITROUS OXIDE CHARACTERISTICS

Nitrous oxide is a low cost non-toxic, non-flammable, non-explosive, non-harmful , and easy to purchase or manufacture propellant (Merrill, 2008; Junior, 2009). It remains stable under normal conditions and is compatible with materials with simple structure; it is storable as a compressed gas or a liquid in wide temperature ranges, theoretically limited by its triple point (-90.8°C). However its storage is recommended in temperatures between –34°C to 60°C. N_2O decomposes exothermically with aid of a catalyst or thermally, and can be stored in saturation conditions in about 50.8 bar at 20°C. A comparison of nitrous oxide with other monopropellants is shown in Table 1.

Nitrous oxide starts decomposition between 520°C to 850°C, forming mostly nitrogen (N_2) and oxygen (O_2), and releasing heat. Due to the high amount of energy needed to initiate the decomposition, catalysts are used to reduce the activation energy of the reaction, then yielding a lower decomposition temperature, around 250°C. However, the system must be preheated to this temperature to start the exothermic decomposition and attain a self-sustaining reaction. Reaching this condition, the adiabatic decomposition temperature is around 1640°C, producing a theoretical specific impulse of 206 s (Zakirov *et al.*, 2001).

NITROUS OXIDE DECOMPOSITION

The chemical mechanism of the gas phase decomposition of nitrous oxide starts with:

$$N_2O + M \rightarrow N_2 + O + M \quad (1)$$

where M is a third body (Karabeyoglu *et al.*, 2008). The O radical reacts with N_2O according to the two parallel reactions:

$$N_2O + O \rightarrow NO + NO \quad (2)$$

$$N_2O + O \rightarrow N_2 + O_2 \quad (3)$$

The O radical can be eliminated by a termolecular homogeneous recombination:

$$O + O + M \rightarrow O_2 + M \quad (4)$$

or by a heterogeneous recombination:

$$O + \text{wall} \rightarrow \frac{1}{2} O_2 + \text{wall} \quad (5)$$

Secondary reactions involved mainly in the formation of NO are:

$$NO + O + M \rightarrow NO_2 + M \quad (6)$$

$$N_2O + NO_2 \rightarrow N_2 + O_2 + NO \quad (7)$$

$$NO_2 + O \rightarrow NO + O_2 \quad (8)$$

$$NO + N_2O \rightarrow N_2 + NO_2\,. \quad (9)$$

The previous equations can be combined, yielding the following global reactions:

$$N_2O \rightarrow N_2 + \frac{1}{2} O_2 \quad ; \quad \Delta H = -1865\ \text{kJ/kg} \quad (10)$$

$$N_2O \rightarrow NO + \frac{1}{2} N_2 \quad ; \quad \Delta H = +19.17\ \text{kJ/kg}. \quad (11)$$

The reaction involving N_2/O_2 is exothermic and the NO/N_2 reaction is endothermic.

Catalytic reactions offer advantages since they favor the N_2/O_2 self-sustaining reaction and improve the reaction efficiency (Wilson *et al.*, 2012; Lohner *et al.*, 2007; Zakirov and Zhang, 2008). The catalytic decomposition of nitrous oxide starts

Table 1. Comparison of nitrous oxide with others monopropellants (Zakirov *et al.*, 2001).

Propellant	Nitrous Oxide	Hydrogen Peroxide	Hydrazine
Chemical formula	N_2O	H_2O_2	N_2H_4
Theoretical I_{sp} (s)	206	179	245
Storable	yes	yes	yes
Density (kg/m^3)	745 at 20°C and 52 bar	1347	1004
Vapor pressure (bar)	50.8 at 20°C	0.00345 at 20°C	0.0214 at 26,7°C
Storage temperature (°C)	-34 to 60	-7 to 38	9 to 60
Toxic	no	burns skin	very toxic
Flammable	no	no	yes

I_{sp} in vacuum conditions with nozzle expansion ratio 200; all propellants stored in liquid state; hydrogen peroxide 89 % m/m.

with the adsortion of a N_2O molecule in an active site on the catalyst surface. After that, the N_2 molecule is released while the oxygen atom remains adsorved on the active site. At last, the oxygen atom is released when it combines with another oxygen atom in a nearby active site, forming O_2 or by direct reaction with another N_2O molecule. These steps are described by:

$$N_2O + [] - M \rightarrow N_2O - M \quad (12)$$

$$N_2O - M \leftrightarrow N_2 + O - M \quad (13)$$

$$2O - M \leftrightarrow O_2 + [] - M \quad (14)$$

$$N_2O + O - M \rightarrow N_2 + O_2 + [] - M \quad (15)$$

In reactions (12-15) the O-M term refers to an oxygen radical and the N_2O-M term refers to a N_2O molecule adsorved on the catalyst surface. The []-M term corresponds to an empty site on the catalyst surface. During N_2O decomposition, the adsortion of oxygen radicals coming from the catalytic surface is often considered a limiting factor. For this reason, the decomposition rate of N_2O is intimately linked to the bind energy between the catalyst metal and the oxygen atom (Scherson and Lohner, 2009; Kapteijn *et al.*, 1996).

CATALYST FOR N_2O DECOMPOSITION

A catalyst is used to provide a rapid propellant decomposition, without changing the global reaction. The reactants bind chemically to the surface of the catalyst active sites, which are atoms or groups of atoms available for reaction. When the catalyst is solid, the active material is composed of metal or metal oxides which are dispersed over a high surface area material, increasing the contact area between reactants and active sites (Batta *et al.*, 1962; Vieira *et al.*, 2003). In the case of gaseous nitrous oxide, this is obtained by making it pass through a material with transition metals, such as Ir, Rh, Ru, Ni, Co, Cu or Zr.

There are several characteristics which affect the catalyst performance. The most important one is the decomposition limiting factor. The reaction can be controlled by different processes:

- Chemical kinetics, limited by the speed of N_2O adsorption on the catalyst, that is, the speed of combination of N_2O with a rhodium molecule in an active site and release of oxygen and nitrogen;
- Diffusion in the catalyst pores, where the reaction is limited by diffusion of reactants and products into and out of these microscopic pores; and
- Convection in the diffusion boundary layer around each pellet catalyst, which cannot carry the products and reactants from the surface.

Knowing which of these modes is important will determine how the catalyst bed should be sized (Figueiredo and Ribeiro, 1989).

Tests are needed also to determine the critical temperature range. It is necessary to know when the decomposition reaction changes from slow to fast and complete. The catalyst must be tested to determine the maximum temperature at which it can be used without deterioration of its properties. This should be checked since the active material (Rh, Ru, Ir) to be deposited on the support material surface also will be subjected to high temperatures. Placing these active elements at very high temperatures can cause their vaporization and removal of the support surface, destroying the catalyst (Figueiredo and Ribeiro, 1989).

Being commercially available and having a relatively low cost, supports based on alumina (Al_2O_3) are widely used for catalytic purposes in propulsion. The aluminum oxide is formed at room temperature in γ-Al_2O_3 phase and has an open structure consisting of amorphous aggregates of $Al_{12}O_{18}$. When the temperature increases (around 1200°C) aluminum oxide goes through different phases irreversibly to form α-Al_2O_3 which is the final stage of its internal structure. Although the stages of transition are not well understood, it is well known that for each transition, the available surface area per gram of oxide decreases significantly, causing a decrease in the number of active sites available for decomposition (Arai and Machida, 1996; Vafai, 2005).

Because of its ready availability and decomposition efficiency, a catalyst of rhodium (Rh_2O_3) oxide supported on alumina (Al_2O_3) was chosen to decompose nitrous oxide in a 2 N thruster prototype. Catalyst pellets (cylindrical extrudate) approximately 2 mm diameter and 3 mm length were used, containing 5% by volume of rhodium oxide. The pellets are shown in Fig. 1.

Figure 1. Rh_2O_3/Al_2O_3 catalyst.

THRUSTER DESIGN

The thruster design was based on a performance theoretical study carried out with aid of CEA2 NASA (2012) program. This program initially determines the chemical equilibrium composition and temperature in the chamber for constant pressure and enthalpy, through minimization of the Gibbs free energy. Then it solves the conservation equations for one-dimensional isentropic steady flow along the nozzle considering equilibrium or frozen conditions. As input parameters, the properties of pure nitrous oxide gas were considered at a temperature of 298.16 K, atmospheric pressure of 0.95 bar, 5 bar pressure in the chamber and assuming frozen flow in the nozzle. Data obtained with the NASA CEA2 code are shown in Tables 2 and 3. Table 3 confirms that the main decomposition products are O_2 and N_2. Table 4 shows some dimensions of the nozzle and the mass flow rate of propellant calculated to yield 2 N thrust.

Estimates of the length and diameter of the catalytic chamber were made based on the literature of monopropellant catalytic thrusters (Lohner *et al.*, 2007; Scherson and Lohner, 2009; Cai *et al.*, 2011). Thus, the reference values adopted for initial design of the catalyst chamber were 15 mm in diameter and 70 mm in length (D15L70). All thruster parts (flange injection, chamber, catalyst bed and nozzle) were manufactured in 316 stainless steel. Computational views of thruster (3D drawings) are shown in Fig. 2, whereas Fig. 3a depicts the disassembled thruster and Fig. 3b, thruster mounted in the thrust balance.

Table 2. Thruster theoretical parameters.

Thruster parameters	Thruster chamber	Nozzle throat	Nozzle exhaustion
P (bar)	5.0000	2.7262	0.9505
T (K)	1906.92	1658.40	1293.97
ρ *(kg/m³)*	0.92529	0.58010	0.25924
c_p *(kJ/kg/K)*	1.2403	1.2213	1.1842
γ	1.2961	1.3021	1.3146
Mach	0.000	1.000	1.758
A_e/A_g	-	1.000	1.4343
c^* *(m/s)*	-	1101.8	1101.8
C_F	-	0.7099	1.1076
I_{vac} *(m/s)*	-	1383.0	1520.9
I_{sp} *(m/s)*	-	782.2	1220.4

P = chamber pressure; T = chamber temperature; ρ = density of products; c_p = specific heat of the products; γ = ratio of specific heats of products; *Mach* = Mach number; A_e = exhaustion area; A_g = throat area; c^* = characteristic velocity; C_F = coefficient of thrust; I_{vac} = specific impulse in vacuum; I_{sp} = specific impulse with adapted nozzle.

Table 3. N_2O decomposition products.

Products	Molar fractions
NO	0.06370
O	0.00008
NO_2	0.00004
O_2	0.32988
N_2	0.66327

Table 4. Thruster parameters and dimensions.

Parameters	Values
A_g *(mm²)*	3.61
D_g *(mm)*	2.14
A_e *(mm²)*	5.17
D_e *(mm)*	2.56
$\dot{m}$ *(/s)*	1.63

TEST BENCH

A test bench was used to determine thrust, mass flow rate, temperatures and pressures in the supply line and in the thruster. A schematic of the test bench is showed in Fig. 4. It contains

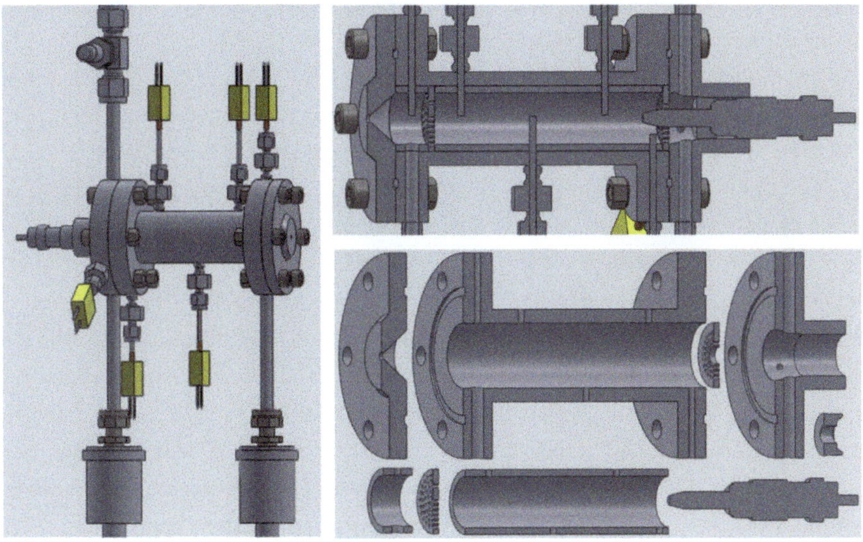

Figure 2. 3D drawings of the thruster with instrumentation and cut views.

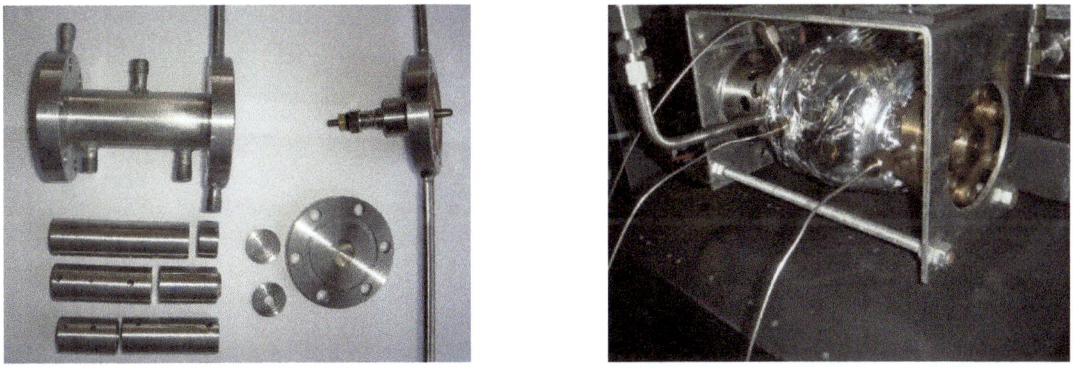

(a) (b)

Figure 3. Thruster photos.

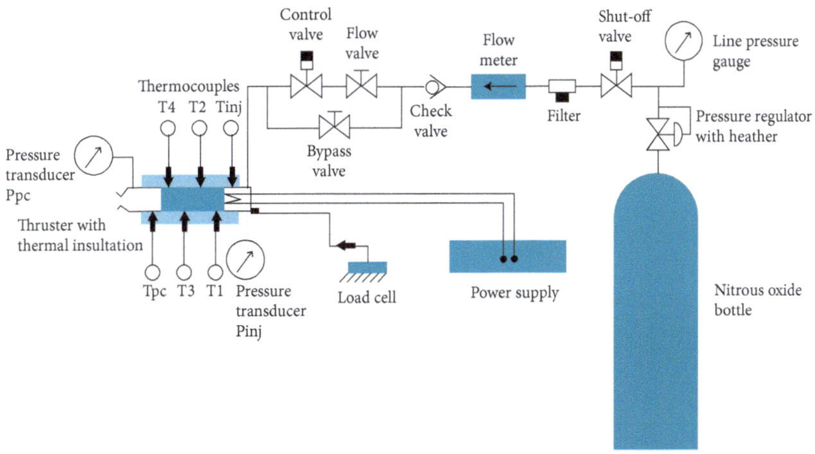

Figure 4. Test bench scheme.

a gas cylinder with nitrous oxide (99.99%), a heated pressure regulator, sintered filters in the gas inlet valves, manual lock valves, two-way solenoid valve with direct action, check valve, pressure transmitters, K-type thermocouples, HBM load cell with 5 N capacity (± 0.01 N), and a Sevenstar nitrous oxidizer mass flow meter. Preheating of the nitrous oxide to start decomposition was performed by a Duratherm Bosch glow-plug placed at the entrance of the catalyst bed, as depicted in Fig. 2. The injection temperature, i.e., the gaseous nitrous oxide temperature at the entrance of the catalyst bed, was adjusted by a temperature controller (Novus 480I) which monitored the glow-plug heating.

A data control system using a National Instruments SC-2345 rack, TCC02 modules (for signals in mV), FT01 modules (with channels 0-10 V), Compact Daq USB module (extra channels of analog inputs and communication module) and a PCI-6221 data acquisition board were used for observations of the experiments variables in real time. A Labview control software monitored variables of pressure, mass flow rate, thrust, temperature and operation of the glow-plug. For this purpose, analog and digital input/output were used, providing data on a computer in graphical form and saving in its database.

RESULTS AND DISCUSSION

A pressure of 7.5 bar at the exit of the regulator valve was adopted to provide the total pressure drop on the various devices installed before the thruster. Once the temperature (thermocouple T_{inj}) measured in the glow-plug heater section reached 220°C the injection of nitrous oxide was performed. This is evidenced in Fig. 5 which shows that a pulse with a mass flow rate (Q) of 1.6 g/s increased rapidly the catalytic bed temperature, thus indicating the occurrence of catalytic decomposition.

Locations of T_{inj}, T_1, T_2, T_3 and T_4 thermocouples along the catalytic chamber can be seen in Figs. 2 and 4. T_{pc} is the measured temperature after the catalytic bed (post-chamber). Prior to testing, pre-heating was kept for several minutes for thermal stabilization of the catalyst bed.

After the preheating period, the glow plug temperature controller was turned off. The mass flow rate was gradually increased to the nominal flow rate and heating was maintained only by the decomposition of N_2O flow. Under these conditions, it was possible to apply multiple sequences of pulses with different duty cycles, i.e., periods with mass flow on and off. Very long shots could compromise the thruster and thermocouples, due to

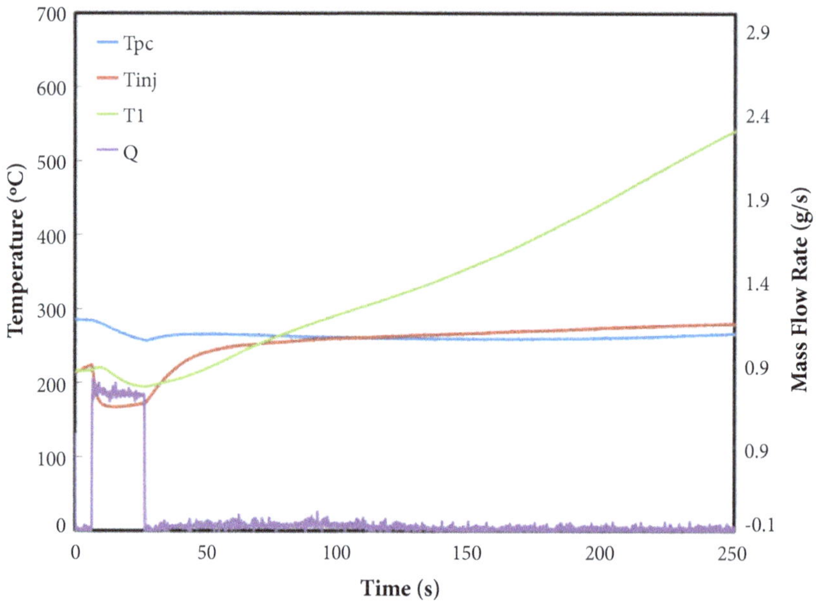

Figure 5. Initial heating using N_2O.

the high temperatures that could be reached in the bed. During intervals, there occurred a temperature uniformization along the thruster, especially in the catalytic bed.

Figure 6 shows curves of pressures, temperatures and mass flow rates measured in the thrust balance. A sequence of 5 pulses with 25 s on and 10 s off is shown, followed by sequences of 3 pulses with 15 s on and 10 s off. It is observed that the injection pressures (P_{inj}) and post-chamber (P_{pc}) reached 5.29 bar and 5.15 bar, respectively. The temperature at the end of the catalytic bed (T_4) reached 1400°C with an average mass flow rate (Q) of 1.65 g/s. The thrust (F) measured on the load cell reached 1.93 N, close to the design value of 2 N.

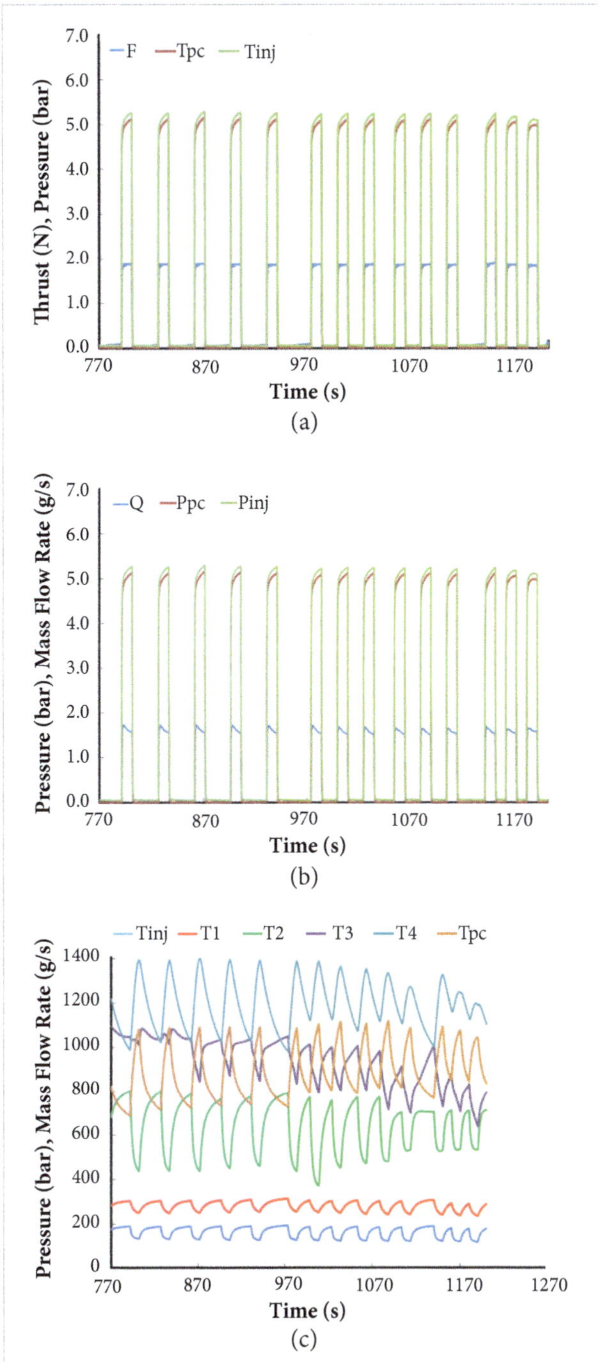

Figure 6. D15L70 thruster duty cycle testing.

Figure 7 shows the gas temperature evolution inside the catalytic chamber of the thruster. A MATLAB algorithm using a cubic interpolation method was developed to determine the temperature distributions during the thermal-catalytic decomposition process.

Taking as reference the pulse in the interval of 860 s to 900 s in Fig. 6c, it was observed that the highest temperature (1400°C) was measured by the thermocouple located at 62 mm from the injection, as shown in Figs. 2 and 4. This corresponds to the last catalytic bed thermocouple (T_4). The maximum temperature reached at the end of the catalytic chamber indicates a bed with optimum length. The hot expanding gases from decomposition pass through the post-chamber and follow to the nozzle for generating thrust.

Pulses present similar profiles for several sequences and in different periods. However, the fatigue of the catalyst can be observed in the range of 1136 s to 1200 s in Fig. 6, where pulses are deliberately shown with a decrease in catalyst efficiency. This is due to the pellet contraction and loss of rhodium oxide. The contraction of the pellets is caused by the low calcination temperatures and phase change of alumina, as was previously described by Arai and Machida (1996). The rhodium oxide loss is due to the high temperatures reached in the catalytic bed. The decrease in pellet volume also creates preferential paths for N_2O flow which pass through the catalytic bed without decomposition. Besides, it causes a decrease in the catalyst specific area, reducing the area of active sites available for catalytic reaction. If the surface available for decomposition is internal to the pellets, with the alumina retraction the active sites become inaccessible for N_2O, preventing its decomposition. However, there was no clustering of catalyst grains or weight loss.

There is also a loss of efficiency concerning the reduction of the adsorption capacity of the molecules of N_2O, as seen in the last three pulses in Figs. 6, 8 and 9. This may be explained by the adsortion of water and oxygen which reduces the number of active sites available for N_2O decomposition (Esclapez, 2011). This factor, jointly with the retraction of alumina and heat losses, can decrease the reaction temperature in

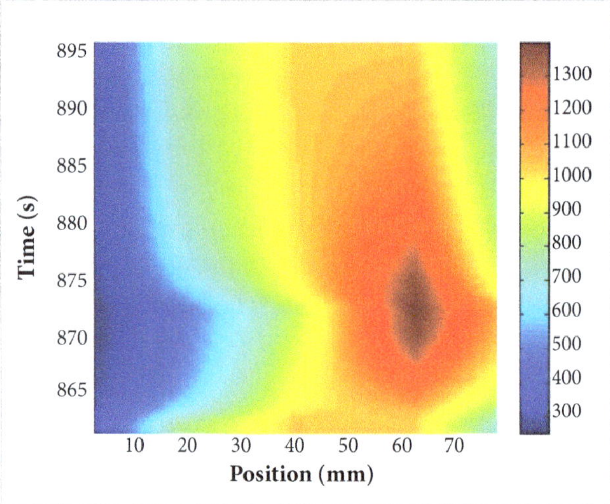

Figure 7. Gas temperature (°C) evolution along the catalytic bed and post-chamber section.

the bed. It should be noted that using shorter pulse lengths would provide lower temperatures and a slower degradation of the catalyst.

Figure 8 shows the specific impulse, $Isp = F/\dot{m}$, and the characteristic velocity, $c^* = A_t P_c /\dot{m}$, obtained experimentally. The pulses shown in Fig. 6 were taken as reference, since they were obtained with the thruster in full operation at the start of degradation of the catalyst.

The average specific impulse obtained experimentally was about 1120 m/s, which is close to the theoretical value, 1220.4 m/s. Therefore, the average specific impulse efficiency, $\eta Isp = I_{sp,exp}/I_{sp,theo}$, was around 91.7 %. The characteristic velocity efficiency, $\eta_{c*} = c^*_{\text{exp}} / c^*_{\text{theo}}$, is shown in Fig. 9. The characteristic velocity value obtained in the experiments was about 970 m/s, whereas the average theoretical value is about 1102 m/s, giving an average efficiency of about 88%. Considering the characteristic velocity efficiency obtained experimentally, it is observed that the catalytic decomposition of nitrous oxide in the bed was not complete.

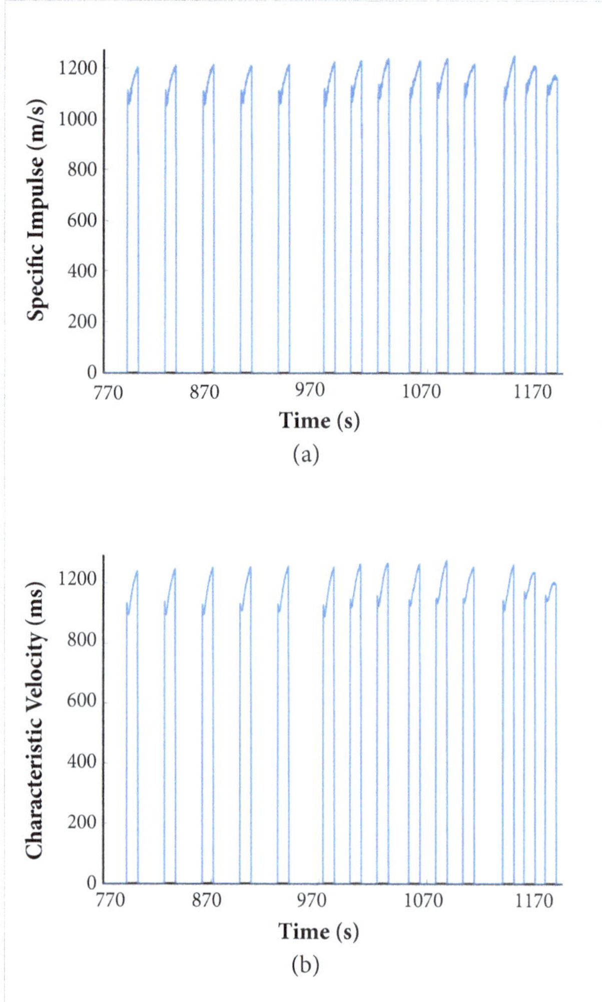

Figure 8. Specific impulse (a) and characteristic velocitie(b).

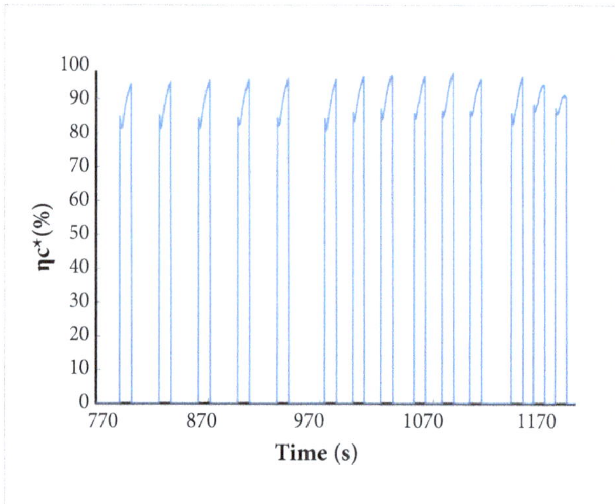

Figure 9. Characteristic velocity efficiency.

CONCLUSIONS

An experimental investigation of a monopropellant 2 N thermal-catalytic prototype thruster using gaseous nitrous oxide was presented. The propellant was decomposed by a catalyst of rhodium oxide supported on alumina, with pre-heating of both catalyst and propellant by a glow-plug system.

Relatively high decomposition efficiency (~ 88 %) was observed prior to catalyst degradation. Changes in the catalytic support preparation are required to reduce the volume loss of the support under the test conditions and to improve catalyst performance for longer firing times. Other factors such as optimization of catalytic bed dimensions, improvements in the injector design and better thermal insulation may increase the thruster's efficiency.

ACKNOWLEDGEMENTS

The authors acknowledge the Brazilian *Conselho Nacional de Desenvolvimento Científico e Tecnológico* (CNPq) for providing a doctoral scholarship to the first author and a research grant to the second one. The catalyst preparation was performed with the kind help of Dr. Ricardo Vieira from Brazilian *Instituto Nacional de Pesquisas Espaciais* (INPE).

REFERENCES

Amaral, L., 1995, "Química", Loyola, São Paulo, 23-24.

Arai, H. and Machida, M., 1996, "Thermal stabilization of catalyst supports and their application to High Temperature catalytic combustion". Applied Catalysis A: General, Vol. 138, No. 2, pp. 161-176. doi: 10.1016/0926-860X(95)00294-4.

Batta, I., Solymost, F. and Szabo, Z., 1962, "Decomposition of nitrous oxide on some doped cupric oxide catalysts". Journal of Catalysis, Vol. 1, No. 2, pp. 103-112. doi: 10.1016/0021-9517(62)90014-3.

Brown, C.D., 1995, "Spacecraft Propulsion". Ohio: Series Editor-in-Chief. AAIA Education Series. Air Force Institute Technology, pp. 224.

Cai, G., Sun, W., Fang, J., Li, M., Cong,Y. and Yang, Z., 2011, "Design and performance characterization of a sub-Newton N_2O monopropellant Thruster". Aerospace Science and Technology, Vol. 23, No. 1, pp. 439-451. doi: 10.1016/j.ast.2011.10.003.

Esclapez, S.P., 2011, "N_2O Decomposition rhodium/ceria catalysts: From principles to pratical application", PhD Thesis, Inorganic Chemistry Department, Alicante University, Spain, pp. 201.

Figueiredo, J.L. and Ribeiro, F.R., 1989, "Catálise Heterogênea". Calouste Gulbenkian Foundation, Porto, Portugal, pp. 353.

Junior, J.A., 2009, "Desenvolvimento de um Propulsor Catalítico para Satélites usando Óxido Nitroso como Propelente", Master Thesis, Instituto Nacional de Pesquisas Espaciais - INPE, Cachoeira Paulista, Brazil, pp. 168.

Humble, R.W., Henry, G.N. and Larson, W.J., 1995, "Space Propulsion Analysis and Designer". MacGraw-Hill Companies, Inc., Primis Custom Publishing.

Kapteijn, F., Mirasol, J.R. and Moulijn, J.A., 1996, "Review: Heterogeneous catalytic decomposition of nitrous oxide". Applied Catalysis B: Environmental 9, pp. 25-64. doi: 10.1016/0926-3373(96)90072-7.

Karabeyoglu, A., Dyer, J., Stevens, J. and Cantwell, B., 2008, "Modeling of N_2O Decomposition Events". American Institute of Aeronautics and Astronautics, 44th AIAA/ASME/SAE/ASEE Joint Propulsion Conference & Exhibit, Hartford.

Lohner, K., Dyer, J., Doran, E., Dunn, Z., Krieger, B., Decker, V., Wooley, E., Sadhwani, A., Cantwell, B. and Kenny, T., 2007, "Design and Development of a Sub-Scale Nitrous Oxide Monopropellant Gas Generator", American Institute of Aeronautics and Astronautics, Stanford University, Stanford, USA.

Makled, A.E. and Belal, H., 2009, "Modeling of Hydrazine Decomposition for Monopropellant Thrusters". 13th International Conference on Aerospace Sciences & Aviation Technology, ASAT - 13, Military Technical College, Kobry Elkobbah, Cairo, Egypt.

Merril, C., 2008, "Nitrous Oxide Explosive Hazards", Air Force Research Laboratory, Edwards AFB, CA.

NASA, "Chemical Equilibrium with Applications (CEA)". Cleveland, OH: Glenn Research. Center NASA, Retrieved on March 5, 2012, from http://www.grc.nasa.gov/WWW/CEAWeb/ceaguiDownloadwin. htm.

Neto, T.G.S., 2011, "Handout catalysis". Instituto Nacional de Pesquisas Espaciais, Curso de Engenharia e Tecnologias Espaciais.

Plumlee, D. and Steciak, J., 2004, "Development of a monopropellant micro-nozzle and Ion Mobility spectrometer in LTCC", In: Ceramic Interconnect Technology Workshop, Denver, CO.

Scherson, Y. and Lohner, K., 2009, "A Monopropellant Gas Generator Based on N_2O Decomposition for "Green" Propulsion and Power Applications". American Institute of Aeronautics and Astronautics, 45th AIAA/ASME/SAE/ASEE Joint Propulsion Conference & Exhibit, Denver, Colorado, EUA.

Sanscrainte, W., 1961, "Hydrogen Peroxide Attitude Control Systems". Planetary and Space Science, Vol. 4, pp. 184-193. doi: 10.1016/0032-0633(61)90131-3.

Stark, J.P.W.J. and Swinerd, G.G., 2003, "Mission Analysis", In: Fortescue, P., Stark, J. and Swinerd, G., (Eds.), Spacecraft Systems Engineering, 3rd ed., John Wiley & Sons, pp 111-167.

Sutton, G.P., 2001, "Rocket propulsion elements: an introduction to the engineering of rockets". 7. ed. John Wiley & Sons, pp. 751.

Vafai, K., 2005, "Handbook of Porous Media". 2nd Edition. Taylor & Francis Group, LLC, London, UK, pp. 742.

Vieira, R., Cuong, P., Keller, N. and Ledoux, M.J., "Novos Materiais à base de Nanofibras de Carbono como Suporte de Catalisador na Decomposição da Hidrazina". Química Nova, Vol. 26, No. 5, pp. 665-669. doi: 10.1590/S0100-40422003000500008.

Wilson, M.D., Eilers, S.D. and Whitmore, S.A., 2012, "Catalytic Decomposition of Nitrous Oxide Monopropellant for Hybrid Motor Re-Ignition". American Institute of Aeronautics and Astronautics, 48th AIAA/ASME/SAE/ASEE Joint Propulsion Conference & Exhibit, Atlanta, Georgia, USA.

Zakirov, A.V. and Zhang, H., 2008, "A model for the operation of nitrous oxide monopropellant". Aerospace Science and Technology, Vol. 12, No. 4, pp.318-323. doi: 10.1016/j.ast.2007.08.003.

Zakirov, V., Sweeting, M., Goeman, V. and Lawrence, T., 2000, "Surrey research on nitrous oxide catalytic decomposition for space applications". 14th AIAAUSU Conference on Small Satellites, pp. 1–9.

Zakirov, V. and Li, L., 2004, "Small Satellite Propulsion Challenges". European Conference for Aerospace Sciences (EUCASS), Tsinghua University, P.R. China.

Zakirov, V., Sweeting, M., Lawrence, T. and Sellers, J., 2001, "Nitrous oxide as rocket propellant". Acta Astronautica, Vol. 48, No. 5-12, pp. 353-362. doi: 10.1016/S0094-5765(01)00047-9.

Reliability Analysis for Aviation Airline Network Based on Complex Network

Dong Bing[1,2]

ABSTRACT: In order to improve the reliability of aviation airline network, this paper presents an empirical analysis on the airline network structure of an aviation company in China from the perspective of complex network, and the calculation result of the statistical features and degree distribution of the network, proves that the network is a small-world network and a scale-free network. Four indicators, i.e. degree, closeness, vertex betweenness and flow betweenness, are utilized for aviation network centralization so as to distinguish the most appropriate method. The influence of nodes in local network is to be measured through the indicators. The results show that vertex betweenness can achieve the best aviation network centralization effect. Specifically, the centrality degree reaches 95.87%. On this basis, the network reliability is analyzed to discover that when two nodes with maximum degree or maximum betweenness are removed, the network performance is reduced by a half. Eventually, countermeasures are proposed for further improvement according to the results. In other words, complex network method is feasible used to analyze the topological structure and statistical features of aviation network. Based on this, a study is conduced to the network reliability and suggestions are proposed for optimizing the aviation network.

KEYWORDS: Complex network, Aviation network, Scale-free property, Reliability.

INTRODUCTION

Aviation network refers to an airline system constituted by airlines connected in a certain way in a district, serving as a basis for the production and development of the airline company. In the studies on aviation network with network research methods, it is found that the aviation network has relevant statistical features (Guimera and Amaral, 2004; Guimera *et al.*, 2005; Barrat *et al.*, 2004) of "small-world network". However, most studies (Barrat *et al.*, 2005) focus on the analysis on physical statistical features of aviation network structure and the evolution of overall topological structure, while only a few studies are made on the analysis of route network with social network methods (Porta *et al.*, 2006). Since there are significant differences between aviation network nodes, it is particularly necessary to conduct comparative analyses on relevant nodes and studies on centrality of aviation network. The concept of network centrality can be traced back to the idea of applied statistics in the 19th Century (Gaertler and Wagner, 2001). Typically, different centrality indicators are required for centralization for different types of networks, and the multiple centrality study method needs to be applied in combination with parameters. In China, some scholars adopt other theories and methods to study the centrality of aviation network (Dang and Li, 2011). For instance, some scholars use rank-size model to measure the air transport concentration degree so as to assess the position of hub airports; some scholars mainly adopt the dominant flow method supplemented by squared Euclidean distance method and distance-based cluster method to analyze the level and change of major cities in China in domestic passenger aviation network, based on the air passenger statistical data; moreover, some scholars (Porta *et al.*, 2006) employ the Geographic Information System (GIS) method to study the spatial pattern

1.Southwest Jiaotong University – Chengdu/Sichuan – China 2.Civil Aviation Flight University of China – Guanghan – Deyang/Sichuan – China
Author for correspondence: Dong Bing | School of Traffic and Transportation | Southwest Jiaotong University | 610031 Chengdu/Sichuan – China | Email: dbcafuc@126.com

of the domestic aviation network airport system on the basis of air flow data.

With the development of the civil aviation industry in China, the air transport network keeps booming in scale. Nevertheless, it still suffers imperfection, low reliability and low operation efficiency. On the airline company's side, while planning aviation network, network planners basically conduct decision analysis according to experience. They only take the demand for a single airline as the primary indicator for assessing the necessity of launching an airline. Besides, they often select airlines that are basically the same in route of airlines benefiting other companies while neglecting the network reliability and its overall synergistic effect. The safety and reliability of aviation network exert on an important impact on market competitiveness and economic benefits of an airline company. Therefore, the aviation network should be planned in a systematical manner to improve its overall synergistic effect. A complex network method is proposed to analyze the topological structure and statistical features of aviation network by taking China Southern Airlines (CSA) as an example. Based on this, a study is conduced to the network reliability and suggestions are proposed for optimizing the aviation network of CSA.

COMPLEX NETWORK PROCESSING METHODS AND RELATED RESEARCH

As a small world model and a scale-free network model were proposed in the end of the 20th century, the complex network gradually became the research hotspot in different discipline. In order to facilitate the study of complex network effectively, all kinds of research software are introduced, such as Pajek, Ucinet, NetworkX and NetMiner 3.

In this paper, Ucinet is used for airline network. Ucinet is a social network analysis program developed by Steve Borgatti, Martin Everett and Lin Freeman. The program is distributed by analytic technologies. The software Ucinet involves, in the network analysis, programs such as community discovery and region analysis, ego network analysis and the hole structure analysis and so on. It also contains a large number of analysis programs, such as cluster analysis, multidimensional scaling, singular value decomposition, factor analysis and correspondence analysis, the role and status analysis, including structure, and the role and regular equivalence.

In this paper, we take airline passenger flow data as samples, the city for the network nodes, routes between cities as the network edge, aviation passenger flow between cities as the mapping relationship between node and nodes in the network structure and construction of air traffic network, as shown in Fig. 1:

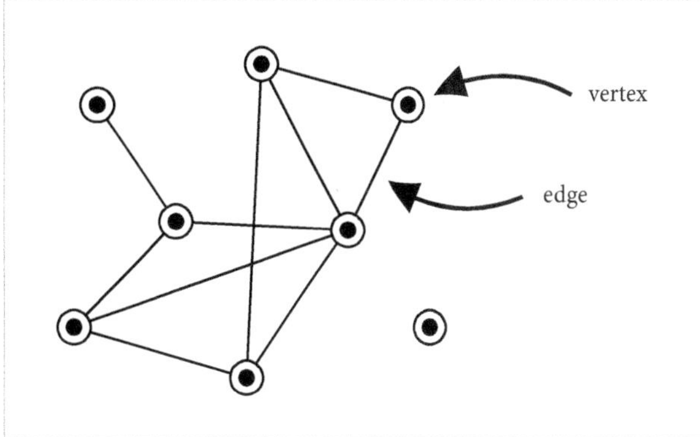

Figure 1. Network structure and construction of air traffic network.

Based on Ucinet software, two kinds of simulation systems were introduced (Hongguang and Liping, 2012), the deliberate targeting system and random interference system are designed, and some simulation experiments are done. The structure diagrams of air passenger flow network are plotted with the software (Dang and Li, 2010) and analyzed from the perspective of structural characteristics, degree of distribution and network centrality. However, traditional research methods fail to thoroughly identify the complexity of aviation network spatial relations between airports in a proper way. Centrality tests for network nodes are important means for judging the importance of nodes in the network, adjusting the aviation layout, and optimizing resource allocation, which, in particular, is greatly significant for safety of aviation network.

EMPIRICAL DATA ACQUISITION AND PROCESSING

In this paper, the domestic and international flight data from CSA´s data centre, in 2010, are taken as samples. Usually, cargo flights are arranged at night and the characteristics of transport cargo flights are different from passenger flights. In this article, cargo flights will not be considered. According to the model, let the airport as network node, the direct airline as network edge, and the number of navigable flights among airports as weight of edge, which constituted a weighted aviation network of CSA. Furthermore, adjacency matrix $(K_{ij})_{n \times n}$ (n refers to the number of nodes) is used in order to represent the aviation network, K_{ij}, in the matrix, refers to the number of flights from Airport i to Airport j. Due to data limitation, the network in this paper is an undirected one. That is, out-degree and in-degree are not involved.

STATISTICAL FEATURE AND THE DEGREE OF DISTRIBUTION

According to the statistics, there are 187 nodes and 1,245 edges. It is thus evident that the aviation network is concentrated and its structure is relatively complex. To show network hierarchical relationship, a backbone aviation network of CSA is built. The node weight threshold is taken as 500 in the weighted network. According to the statistics, there are 99 nodes and 369 edges.

In the complex network theory (Newman, 2003), the indicators reflecting the statistical features of network structure are mainly degree of nodes, average degree, average path length, density, clustering coefficient, etc.

The relation matrix (a_{ij}) is built for the aviation network, in which a_{ij} indicates the relation of flight numbers between city i and city j: if there are any flights between city i and city j, $a_{ij} = 1$; otherwise, $a_{ij} = 0$. Upon calculation, the statistical feature indicators of 2010 aviation network of CSA are listed in Table 1.

Table 1. Statistical features of 2010, CSA aviation network.

	Project	Statistical Value
1	Number of Nodes	187
2	Degree of Node	1250
3	Average Degree	6.684
4	Average Path Length	2.558
5	Network Density	0.0358
6	Clustering Coefficient	0.592

The degree of a node in an aviation network refers to the number of airports having direct flights with this airport (node); greater degree of a node means greater importance to some extent. Table 2 shows CSA ten airports with top degrees in 2010, among which Guangzhou ranks the first of airport degree as an airline hub of CSA. The average degree of a network refers to the average value of all degrees of nodes in the network. According to Table 1, the average degree is 6.684, indicating that, in average, each city node is connected to other 6.684 city nodes.

Table 2. Top 10 degree values of CSA airports, in 2010.

	Airport	degree
1	Guangzhou	106
2	Shenzhen	48
3	Beijing Capital	48
4	Urumqi	46
5	Changsha	42
6	Dalian	39
7	Zhengzhou	38
8	Shenyang	37
9	Pudong	34
10	Wuhan	32

Generally, the distance between two nodes is defined as the number of edges of the shortest path between two nodes, and the average path length of a network is the average value of the distances between all node pairs. In an aviation network, the distance describes the path from one airport to another one using the minimum transit times; the shorter the distance is, the less transit times are required. Meanwhile, the average path length stands for the depth of the air transport, which is a property of the transport shortcut in the integral network; the shorter the average path length is, the less transit times are required between any two airports, bringing more convenience for the passengers. In 2010, the average path length of CSA aviation network is 2.558, which means only 1.558 transit times are required for transporting from one airport to another, favorably meeting the air transport demand.

An undirected network density is defined as the ratio of actual connection number to the maximum possible connection numbers in the figure. In an aviation network, it describes the ratio of actual number of opened segments to the number of all possible segments. The density indicates the closeness of the air connections among all cities in the network. The value is taken between 0 and 1; if the value is closer to 1, the network structure is more perfect and the connections of the air transport are closer. In 2010, the density of aviation network is 0.0358, which is relatively small, indicating that the connections are not very strong among CSA airports.

The clustering coefficient of a node in the network stands for the ratio of the actual connection number to the maximum possible connection edges between this node and its adjacent nodes. In an aviation network, the clustering coefficient of a node indicates the average cluster degree of the local network comprised of the airport and its adjacent airports. Higher clustering coefficient means greater cluster degree of the local network, and smaller impact of this node on the adjacent airports; on the contrary, lower value means more dependence of the adjacent airports on this node. Guangzhou Airport has a clustering coefficient of 0.06, which is the smallest of all, indicating that the adjacent airports are highly dependent on Guangzhou Airport, and large numbers of flights will be affected if failure occurs in Guangzhou Airport. The clustering coefficient of the integral network is the average value of the clustering coefficients of all city nodes. As shown in Table 1, CSA aviation network, in 2010, has relatively small average shortest path length and large clustering coefficient; hence this aviation network belongs to a small-world network.

Betweenness is generally defined as the capability of a node for controlling the connection of other node pairs, i.e., the effect of a node acting as a bridge between other node pairs. Higher betweenness of a node indicates stronger effect of the node as a bridge, and more important role in the network. See Table 3 for the betweennesses of top 10 airports of CSA Aviation Network in 2010, based on which Guangzhou Airport, as a CSA hub, has the maximum degree and betweenness, taking up the most important position in the network. Secondly, Urumqi, Beijing Capital and Shenzhen Airports also play prominent roles as bridges. Specifically, Pudong Airport ranks third in the betweenness value; although its degree value is not particularly high, it is still a significant transit in the network. If a fail occurs in important transit airports mentioned above, the connections will be greatly affected between other nodes.

Table 3. Betweenness values of top 10 airports of CSA aviation network, in 2010.

	Airport	Betweenness Values
1	Guangzhou	51.414
2	Urumqi	25.782
3	Pudong	9.61
4	Beijing Capital	8.418
5	Shenzhen	6.21
6	Harbin	5.828
7	Shenyang	5.68
8	Kunming	4.811
9	Dalian	4.571
10	Changsha	4.224

The degree distribution of the nodes in the network can be described in the power-law distribution function;

$$p(k)=k^{-\alpha} \qquad (1)$$

The power-law distribution coefficient α shows the degree distribution characteristics of a network(Newman, 2003). The distribution $p(k)$ is the probability of a randomly selected node of degree k.

Power-law distribution is also known as scale-free distribution; a scale-free network is a network whose degree distribution follows a power law. In the log-log plot, a straight line with negative slope can be obtained by conducting a linear fitting for the degree distribution, and the absolute value of slope is the power exponent. If the absolute value is relatively small, this network is scale-free. Correlation coefficient indicates the fitting degree of the curve; higher correlation coefficient means the curve is more favorably fitted, explaining actual problems more adequately. The degree distribution of undirected network is discussed here. By applying a double-segment fitting in the log-log plot, we obtained the power exponent and correlation coefficient variance of the degree distribution (Table 4), and degree distribution (Fig.2) of CSA aviation network in, 2010.

Based on Table 3 and Fig. 2 , at the significance level of α=0.0001, Segment 1 and Segment 2 both have excellent fitting, with correlation coefficient exceeding 0.96. Since Segment 1 and Segment 2 both have relatively small power exponents, the degree distribution of CSA aviation network is subject to double-segment power-law distribution, and thus this network is scale-free. Hence, the nodes in the aviation network are heterogeneous, certain nodes (hubs) have large numbers of connections, playing the dominant role in the network, while other large number of nodes only have small numbers of connections and are located on the edge of the network, according to the Matthew Effect.

Table 4. Feature values of degree distribution of domestic flight network.

	Project		Statistical Value
1	Number of Nodes(N)		187
2	Degree of Nodes(k)		1250
3	Fitting of Segment 1	Power Exponent(|α|)	0.55094
		Correlation Coefficient(|ACC|)	0.55094
4	Fitting of Segment 2	Power Exponent(|α|)	1.53157
		Correlation Coefficient(|ACC|)	0.96051

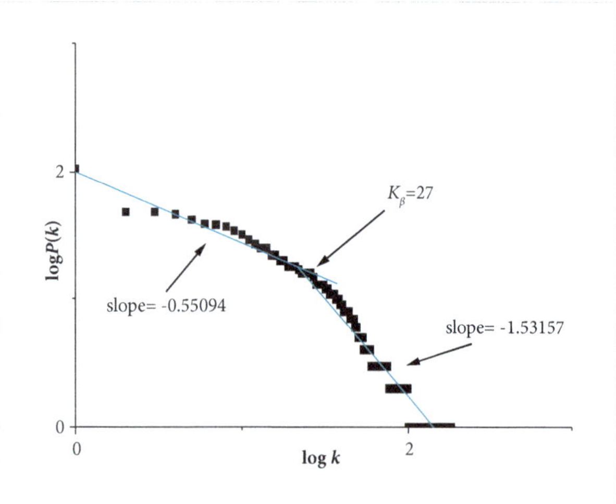

Figure 2. Node degree distribution of CSA aviation network, in 2010.

COMPARISON OF CENTRALITY DEGREE UNDER DIFFERENT INDICATORS

Relevant network centrality indicators serve as the basis for measuring the centrality degree of the network. On the assumption that centrality indicators have been defined in Network C_A with n nodes, the centrality degree of the network will be defined as follows (Costenbader and Valente, 2003):

$$C_A^g = \frac{\sum_{x \in W} (C_A^* - C_A(x))}{(n-1)\max_{x \in W}(C_A^* - C_A(x))} \tag{2}$$

Where, W refers to the whole network, $C_A^* = \max_{x \in W} C_A(x)$ represents the centrality value of the node with the largest centrality degree.

According to the equation, if the centrality of all nodes is the same, namely the network has no center, then $C_A^* = 0$. In case the centrality degree of only one node is of 1 and that of other nodes is 0, C_A^g will be greater, and the handful of center nodes will be more prominent, which shows that the larger the centrality difference between network nodes, the higher the centrality indicators of the handful of center nodes. Thus, the accuracy of center nodes will be higher, and so will the centrality degree.

Different network centrality (Friedkin, 1991; Newman, 2005) degrees can be figured out by substituting degree, closeness, vertex betweenness and flow betweenness into the following equations respectively.

$$\begin{aligned}
C_D^g &= \frac{\sum_{x \in W} [n-1-C_D(x)]}{(n-1)(n-2)} \\
C_C^g &= \frac{\sum_{x \in W} [n-1-C_C(x)]}{(n-1)(n-2)/(2n-3)} \\
C_B^g &= \frac{\sum_{x \in W} [(n-1)(n-2)-2C_B(x)]}{(n-1)^2(n-2)} \\
C_{FB}^g &= \frac{\sum_{x \in W} [n(n-1)-4C_{FB}(x)]}{n(n-1)^2}
\end{aligned} \tag{3}$$

Where $C_D(x)$, $C_C(x)$, $C_B(x)$, and $C_{FB}(x)$ represents the degree, closeness, vertex betweenness and flow betweenness indicator values of network nodes, respectively. In contrast, the degree indicator is more suitable for measuring the influence of nodes in local network. In global scope, however, the closeness indicator needs to be referenced. The two indicators are only applicable to static network analysis, while the betweenness indicator is more suitable for analysis of dynamic network. When the degree, closeness, vertex betweenness and flow betweenness indicators are used for aviation network centralization based on the formula above, the centrality degree under the different indicators is shown in Table 3.

Obviously, the centrality degree varies with the selected centrality indicators for network centralization. The centrality degree of closeness is relatively low, which indicates that closeness is not suitable for aviation network centralization. Moreover, the top 10 nodes in the aviation network are selected to catch the distribution of indicator values for each node, as shown in Fig. 2. Therefore, degree and closeness fail to obviously distinguish nodes, while the difference of distributions of betweenness is relatively large. Through calculating the centrality degree of each indicator, and comparing the cumulative distribution of centrality data of top 10 nodes, vertex betweenness is the most suitable indicator for aviation network centralization.

NETWORK RELIABILITY ANALYSIS

There are a lot of network attack means in reality (Barrat *et al.*, 2005; Holme *et al.*, 2002; Kai-Quan *et al.*, 2012; Li and Cai, 2012), where the random attack and the hostile attack are relatively representative, and the hostile attack is very destructive. As it was demonstrated before, the aviation network of CSA is a scale-free one, with associate scale-free properties, such as "Stable and Fragile" in attacks, which means it has a very strong flexibility in random attacks or unexpected malfunctions while it is very fragile in hostile attacks (Xiaohuan Wu *et al.*, 2013).

The most important indicators to characterize the network topologic structure are the average path length and the clustering coefficient. The average path length in the aviation network represents air transport depth, the clustering coefficient represents air transport width, and the network efficiency represents overall coordination of the network. With a smaller average path length of the network, a bigger clustering coefficient, and higher network efficiency, it is indicated that the network has a better performance and a stronger fault-tolerant capability. Therefore, this paper will have the reliability analysis on the current network of CSA with the three indicators, and the calculation formulas are as follows:

$$L = \frac{2}{N(N-1)} \sum_{i \geq j} d_{ij} \tag{4}$$

In Eq.(4) N stands for network node number, d_{ij} stands for the shortest distance from Node i to Node j.

$$C_i = 2E_i/(k_i(k_i-1)) \tag{5}$$

In Eq. (5) k_i stands for the number of edges directly connecting with airport i, and E_i stands for the number of existing connecting edges between airports in number of k_i.

$$E = \frac{1}{N(N-1)} \sum_{i,j \in N, i \neq j} \frac{1}{d_{ij}} \tag{6}$$

In Eq.(6) with $0 \leq E \leq 1$. When $E=1$, the network is completely connected; and when $E=0$, all nodes in the network are isolated.

Airport nodes in the aviation network of CSA are sorted according to the degree and the betweenness value in descending order, and then the airport based on a relatively big degree and that, based on a relatively big betweenness, are removed orderly, the average path length, the clustering coefficient, and the network efficiency of the network are calculated respectively, and changes of the average path length, the clustering coefficient and the network efficiency corresponding to the decrease of the airport number are counted and compared, then the contents indicated in Fig.4, Fig. 5 and Table 5 are respectively obtained.

As indicated by Figs. 3 and 4, the fluctuation of average path length is relatively strong when several nodes are removed, and the fluctuation of average path length based on the remove policy of degree priority is stronger than the one based on the policy of

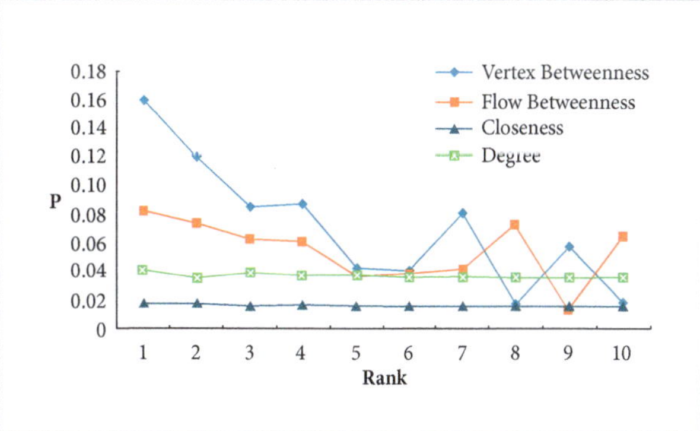

Figure 3. The cumulative distribution of centralized data under four indicators of the top 10 nodes.

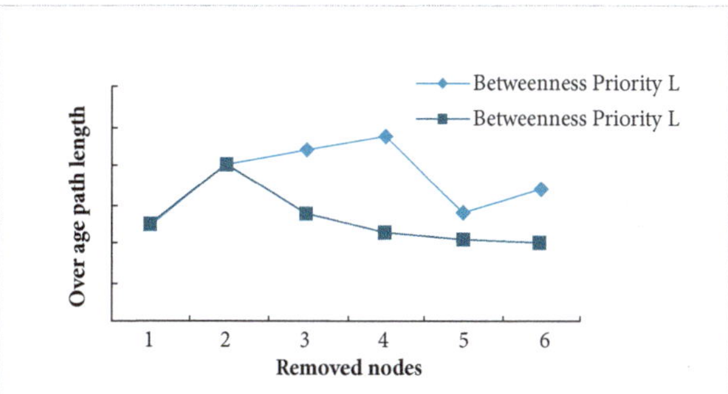

Figure 4. Comparison of average path length changes.

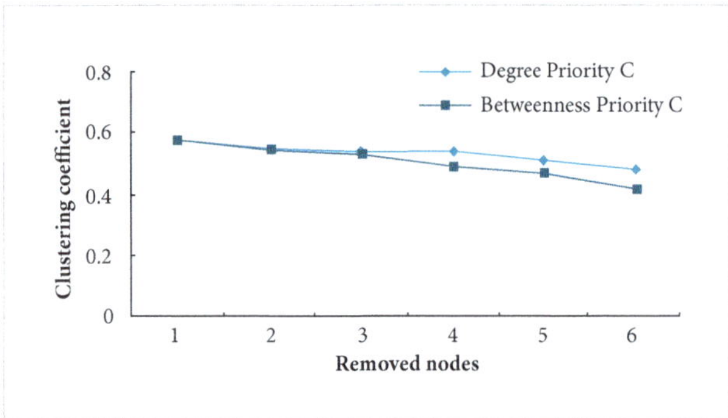

Figure 5. Comparison of clustering coefficient changes.

Table 5. Centrality degree under the different indicators of CSA aviation network.

	Indictor	Centrality Degree
1	Degree CD	83.12%
2	Closeness CC	42.50%
3	Vertex Betweenness CB	95.87%
4	Flow Betweenness CFB	93.12%

priority of betweenness. The average path length increases to a largest extent when Guangzhou Airport node with the greatest degree and the highest betweenness is removed, then it decreases afterwards because the network is divided into multiple connected sub-graphs, among which no connection exists. The average path length decreases after increasing, which is a shortcoming for measuring the network performance.

However, overall, it shows that after the node with the greatest degree or the highest betweenness has been removed, the average path length fluctuates widely, which causes the network instability. As it is indicated in Fig. 4, the clustering coefficient decreases after several nodes are removed, and the decrease of the one based on the remove policy of degree priority is slightly larger than the one based on the betweenness priority; therefore, the clustering coefficient decreases and the clustering degree becomes smaller when the node with the greatest degree or highest betweenness has been removed from the network.

As shown in Table 6, when 1~5 nodes with the greatest degree and highest betweenness are removed, the network efficiency suffers a relatively great impact and the backbone network suffers an even greater one, because both of the removed airlines and the flights are of huge numbers in the original network. Especially, when 5 nodes are removed, the efficiency of the original network is decreased by more than 30% and the backbone network decreased by over 70% and the whole network is almost paralyzed. In addition, the decrease based on the remove policy of degree priority is larger than that based on the betweenness priority. In order to understand the network change condition deeply, taking the test of removing two cities one-time for example, we conduct specific analysis on the network performances, based on the two kinds of remove policies before and after the test. Before the test, backbone network of the CSA

Table 6. Comparison of network efficiency changes based on two different remove policies.

Test Procedure	Based on Degree Priority							Based on Betweenness Priority						
	Original Network			Backbone Network			Impacted Flights	Original Network			Backbone Network			Impacted Flights
	Node	Edge	Efficiency	Node	Edge	Efficiency		Node	Edge	Efficiency	Node	Edge	Efficiency	
Before Test	187	1245	—	99	369	—	—	187	1245	—	99	369	—	—
Remove 1 Point	186	1035	-11.21	98	267	-34.68	18.6	186	1035	-10.35	98	267	-34.68	18.6
Remove 2 Points	185	939	-18.36	97	228	-51.46	28.4	185	945	-17.54	97	237	-48.73	25.42
Remove 3 Points	184	847	-25.50	96	188	-60.78	37.9	184	882	-23.26	96	217	-56.83	34.26
Remove 4 Points	183	761	-30.64	95	160	-69.47	40.1	183	790	-29.01	95	177	-64.30	39.62
Remove 5 Points	182	685	-33.73	94	136	-76. 31	43.3	182	700	-31.43	94	142	-72.46	40.05

contains 99 nodes and 369 routes; when two airports with the greatest degree value (Guangzhou Airport and Shenzhen Airport) have been removed according to the first policy, 306 routes disappear, the original network efficiency is decreased by 18.36%, the backbone network efficiency is decreased by 51.46%, and the network performance is decreased by half. When two airports with the highest betweenness (Guangzhou Airport and Urumqi Airport) are removed according to the second policy, 300 routes disappear, the original network efficiency is decreased by 17.54%, and the backbone network efficiency is decreased by 48.73%. Its decrease is slightly smaller than that of the remove policy based on the degree priority.

When two airports with the greatest degree are removed, the number of routes of the whole network in south-to-north direction is decreased substantially; when two airports with the highest betweenness are removed, most airports in the western area become isolated nodes, holding up flights, and the overall network performance is decreased by half or so. However, the network performance decreases approximately by about 25% (Dang and Li, 2011) when two nodes are removed in the America aviation network, indicating that the southern aviation network is relatively fragile when facing a selective attack and the overall network reliability is expected to be enhanced. Apart from the selective attack, the current network layout of CSA can cause heavy strike to the aviation transportation when facing random attacks, including natural disasters. Therefore, while enlarging the scale of network development, CSA should also develop its aviation network into a multi-hub system. Judging from the degree value and betweenness value of the airport node, number of airlines removed from the network, and the network efficiency decrease condition, in addition to taking Guangzhou Airport as the core hub, CSA can further plan and revise its aviation network by defining Beijing as the important hub between Europe, America and the inland of China, defining Urumqi as the regional hub between the middle Asia and the inland of China. On the other hand, the Company can promote connections in east-to-west direction, for example, establishing connections between the west and the east centered in Zhengzhou to fill up the blank in this direction. In this way, the overall performance and reliability of the network can be enhanced, ensuring fluent running of the air transport system.

CONCLUSION

In this paper, complex network theory is applied, an aviation network structure model is built for CSA, and its structure is analyzed. It is discovered that Guangzhou Baiyun Airport is the hub of the aviation network of CSA, the majority of airlines are in the south and north directions, the airlines in Western Region are distributed around Urumqi in a radial way. Moreover, analysis is conducted on the distribution of statistical features and degree distribution of the network, and it is proven that the aviation network of CSA is a small-world network and a scale-free network. Based on this, the reliability of the network is analyzed according to the degree and betweenness of nodes in the network. The results show that the backbone network performance will be reduced to a half, once 2 nodes are removed and it basically breaks down once 5 nodes are removed. The overall reliability of the network is far from high. Therefore, CSA should put more efforts in the overall programming of the aviation network as well as the construction and management of the aviation hub, seek to develop as an multi-hub airport, launch more flights in west and east directions, reasonably allocate air transport resources, and improve the overall performance of the network so as to meet the demand of sustained and healthy development of air transport of CSA.

The Complex network method is used to analyze the topological structure and statistical features of aviation network. Based on this, the suggestions are proposed for optimizing the aviation network from the analysis of the network reliability the application of the methodology (airport planning, fleet sizing, route planning, etc) isn't mentioned in the paper. They will be discussed in other papers.

ACKNOWLEDGEMENTS

This work is supported by the Joint Funds of the National Natural Science Foundation of China and the Civil Aviation Administration of China (Grant No.U1233105), partially supported by grant J2010-03 of Science and Technology Fund of CAFUC.

REFERENCES

Barrat, A., Barthélemy, M., Pastor-Satoras, R. and Vespignani, A., 2004, "The Architecture of Complex Weighted Networks", Proceedings of the National Academy of Sciences of the United States of America, Vol. 101, No.11, pp. 3747-3752.

Barrat, A., Barthélemy, M. and Vespignani, A., 2005, "The Effects of Spatial Constraints on the Evolution of Weighted Complex Networks", Journal of Statistical Mechanic, doi:10.1088/1742-5468/2005/05/P05003.

Costenbader, E. and Valente, T.W., 2003, "The Stability of Centrality Measures When Networks Are Sampled", Social network, Vol. 25, pp.238-307.

Friedkin, N.E., 1991, "Theoretical foundations for centrality measures", American Journal of Sociology, Vol. 96, pp. 1478-1504.

Gaertler, M. and Wagner, D., 2001, "Algorithms for representing network centrality, groups and density and clustered graph representation", Convolution and Self- organization in Dynamical Networks, Vol. 3, pp.1-7.

Guimera, R.and Amaral, L.A.N., 2004, "Modeling the world-wide airport network", The European Physical Journal B, Vol. 38, pp. 381-385.

Guimera, R., Mossa, S., Turtschi, A. and Amaral, L.A.N., 2005, "The Worldwide Air Transportation Network: Anomalous Centrality, Community Structure, and Cities' Global Roles", Proceedings of the National Academy of Sciences of the United States of America, Vol. 102, pp. 7794-7799.

Holme, P., Kim, B.J., Yoon, C.N. and Han, S.K., 2002, "Attack Vulnerability of Complex Networks", Physical Review E, Vol. 65, pp.1-14.

Hongguang, Y. and Liping, Z., 2012, "Research on Robustness of China's Aviation Network Based on Simulation Analysis", Journal of Wuhan University of (Technology Transportation Science & Engineering), Vol. 36, pp. 42-46.

Kai-Quan, C., Jun, Z., Wen-Bo, D. and Xian-Bin, C., 2012, "Analysis of the Chinese Air Route Network as a Complex Network", Chinese Physics B, Vol. 21, pp. 31-37.

Li, W. and Cai, X., 2012, "Statistical Analysis of Airport Network of China", Physical Review E, Vol. 69, 6106-6112.

Newman, M.E.J., 2003, "The Structure and Function of Complex Networks", SIAM Review, Vol. 45, pp. 167-225.

Newman, M.E.J., 2005, "A measure of Betweenness Centrality Based on Random Walks", Social Networks, Vol. 27, pp. 39-54.

Porta, S., Crucitti, P. and Latora, V., 2006, "The Network Analysis of Urban Streets: A Dual Approach", Physica A: Statistical Mechanics and its Aplications, Vol. 369, pp. 853-866.

Xiaohuan, W., Jinfu, Z., Weiwei, W. and Qiang, G., 2013, "Interval Robust Optimization of Airline Network Designing", Journal of Southwest Jiaotong University, Vol. 48, pp. 559-564.

Dang, Y. and Li, W., 2010, "Air passenger Flow Structure Analysis with Network View", Journal of Transportation Systems Engineering and Information Technology, Vol. 10, pp.167-174.

Dang, Y. and Li, W., 2011, "Comparative analysis on weighted network structure of air passenger now of china and US", Journal of Transportation Systems Engineering and Information Technology, Vol.11, pp.156-162.

Permissions

All chapters in this book were first published in JATM, by Departamento de Ciência e Tecnologia Aeroespacial (DCTA); hereby published with permission under the Creative Commons Attribution License or equivalent. Every chapter published in this book has been scrutinized by our experts. Their significance has been extensively debated. The topics covered herein carry significant findings which will fuel the growth of the discipline. They may even be implemented as practical applications or may be referred to as a beginning point for another development.

The contributors of this book come from diverse backgrounds, making this book a truly international effort. This book will bring forth new frontiers with its revolutionizing research information and detailed analysis of the nascent developments around the world.

We would like to thank all the contributing authors for lending their expertise to make the book truly unique. They have played a crucial role in the development of this book. Without their invaluable contributions this book wouldn't have been possible. They have made vital efforts to compile up to date information on the varied aspects of this subject to make this book a valuable addition to the collection of many professionals and students.

This book was conceptualized with the vision of imparting up-to-date information and advanced data in this field. To ensure the same, a matchless editorial board was set up. Every individual on the board went through rigorous rounds of assessment to prove their worth. After which they invested a large part of their time researching and compiling the most relevant data for our readers.

The editorial board has been involved in producing this book since its inception. They have spent rigorous hours researching and exploring the diverse topics which have resulted in the successful publishing of this book. They have passed on their knowledge of decades through this book. To expedite this challenging task, the publisher supported the team at every step. A small team of assistant editors was also appointed to further simplify the editing procedure and attain best results for the readers.

Apart from the editorial board, the designing team has also invested a significant amount of their time in understanding the subject and creating the most relevant covers. They scrutinized every image to scout for the most suitable representation of the subject and create an appropriate cover for the book.

The publishing team has been an ardent support to the editorial, designing and production team. Their endless efforts to recruit the best for this project, has resulted in the accomplishment of this book. They are a veteran in the field of academics and their pool of knowledge is as vast as their experience in printing. Their expertise and guidance has proved useful at every step. Their uncompromising quality standards have made this book an exceptional effort. Their encouragement from time to time has been an inspiration for everyone.

The publisher and the editorial board hope that this book will prove to be a valuable piece of knowledge for researchers, students, practitioners and scholars across the globe.

List of Contributors

Claudio Antonio Federico
Instituto de Esudos Avançados - São José dos Campos/SP - Brazil
Trevo Coronel Aviador, José Alberto Albano do Amarante,1 | CEP 12.228-001, São José dos Campos/SP - Brazil

Odair Lélis Gonçalez
Instituto de Esudos Avançados - São José dos Campos/SP - Brazil

Evaldo Simões da Fonseca
Instituto de Radioproteção e Dosimetria - Rio de Janeiro/RJ - Brazil

Karla Cristina de Souza Patrão
Instituto de Radioproteção e Dosimetria - Rio de Janeiro/RJ - Brazil

Marlon Antonio Pereira
Instituto de Esudos Avançados - São José dos Campos/SP - Brazil

Linda Viola Ehlin Caldas
Instituto de Pesquisas Energéticas e Nucleares - São Paulo/SP - Brazil

Susane Ribeiro Gomes
Instituto Tecnológico de Aeronáutica - São José dos Campos/SP - Brazil
Praça Eduardo Gomes, 50 | CEP 12.228-901 São José dos, Campos/SP - Brazil

Leopoldo Rocco Junior
Instituto Tecnológico de Aeronáutica - São José dos Campos/SP - Brazil

José Atílio Fritz Fidel Rocco
Instituto Tecnológico de Aeronáutica - São José dos Campos/SP - Brazil

Koshun Iha
Instituto Tecnológico de Aeronáutica - São José dos Campos/SP - Brazil

Flaviano Willians Fernandes
Instituto Tecnológico de Aeronáutica - São José dos Campos/SP - Brazil

Tiago Moreira Bastos Campos
Instituto Tecnológico de Aeronáutica - São José dos Campos/SP - Brazil

Luciana de Simone Cividanes
Instituto Tecnológico de Aeronáutica - São José dos Campos/SP - Brazil

João Paulo Barros Machado
Instituto Tecnológico de Aeronáutica - São José dos Campos/SP - Brazil

Evelyn Alves Nunes Simonetti
Instituto Tecnológico de Aeronáutica - São José dos Campos/SP - Brazil

Gilmar Patrocínio Thim
Instituto Tecnológico de Aeronáutica - São José dos Campos/SP - Brazil
Praça Marechal Eduardo Gomes, 50 - Vila das Acácias | CEP 12.228-900 São José dos Campos/SP - Brazil

Thiago Hanna Both
Universidade Federal do Rio Grande do Sul - Porto Alegre/RS - Brazil
Avenida Bento Goncalves, 9500, Campus do Vale, Bloco IV, CEP: 91.509-900 Porto Alegre/RS - Brazil

Dalton Colombo
Universidade Federal do Rio Grande do Sul - Porto Alegre/RS - Brazil

Ricardo Vanni Dallasen
Universidade Federal do Rio Grande do Sul - Porto Alegre/RS - Brazil

Gilson Inacio Wirth
Universidade Federal do Rio Grande do Sul - Porto Alegre/RS - Brazil
Avenida Bento Goncalves, 9500, Campus do Vale, Bloco IV, CEP: 91.509-900 Porto Alegre/RS - Brazil

Marcelo Sampaio
Instituto Nacional de Pesquisas Espaciais - São José dos Campos/SP - Brazil
Avenida dos Astronautas 1758, São José dos Campos/SP | CEP: 12.227-010 - Brazil

Heitor Evangelista
Universidade do Estado do Rio de Janeiro - Rio de Janeiro/RJ - Brazil

Roberto d'Amore
Instituto Tecnológico de Aeronáutica - São José dos Campos/SP - Brazil

Nasser Ribeiro Asad
Universidade do Estado do Rio de Janeiro - Rio de Janeiro/RJ - Brazil

Lídia Maria Buarque de Oliveira Asad
Universidade do Estado do Rio de Janeiro - Rio de Janeiro/RJ - Brazil

Adriano Caldeira de Araújo
Universidade do Estado do Rio de Janeiro - Rio de Janeiro/RJ - Brazil

Ana Paula Hagge Brasil
Universidade do Estado do Rio de Janeiro - Rio de Janeiro/RJ - Brazil

Nelson Veissid
Instituto Nacional de Pesquisas Espaciais - São José dos Campos/SP - Brazil

Valeri Vlassov
Instituto Nacional de Pesquisas Espaciais - São José dos Campos/SP - Brazil

Alessandra Pacini
Universidade do Vale do Paraíba - São José dos Campos/SP - Brazil

Monique Thérèze Schulz Fontoura
Universidade do Estado do Rio de Janeiro - Rio de Janeiro/RJ - Brazil

Fredy Marcell Villanueva
School of Astronautics - Beihang University - Beijing - China
Beihang University | School of Astronautics | Haidian District, 37 Xueyuan Road - New Main Building, B923 | Beijing 100191 - China

He Linshu
School of Astronautics - Beihang University - Beijing - China

Xu Dajun
School of Astronautics - Beihang University - Beijing - China

Cesar Celis
Pontifícia Universidade Católica do Rio de Janeiro - Rio de Janeiro/RJ - Brazil
Departamento de Engenharia Mecânica | Pontifícia Universidade Católica do Rio de Janeiro - Rua Marquês de São Vicente 225, Rio de Janeiro/RJ | CEP: 22453-900 - Brazil

Vishal Sethi
Cranfield University, Cranfield - United Kingdom

David Zammit-Mangion
Cranfield University, Cranfield - United Kingdom

Riti Singh
Cranfield University, Cranfield - United Kingdom

Pericles Pilidis
Cranfield University, Cranfield - United Kingdom

Mohamed Kameche
Centre de Développement des Satellites - Bir Eldjir/Oran - Algeria
Centre de Développement des Satellites Agence Spatiale Algérienne | BP 4065, Ibn Rochd USTO | 31130 | Oran/ Algeria

Haider Benzeniar
Centre de Développement des Satellites - Bir Eldjir/Oran - Algeria

Ayhane Bey Benbouzid
Centre de Développement des Satellites - Bir Eldjir/Oran - Algeria

Redha Amri
Centre de Développement des Satellites - Bir Eldjir/Oran - Algeria

Nadir Bouanani
Centre de Développement des Satellites - Bir Eldjir/Oran - Algeria

Victor Dmitriev
Universidade Federal do Pará - Belém/PA - Brazil

Fernando Gomes
Universidade Federal do Pará - Belém/PA - Brazil
Departamento de Engenharia Elétrica, Universidade Federal do Pará | Avenida Augusto Correa, 1 | CEP: 66.075-110 Belém/PA - Brazil

Clerisson Nascimento
Universidade Federal do Pará - Belém/PA - Brazil

Cleverson Bringhenti
Instituto Tecnológico de Aeronáutica - São José dos Campos/SP - Brazil
Instituto Tecnológico de Aeronáutica | Departamento de Turbomáquinas | Pr. Marechal Eduardo Gomes, 50 Vila das Acácias | CEP: 12.228-900 - São José dos Campos/ SP - Brazil

Jesuíno Takachi Tomita
Instituto Tecnológico de Aeronáutica - São José dos Campos/SP - Brazil

João Roberto Barbosa
Instituto Tecnológico de Aeronáutica - São José dos Campos/SP - Brazil

Luiz Aldo Leite das Neves
Universidade de Taubaté - Taubaté/SP - Brazil
Instituto de Aeronáutica e Espaço - São José dos Campos/SP - Brazil
Praça Marechal Eduardo Gomes, 50 - Vila das Acácias | CEP 12.228-904 São José dos Campos/SP - Brazil

Luiz Eduardo Nicolini do Patrocínio Nunes
Universidade de Taubaté - Taubaté/SP - Brazil

Valesca Alves Corrêa
Universidade de Taubaté - Taubaté/SP - Brazil

Mirabel Cerqueira Rezende
Instituto de Aeronáutica e Espaço - São José dos Campos/SP - Brazil

Zhang Jing
Taiyuan University of Technology - College of Mechanical Engineering - Taiyuan - China
79 - Yingze West Street - Wanbolin District | Zip Code: 030024 - Taiyuan - China

Guo Hong-Wei
Harbin Institute of Technology - School of Mechatronics Engineering Harbin - China

Liu Rong-Qiang
Harbin Institute of Technology - School of Mechatronics Engineering Harbin - China

Deng Zong-Quan
Harbin Institute of Technology - School of Mechatronics Engineering Harbin - China

Rosângela de Oliveira Colabone
Academia da Força Aérea - Pirassununga/SP - Brazil
Rua 11, N. 3.015 | CEP: 13.504-102 - Rio Claro/SP - Brazil

Antonio Luiz Ferrari
Academia da Força Aérea - Pirassununga/SP - Brazil

Francisco Arthur da Silva Vecchia
Universidade de São Paulo - Escola de Engenharia de São Carlos - São Carlos/SP - Brazil

Adriano Rogério Bruno Tech
Universidade de São Paulo - Faculdade de Zootecnia e Engenharia de Alimentos - Pirassununga/SP - Brazil

Jéssica Garcia de Azevedo
Instituto de Aeronáutica e Espaço - São José dos Campos/SP - Brazil

Artur Cristiano Arantes Filho
Instituto de Aeronáutica e Espaço - São José dos Campos/SP - Brazil
Instituto de Aeronáutica e Espaço | Praça Marechal Eduardo Gomes, 50 - Vila das Acácias CEP: 12.220-098 - São José dos Campos/SP - Brazil

Luis Eduardo Vergueiro Loures da Costa
Instituto de Aeronáutica e Espaço - São José dos Campos/SP - Brazil

Amaury Caruzzo
Instituto Tecnológico de Aeronáutica - São José dos Campos/SP - Brazil
Praça Marechal Eduardo Gomes, 50, Vila das Acácias | CEP: 12.228-900 - São José dos Campos/SP - Brazil

Mischel Carmen Neyra Belderrain
Instituto Tecnológico de Aeronáutica - São José dos Campos/SP - Brazil

Gilberto Fisch
Instituto de Aeronáutica e Espaço - São José dos Campos/SP - Brazil

Daniel Ferreira Manso
Instituto de Aplicações Operacionais - São José dos Campos/SP - Brazil

Bento Silva de Mattos
Instituto Tecnológico de Aeronáutica - São José dos Campos/SP - Brazil
Praça Marechal Eduardo Gomes, 50 - Vila das Acácias | CEP 12.228-900, São José dos Campos/SP - Brazil

Ney Rafael Secco
Instituto Tecnológico de Aeronáutica - São José dos Campos/SP - Brazil

Eduardo Francisco Salles
Instituto Tecnológico de Aeronáutica - São José dos Campos/SP - Brazil

Luiz Eduardo Medeiros
Instituto de Aeronautica e Espaco - Sao Jose dos Campos/SP - Brazil

Roberto de Oliveira Magnago
Universidade do Estado do Rio de Janeiro - Resende/RJ - Brazil
Departamento de Matematica, Fisica e Computacao/Faculdade de Tecnologia/Universidade do Estado do Riode Janeiro | Rodovia Presidente Dutra km 298 - Polo Industrial | CEP 27.537-000 Resende/RJ - Brazil

Gilberto Fisch
Instituto de Aeronautica e Espaco - Sao Jose dos Campos/SP - Brazil

Edson Roberto Marciotto
Instituto de Aeronautica e Espaco - Sao Jose dos Campos/SP - Brazil

Natália Beck Sanches
Instituto Tecnológico de Aeronáutica - São José dos Campos/SP - Brazil

Ricardo Pedro
Instituto Tecnológico de Aeronáutica - São José dos Campos/SP - Brazil

Milton Faria Diniz
Instituto de Aeronáutica e Espaço - São José dos Campos/SP - Brazil

Elizabeth da Costa Mattos
Instituto de Aeronáutica e Espaço - São José dos Campos/SP - Brazil

Silvana Navarro Cassu
Instituto de Aeronáutica e Espaço - São José dos Campos/SP - Brazil

Rita de Cássia Lazzarini Dutra
Instituto de Aeronáutica e Espaço - São José dos Campos/SP - Brazil
Praça Marechal Eduardo Gomes, 50 - Vila das Acácias | CEP 12.228-904 São José dos Campos/SP - Brazil

Otavio Luiz Bogossian
Instituto Nacional de Pesquisas Espacias - São José dos Campos/SP - Brazil
Avenida dos Astronautas, 1.758 - Jardim da Granja | CEP 12.227-010, São José dos Campos/SP - Brazil

Geilson Loureiro
Instituto Nacional de Pesquisas Espacias - São José dos Campos/SP - Brazil

Roberto Vieira Fonseca Lopes
Instituto Nacional de Pesquisas Espacias - São José dos Campos/SP - Brazil

Edgardo Roggero
Comisión Nacional de Actividades Espaciales - Buenos Aires - Argentina

Luiz Alberto de Andrade
Instituto de Aeronáutica e Espaço - São José dos Campos/SP - Brazil
Instituto de Aeronáutica e Espaço - Divisão de Materiais | Praça Marechal Eduardo Gomes, 50, Vila das Acácias, São José dos Campos/SP | CEP: 12228-904 - Brazil

Luan Silva Carvalho dos Santos
Instituto de Aeronáutica e Espaço - São José dos Campos/SP - Brazil

Adriana Medeiros Gama
Instituto de Aeronáutica e Espaço - São José dos Campos/SP - Brazil

Patrícia Helena de Oliveira Teixeira
Faculdade de Tecnologia de São José dos Campos - São José dos Campos/SP - Brazil
Instituto Tecnológico de Aeronáutica - São José dos Campos/SP - Brazil
Avenida Cesare Mansueto Giulio Lattes, s/n, Eugênio de Melo | CEP: 12.247-014 - São José dos Campos/SP - Brazil

Adriano Gonçalves dos Reis
Faculdade de Tecnologia de São José dos Campos - São José dos Campos/SP - Brazil Instituto Tecnológico de Aeronáutica - São José dos Campos/SP - Brazil

Andreas Janssen
Fraunhofer Institute for Production Technology - Aachen - Germany

Arineiza C. Pinheiro
Instituto de Ciências Matemáticas e de Computação/USP - São Carlos/SP - Brazil

Adenilso Simão
Instituto de Ciências Matemáticas e de Computação/USP - São Carlos/SP - Brazil

Ana Maria Ambrosio
Instituto Nacional de Pesquisas Espaciais - São José dos Campos/SP - Brazil
Avenida dos Astronautas, 1.758, CEP: 12.227-010 -São José dos Campos/SP | Brazil

Hu Fuwen
North China University of Technology - Beijing - China
College of Mechanical and Electrical Engineering | North China University of Technology | 100144 | Beijing-China

Douglas Felipe da Silva
Instituto Nacional de Pesquisas Espaciais - São José dos Campos/ SP - Brasil
Instituto Nacional de Pesquisas Espaciais - Av. dos Astronautas, 1.758, São José dos Campos/SP | CEP: 12.227-010 | Brazil

Issamu Muraoka
Instituto Nacional de Pesquisas Espaciais - São José dos Campos/ SP - Brasil

Ezio Castejon Garcia
Instituto Tecnológico de Aeronáutica - São José dos Campos/ SP - Brazil

Carlos Henrique Marchi
Universidade Federal do Paraná - Curitiba/PR - Brazil

Luciano Kiyoshi Araki
Universidade Federal do Paraná - Curitiba/PR - Brazil
Departamento de Engenharia Mecânica | Caixa Postal 19040 | CEP: 81.531-980 Curitiba/PR - Brazil

Luciano Hennemann
Instituto Nacional de Pesquisas Espaciais - Cachoeira Paulista/SP - Brazil
Laboratório Associado de Combustão e Propulsão - Instituto Nacional de Pesquisas Espaciais | Caixa Postal 1, Cachoeira Paulista/SP | CEP: 12.630-000 - Brazil

José Carlos de Andrade
Instituto Nacional de Pesquisas Espaciais - Cachoeira Paulista/SP - Brazil

Fernando de Souza Costa
Instituto Nacional de Pesquisas Espaciais - Cachoeira Paulista/SP - Brazil

Dong Bing
Southwest Jiaotong University - Chengdu/Sichuan - China
Civil Aviation Flight University of China - Guanghan - Deyang/Sichuan - China
School of Traffic and Transportation | Southwest Jiaotong University | 610031 Chengdu/Sichuan - China